Lecture Notes in Computer Science   10387

Commenced Publication in 1973
Founding and Former Series Editors:
Gerhard Goos, Juris Hartmanis, and Jan van Leeuwen

Ying Tan · Hideyuki Takagi
Yuhui Shi (Eds.)

Data Mining
and Big Data

Second International Conference, DMBD 2017
Fukuoka, Japan, July 27 – August 1, 2017
Proceedings

 Springer

Editors
Ying Tan
Peking University
Beijing
China

Hideyuki Takagi
Kyushu University
Fukuoka
Japan

Yuhui Shi
Southern University of Science
and Technology
Shenzhen
China

ISSN 0302-9743   ISSN 1611-3349   (electronic)
Lecture Notes in Computer Science
ISBN 978-3-319-61844-9   ISBN 978-3-319-61845-6   (eBook)
DOI 10.1007/978-3-319-61845-6

Library of Congress Control Number: 2017945275

LNCS Sublibrary: SL3 – Information Systems and Applications, incl. Internet/Web, and HCI

Printed on acid-free paper

This Springer imprint is published by Springer Nature
The registered company is Springer International Publishing AG
The registered company address is: Gewerbestrasse 11, 6330 Cham, Switzerland

Preface

This volume (LNCS vol. 10387) constitutes the proceedings of the Second International Conference on Data Mining and Big Data (DMBD 2017), which was held in conjunction with the 8th International Conference on Swarm Intelligence (ICSI 2017), from July 27 to August 1, 2017, in Fukuoka, Japan.

The Second International Conference on Data Mining and Big Data (DMBD 2017) (IEEE Conference Record 41362) served as an international forum for researchers and practitioners to exchange the latest advances in theories, technologies, and applications of data mining and big data. The theme of DMBD 2017 was "Serving Life with Data Science." DMBD 2017 was the second conference in the series after the successful first event (DMBD 2016) at Bali Island, Indonesia, during June 25–29, 2016.

Data mining refers to the activity of going through big data sets to look for relevant or pertinent information. This type of activity is a good example of "looking for a needle in a haystack." The idea is that businesses collect massive sets of data that may be homogeneous or automatically collected. Decision-makers need access to smaller, more specific pieces of data from these large sets. They use data mining to uncover the pieces of information that will inform leadership and help chart the course for a business. Big data contains a huge amount of data and information and is worth researching in depth. Big data, also known as massive data or mass data, refers to the amount of data involved that are too great to be interpreted by a human. Currently, the suitable technologies include data mining, crowdsourcing, data fusion and integration, machine learning, natural language processing, simulation, time series analysis, and visualization. It is important to find new methods to enhance the effectiveness of big data. With the advent of big data analysis and intelligent computing techniques we are facing new challenges to make the information transparent and understandable efficiently. DMBD 2017 provided an excellent opportunity and an academic forum for academics and practitioners to present and discuss the latest scientific results, methods, and innovative ideas and advantages in theories, technologies, and applications in data mining, big data, and intelligent computing. The technical program covered all aspects of data mining, big data, and swarm intelligence as well as intelligent computing methods applied to all fields of computer science, machine learning, data mining and knowledge discovery, robotics, big data, scheduling, parallel realization, etc.

DMBD 2017 took place in the center of the historical Fukuoka City. Fukuoka is the fifth largest city in Japan with 1.55 million inhabitants and is the seventh most liveable city in the world according to the 2016 Quality of Life Survey by *Monocle*. Fukuoka is the northern end of Kyushu Island and is the economic and cultural center of Kyushu Island. Because of its closeness to the Asian mainland, Fukuoka has been an important harbor city for many centuries. Today's Fukuoka is the product of the fusion of two cities in the year 1889, when the port city of Hakata and the former castle town of Fukuoka were united into one city called Fukuoka. The participants of ICSI 2017 enjoyed traditional Japanese dances, the local cuisine, beautiful landscapes, and the hospitality of the Japanese people in modern Fukuoka, whose sites are part of UNESCO's World Heritage.

DMBD 2017 received 96 submissions from about 231 authors in 34 countries and regions (Algeria, Australia, Bangladesh, Brazil, Brunei Darussalam, Bulgaria, China, Colombia, Ecuador, France, Germany, Hong Kong SAR China, India, Indonesia, Iran, Japan, Malaysia, The Netherlands, Pakistan, Poland, Portugal, Romania, Russia, Serbia, Slovakia, South Africa, South Korea, Spain, Chinese Taiwan, Thailand, Turkey, USA, UK, and Vietnam) across six continents (Asia, Europe, North America, South America, Africa, and Oceania). Each submission was reviewed by at least two reviewers, and on average 2.6 reviewers. Based on rigorous reviews by the Program Committee members and reviewers, 53 high-quality papers were selected for publication in this proceedings volume with an acceptance rate of 55.21%. The papers are organized in 13 cohesive sections covering major topics of data mining and big data.

On behalf of the Organizing Committee of DMBD 2017, we would like to express sincere thanks to the Research Center for Applied Perceptual Science of Kyushu University and the Computational Intelligence Laboratory of Peking University for their sponsorship, to the IEEE Computational Intelligence Society for its technical sponsorship, to the Japan Chapter of IEEE Systems, Man and Cybernetics Society for its technical co-sponsorship, as well as to our supporters the International Neural Network Society, World Federation on Soft Computing, IEEE Beijing Section, Beijing Xinghui Hi-Tech Co., and Springer. We would also like to thank the members of the Advisory Committee for their guidance, the members of the international Program Committee and additional reviewers for reviewing the papers, and the members of the Publications Committee for checking the accepted papers in a short period of time. We are particularly grateful to Springer for publishing the proceedings in the prestigious series *Lecture Notes in Computer Science*. Moreover, we wish to express our heartfelt appreciation to the plenary speakers, session chairs, and student helpers. In addition, there are still many more colleagues, associates, friends, and supporters who helped us in immeasurable ways; we express our sincere gratitude to them all. Last but not the least, we would like to thank all the speakers, authors, and participants for their great contributions that made DMBD 2017 successful and all the hard work worthwhile.

May 2017   Ying Tan
                                                       Hideyuki Takagi
                                                          Yuhui Shi

Organization

General Co-chairs

Ying Tan   Peking University, China
Hideyuki Takagi   Kyushu University, Japan

Program Committee Chair

Yuhui Shi   Southern University of Science and Technology, China

Advisory Committee Co-chairs

Russell C. Eberhart   IUPUI, USA
Gary G. Yen   Oklahoma State University, USA
Hisao Ishibuchi   Osaka Prefecture University, Japan

Technical Committee Co-chairs

Kay Chen Tan   National University of Singapore, Singapore
Xiaodong Li   RMIT University, Australia
Nikola Kasabov   Auckland University of Technology, New Zealand
Ponnuthurai N. Suganthan   Nanyang Technological University, Singapore

Plenary Session Co-chairs

Mengjie Zhang   Victoria University of Wellington, New Zealand
Andreas Engelbrecht   University of Pretoria, South Africa

Invited Session Co-chairs

Yan Pei   University of Aizu, Japan
Chaomin Luo   University of Detroit Mercy, USA

Special Sessions Co-chairs

Shangce Gao   University of Toyama, Japan
Ben Niu   Shenzhen University, China
Qirong Tang   Tongji University, China

Tutorial Co-chairs

Milan Tuba   John Naisbitt University, Serbia
Andreas Janecek   University of Vienna, Austria

Publications Co-chairs

Swagatam Das   Indian Statistical Institute, India
Xinshe Yang   Middlesex University, UK

Publicity Co-chairs

Yew-Soon Ong   Nanyang Technological University, Singapore
Carlos Coello   CINVESTAV-IPN, Mexico
Yaochu Jin   University of Surrey, UK
Shi Cheng   Shanxi Normal University, China
Bin Xue   Victoria University of Wellington, New Zealand

Finance and Registration Co-chairs

Chao Deng   Peking University, China
Suicheng Gu   Google Corporation, USA

Local Arrangements Chair

Ryohei Funaki   Kyushu University, Japan

Conference Secretariat

Xiangyu Liu   Peking University, China

ICSI 2017 International Program Committee

Mohd Helmy Abd Wahab   Universiti Tun Hussein Onn Malaysia, Malaysia
Miltos Alamaniotis   Purdue University, USA
Tomasz Andrysiak   UTP Bydgoszcz, Poland
Duong Tuan Anh   HoChiMinh City University of Technology, Vietnam
Carmelo J.A. Bastos Filho   University of Pernambuco, Brazil
David Camacho   Universidad Autonoma de Madrid, Spain
Vinod Chandra S.S.   Kerala University, India
Yenming J. Chen   National Kaohsiung First University of Science and
                               Technology, Taiwan
Shi Cheng   Shaanxi Normal University, China
Jose Alfredo Ferreira Costa Federal University
Bogusław Cyganek   AGH University of Science and Technology, Poland
Ke Ding   Tencent Corporation

Teresa Guarda	Universidad Estatal de Peninsula de Santa Elena, Ecuador
Cem Iyigun	Middle East Technical University, Turkey
Dariusz Jankowski	Wrocław University of Technology, Poland
Mingyan Jiang	Shandong University, China
Chen Junfeng	Hohai University, China
Imed Kacem	Université de Lorraine, France
Kalinka Kaloyanova	University of Sofia, Bulgaria
Germano Lambert-Torres	PS Solutions
Wei-Po Lee	National Sun Yat-sen University, Taiwan
Bin Li	University of Science and Technology of China
Michael Li	Central Queensland University, Australia
Churn-Jung Liau	Academia Sinica at Taipei, Taiwan
Andrei Lihu	Politehnica University of Timisoara, Romania
Wenjian Luo	University of Science and Technology of China
Wojciech Macyna	Wroclaw University of Technology, Poland
Guojun Mao	Central University of Finance and Economics
Vasanth Kumar Mehta	SCSVMV University
Mohamed Arezki Mellal	M'Hamed Bougara University
Sanaz Mostaghim	Institute IWS
Sheak Rashed Haider Noori	Daffodil International University, Bangladesh
Somnuk Phon-Amnuaisuk	Universiti Teknologi, Brunei
Gerald Schaefer	Loughborough University, UK
Manik Sharma	DAV University
Ivan Silva	University of São Paulo, Brazil
Pramod Kumar Singh	ABV IIITM Gwalior, India
Hung-Min Sun	National Tsing Hua University, Taiwan
Ying Tan	Peking University, China
Paulo Trigo	ISEL
Milan Tuba	John Naisbitt University, Serbia
Agnieszka Turek	Warsaw University of Technology, Poland
Gai-Ge Wang	Jiangsu Normal University, China
Guoyin Wang	Chongqing University of Posts and Telecommunications, China
Lei Wang	Tongji University, China
Zhenzhen Wang	Jinling Institute of Technology, Nanjing, China
Ka-Chun Wong	City University of Hong Kong, SAR China
Michal Wozniak	Wroclaw University of Technology, Poland
Bo Xing	University of Johannesburg, South Africa
Yingjie Yang	De Montfort University, UK
Jie Zhang	Newcastle University, UK
Qieshi Zhang	Shaanxi Normal University, China

Additional Reviewers for ICSI 2017

Diván, Mario José
Fernandes, Bruno
Guo, Yuqian
Habib, Md. Tarek
Junyi, Chen
Li, Xiangtao

Lima, Emerson
Lin, Jiecong
Saganowski, Łukasz
Yan, Shankai
Zhang, Jiao

Contents

Classification

Schedule and Sequence Analysis

Big Data

Data Analysis

High Performance Computing

Knowledge Base and its Framework

Fuzzy Control

Association Analysis

A Process for Exploring Employees' Relationships via Social Network and Sentiment Analysis

Jeydels Barahona[1] and Hung-Min Sun[2,3(✉)]

[1] Institute of Information Systems and Applications,
National Tsing Hua University, Hsinchu, Taiwan
[2] Department of Computer Science, National Tsing Hua University,
Hsinchu, Taiwan
hmsun@cs.nthu.edu.tw
[3] Research Center for Information Technology Innovation,
Academia Sinica, Taipei, Taiwan

Abstract. The proposed study is to analyze and visualize properties of the social network constructed from a dataset based on Enron mail dataset, and utilize sentiment analysis as an additional source of information to study employees' relationships in a company. We concluded that when social network analysis is used in conjunction with emotion detection, it is possible to see the positive or negative areas where the company must work to promote a healthy organizational culture and uncover possible organizational issues in a timely manner.

Keywords: Sentiment Analysis · Social computing · Text mining · Organizational aspects

1 Introduction

Sentiment Analysis (SA) is related to data mining, which is implemented from structured data. Companies produce unstructured data such as emails logs, and emails provide valuable information about what entities are involved directly and indirectly in making decisions. One example of this kind of valuable data is the Enron dataset. Enron corporation filed for bankruptcy in 2001 and the Enron dataset was then collected and posted to the web. This research is aimed at building a social network out the communication flow among the entities in the Enron dataset to find the most influential people in the company, and extract the sentiment associated to the email content. Besides, we will also discuss about the need of adding a neutral class and its relevance to this domain.

2 Related Work

Some previous work has worked on Enron dataset. [2] demonstrated that SNA can be used to assess the stability of organizations, since it can help in the identification of culture changes that may be hard to identify at individual level.

© Springer International Publishing AG 2017
Y. Tan et al. (Eds.): DMBD 2017, LNCS 10387, pp. 3–8, 2017.
DOI: 10.1007/978-3-319-61845-6_1

Moreover, many companies use opinion mining and SA for their research nowadays. For instance, companies make their systems continuously gather information from the Web and use the result of SA to develop marketing strategies [4]. Since SA is not often used with internal data such as email communications among employees, this presents an opportunity for linking both SNA and SA and extract new interesting information for supporting the decision-making process in companies from the organizational point of view.

3   Methodology

3.1   The Enron Dataset

We used the version released in May 7th 2015 of Enron dataset. For our study, we explored the data contained in the "inbox" subfolder in the 150 folders.

3.2   Data Transformation

To transform data into a more manageable format for text mining, we first searched for folders named "inbox" of the Enron corpus and add contents to output a file named "enron.mbox". Second, we converted the mbox data to JSON format. Finally, we import the JSON file to MongoDB [9].

3.3   Extracting the Email Addresses

We extracted only the From and To fields of each message from MongoDB. Each sender-receiver pair was taken and coupled back with the subject field and data field. This resulted in a filtered data collection with each record represented by the fields: From, To and Email-ID.

3.4   Preprocessing the Email Addresses

The network will be structured with employees as nodes and each sent email as an edge [6]. We filtered the interactions leaving out email addresses that lack the string "@enron" as our main focus is on internal communication, and we limited our scope to only the most active users who sent the most messages to others. Lastly, we kept only the interactions among the 371 selected entities and about 10 thousand emails sent among them.

3.5   Preprocessing the Messages

Four preprocessing steps of text classification are tokenization, stop-word removal, lowercase conversion, and stemming [1]. We used the "tm" package in R [5] to preprocess the emails. The main structure for managing documents in tm is a so-called Corpus. For the Enron Emails, the Corpus will be formed by the collection of emails belonging to the 371 top users.

3.6 Feature Selection

TF-IDF composed of two scores, term frequency and inverse document frequency is a common metric used for text categorization, and the multiplication of the two scores allow us to find terms that are important in a document [7]. We applied TD-IDF to find the most relevant terms in the documents. Among the most frequent terms are: energy, gas, company, trading and time.

3.7 Sentiment Analysis

We are finding the quadruple (s, g, h, t) in this step, where s represents the sentiment, g represents the target object for which the sentiment is expressed, h represents the one expressing the sentiment, and t represents the time at which the sentiment was expressed. SA was performed by reading the content of the emails and manually annotating a subset of 1000 messages. The introduction of a neutral category was necessary as many of the messages were plain communications. As positive emails we considered socialization opportunities and any other factor that promotes a healthy organizational culture. As for negative emails, we included those that denoted delays in processes, financial audit concerns, problems in the market and court issues. Finally, we used a baseline algorithm proposed by [3] to classify emails. The Rtexttools package developed by [8] along with the e1071 package were used to implement the method using the annotated email collection 80% of the data as the training set and 20% as the testing set. The classifier product of this process was used later for assigning labels to the rest of emails in the study.

3.8 Social Network Analysis on the Enron Emails

After preceding steps, we moved on to build the social network. Basically, the data resulting from both processes was merged into a single edge table containing the following attributes: id, source, target, emailId and sentiment. The resulting social graph will be analyzed using SNA centrality metrics. The visualization and analysis process will be done through Gephi.

4 Results and Discussion

4.1 The Documents in the Enron Corpus

After preprocessing, there were of 371 employees who exchanged about 10,700 emails. To identify the most prominent terms, Tm package was used to complete hierarchical clustering using terms similarity and frequency only. For our data, "attached", "contact" are large clusters. However, the remaining clusters are even more important as they give insights about specific messages on "Energy", "market" describing Enron as a company that traded Energy.

4.2   Enron Emails Sentiment Classification

For machine learning purposes, there are different classifiers available that perform well for Natural Language Processing tasks such as SA. Classifiers with better performance: Naive Bayes, Support Vector Machine, Maximum Entropy and Ensemble classifiers [7] were trained and tested. We progressively implemented the algorithms with 400 emails and increased until 800 in the testing set. For all the classifiers, both precision and recall were lower than the accuracy. Moreover, the trained SVM classifier has better performance, so it was applied to rest of the unlabeled data to predict the sentiment. The prediction resulted in most of the data being neutral, followed by negative and positive emails respectively.

The imbalance in the classes in the training set affected specific measures such as precision and recall. In information retrieval, the goal is to identify a small number of matches from a large number of documents, in this case emails that fall in a specific sentiment. This task makes getting a higher precision more difficult than getting a good recall. The precision for the neutral class was 81% and recall 97%. These results were acceptable but they decreased significantly for both the positive and negative classes.

Our next step consisted in visualizing the data by sentiment over time. The analysis in time of the sentiment is helpful to show the specific time in which a particular sentiment spreads. Positive emails were sent between the months of August and October, followed by a sudden decrease in November.

Concerning neutral emails, there is a low frequency between March and July. Neutral emails are still relevant because they refer to basic activities such as asking for reports and meetings schedules. After July 2001 there was a massive growth in the quantity of neutral emails, most likely due to report inquiries from different areas in the company to address specific issues. Lastly, we see a sudden peak again in December because Enron Corp. filed for bankruptcy on December 2, 2001.

The email flow was similar for negative emails. Most of them were sent between August 2001 and January 2002. This is probably related to situations affecting the market and Enron's profit in this period. There was a significant increase in negative emails in October 2001, which may obey to Enron reporting a 638-million-dollar third quarter loss.

4.3   Social Network Analysis

A major difference with other studies on the Enron email network is that we are not simply considering the 150 folders in the dataset as the nodes of the network, but instead we fetched the data in the inbox folder for these 150 users and mapped all the sender-recipient relationships available irrespective of the folder where the email belongs to.

Degree Centrality. The social network built out of the emails is not very dense. Here a node's in-degree would be the number of emails an employee is receiving, whereas out-degree would be the emails this person is sending to others. For the network in this study, the nodes with the highest degree included people such as Jae Black, Mary Cook, David Forster and Veronica Espinoza, which means they have the most ties with other employees (Fig. 1).

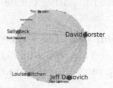

Fig. 1. Degree in Enron's network   **Fig. 2.** Betweenness in Enron's network

Betweenness Centrality. Betweenness can be used to find out who the actors in the network are and how influential their role is at the moment of making information flow through the network. Some actors' relevance to the network was more evident when exploring betweenness centrality in the graph, which includes people like Sally Beck, David Forster and Jeff Dasovich (Fig. 2).

Closeness Centrality. A node's closeness centrality is the sum of graph-theoretic distances from all other nodes. Nodes with low closeness scores are well-positioned to obtain novel information earlier. For the Enron emails, most of the nodes had a closeness between 0.30 and 0.55.

4.4 Sentiment Diffusion in Social Network

The initial results of the classification process showed that there were more negative emails than positive in the data by about 4%. However, once the respective emails were associated with their senders and recipients, mails with a positive sentiment were sent to more people and negative emails to a fewer number.

Positive Sentiment Subgraph. Positive emails as visualized with betweenness centrality were primarily sent by three important nodes: Louise Kitchen, Jeff Dasovich and David Forster. They all brought success for Enron (Fig. 3).

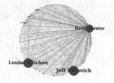
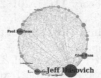

Fig. 3. Positive sentiment in Enron's network   **Fig. 4.** Negative sentiment in Enron's network

Negative Sentiment Subgraph. The sub-network for negative sentiment shows that most negative emails came from Paul Kaufman, Glen Hass and Jeff Dasovich, Enron's governmental affairs executive. Jeff Dasovich appeared the graph for positive sentiment as well. Here we conclude that Jeff Dasovich was a major bridge for negative issues concerning Enron. All of the 378 most active nodes received negative emails at some point (Fig. 4).

Neutral Sentiment Subgraph. As for neutral emails, they are still relevant because we can also see who are the people are participating the most in meetings or reporting new information.

5   Conclusion

We proposed a process for exploring employees' relationships based on email communications at Enron through SNA and SA. Our results show that there were conflicting situations and distress related to legal issues and employees' satisfaction. This reflected the different sub-networks shown in this work and the analysis of the emails by sentiment over time. We conclude that visualizing the diffusion of sentiment throughout the network provides a meaningful information for organizational purposes.

Acknowledgment. This research was supported in part by the Ministry of Science and Technology, Taiwan, under the Grant MOST 103-2221-E-007-073-MY3.

References

1. Uysal, A.K., Gunal, S.: The impact of preprocessing on text classification. Inf. Process. Manage. **50**(1), 104–112 (2014)
2. Collingsworth, B., Menezes, R., Martins, P.: Assessing organizational stability via network analysis. In: 2009 IEEE Symposium on Computational Intelligence for Financial Engineering, Nashville, TN, pp. 43–50 (2009)
3. Pang, B., Lee, L.: Opinion mining and sentiment analysis. J. Found. Trends Inform. Retrieval **2**(January), 1–135 (2008)
4. Cambria, E., Schuller, B., Xia, Y., Havasi, C.: New avenues in opinion mining and sentiment analysis. IEEE Intell. Syst. **28**, 15–21 (2013)
5. Feinerer, I., Hornik, K., Meyer, D.: Text mining infrastructure. R. J. Stat. Software **25**(5), 1–54 (2008)
6. Diesner, J., Frantz, T.L., Carley, K.M.: Communication networks from the enron email corpus "It's Always About the People. Enron is no Different". Comput. Math. Organ. Theory **11**(3), 201–228 (2005)
7. O'Keefe, T., Koprinska, I.: Feature selection and weighting methods in sentiment analysis. In: Proceedings of the 14th Australasian document computing symposium, Sydney, pp. 67–74 (2009)
8. Jurka, T.P., Collingwood, L., Boydstun, A.E., Grossman, E., van Atteveldt, W.: RTextTools: Automatic Text Classification via Supervised Learning. http://CRAN.R-project.org/package=RTextTools
9. Russell, M.A.: Mining the Social Web: Analyzing Data from Facebook, Twitter, LinkedIn, and Other Social Media Sites. O'Reilly Media Inc. (2011)

Mining Relationship Between User Purposes and Product Features Towards Purpose-Oriented Recommendation

Sopheaktra Yong[✉] and Yasuhito Asano

Department of Social Informatics, Graduate School of Informatics, Kyoto University, Yoshida-Honmachi, Sakyo-ku, Kyoto 606-8501, Japan
sopheaktra.yong@outlook.com, asano@i.kyoto-u.ac.jp

Abstract. To help in decision making, buyers in online shopping tend to go through each product features, functionalities, etc. provided by vendors and reviews made by other users, which is not an effective way when confronting loaded of information and especially if the buyers are beginner users who have limited experience and knowledge. To deal with these problems, we propose a framework of purpose-oriented recommendation which present a ranking of products suitable for a designated user purpose by identifying important product features to fulfill the purpose from user reviews. As technical foundation for realizing the framework, we propose several methods to mine relation between user purposes and product features from the online reviews. The experimental results employing reviews of digital cameras in Amazon.com show the effectiveness and stability of proposed methods with acceptable rate of precision and recall.

Keywords: Recommendation system · Review analysis · Bootstrapping

1 Introduction

Nowadays, increasing large numbers of consumers have gradually shifted their purchasing behavior to online shopping sites. Online retailers like Amazon has set up a platform where vendors could upload their products together with descriptions illustrated the functions and features of the product. Consumers often browse through other users' feedbacks to help in their decision-making. However, it is tough sometimes to go through all the reviews since some products might have hundred comments. Many sites have developed various recommender systems to help users choosing suitable products for their needs. The common techniques are based on the similar products a consumer used to buy previously or based on other consumers who share similar interest or purchasing history. However, it is still not sufficient for users since the system does not recommend them according to their needs particularly when it comes to the commodities designed for variety of usage purposes such as PC, smartphone, digital camera etc. For instance, there are various purposes associating with digital camera such as shooting portrait, landscape, wildlife, sports, car racing, etc. Generally, different product features are required for different purposes. Camera makers could not describe

© Springer International Publishing AG 2017
Y. Tan et al. (Eds.): DMBD 2017, LNCS 10387, pp. 9–21, 2017.
DOI: 10.1007/978-3-319-61845-6_2

them all in product description or advertisement. Various recommended techniques with an emphasis on product features including methods to automatically extract product features from the reviews and techniques for ranking products based on features have been proposed in [1–5]. However, it is still very difficult for novice user to choose the suitable camera since they could not tell which camera features are good for their specific needs.

For this reason, we propose a framework of purpose-oriented recommendation which can greatly improve the convenience by recommending products automatically based on consumer's purposes. In this paper, we describe a methodology to mine hidden knowledge from the online reviews, particularly, to extract relationship between user purposes and product features as the technical foundation to realize the proposed framework.

Our Contribution. The contribution of our paper can be summarized as follows:

- We propose a purpose-oriented recommendation framework to suggest products suitable for users especially novice users who do not have much knowledge about the products and a methodology as the technical foundation for realizing the purpose-oriented recommendation framework.
- We present a bootstrapping-based methodology employing state-of-the-art Natural Language Processing (NLP) techniques to extract the relationship between user purposes and product features from online reviews.
- We Introduce (1) a conditioning approach for filtering new seeds by computing the word semantic similarity to lessen the common problem of bootstrapping – semantic drift and (2) an ontology-based technique to generate more seeds for bootstrapping process.
- We tested and confirmed the effectiveness of the proposed methods through experiments on very popular and large Amazon review dataset and user evaluation.

The rest of the paper is organized as follows: our framework of purpose-oriented recommendation is illustrated in Sect. 2. Section 3 presents the methodology for extracting the relationship between product features and user purposes. Section 4 contains experimental results. Brief reviews of related works conducted in this area are summarized in Sect. 5 followed by the conclusion in Sect. 6.

2 Purpose-Oriented Recommendation Framework

As we described above, the users tend to face difficulties when deciding which products they should purchase for their specific needs if the products contains lots of technical features and functionalities. Figure 1 illustrates the flow of the recommendation framework we wish to build in future work.

First, the system will present a list of all possible purposes for a particular product type, e.g. digital camera. In this study, we assume that the users are aware of their purposes and asked to choose from one or more purposes from the purpose list presented by the system. After that, the system will output the features which are most relevant and associated with the chosen purposes. Here, the system also analyses the optimal or minimum requirement for each product features to achieve the

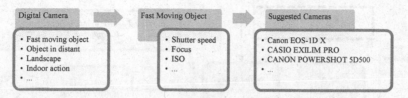

Fig. 1. Purpose-oriented recommendation framework

corresponding purposes. For example, at least a burst rate of 6 frame-per-second is required to capture a flying bird. These kinds of information could help users to not only pick the right camera but also educate them to set the right setting for shooting a specific target. Finally, according to the relationship between product features and purposes, the system will recommend suitable products to the users.

In order to realize the framework, the most crucial part of the system is to find the association between product features and purposes which expressed in the users' reviews. Next section, we will describe the methodology in detail.

3 Methodology

The proposed methodology is composed of three main components, Component I - NLP Preprocess, Component II - Initial Seeds Generation, and Component III - Purpose-Feature Relationship Extraction Bootstrapping Model. The method extracts keywords representing purposes from the online reviews and connect them with product features. In this study, product features as the input are manually selected. They can also be extracted from several online shopping websites such as cnet.com, dpreview.com, amazon.com, etc. There have been several studies on identifying and extracting product features automatically, which is not the focus of this research. Another input of the system is text data of consumer reviews. In Component I, the customer reviews are preprocessed using sets of NLP tools. After that, purposes are extracted using a combined technique of Labeled LDA [6] and Word2Vec [7] by Component II. Then starting with a few initial seeds, Component III iteratively learns patterns to extract more seeds. Figure 2 illustrates the overview of our method.

3.1 NLP Preprocess (Component I)

NLP is applied for linguistic preprocessing and raw text analysis such as removing stop words, sentence tokenizer, and part-of-speech tagging. The tool used in this paper is NLTK: The Natural Language Toolkit [8]. A typical review contains several sentences. Here, we employ Punkt Sentence Tokenizer for splitting reviews into sentences.

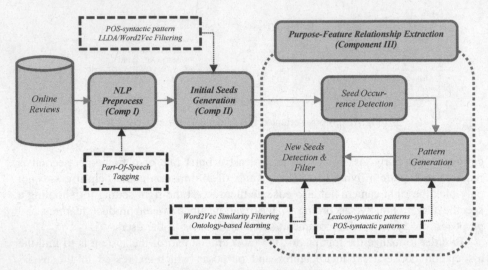

Fig. 2. Overview of methodology

3.2   Initial Seeds Generation (Component II)

Choosing of seeds is arguably the most critical step in bootstrapping. However, most of the previous researchers either chose manually [9] or picked the most frequently occurring words in their corpus [10] that they have identified belong to the category. In this study, we propose a technique which combine topic modeling and word embedding technique to generate initial seeds for bootstrapping model. Our initial seeds generation process is composed of three steps: candidate sentences extraction, term extraction, and term filtering.

We assume that consumers described their usage experiences or purposes inside the reviews. They often exemplified how they use the product or specific function to perform or to achieve something. Table 1 shows some sample reviews containing intended use or purposes of digital cameras' shutter speed functionalities. Mining this type of reviews would enable us to detect the purposes expressed by the consumers. We find "candidate sentences" which contains product features and purpose-oriented expression, by the procedures explained below.

The process of generating initial seeds is described as follows:

1. Candidate Sentences Extraction: Candidate sentences are sentences containing product features as well as purpose-related expression. Sentences containing about product features often have expression whether such features are good or bad for a particular purpose. For example, there is a statement - *"The unlimited continuous shooting mode, which is perfect for catching fast-action sports shots"*. Here, the phrase *"Catching fast-action sports shots"* is the purpose of using the *"shutter"* feature. This piece of review tells us that if we consider buying a camera for shooting a fast-moving object, we should choose the camera with the high *"shutter"* speed. To detect the candidate sentences from the reviews, we manually define some syntactic-sentence patterns used to express purposes. An example of the pattern is

Table 1. Customer reviews from Amazon.com. Product features and corresponding purposes are highlighted.

Review #1: The unlimited continuous shooting mode, which is perfect for catching fast-action sports shots, solves the problem of slow shutter speeds and is not found in any of the other camera's in the s3's class.
Feature: shutter, Purpose: catching fast-action

Review #2: I especially liked the panoramic feature for the beautiful Alaskan landscapes and the burst feature for shooting wildlife in action.
Feature: shutter, Purpose: shooting wildlife

Review #3: The shutter speed allows her to take pictures of friends in motion [dancing, skiing and other sporting events] amazing window on back for best picture view.
Feature: shutter, Purpose: dancing, skiing, sporting events

<feature> + Verb + Adjective + Preposition + Noun/Noun Phrase. Then we build the set of candidate sentences by extracting any sentences following these patterns.

2. Term Extraction: Once we have a list of candidate sentences, we extract the "purposes" from this set using topic modeling – Labeled LDA. Unlike the conventional LDA, Labeled LDA is a generative model which constrains the topic model to use only those topics corresponding to a document's label set [6]. In our setting, a list of product features with their corresponding candidate sentences are prepared to train the model. The product features are trained as the topics (labels) and the candidate sentences are treated as documents. Thus, The Labeled LDA outputs the important terms per product feature. These terms are considered as keywords or key phrases representing purposes. An example of this component output is {shutter speed: kids, wildlife, action_shots, recording, getting}, where "shutter speed" is the product feature (topic) and "kids, wildlife, action_shots, recording, getting" are the top keywords representing purposes.

3. Term Filtering: The list of extracted keywords might contain unrelated words since the Labeled LDA output the list based on word frequency only. Therefore, we propose a filtering approach by calculating word semantic relationship using a recent technique Word2Vec. It is a two-layer neural net that processes text. It can convert words into vectors by learning a corpus. Word2Vec can compute the word semantic similarity by calculating their vectors. The corpus we used to train Word2Vec is the consumer reviews. In our methodology, we utilize Word2Vec to filter out the anomaly words from the Labeled LDA's output. For example, Word2vec is able to remove the unrelated words - "recording" and "getting" from the Labeled LDA output above.

3.3 Extracting Purpose-Feature Relationship (Component III)

To obtain more purpose-related sentences or purposes, we adopt a bootstrapping method as in [9] by detecting more sentence patterns for describing purposes to extract appropriate nouns and noun phrases. Because a simple bootstrapping method tends to produce noises in the results, we also proposed filtering methods to reduce those noises.

Our bootstrapping mainly iterates between the following two phases: (1) pattern generation and (2) seed extraction and filter. The algorithm begins with initial seeds and then iterates through the phases until it could not extract any more patterns or generate new seeds with the word similarity compared to the seeds pool higher than a given threshold τ_2.

Pattern Generation. In the pattern generation phase, we would like to detect the extraction patterns in the form of lexicon and POS pattern which connect between product feature and user purpose in the reviews. The process starts with a few feature-purpose seed tuples obtained from Component II. For every seed tuple $<x, y>$, e.g.$<x = $ shutter, $y = \{kids,\ wildlife,\ ...\}>$, we first retrieve all the sentences containing these terms x, y. Next, all "connecting text" linking terms x and y $(kids|\ wildlife|...)$ are extracted into the candidate list. Since we also consider detecting POS syntactic patterns, the "connecting text" is parsed and tagged into the part-of-speech. As shown in the first sentence in Table 2, the phrase *"is good for"* (or POS-tagged *"VB JJ IN"*) is the pattern connecting between product feature *(shutter speed)* and purposes *(kids)*. The strength of each pattern is computed by how frequently they are used to connect the seed tuples. Finally, we augment the extraction pattern if any pattern in the candidate list appeared more than a specific threshold τ_1 in the reviews.

Table 2. Sample sentences and extracted patterns

Sentences	Lexicon patterns	POS patterns
Shutter speed is good for kids	*is good for*	VB JJ IN
I use burst feature for shooting wildlife	*for shooting*	IN VBG

Seed Extraction and Filter. Most common bootstrapping models are pattern-based approaches in which new seeds are accepted based on the extraction patterns. This could lead to a well-known flaw in bootstrapping known as "semantic drift". To avoid this problem, we adopt Word2Vec word similarity as the filtering methods so that the current seeds do not wander away from the original semantic meaning of the initial seeds. The process is simply described as following;

- First, our method retrieves from the corpus the set of words W that match any of the extraction patterns P, e.g. Shutter speed <u>is good for</u> "**sport events**". The word W could be expressed in multi-word terms as Noun or Noun Phrase *(NP)*.

$$NP: \{(<VBG>|<JJ>|<JJR>|<JJS>)*(<NN>|<NNS>)+\}.$$

- The term W is augmented to the seed list only if its similarity exceeds the predefined similarity threshold τ_2 in average.

In the next section, we will introduce an ontology-based technique to generate more seeds in complement to the conventional pattern-based bootstrapping approach.

Ontology-based Approach for Seed Extraction. As described in previous section, we could extract new seeds or purposes only if the terms are expressed in the extraction patterns. Therefore, we could miss some interesting purposes because not all the customers would use the same sentences to express their purposes or experiences. Let's look at the 3rd review in Table 1. The terms *"dancing, skiing and other sporting events"* share the hyponym relation which means that if one of them is classified as the purpose, so does the rest. For instance, the context pattern of *"allows her to take pictures of friends in"* is not an extraction pattern, thus the terms *"dancing, skiing and other sporting events"* would not have a chance to be detected. In this study, we propose an additional approach and in complement to the pattern-based bootstrapping in detecting new seeds or purposes by mining the new purposes if they have taxonomic relations with any in the seed list.

Taxonomic relations are the most important semantic relations in a domain ontology, the extraction of which has been well studied in the field of lexicon building. We adopt the lexicon-syntactic patterns [11] for taxonomic relation extraction. A total of five lexicon-syntactic patterns are used and described in Table 3. The process of extracting taxonomic relations and generating new seeds for bootstrapping is described as following:

Table 3. The lexicon-syntactic patterns of taxonomic relation extraction

1. NP_0 such as NP_1 {, NP_2, ... (and/or) NP_n}	Hyponym (NP_i, NP_0)
2. NP_1 is a kind of NP_0	Hyponym (NP_i, NP_0)
3. NP{, NP}*{,} or other NP_0	Hyponym (NP, NP_0)
4. NP{, NP}*{,} and other NP_0	Hyponym (NP, NP_0)
5. NP_0, including {NP}* or/and NP	Hyponym (NP, NP_0)

- First, we (1) retrieves from the corpus the set of NPs that match any of the predefined patterns and (2) scan whether at least one of the NP set is already in the seed list.
- Next, the rest of the NP in the set is considered as seed candidates and are augmented as new seeds if its word semantic similarity exceeds the predefined similarity threshold $\tau2$ in average against the other terms in seed list.

Adopting this approach could help us extracting both more precise and general purposes which is more convenient for novice users to comprehend the meaning. It could also assist us to classify the purposes into hierarchy straightforwardly.

4 Experimental Results

In this section, we will illustrate our experimental results and case studies. We conducted our experiments on customer reviews on digital cameras from Amazon.com. The dataset is made available via SNAP, Stanford University [12]. The dataset contains 203,773 reviews and were collected from May 1996 to July 2014.

4.1 Extraction Purpose-Feature Relationship Using Bootstrapping Approach

In this experiment, we manually select 8 camera features and analyze the performance of each model separately by evaluating the number of purposes it could generate, precision, and (relative) recall of all various models i.e. Lexicon, POS, LP, O, and LPO. The first three models are the conventional pattern-based whereas the LPO is the proposed model which combine the pattern-based and ontology-based together. Each model description is as follow:

Lexicon: Generic patterns are only learnt from lexicon-syntactic patterns

POS: Generic patterns are only learnt from POS-syntactic patterns

LP: Generic patterns are learnt from new seeds generated from lexicon and POS syntactic patterns per iteration

O: Generic patterns are ignored in this context. New seeds are only generated from the ontology-based approach (the hyponym relation in Table 3)

LPO: A combination of all models

Table 4 shows initial seeds for each camera feature and its selective corresponding new seeds generated by the proposed model – LPO. Note that some camera features' initial seeds are manually modified prior the bootstrapping process. Table 5 shows the total number of new seeds, precision, and relative recall generated by each model separately. LPO could generate the most seeds. Particularly, a combination of Ontology learning and LP increase the number of seeds than individual model. It concludes that by discovering new seeds via ontology relation during the bootstrapping process could increase the number of seeds accordingly.

Table 4. Initial seeds used for each camera feature and outputs from proposed model. The incorrect purposes are bold.

	Shutter	ISO	Weight	IS	Screen	Flash	Pixel	Focus
Seeds	kids sports wildlife	night light bright	vacation trip carrying	wildlife capturing	reviewing previewing low_angle	light indoor night	printing uploading	kids macros moving
Output	skiing action shots baseball sporting pets...	light **condition** dim lit darkness dusk...	bicycle_ touring hiking adventures walks...	kids animals fish candids sports...	**angle** viewing **inwards** **eye_piece** composing framing...	castle dim_light dark_room museums ...	publishing share transfer editing email...	moving_targets insects portraits action shots...

Although it is difficult to get all the correct instances (all purposes) for each camera feature, it is still possible to compare the recall of a method relative to another method recall. Following the Pantel et al. [13], the relative recall $R_{A/B}$ of method A given method B can be calculated as:

Table 5. Precision, relative recall, and F-score of each method

Methods	#New seeds	Precision	Rel recall	F-score
Lexicon	94	0.84	0.40	0.54
POS	240	0.72	0.87	0.79
LP	235	0.71	0.84	0.77
O	12	0.80	0.05	0.09
LPO	263	0.75	0.99	**0.85**

$$R_{A/B} = \frac{R_A}{R_B} = \frac{\frac{C_A}{C}}{\frac{C_B}{C}} = \frac{C_A}{C_B} = \frac{P_A \times |A|}{P_B \times |B|}, \tag{1}$$

where R_A is the recall of A, C_A is the number of correct seeds extracted by A, C is the (unknown) total number of correct seeds in the corpus, P_A is A's precision in our experiments, and $|A|$ is the total number of seeds discovered by A. We can compute the relative recall by comparing the number of seeds extracted by each model with the number of seeds retrieved by all the models together. The total number of seeds is 272 keywords in which 199 keywords are corrected. We asked a professional photographer who has experiences for more than 10 years to annotate and evaluate the correctness of our extracted purposes. Therefore, we could calculate the performance for all models as shown in Table 5. The lexicon model generates a relatively small number of purposes (94 keywords) with the best precision while the POS and LP model could generate much more purposes (235–240 keywords) but the precision is dropping. It might be implied that we could get more purposes but the it might also contain a lot of noises in the output. However, our proposed model (LPO) which combines pattern-based and ontology-based together can not only generate more purposes but also achieve better accuracy comparing to individual components. In this experiment the threshold $\tau 1$ and $\tau 2$ are set to 3 and 0.5 respectively. However, the result might seem a bit bias since the evaluation is performed by only one subject.

4.2 Subjective Evaluation

A subjective evaluation is further conducted to examine and emphasize the accuracy and effectiveness of the proposed method – LPO from a variety of users. As aforementioned, there were 263 keywords generated using LPO model which is inconvenient and time-consuming for users to evaluate the result. Therefore, we categorize purposes and ask users to enumerate important features for each purpose category. We manually defined nine categories, while this categorization could be done automatically in the future work by mining both taxonomic and non-taxonomic relations as in [14] (Table 6).

Questionnaire Setup. We created 10 questions in total [15]. Question #1 asks about the respondent's experiences in photography skill. Question #2 – Question #10 survey respondent's knowledge or experiences for which relevant camera features they are using for each purpose. Relevant camera features per purpose is defined by selecting camera features receiving 50% votes from all respondents.

Table 6. Sample of purpose categories and individual keywords

Purposes	Keywords
Macros	flower, insects, foods, …
Moving targets	animals, kids, freezing, motion, action shots, sport events, soccer, …
Traveling	trips, touring, backpack, adventures, outings, …
Low light condition	dimmer environments, backlit daylight, dark room, …
Bright condition	sunlight, daylight, outdoor, portraits at outdoor, …
…	…

Result. We received 31 respondents in which the majority of respondents are beginner users (less than 1 year) and professional users (more than 5 years). The responses from them could possibly be viewed as subjective evaluations with different opinions and diversities. Since we asked people to rank the camera features for different purposes, we could obtain the ground truth of relationship between user purposes and features and able to calculate the accuracy of each method as shown in Tables 7 and 8. Though the precision of pattern-based models and LPO is the same, but the number of individual keywords extracted by the LPO is higher based on evaluation in Sect. 4.1. Therefore, we could conclude that our proposed model could extract more good purposes (better recall) than the pattern-based and ontology-based separately with the best accuracy of F-score 0.67.

Table 7. Category-based precision and recall of proposed model – LPO

Purpose category	Ground truth	Proposed model	P	R
Macros/Closeups	Focus, IS, ISO	Focus, IS	1	0.67
Landscape/Scenery	ISO, Focus, IS	Focus, Shutter, IS	0.67	0.67
Moving targets	Shutter, ISO, Focus, IS	Focus, Shutter, IS	1	0.75
Portraits	Focus, IS, ISO	Focus, IS	1	0.67
Viewing and framing	Screen	Screen	1	1
Uploading, editing, and printing	N/A	Pixel	0	N/A
Low light condition	ISO, Flash	ISO, Flash	1	0.5
Bright condition	ISO, Shutter	ISO, Flash	0.5	0.5
Traveling	Shutter, Focus, IS	Weight	0	0
Average			**0.69**	**0.66**

Table 8. Precision, recall, and F-score of each method (Category-based)

Methods	P	R	F-score
Lexicon	0.69	0.54	0.61
POS	0.69	0.61	0.65
LP	0.69	0.61	0.65
O	0.11	0.06	0.08
LPO (Proposed Model)	**0.69**	**0.66**	**0.67**

5 Related Work

This paper deals with various fields of studies, including feature-based ranking system, mining semantic relations using bootstrapping, and ontology learning. We will take a look at some of the previous works related to each field of study.

Feature-Based Ranking System. Mining of opinions from reviews has become a popular area of research. Users often express their usage experiences in reviews. Regarding the feature extraction techniques, it could be classified into supervised and unsupervised approach. An example of supervised techniques, Wong and Lam [16] employ the hidden Markov model and conditional random fields as the underlying learning method for extracting product features from auction websites. In contract, product features can also be extracted automatically using unsupervised technique presented by Hu and Liu [17]. They made assumptions that product features must be noun or noun phrases. This utilizes the association rule mining algorithm to discover all frequent item sets within a target set of reviews. Regarding the problem of feature-based summarization, Kangale et al. [3] provided a solution which could extract user opinions from reviews and generate rating as well as review summary of each product feature. Zhang et al. [1] proposes a methodology to rank products based on their features using online reviews. First, they manually define a set of product features that are of interest to the customers. Next, they detect subjective and comparative sentences in reviews using predefined structural patterns. Using these sentences, they construct a feature-specific product graph and apply page-rank like algorithm to rank products based on the product features. Our work differs from theirs in the following aspects: (1) we assume that most of the consumers do not have much knowledge about product features and attributes and thus, they could not prioritize which product feature is important or relevant and (2) instead of generating a review summary on each product feature or product as a whole, we are interested only on extracting the semantic relationship between product features and user purposes.

Mining Semantic Relations with Bootstrapping. Most common approach of bootstrapping to extract the semantic relations is pattern-based approach. Agichtein and Gravano [9] proposed "Snowball" algorithm to identify and extract structured relations between named entities from unstructured text. The system is given with some initial seeds then searches sentences in which terms in a seed tuple occur closely to each other. It then analyzes the connecting text and surrounding context and generates an extraction pattern. Pantel et al. [13] proposed an approach based on an edit-distance technique to learn lexicon-POS patterns and could obtain both good performance and efficiency. Later Espresso [18] has been proposed by the same author to infer patterns for harvesting binary semantic relations (is-a, part-of, succession, reaction, and production). It also describes refining techniques to deal with wide variety of relations by measuring the strength of association between patterns and seeds using pointwise mutual information. In our work, we propose an additional filtering approach for accepting new seeds by measuring the word semantic similarity to avoid the semantic drift.

Ontology Learning. Because we have introduced an ontology learning approach to generate more seed instances in conventional bootstrapping, we will take a look at some studies related to ontology. Hearst [11] pioneered using patterns to extract hyponym (is-a) relations. Manually building a few lexicon-syntactic patterns, Hearst sketched a bootstrapping algorithm to learn more patterns from instances, which has served as the model for most subsequent pattern-based algorithms. We have adopted this approach in our study by extracting hyponym relation of generated seeds using some predefined lexicon-syntactic patterns listed in Table 3. Chen et al. [14] proposed an ontology learning framework to extract customer needs of digital cameras. To detect the ontology relation, the authors detect taxonomic and non-taxonomic relations from the customer reviews. Taxonomic relations are extracted using several methods including string matching, lexicon-syntactic patterns, and WordNet taxonomic relations. They proposed a word property-based method for extracting non-taxonomic relations by measuring the support and confidence of the co-occurrence noun phrases from the corpus.

6 Conclusion and Future Work

In this paper, we proposed a methodology for extracting semantic relations between product features and purposes. This work is necessarily important in order to realize the proposed framework of purpose-oriented recommendation. The output of the framework could help educating novice users to comprehend the product features which should be taken into account for their specific needs or purposes. The result of experiments indicates that the combination of pattern-based and ontology-based approach could extract number of good purposes and obtain a good result of precision and recall.

For the future research, we are considering to quantify the optimal setting or requirements of product features for achieving a specific purpose. For instance, if the user is interested in buying a digital camera for shooting their busy kids, then it is desired to not only rank the relevant features but also the optimal setting configuration (shutter speed of 1/500 s, 5 fps burst rate, ISO 1600, continuous AFC etc. is at least required to capture sharp image). Such quantification not only tells users which camera they should purchase considering the important features but also educates them to set the right setting for achieving their goal when they use the product.

Acknowledgment. This work was supported by JSPS KAKENHI Grant Number 15K00423 and the Kayamori Foundation of Informational Science Advancement.

References

1. Zhang, K., Narayanan, R., Choudhary, A.: Voice of the customers: Mining online customer reviews for product feature-based ranking. In: Proceeding of 3rd Workshop on Online Social Networks (WOSN), pp. 1–9 (2010)
2. Hu, M., Liu, B.: Mining opinion features in customer reviews. In: Proceeding of 19th National Conference on Artificial Intelligence, pp. 755–760 (2004)

3. Kangale, A., Kumar, S.K., Naeem, M.A., Williams, M., Tiwari, M.K.: Mining consumer reviews to generate ratings of different product attributes while producing feature-based review-summary. Int. J. Syst. Sci. **47**, 3272–3286 (2016)
4. Kamal, A.: Review mining for feature based opinion summarization and visualization. arXiv Preprint arXiv:1504.03068 (2015)
5. Uchida, S., Yamamoto, T., Kato, M.P., Ohshima, H., Tanaka, K.: Object search by experience attributes. DBSJ Japanese J. **14-J**(7), 1–7 (2016)
6. Ramage, D., Hall, D., Nallapati, R., Manning, C.D.: Labeled LDA: a supervised topic model for credit attribution in multi-labeled corpora. In: Proceeding of EMNLP 2009, pp. 248–256 (2009)
7. Mikolov, T., Chen, K., Corrado, G., Dean, J.: Efficient estimation of word representations in vector space. In: Proceeding of Workshop at International Conference on Learning Representations (ICLR) (2013)
8. Loper, E., Bird, S.: NLTK: the natural language toolkit. In: Proceeding of ETMTNLP 2002, vol. 1, pp. 63–70 (2002)
9. Agichtein, E., Gravano, L.: Snowball: extracting relations from large plain-text collections. In: Proceeding of 5th ACM Conference on Digital libraries, pp. 85–94 (2000)
10. Thelen, M., Riloff, E.: A bootstrapping method for learning semantic lexicons using extraction pattern contexts. In: Proceeding of EMNLP 2002, pp. 214–221 (2002)
11. Hearst, M.A.: Automatic acquisition of hyponyms from large text corpora. In: Proceeding of 14th Conference on Computational Linguistics, pp. 539–545 (1992)
12. McAuley, J., Pandey, R., Leskovec, J.: Inferring networks of substitutable and complementary products. In: Proceeding 21th KDD, pp. 785–794
13. Pantel, P., Ravichandran, D., Hovy, E.: Towards terascale knowledge acquisition. In: Proceeding of 20th International Conference on Computational Linguistics, pp. 771–777 (2004)
14. Chen, X., Chen, C.-H., Fai Leong, K., Jiang, X.: An ontology learning system for customer needs representation in product development. Int. J. Adv. Manuf. Technol. **67**, 441–453 (2013)
15. Online Survey about Digital Cameras, https://www.surveymonkey.com/r/RMW9C7T
16. Wong, T.-L., Lam, W.: Learning to extract and summarize hot item features from multiple auction web sites. Knowl. Inf. Syst. **14**, 143–160 (2008)
17. Hu, M., Liu, B.: Mining and summarizing customer reviews. In: Proceeding of 10th KDD. pp. 168–177 (2004)
18. Pantel, P., Pennacchiotti, M.: Espresso: leveraging generic patterns for automatically harvesting semantic relations patrick. In: Proceeding of Joint Conference of 21st COLING and 44th ACL, pp. 113–120 (2006)

Finding Top-k Fuzzy Frequent Itemsets from Databases

Haifeng Li[✉], Yue Wang, Ning Zhang, and Yuejin Zhang

School of Information, Central University of Finance and Economics,
Beijing, China
{mydlhf,wangyue,zhangning,zhangyuejin}@cufe.edu.cn

Abstract. Frequent itemset mining is an important in data mining. Fuzzy data mining can more accurately describe the mining results in frequent itemset mining. Nevertheless, frequent itemsets are redundant for the users. A better way is to show the top-k results accordingly. In this paper, we define the score of fuzzy frequent itemset and propose the problem of top-k fuzzy frequent itemset mining, which, to the best of our knowledge, has never been focused on before. To address this problem, we employ a data structure named *TopKFFITree* to store the superset of the mining results, which has a significantly reduced size in comparison to all the fuzzy frequent itemsets. Then, we present an algorithm named *TopK-FFI* to build and maintain the data structure. In this algorithm, we employ a method to prune most of the fuzzy frequent itemsets immediately based on the monotony of itemset score. Theoretical analysis and experimental studies over 4 datasets demonstrate that our proposed algorithm can efficiently decrease the runtime and memory cost, and significantly outperform the naive algorithm *Top-k-FFI-Miner*.

Keywords: Fuzzy frequent itemset · Data mining · Top-k fuzzy frequent itemset · Quantitative database

1 Introduction

Tin he fuzzy set [15] can be used in network applications to make a better decision. Many fuzzy mining algorithms, such as classification [13], clustering [14], web mining [8], neural networks [1], have been proposed. Frequent itemset mining is a very important method in data mining, which together with the fuzzy techniques, can be more powerful to obtain the accurate mining results [7]. Several methods were proposed to improve the performance. [4] focused on the development of a general model to discover association rules, which can be

This research is supported by the National Natural Science Foundation of China (61100112,61309030), Beijing Higher Education Young Elite Teacher Project (YETP0987), State Key Program of the National Social Science Foundation of China(13AXW010), Discipline Construction Foundation of Central University of Finance and Economics(2016XX05).

© Springer International Publishing AG 2017
Y. Tan et al. (Eds.): DMBD 2017, LNCS 10387, pp. 22–30, 2017.
DOI: 10.1007/978-3-319-61845-6_3

used in relational databases that contain quantitative data. [5] addressed the problem of the increasing use of very large quantitative-value databases, and proposes a new fuzzy mining algorithm based on the *AprioriTid* approach to find fuzzy association rules from given quantitative transactions; each item only used the linguistic term with the maximum cardinality in later mining processes, and thus made the number of fuzzy regions to be processed the same as that of the original items. [6] introduced a *FP-tree* based data structure *FUFP-tree*, which made the tree updating process easier; an incremental algorithm was also proposed for reducing the execution time in reconstructing the tree when new transactions were inserted, and thus can achieve a trade-off between runtime and space complexity. [10] designed a novel tree structure called *CFFP-tree* to store the related information in the fuzzy mining process, where each node maintained the membership value of the contained item and the membership values of its super-itemsets in the path; moreover, the *CFFP-tree* was built by the proposed algorithm CFFP-growth. The author then proposed a simple tree structure called the *UBFFP-tree* [9]; the designed two-phase fuzzy mining approach can easily derive the upper-bound fuzzy counts of itemsets using the tree; thus, it can prune the unpromising itemsets in the first phase, and then finds the actual fuzzy frequent itemsets in the second phase. [11] developed a fuzzy frequent itemset algorithm *FFI-Miner* to mine the complete set of fuzzy frequent itemsets without candidate generation. *FFI-Miner* used a novel fuzzy-list structure to keep the essential information for later mining process; plus, it employed an efficient pruning strategy to reduce the search space, and thus reduce the runtime cost.

Motivation: The mined fuzzy frequent itemsets are massive when the threshold is small. Traditional frequent itemset mining methods use the itemset representation, such as the maximal itemsets [2], the closed itemsets [12] and the non-derivable itemsets [3] to reduce the itemset count, which can not only improve the performance, but also enable the user to understand the mining results easier. But on the other hand, Specifying an adaptive threshold is much difficult for users, a much reasonable method is to find the top-k fuzzy frequent itemsets. In this paper, we focus on the problem of how to discover the top-k fuzzy frequent itemsets with an effective and efficient method over quantitative databases.

The rest of this paper is organized as follows: In Sect. 2 we present the preliminaries of the fuzzy itemset mining and define the top-k fuzzy frequent itemset, and state the problem addressed in this paper. Section 3 presents the data structures, and illustrates our algorithm in detail. Section 4 evaluates the performance with theoretical analysis and experimental results. Finally, Sect. 5 concludes this paper.

2 Preliminaries and Problem Statement

2.1 Preliminaries

Let $\Gamma = \{i_1, i_2, \cdots, i_m\}$ be a set of m distinct items in a quantitative database $QD = \{QT_1, QT_2, \cdots, QT_n\}$, where each $QT_i \subseteq \Gamma$ is called a quantitative transaction, which contains items with the quantities. An example is shown in Table 1.

Table 1. A quantitative database

TID	QTransaction
1	A:5, C:10, D:2, E:9
2	A:8, B:2, C:3
3	A:5, B:3, C:9
4	A:5, B:3, C:10, E:3
5	A:7, C:9, D:3
6	A:5, B:2, C:10, D:3
7	A:5, B:2, C:5
8	A:3, C:10, D:2, E:2

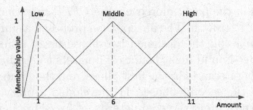

Fig. 1. The membership functions of linguistic 3-terms

The membership functions μ are used to fuzzy up the database, which is shown in Fig. 1.

(1) The quantitative value of each item will be converted to a fuzzy value; thus, each item Z can be represented by a special notation $\Sigma \frac{\mu_i}{z_i}$, where μ_i denotes the membership grade, and z_i denotes the fuzzy region of Z. As an example, item $A : 5$ in transaction 1 can be denoted as $\frac{0.2}{A.L} + \frac{0.8}{A.M}$, in which 0.2 is the membership grade of item A when computing with membership function $\mu.Low$.

(2) The membership grades will be summed up in the quantitative database according to different fuzzy areas, and the area with maximal summed membership grade will be choose as the final one. Again, taking item A as an example, the summary of $A.L$ is 1.6, $A.M$ is 5.8, and $A.H$ is 0.6. Then, we regard $A.M$ as the final fuzzy area of item A, with summed membership grade 5.8 and occurrence number 8. Other items are $B.L$, $C.H$, $D.L$ and $E.L$.

(3) The items in quantitative database will be updated by the fuzzy areas. For an instance, item $A : 5$ in transaction 1 will be replaced by $\frac{0.8}{A.M}$. The updated fuzzy database FD is shown in Table 2.

An itemset $X = \{x_1, x_2, \cdots, x_k\}$ is a k-items set, and we use $|X|$ to denote the size of X, that is, $|X| = k$. If X is covered by a fuzzy transaction FT, then we regard the minimum membership grade of x_i as the membership grade of X.

Table 2. Fuzzy database

TID	FTransaction
1	$\frac{0.8}{A.M}, \frac{0.8}{C.H}, \frac{0.8}{D.L}$
2	$\frac{0.6}{A.M}, \frac{0.8}{B.L}$
3	$\frac{0.8}{A.M}, \frac{0.6}{B.L}, \frac{0.6}{C.H}$
4	$\frac{0.8}{A.M}, \frac{0.6}{B.L}, \frac{0.8}{C.H}, \frac{0.6}{E.L}$
5	$\frac{0.8}{A.M}, \frac{0.6}{C.H}, \frac{0.6}{D.L}$
6	$\frac{0.8}{A.M}, \frac{0.8}{B.L}, \frac{0.8}{C.H}, \frac{0.6}{D.L}$
7	$\frac{0.8}{A.M}, \frac{0.8}{B.L}$
8	$\frac{0.4}{A.M}, \frac{0.8}{C.H}, \frac{0.8}{D.L}, \frac{0.8}{E.L}$

The summed grade in fuzzy database FD is called the support of X, denoted $\Lambda(X)$. Also, we use $ct(X)$ to denote the count of occurrence of X. Given a minimum support λ, X is called frequent if $\Lambda(X) \geq \lambda * |FD|$.

2.2   Problem Statement

Intuitively, on the one hand, if an itemset has a larger size, it is much more useful; on the other hand, the larger the itemset, the more important it is. In this section, we define the score of a fuzzy frequent itemset X as $\varsigma(X) = |X| * \Lambda(X)$, based on which the problem is introduced. Given the minimum support λ, the parameter k, and the membership functions μ, the problem in this paper is to discover the top-k frequent itemsets based on their scores over quantitative databases.

3   Top-k Fuzzy Frequent Itemset Mining Method

3.1   A Naive Method

An intuitive method to find the k most valuable fuzzy frequent itemset is simple. First, we can use an existing method to discover all the fuzzy frequent itemsets, then we compute the scores and sort all the fuzzy frequent itemsets with descendant order. Finally, we get the first k itemsets as the mining results. In this paper, we use the state-of-the-art algorithm to discover the fuzzy frequent itemsets. This method will be used in our experiments as the evaluation method.

3.2   Data Structure

TopKFFITree. We build an in-memory prefix-tree named *TopKFFITree* to store the super set of the fuzzy itemsets, in which each itemset is denoted by the tree nodes. A node n_X is a 3-tuple $<X, sup, score>$. X denotes the fuzzy itemset. *sup* is the support of X in the database. *score* is the score of X. Each node has a pointer to its parents except the root node. As a result, the child

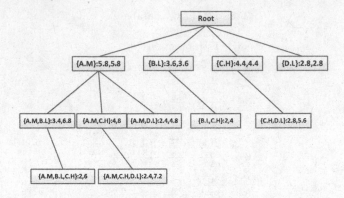

Fig. 2. TopKFFITree for $\lambda = 0.2$

node represents an itemset that covers the parent node. Using the database from Table 2, we show the *TopKFFITree* in Fig. 2 when the minimum support is set to 0.2. For an instance, the itemset $D.L$ has support 2.8, and has the score 2.8 either.

3.3 Pruning Strategies

Theorem 1. *For three fuzzy itemsets X, Y and Z, which satisfy $X \in Y \in Z$ and $|X| = |Y| - 1$, if $\varsigma(X) > \varsigma(Y)$, then $\varsigma(Y) > \varsigma(Z)$.*

Proof. For any two itemsets I and J s.t. $I \in J$, then $\Lambda(I) \leq \Lambda(J)$, we use $T(|I|) = \varsigma(I) - \varsigma(J)$, that is, $T(|I| = |I| * \Lambda(I) - |J| * \Lambda(J) = |I| * \Lambda(I) - (|I| + 1) * \Lambda(J) = |I| * (\Lambda(I) - \Lambda(J)) - \Lambda(J)$. Thus, $T'(|I|) = \Lambda(I) - \Lambda(J) \leq 0$. If $\varsigma(X) > \varsigma(Y)$, that is, $T(|X|) > 0$, given $|X| < |Y|$, then we can get $T(|Y|) = \varsigma(Y) - \varsigma(Z) > 0$ based on $T'(X) \leq 0$; thus, $\varsigma(Y) > \varsigma(Z)$.

Theorem 1 shows that when score of an itemset is partially monotonously consistent the size of the itemset. That is, when an itemset X has a larger score than its superset Y, the scores of all the supersets of Y are smaller than that of Y. This supplies us a method to prune the support computation of itemsets. Once we find an itemset has the smaller score that its subset, we can prune all its superset directly.

3.4 TopK-FFI Algorithm

We propose the *TopK-FFI* algorithm to bottom-up construct the *TopKFFITree*. The basic idea is to achieve the superset of our mining results from the *TopKF-FITree*. To conduct this process, we will consider a key problem, that is, how to find the top-k results from the *TopKFFITree*? To address this problem, we generate the itemsets with breadth manner, and consider 2 conditions. First, if the size of all the fuzzy frequent itemsets is smaller than k, we regard all these

---

Algorithm 1. TopK-FFI Algorithm

Require: *root*: root of *TopKFFITree*; *k*: the size of mining results; λ: minimum support; *FFIC*: Collection of TopKFFITree nodes;

1: **for** each item X **do**
2:     compute the support $\Lambda(X)$ and score $\varsigma(X)$;
3:     **if** $\Lambda(X) \geq \lambda$ **then**
4:         generate new child node n_X of root;
5:         add X into *FFIC*;
6: **if** $|FFIC| < k$ **then**
7:     **for** each child node n_X of root **do**
8:         CALL Explore(n_X, k, λ, *FFIC*);
9: return *FFIC*;

Algorithm 2. Explore Algorithm

Require: n_X: node of *TopKFFITree*, denotes itemset X; k: the size of mining results; λ: minimum support; *FFIC*: Collection of TopKFFITree nodes;

1: **for** each frequent node n_X's right sibling node n_Y **do**
2:     get $Z = X \cup Y$
3:     compute the support $\Lambda(Z)$ and the score $\varsigma(Z)$;
4:     **if** $\Lambda(Z) > \lambda$ **then**
5:         generate new child node n_Z of n_X;
6:         add Z into *FFIC*;
7: **if** |FFIC|>k **then**
8:     return;
9: **for** each child node n_Y of n_X **do**
10:     **if** $\varsigma(Y) > \varsigma(X)$ **then**
11:         CALL Explore(n_Y, k, λ, *FFIC*);

---

itemsets as the mining results. Second, if the size of all the fuzzy frequent itemsets is larger than k, we will generate the itemsets until their scores are smaller than the scores of 1-itemsets they cover; we suppose the size is p, then (1) if $p < k$, we will continuously generate the supersets until $p >= k$; (2) if $p > k$, then we will sort the generated itemsets based on their scores with descendant order, and get the first k itemsets. Note that both two aspects may not generate the exact k results since we have to generate all the itemsets when we begin to explore a level of the *TopKFFITree*. To perform the sorting, we employ a vector, named *FFIC* to maintain the mining results in the *TopKFFITree*. Algorithm 1 shows the detail of our method.

4   Experimental Results

In this section, we evaluate the performance of our algorithm *TopK-FFI*. We employed the state-of-the-art algorithm *FFI-Miner* [11] to discover the fuzzy frequent itemsets, and we call this method the *Top-k-FFI-Miner*, which was used as the evaluation method. The minimum support and the parameter k are the parameters to evaluate the mining performance.

4.1   Running Environment and Datasets

We implemented the algorithms with Python 2.7 running on Microsoft Windows 7 and performed on a PC with a 3.60 GHZ Intel Core i7-4790M processor and 12 GB main memory. We used 2 synthetic datasets and 2 real-life datasets as the evaluation datasets. The detailed data characteristics are shown in Table 3.

Table 3. Fuzzy dataset characteristics

DataSet	Transaction count	Average size	Min size	Max size	Items count	Transaction correlation
T25I15D100K	100 000	26	4	67	1000	38
T40I10D100K	100 000	39	4	77	1000	25
KOSARAK	990 002	8	1	2498	41 270	5159
ACCIDENTS	340 183	33	18	51	468	14

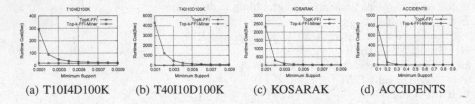

(a) T10I4D100K   (b) T40I10D100K   (c) KOSARAK   (d) ACCIDENTS

Fig. 3. Runtime cost vs. minimum support

4.2   Effect of Minimum Support

In this section, we fixed the k parameter to 1000, and evaluate the performance when the minimum support was changed. Figures 3 and 4 presented the runtime cost over different datasets. As can be seen, our algorithm *TopK-FFI* achieved significantly better computing performance over the naive algorithm *Top-k-FFI-Miner*. We can clearly see that no matter how the minimum support changed, the runtime of our algorithm was similar. The runtime of *Top-k-FFI-Miner*, however, when the minimum support turned smaller, increased greatly. This is due to the fact that our algorithm only to maintain k, which here is 1000, or a little more nodes in the *TopKFFITree*; thus, most part of the fuzzy frequent itemsets will be pruned, which result in the significantly reduction of computing cost and memory cost. Note that no matter how dense a database was, the mining performance was stable when we fixed the k. As can be seen from in Figs. 3(d) and 4(d), even though the *ACCIDENTS* dataset was dense, our algorithm kept almost unchanged computing cost and memory cost, in comparison to that, *Top-k-FFI-Miner* had a much worse performance when the minimum support turned smaller.

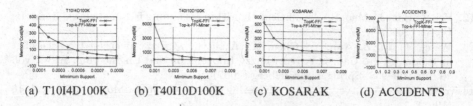

(a) T10I4D100K   (b) T40I10D100K   (c) KOSARAK   (d) ACCIDENTS

Fig. 4. Memory cost vs. minimum support

5   Conclusions

In this paper, we made a research on how to discover the top-k fuzzy frequent itemsets over databases, which, to our best knowledge, has never been studied before. We designed an in-memory data structure named *TopKFFITree* to store the itemsets. An efficient *TopK-FFI* algorithm was presented to build and maintain the *TopKFFITree*. We showed that the score of the itemset was partially monotonous when the size was increased; thus, a prune strategy was employed in our algorithm, which can significantly reduce the search space. Our extensive experimental results shown that the algorithm outperformed the baseline algorithm significantly.

References

1. Buckley, J.J., Hayashi, Y.: Fuzzy neural networks: a survey. Fuzzy Sets Syst. **66**(1), 1–13 (1994)
2. Burdick, D., Calimlim, M., Gehrke, J.: MAFIA: a maximal frequent itemset algorithm for transactional databases (2001)
3. Calders, T., Goethals, B.: Mining all non-derivable frequent itemsets. In: Elomaa, T., Mannila, H., Toivonen, H. (eds.) PKDD 2002. LNCS, vol. 2431, pp. 74–86. Springer, Heidelberg (2002). doi:10.1007/3-540-45681-3_7
4. Delgado, M., Marin, N., Sanchez, D., Vila, M.A.: Fuzzy association rules: general model and applications. IEEE Trans. Fuzzy Syst. **11**(2), 214–225 (2003)
5. Hong, T., Kuo, C., Wang, S.: A fuzzy aprioritid mining algorithm with reduced computational time. Appl. Soft Comput. **5**(1), 1–10 (2004)
6. Hong, T., Lin, C., Yulung, W.: Incrementally fast updated frequent pattern trees. Expert Syst. Appl. **34**(4), 2424–2435 (2008)
7. Kuok, C.M., Fu, A., Wong, M.H.: Mining fuzzy association rules in databases. SIGMOD Rec. **27**(1), 41–46 (1998)
8. Lin, C.W., Hong, T.P.: A survey of fuzzy web mining. WIREs Data Min. Knowl. Disc. **3**(3), 190–199 (2013)
9. Lin, C., Hong, T.: Mining fuzzy frequent itemsets based on UBFFP trees. J. Intell. Fuzzy Syst. **27**(1), 535–548 (2014)
10. Lin, C., Hong, T., Wenhsiang, L.: An efficient tree-based fuzzy data mining approach. Int. J. Fuzzy Syst. **12**(2), 150–157 (2010)
11. Lin, J.C.W., Li, T., Fournier-Viger, P., Hong, T.P.: A fast algorithm for mining fuzzy frequent itemsets. J. Intell. Fuzzy Syst. **29**(6), 2373–2379 (2015)

12. Pei, J., Han, J., Mao, R.: An efficient algorithm for mining frequent closed itemsets, Closet (2000)
13. Wang, T., Li, Z., Yan, Y., Chen, H.: A survey of fuzzy decision tree classifier methodology. In: Cao, B.Y. (ed.) Fuzzy Information and Engineering. AISC, vol. 40, pp. 959–968. Springer, Heidelberg (2007)
14. Yang, M.S.: A survey of fuzzy clustering. Math. Comput. Model. **18**(11), 1–16 (1993)
15. Zadeh, L.A.: Fuzzy sets and systems. Int. J. General Syst. (2007)

Association Rule Mining in Healthcare Analytics

S. Anand Hareendran[1(✉)] and S.S. Vinod Chandra[2]

[1] Department of Computer Science, MITS, Kochi, Kerala, India
anandhareendrans@mgits.ac.in
[2] Computer Center, University of Kerala, Thiruvananthapuram, Kerala, India
vinod@keralauniversity.ac.in

Abstract. Big data analytics examines large amounts of data to uncover hidden patterns, correlations and other insights. In this work, a novel association rule-mining algorithm is employed for finding various rules for performing valid prediction. Various traditional association mining algorithms has been studied carefully and a new mining algorithm, Treap mining has been introduced which remedies the drawbacks of the current Association Rule Learning (ARL) algorithms. Treap mining is a dynamic weighted priority model algorithm. As it works on dynamic priority, rule creation happens in least time complexity and with high accuracy. When comparing with other association mining algorithms like Apriori and Tertius, we could see that Treap algorithm mines the database in an $O(n \log n)$ when compared to Apriori's $O(e^n)$ and Tertius's $O(n^2)$. A high precise mining model for the post Liver Transplantation survival prediction was designed using the rules mined by Treap algorithm. United Nations Organ Sharing dataset was used for the study. Rule accuracy of 96.71% was obtained while using Treap mining algorithm where as, Tertius produced 92% and Apriori created 80% valid results. The dataset has been tested in dual environment and significant improvement has been noted for Treap algorithm in both cases.

Keywords: Treap algorithm · Association mining · Survival prediction · Apriori Algorithm · Priority model

1 Introduction

Association rules are *if-then* rules, which help to uncover the vast relationship between seemingly unrelated data. It uses a combination of statistical analysis, machine learning and database management to exhaustively explore the data to reveal the complex relationships that exists. To do so, a complete study and analysis of the entire database is required. This is the primary challenge of all mining tasks, the huge amount of time that is required to virtually finish the mining. Another challenge of mining is incremental mining, where data is mined over an already mined pattern. This can make the discovered relations totally inefficient.

There are two main concept related to association rule mining. They are support and confidence. Let us define both terms

© Springer International Publishing AG 2017
Y. Tan et al. (Eds.): DMBD 2017, LNCS 10387, pp. 31–39, 2017.
DOI: 10.1007/978-3-319-61845-6_4

Support: The rule $x \rightarrow y$ holds with support s, if $s\%$ of transactions in the database D contains all items of x and y, i.e. x union y. Rules that have a greater s than a user defined support is said to have minimum support.

Confidence: The rule $x \rightarrow y$ holds with confidence c, if there is $c\%$ of the transactions in D that contains x also contains y. Rules that have a greater c than a user defined confidence is said to have minimum support.

In this work, a novel approach of rule mining is put forward. A priority based elimination scheme is employed for the process. The data structure used is Treap. The work basically is carried out in four different procedure calls. Treap is used as it shows both the properties of tree and heap data structures. This helps to organize the data in priority calculation, which in turn helps in mining high priority itemsets. From the items the association rules are created.

Medical databases contain large volume of data about patients and their clinical information. For extracting the features and their relationships from such huge database, various data mining techniques need to be employed. This scope has been utilized in this work. Liver transplantation data was collected from the United Nations Organ Sharing (UNOS) registry and long-term survival prediction rules has been generated using ARL algorithms.

Liver Transplantation (LT) is considered as the only viable treatment for end stage liver diseases. In transplantation, the physician or the medical experts decide the priority of allocating of resource according to the disease severity of patient and MELD (Model for End Stage Liver Disease) score. MELD score, which yields a numeric value based upon serum creatinine, bilirubin and INR has been successful in prognosticating 90 day mortality for these patients, and has proven to be a method of liver allocation. However, a careful look at the parameters of the MELD score reveals the limitations and resultant caution that should be given to ostensibly objective data. Creatinine and INR are labile especially in the setting of patients with advanced liver disease that are prone to alteration not only by the inherent disease state but also iatrogenic interventions. The implications of these interventions have significant medical and moral consequences as they not only determine immediate treatment but also which patients are allocated the precious life-extending resource of organ transplant. So a high accuracy model for prediction need to be in place, which can overcome the local maxima to a global maxima solution. Using the dynamic priority Treap mining, set of valid association rules were able to be generated, which in turn was the major input for building a long term survival prediction model. The association rules pointed out the impact of various serum attributes and the relations among them. Artificial neural network based prediction model has also been developed using this association mining rules, which in turn out performed the MELD score prediction.

2   Related Research

Fast Algorithms for Mining Association Rules [Agrawal and Srikant, 1993] throws light to the problems of discovering association rules between items in a large database of sales transactions using Apriori Algorithm. In Advanced Version of Apriori Algorithm

(Suneetha and Krishnamoorti 2010) the authors discuss the limitations of the original Apriori algorithm and present a modified version of Apriori algorithm to increase the efficiency of rules generated. High dimension oriented Apriori Algorithm (Lei Ji et al. 2006) adopts a new method to reduce the redundant generation of sub-itemsets during pruning the candidate itemsets, which can obtain higher efficiency of mining than that of the original algorithm when the dimension of data is high. Theoretical proof and analysis are also provided for strong validation. Tertius is a mining algorithm, which works with the help of background knowledge (Flach and Lachiche 2001). It is an inductive logic programming system that performs confirmatory induction, i.e., it looks for the n clauses that have the highest value of a confirmation evaluation. In order to compute its confirmation measure, Tertius needs to populate two contingency tables for each clause, one for observed values and one for expected values. Based on a time series sequence of clinical data (Paramanto et al. 2006) conducted a study with recurrent neural networks using time series sequence of medical data in 2001. They used Back Propagation Through Time (BPTT) algorithm and achieved a better survival rate with 6-fold cross validation. (Zhang et al.) performed a study for the comparison between MELD and Sequential Organ Failure Assessment scores. They used a MLP model for liver patients with Benign End-Stage Liver Diseases. In 2013, (Cruz-Ramirez et al.) introduced a Radial Basis Function network model using multi objective evolutionary algorithm to address the liver allocation and survival prediction.

3 Treap Mining Algorithm

Treap is a data structure, which has the properties of both tree and heap. A node in a Treap is like a node in a Binary Search Tree (BST). In BST, it has a data value, x, but in Treap it has a unique numerical priority, p, in addition to, x [20]. The nodes in a Treap also obey the heap property; that is, at every node u, except the root, u.parent.p<u.p. By the way in which the root and its child are related, treap can be classified as Mini Treap (minimum priority) and Max Treap (maximum priority).

The basic input to the algorithm is sets of items (variables) say n and m represents the number of transactions and S represents the support. Major data structures used in this algorithm are Array and Treap. The basic design of the Treap mining algorithm in pseudo code format is given in Algorithm 1.

Algorithm 1. Rule Mining Treap

Input: Database D, with n itemsets and m transactions. Let the support be S
Output: Association Rules
Data structures Used: Array, Treap
        Step 1: for each item set n ϵ D, calculate priority
        Call Priority Procedure (D, S)
        Step 2: For i=1 to n in Priority Array P[n]
        Call BuildTreap Procedure
        Step 3: BSF traversal in Min Treap
        Call Traversal Procedure
        Step 4: For each mined item
        Call GenerateRule Procedure
        Step 5: Output rulesR

Priority Procedure: In a database even if an item set is not frequent there is chance that it can be of some importance in rule generation. In all association algorithms such infrequent items are pruned off without much analysis. In this proposed work, the priority of all item sets is found and it is analyzed with the frequency. After this analysis, if the item set is still invalid it is eliminated or else it is considered for rule creation. The algorithm begins by scanning the database and frequency, **f** is calculated for all item set. Partial priority of item set, which is the component of priority, is found out using the equation

Partial priority p' for each itemset,

$$pi' = S * \frac{fi}{\sum f} \tag{1}$$

The calculated partial priority is added up with the normalized frequency to calculate the weight. Normalization is important since it is a variance maximizing exercise and it projects our original data onto directions which maximize the variance. Thus even if we have a big difference in the frequency range, it brings down the range to a favorable boundary. Here Z-score normalization technique is used.

$$\text{Normalized,} \quad e_i = \frac{e_i - E'}{std(E)} \tag{2}$$

where

$$std(E) = \sqrt{\{\frac{1}{n-1} \sum_{i=1}^{n} (e_i - E')^2\}} \tag{3}$$

and

$$E' = \frac{1}{n} \sum_{i=1}^{n} e_i \tag{4}$$

Now the threshold is set and the priority of each itemset is found out. The weight added is the maximum allowed that can be given to any item set in that database D to maintain threshold.

The maximum possible weight (W_{mp}): Let **Y** be a p-itemset, and **X** is a superset of **Y** with the k-itemset (p < k). The maximum possible weight for any k-itemset containing **Y** is defined as

$$W_{mp}(Y, k) = \frac{1}{k}\left(\sum_{ij \in y, j=1}^{n} w_j + \sum_{l=1}^{k-n} w_l\right) \tag{5}$$

In Eq. (5), the first part is the average weight for the p-itemsets and the second part is the average weight of the (k-p) maximum remaining itemsets.

BuildTreap Procedure: Min Treap is the data structure used in this work. This is a tree, which satisfies the Min Heap Property. The basic property of Min Treap is the root node has a smaller priority than its children. The BuildTreap procedure obtains its input from the Priority procedure. The algorithm scans each element from the list and builds a Min Treap according to the priority. Whenever a new element is added the BuildTreap subroutine is recursively called in-order to maintain the Treap structure. The process is very similar to that of heap creation. By MinTreap procedure, Treap for each transaction is build and the leaf node will be the one with highest priority. So, by employing a depth first search we can get to the nodes with highest priority.

Traversal Procedure: This is a depth first Treap traversal where the algorithm tries to find the most frequent items with highest priority from the Treap. Search continues till the leave node returns the priority value and the node value. If the priority is greater than the threshold, parent and the sibling of the node is returned. If there is no sibling, then the sibling of the parent is retuned by the subroutine. The search moves in a bottom up approach until the priority has reached its minimum threshold.

GenerateRule Procedure: This is the final step of the algorithm, here all those rules above the threshold T is calculated. Initially the rule set R is initialized to NULL. For each frequent item obtained from the Treap traversal, the ratio of frequent item to its supporting item sets is found out. If it is greater than the threshold, the rule is added to R. This process is continued until all the frequent items are visited for rule creation.

3.1 Analysis

In the proposed work, four procedures are called back to back, thus the total complexity will be the upper bound of these procedure calls. Priority calculation procedure always takes a linear order as the amount of time taken for calculation is directly proportional to the number of input elements. BuildTreap procedure works in a similar way as heap sort. As we know that the total running time of a heap sort is O (n log n). In case of treap, we can build it more quickly, since we may not have to extract all the elements from the tree. Let us assume n takes the form of,

$$n = 2^{h+1} - 1 \text{ where h is the height of the tree.} \tag{6}$$

At the bottom most level there are 2^h nodes, the one at the next to bottom level has 2^{h-1} node. In general, at level j from the bottom there are 2^{h-j} nodes and it need to be shifted down j levels to satisfy the heap property. Thus time complexity proportional to this is given by,

$$T(n) = \sum_{j=0}^{h} j.2^{h-j} = \sum_{j=0}^{h} j\frac{2^h}{2^j} \tag{7}$$

Factor out the 2^h term and taking derivative, we get

$$\sum_{j=0}^{\infty} j\, x^{j-1} = \frac{1}{(1-x^2)} \tag{8}$$

Multiplying x on both sides and substitute value x = 1/2

$$\sum_{j=0}^{\infty} \frac{j}{2^j} = \frac{\frac{1}{2}}{(1-\frac{1}{2})^2} = 2 \tag{9}$$

Substituting the value of Eq. 8 in Eq. 7, we get

$$T(n) = 2^h . \sum_{j=0}^{h} \frac{j}{2^j} \leq 2^h . \sum_{j=0}^{\infty} \frac{j}{2^j} \leq 2^h.2 = 2^{h+1} \tag{10}$$

So we can conclude, $T(n) \leq n + 1 \in O(n)$
Similarly for rule generation for n transactions with m itemsets can be given as,

$$\sum_{k=1, j=m_{c_k}}^{k=m} \sum_{i=1}^{j} j_{C_i} \tag{11}$$

$$\sum_{k=1}^{m} m \sum_{i=1}^{j} j_{C_i} \tag{12}$$

$$\cong O(n)\text{as } n \to \infty \tag{13}$$

The analysis part clearly describes that the algorithm works in the linear order, O (n) in the best case scenario, and even keeps the algorithm steady in O(n log n) in worst case scenarios, which is far better than the competitor algorithms of the same genre.

4   Results and Discussion

Treap mining algorithm has been tested with the UNOS dataset. The various associations among the attributes from the database are being calculated to find the priority. This priority of attributes also plays an important role while modeling the prediction system. Association mining algorithms were run on dual environment, the one before preprocessing and ranking (which contained 197 attributes) and one after ranking (which had 27 attributes). This was done in-order to find the rule prediction accuracy. Associations mean how much the attributes in that rule are inter-related. By checking the rule accuracy, the attributes of high affinity can be selected. Figure 1 shows the number of rules generated while applying the association mining algorithms to the 197 attributes.

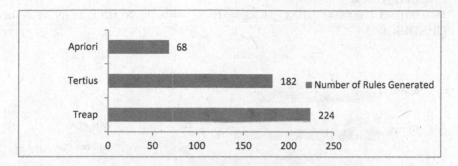

Fig. 1. Number of rules generated before ranking

For finding the relevance of attributes, the 197 attributes were ranked and dimensionality of data was reduced. These attributes were fed to the Weka ranker. With the 27 attributes obtained from Weka ranker, the association mining algorithms were run again to compare the prediction accuracy. The results showed that the rules obtained while using 27 attributes were the subset of the rules generated while working with 197 attributes. The accuracy of Treap was increased to 96.71% and Tertius showed a steep performance improvement to 92.18% and even though Apriori could generate only ten rules due to memory overflow, the accuracy was found to be 79.87%. Increasing the threshold can increase the rule accuracy and support. Figure 2 shows the comparison of three mining algorithms with respect to Number of rules generated and rule accuracy.

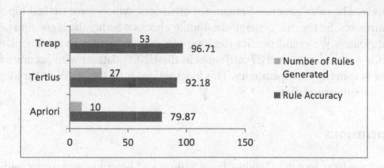

Fig. 2. Number of rules generated and rule accuracy

Rule accuracy improvement graph is shown in Fig. 3.
The sample rules obtained are shown below.

1. /* 0.001780 0.000000 */FINAL_MELD_OR_PELD = PELD ==> NON_HRT_
DON = N

2. /* 0.001485 0.000000 */EXC_HCC = HCC ==> NON_HRT_DON = N or MALIG_TRR = N
3. /* 0.001208 0.000000 */EXC_HCC = HCC ==> NON_HRT_DON = N or ENCEPH_TCR = Y
4. /* 0.001075 0.000000 */EXC_HCC = HCC ==> NON_HRT_DON = N or GENDER = M

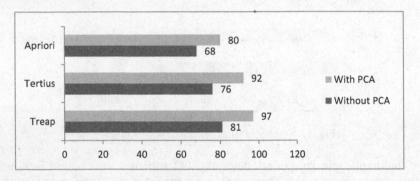

Fig. 3. Rule accuracy comparison

From the comparisons, we could see that association rule mining algorithms can be used in predicting long-term survival rates with high accuracy level. Among the rule mining algorithms, Treap mining was giving more consistent results. Without using preprocessing and ranker, the rule accuracy was 68% and with Weka ranker, the rule accuracy was 80% while using Apriori algorithm. The rule accuracy was 92% with Weka ranker and 76% without ranker for Tertius algorithm. The most powerful association rule mining algorithm, Treap produced 81% rule accuracy without ranking and 97% with Weka ranker. The rules generated contained attributes, which were ranked top among the 197 attributes. So the rule generations double checks whether the algorithm selected the top attributes. We could predict the long-term survival after liver transplantation successfully with the extracted 27 attributes in the UNOS dataset with accurate results in terms of Sensitivity and Specificity. This associations form the platform for long-term survival prediction modeling.

5   Conclusions

Health and medical sector is having large volume of heterogeneous data and hence powerful data mining techniques are needed for predictions and decision-making. One of the key areas in medicine is the prediction of suitability and survival rate after organs transplantation. In this work a new association mining algorithm, Treap is being discussed and its scope in predicting valid associations among the various available attributes. Treap mining algorithms when compared with other traditional algorithms produces valid results in least time complexity, O (n log n). While using Treap mining in creating association rules for designing a survival prediction model, the mining results produced were far superior to other algorithms of the same genre. Treap mining algorithm produced 97% accurate association predictions and there by acting as the basic input to survival prediction model.

References

1. Boney, L., Tewfik, A.H., Hamdy, K.N.: Minimum association rule in large Database. In: Proceedings of Third IEEE International Conference on Computing, pp. 12–16 (2006)
2. Agarwal, R., Srikant, R.: Fast Algorithms for mining association rules. In: Proceedings of VLDB, pp. 487–499 (1994)
3. Bodon, F.: A fast apriori implementation. In: Proceedings of IEEE ICDM Workshop on Frequent Item set Mining Implementation, vol. 9 (2003)
4. Borgelt, C.: Recursion pruning for the apriori algorithm. In: Proceedings of 2nd IEEE ICDM Workshop on Frequent Item Set Mining Implementations, vol. 126 (2004)
5. Zaki, M., Parthasarathy, S., Ogihara, M., Li, W.: New algorithms for fast discovery of association rules. In: Proceedings of 3rd International Conference on Knowledge Discovery and Data Mining, vol. 2, pp. 283–296 (1997)
6. Anandhavalli, Gautaman, K.: Association rule mining in genomics. Int. J. Comput. Theory Eng. **1** (2007)
7. Cooper, C., Zito, M.: Realistic synthetic data for testing association rule mining algorithms for market basket databases. Knowl. Disc. Databases: PKDD **9**, 398–405 (2007)
8. Varde, A.S., Takahashi, M., Rundensteiner, E.A., Ward, M.O., Maniruzzaman, M., Sisson, R.D.: Apriori algorithm and game of life for predictive analysis in materials science. Int. J. Knowl. Based Intell. Eng. Syst. **8**, 116–122 (2004)
9. Wu, H., Lu, Z., Pan, L., Xu, R., Jiang, W.: An improved apriori based algorithm for association rules mining. In: Proceedings of Sixth International Conference on Fuzzy Systems and Knowledge Discovery, pp. 51–55 (2009)
10. Kamath, Wiesner: Malinchoc, Kremers, Therneau, Kosherg.: A model to predict survival in patients with end stage liver disease. Hepatology **1**, 464–470 (2001)
11. Cruz-Ramírez, M., Hervás-Martínez, C., Fernandez, J.C., Briceno, J., De-La-Mata, M.: Predicting patient survival after liver transplantation using evolutionary multi-objective artificial neural networks. Artif. Intell. Med. **58**, 37–49 (2013)
12. Doyle, H.R., Dvorchik, I., Mitchell, S., Marino, I.R., Ebert, F.H., McMichael, J.: Predicting outcomes after liver transplantation. A connectionist approach. Annals of Surgery, pp. 408–419 (1994)

Does Student's Diligence to Study Relate to His/Her Academic Performance?

Toshiro Minami[1($\boxtimes$)], Yoko Ohura[1], and Kensuke Baba[2]

[1] Kyushu Institute of Information Sciences, Dazaifu, Fukuoka, Japan
minamitoshiro@gmail.com , ohura@kiis.ac.jp
[2] Fujitsu Laboratories, Kawasaki, Kanagawa, Japan
baba.kensuke@jp.fujitsu.com

Abstract. It is often pointed out that students' academic performance becomes worse. Lack of professors' teaching ability is often considered its major cause, and universities promote faculty development programs. According to our observation, however, the major cause is rather on student's side, such as lack of motivation, diligence, and other attitudes toward learning. In this paper, we focus on diligence. Diligence is quite important for students to learn effectively. Among various kinds of diligence, we take two kinds of them into consideration; the length of answer text to a questionnaire, and the amount of submitted homework assignments. We investigate how these kinds of diligence of students relate each other, and how they relate to the examination score.

Keywords: Educational Data Mining · Lecture data analytics · Attitudes to learning · Retrospective evaluation

1 Introduction

Universities have been popularized in many countries including Japan. As a result, students' academic performances have been decreasing, and academic staff are required to improve their teaching skills by attending faculty development (FD) programs. As we observe, however, the major cause is rather on student side, such as their lack of motivation, diligence, and attitudes to learning. Therefore, it is more important to motivate students and correcting their attitudes to learning. To this aim, we need to know more about students.

We choose an approach with data analytics for this aim. It consists of two steps: (1) to make a student's learner model which includes attitudes to learning by proposing new concepts and measuring indexes for them, and see what we can find, and (2) to advise the student on the basis of his or her learner model. This approach has an advantage in terms of understandability of humans. Even though the method we are developing is a naive one, we prefer to choose the understandable method rather than the established and more sophisticated methods if they are less understandable.

© Springer International Publishing AG 2017
Y. Tan et al. (Eds.): DMBD 2017, LNCS 10387, pp. 40–47, 2017.
DOI: 10.1007/978-3-319-61845-6_5

As a part of such an approach, we have analyzed the answer texts of a questionnaire, which asked the students to evaluate the class and themselves by looking-back [4]. These text data are appropriate for analyzing the students' attitudes to learning. Diligence is one of such attitudes and is quite important for students learn effectively. In this paper, we deal with two types of diligence; length of the answer text to a questionnaire, and amount of submitted homework assignments. We investigate how these kinds of diligence of students relate each other, and how they relate to the examination score.

Educational data analysis has been conducted mainly in the field of Educational Data Mining (EDM) [8]. For example, Romero et al. [9] studied algorithms to classify students using e-learning system data. Its interest was on predicting the student's outcome. We focus on the student's behavioral tendency in learning, such as eagerness, diligence, or seriousness. Even though the motivation of Ames et al. [1] is close to us, it is different that their underlying data were obtained by asking the students to choose the rate, whereas most of our data are free-texts, and thus, potentially contain information about the students with a wide spectrum of attitudes in more detail than with only the rates.

Most studies in EDM field intend to deal with big data, and the data are obtained automatically as log data from learning management systems. By contrast, the data we deal with are small data, because our target data could be very small [3,4]. We apply statistical and data mining methods carefully, and we have to take care of all data even if they are outliers statistically, because even such data represent some actual students.

The rest of this paper is organized as follows. In Sect. 2, we describe the target data for analysis. In Sect. 3, we analyze the answer-texts of students to the questions about the course and the students. In Sect. 4, we analyze the amount of homework submissions, including a comparative study of two types of diligence. Finally, in Sect. 5, we conclude the discussions and findings in this paper together with some prospects to the future study.

2   Target Data

The data used in this paper originated from the class named "Exercise on Information Retrieval" in 2009 in a junior (two year) college [2–7]. The number of attended students was 35. They were 2nd year students and were going to graduate. It is a compulsory course for librarian certificate. The course consisted of 15 lectures. A homework was assigned in every lecture, which aimed to let the students review what they had learned that day.

The term-end examination of the course aimed to evaluate the skills which were supposed to have learned and trained in the classroom and through doing the homework assignments. We will use the score of term-end examination as the measure for the student's academic performance.

At the end of the course, we asked the students to answer some retrospective questions that evaluate themselves and the lectures/lecturer. For example, (Q1) What did you learn in this class? Did it help you?, (Q2) What are the good points of the lectures?, (Q3) What are the bad points that need to be improved?, (Q4) What score you give to the lectures as a whole?, (Q6) What are your good points in learning attitudes and efforts for the course?, (Q7) What are your bad points that should have to be improved?, (Q8) How do you evaluate of your diligence and eagerness to study?, and (Q11) What score you give to yourself as the evaluation of your own efforts and attitude toward the course.

3   Text-Length Analysis

We take the length of answer-text of a questionnaire, which we call "l-diligence", as the first index for measuring student's "diligence" [7]. It is reasonable to use text-length for measuring diligence because if a student is more diligent to learning, she might try harder to answer the questions and the text should become longer. As the target text for analysis, we choose the answers to the questions (Q2), (Q3), (Q6), and (Q7), which are the answers to evaluation questions on the contrasting topics for evaluation such as good and bad points of the lectures and the student herself.

In this section, we start with investigating the correlation of l-indexes for different questions and how they are related with academic performance in Sect. 3.1, Then we divide the students into 4 types by l-index, and investigate the differences between them in Sect. 3.2.

3.1   Correlation Analysis of l-Indexes for Questions and Their Relation to Academic Performance

We deal with 4 l-indexes for contrasting questions, which we call LG (Lecture-Good), LB (Lecture-Bad), SG (Student/Self-Good), and SB (Student/Self-Bad) for the answer texts to the questions (Q2), (Q3), (Q6), and (Q7), respectively. For a text T, let $|T|$ denote the length (l-index) of T. For example, $|LG|$ denotes the text length of LG, and $|SB|$ the length of SB. We also use L for lecture data text obtained by concatenating the texts of LG and LB. Thus, $|L| = |LG|+|LB|$. Similarly, we use S for SG and SB, G for LG and SG, B for LB and SB. We use All to all the texts. Thus, $|All| = |L|+|S|$ of the student.

Table 1 shows sample length data. The length "0" means that the student did not answer the questions at all. The numbers of students who answered the questions are 29, 23, 22, 21, and 29 for LG, LB, SG, SB, and All, respectively. For L, S, G, and B, the numbers are 29, 22, 29, and 24, respectively. There were no students who answered to SG or SB and did not answer to LG and LB, Only 20 students answered to all questions.

We are interested in whether/how the text lengths relate to the performances of students. Figure 1 shows the correlation between the total length, i.e., $|All|$, of students' answer texts and their examination scores, which we consider the index for their performances. Their correlation coefficient is 0.26, which shows that there is no strong correlation between them.

Table 1. Sample l-diligence data for students from "st01" to "st05"

StID	|LG| (Q2)	|LB| (Q3)	|SG| (Q6)	|SB| (Q7)	|All| (Q2-7)	|L| (Q2,3)	|S| (Q6,7)	|G| (Q2,6)	|B| (Q3,7)
st01	130	0	0	0	130	130	0	130	0
st02	165	0	0	0	165	165	0	165	0
st03	220	123	168	212	723	343	380	388	335
st04	81	136	37	72	326	217	109	118	208
st05	214	227	117	55	613	441	172	331	282

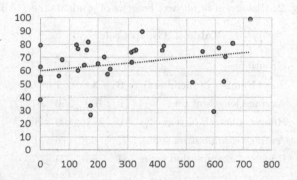

Fig. 1. Correlation between the examination score, (y-axis) and the numbers of characters in all the answer texts of the students (x-axis)

3.2   Analysis of Differences Between Groups

We have found that there is no strong correlation between the student's l-diligence and her performance measured by examination score in our previous study. However, according to Fig. 1, students may be categorized by l-diligence, and the performances may differ from one group to another. We investigate the gaps between the consecutive text lengths.

Firstly, we sort the values of text lengths and get the sequence < 723, 661, 635, 630, 613, 595,..., 0, 0 >. Then we get the values of differences between a value and the next one, as < 62, 26, 5, 17, 18,... 0 >. Note that $723 - 661 = 62$, $661 - 635 = 26$, $635 - 630 = 5$, and so on. The maximum gap locates between 521 of the student st35 and 420 of st11, followed by 70 between 311 of st21 and 241 of st25, and 69 between 415 of st21 and 346 of st14. Further, the minimum non-zero value is 64 of st23. By considering these values, we divide the students into 4 groups using text length: TU (Upper) for those whose text length is greater than 500, TM (Medium) for others with length > 300, TL (Lower) for others with length > 50, which actually means non-zero, and TZ (Zero) for those who did not answer to any questions.

Figure 2 shows the histogram of the text length of student together with the boundary values for dividing the students into TU, TM, TL and TZ groups. Table 2 shows some of the profile information of the groups.

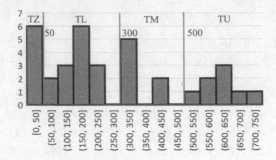

Fig. 2. Histogram of all text lengths of students, i.e., |All|

Table 2. Group profiles

Group	TZ	TL	TM	TU
Size (Number of members)	6	14	7	8
Average length of text	0	160	350	617
Range of length of text	0	177	109	202
Average examination score	57	63	76	67
Range of examination score	41	55	24	70

Regarding the group size, TL is the biggest with 14 members and the other groups are about half of TL. Regarding the range of length of text, TU has the biggest value of 202, followed by TL with 177, and TM with 109. Regarding the average examination score, TM has the maximum score 76, followed by TU with 67, TL with 63, and TZ with 57. Regarding its range, TU with 70 has the maximum, followed by TL with 55, TZ with 41, and TM with 24.

It is interesting to see that TU has the biggest range in its l-diligence, and in its examination score. This result shows that the students who answered with long texts vary a lot in their academic performances. Thus, l-diligence is not effective to estimate the student's academic performance. However, amount of writing, or l-diligence, should be used for measuring some kind of student's diligence because it needs some amount of labor to write a long text, which is different from academic performance in learning.

Another interesting point is, as we can see in Fig. 1 also, that TM has the highest examination score than other groups and its range is the smallest. The students in group TM belong to the middle-upper group in their academic performances. These students are moderate in many aspects. They are the people who are moderate in answering text, and might be moderate in learning, and are moderate in academic performance. The group TL is common with TU in many aspects. The ranges are second biggest both in l-diligence and examination score. Different from TM, the extreme cases TU and TL contain a wide variety students. This may be a generally applicable rule. We need to investigate further in the future.

The group TZ consists of the students who did not answer at all, thus they are non-diligent regarding l-diligence. Probably because they are non-diligent also in learning in general, the average examination score of TZ is the lowest. However, this is not a strong rule because the highest examination score of TZ is 79 of st09. Presumably, some students with high academic performance may not be diligent enough in answering to questions, possibly because answering to questions does not directly relate to studying.

4 Analysis of Homework Submission

In this section, we investigate another candidate index for diligence, the number, or the amount, of homework submissions, or h-diligence. Some sort of diligence is necessary in order to do homework assignment and submit it.

Our experience shows that some students regularly submit homework whenever it is assigned, whereas some others do partially, or do not do at all. We introduce h-diligence as another type of diligence by the number of homework submissions.

Figure 3 shows the correlation between h-diligence and examination score. Their correlation coefficient is -0.05; no strong correlation between them. The figure shows that the range of the examination scores is very wide among the students with high h-diligence. Precisely, the range for the students with h-diligence 13 is 72 (from 27 in minimum to 99 in maximum), and the range for those with h-diligence 12 is 53. It is highly possible that the students with high h-diligence and low examination score are those who do their homework because it is their obligation, not because it is good for them to learn more effectively. This result shows that the teacher has to discriminate false diligence from true one in order to advise students more appropriately.

Figure 4 shows the correlation between h-diligence and l-diligence. The two indexes do not correlate strongly (with negative correlation coefficient -0.17). It is interesting that there are two students who are very low in h-diligence, and l-diligence is very high. Like in Fig. 3, the range of l-diligences for those with h-diligence 13 is very large. It holds also for those with h-diligence 12. Different from these students, the ranges are smaller for other h-diligence values.

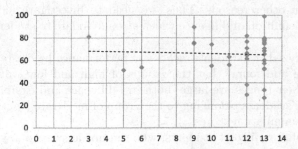

Fig. 3. Correlation between h-diligence (x-axis) and examination score (y-axis)

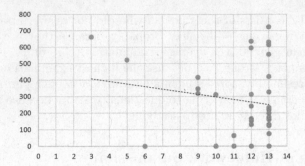

Fig. 4. Correlation between h-diligence (x-axis) and l-diligence (y-axis)

This phenomenon also holds as we see the figure by l-diligence. The students with high l-diligence values range widely in their h-diligence. The ranges are small for those with lower l-diligence values.

According to these observations, the students with high values in one diligence vary a lot in the other diligence and academic performance, whereas those with middle or low diligence indexes vary moderately. Thus, in order to advise students appropriately, the teacher needs to find the students who are seemingly diligent, but actually are not so much diligent, and advise them.

5 Concluding Remarks

In the universities, it is important to improve students' attitudes to leaning in order to improve their academic performance. In this paper, we proposed two indexes which intend to capture students' diligence. The first index (l-diligence) comes from the answer texts to an evaluation questionnaires. The second one (h-diligence) comes from the homework submissions.

We examined the hypotheses that indexes reflect some types of student's diligence, which are correlated with academic performance. The result shows that the diligences in these types do not strongly relate to student's examination score. We have concluded that l- and h-diligence capture the concept superficially. A student submits homework not because he/she is diligent, but because they do so from other reasons. Thus, we need to investigate more deeply in order to find a definition of diligence which strongly relate to student's academic performance.

We also analyzed the students' academic performance by clustering them by using l-diligence and found that the groups of high and low values share the property that their ranges are large, both in l-diligence and examination score. By contrast, the students belonging to the group of moderate length has a small range in their examination scores.

The group of non-diligent students, who did not answer to any questions at all, have some notable amount of range, and thus, non-diligent does not mean they are low performers. Even though, in average, their examination score is lower than other groups, some students have high examination scores. This also indicates that students' diligences we dealt with in this paper and their academic performance do not have an explicit relation.

Our future work includes: (1) a further investigation using other data in order to verify our findings are applicable to them; (2) an investigation for alternative index for measuring diligence from the answer text, such as the number of the occurrences of each word; (3) a formalization of the concept of diligence from the beginning in order to discriminate the true diligence from the false diligence. To find an appropriate formalization, we have to answer such questions as: "What is diligence?", "In what way diligence is important for students?", and "How can a lecture help students with improving diligence?".

Acknowledgments. This work was supported in part by JSPS KAKENHI Grant Number JP15K00310.

References

1. Ames, C., Archer, J.: Achievement goals in the classroom: students' learning strategies and motivation processes. J. Educ. Psychol. **80**(3), 260–267 (1988)
2. Minami, T., Ohura, Y.: An attempt on effort-achievement analysis of lecture data for effective teaching. In: Kim, T., Ma, J., Fang, W., Zhang, Y., Cuzzocrea, A. (eds.) FGIT 2012. CCIS, vol. 352, pp. 50–57. Springer, Heidelberg (2012). doi:10.1007/978-3-642-35603-2_8
3. Minami, T., Ohura, Y.: Towards development of lecture data analysis method and its application to improvement of teaching. In: 2nd International Conference on Applied and Theoretical Information Systems Research (2ndATISR 2012), 14p. (2012)
4. Minami, T., Ohura, Y.: Investigation of students' attitudes to lectures with text-analysis of questionnaires. In: 4th International Conference on E-Service and Knowledge Management (ESKM 2013), 7p. (2013)
5. Minami, T., Ohura, Y.: Towards improving students' attitudes to lectures and getting higher grades -with analyzing the usage of keywords in class-evaluation questionnaire-. In: The Seventh International Conference on Information, Process, and Knowledge Management (eKNOW 2015), pp. 78–83 (2015)
6. Minami, T., Ohura, Y.: How student's attitude influences on learning achievement? -an analysis of attitude-representing words appearing in looking-back evaluation texts. Int. J. Database Theory Appl. (IJDTA), Sci. Eng. Res. Support Soc. (SERSC) **8**(2), 129–144 (2015)
7. Minami, T., Ohura, Y., Baba, K.: How can we assess student's diligence from lecture/self-evaluation -an approach with answer-text analysis of looking-back questionnaire-. In: International Joint Conference on Convergence (IJCC 2017), pp. 1–7 (2017)
8. Romero, C., Ventura, S.: Educational data mining: a survey from 1995 to 2005. Expert Syst. Appl. **33**(1), 135–146 (2007)
9. Romero, C., Ventura, S., Espejo, P., Hervas, C.: Data mining algorithms to classify students. In: 1st International Conference on Educational Data Mining (EDM 2008), pp. 8-17 (2008)

Correlation Analysis of Diesel Engine Performance Testing Data Based on Mixed-Copula Method

Zha Dongye[1], Qin Wei[1(✉)], Zhang Jie[2], and Zhuang Zilong[1]

[1] School of Mechanical Engineering, Shanghai Jiao Tong University, 200240 Shanghai, China
wqin@sjtu.edu.cn
[2] College of Mechanical Engingeering, Donghua University, 201620 Shanghai, China

Abstract. As an important quality index of diesel engine, the power of diesel engine has a direct impact on the quality and competitiveness of products. Considering the diesel engine performance testing process including many control parameters which are strongly coupled with each other, it is difficult to carry out accurate probability distribution description and correlation analysis, this paper presents a correlation analysis method based on mixed copula model to figure the correlations between multi parameters. This paper begins with the analysis of the engine performance test data with characteristic of non-normal, peak and fat tail, and according to the correlation structure of diesel engine power data, the mixed copula function is constructed by using the weighted linear model to describe asymmetric tail behavior and the expectation maximization method to estimate the related parameters. The results showed that the mixed copula function can well describe the power of diesel engine related structure and tail characteristics.

Keywords: Correlation analysis · Performance testing data · Tail dependence · Mixed-Copula method

1 Introduction

The performance, the reliability and the service life of the automobile are affected by the performance of the diesel engine which is taken as the heart of the loading automobile [1]. The diesel engine manufacturer will inspect the assembling quality and performance of the whole diesel engine by testing the performance of the engine bench upon completion of general assembly of the diesel engine. The inspection will test up to 2000 parameters, including environmental parameters (atmospheric pressure, temperature, humidity, etc.), oil pressure, crank angle, starting torque of crank, etc., of which some performance parameters correlate with each other [2]. Ignoring the relation, respectively monitoring each parameter, will lead to the overall performance of diesel engine offset. So the correlation analysis must be implemented first for effectively quality control of diesel engine. Then how to carry out the correlation analysis between the inspected parameters of diesel engine effectively is the crucial problem that should be explored and studied.

© Springer International Publishing AG 2017
Y. Tan et al. (Eds.): DMBD 2017, LNCS 10387, pp. 48–57, 2017.
DOI: 10.1007/978-3-319-61845-6_6

The inspected data of different parameters show different unknown distribution, and most of them have the characteristic of peak and thick tail [3]. These inspected data were seldom explored and utilized before, and there were a few simple distribution statistics and analysis of single testing curve only [4]; With the development of the research, domestic and foreign scholars studied the correlation between the engine testing process parameters. Peng Su [5] used one dimension nonlinear regression method to analyze the correlation between oil pressure and oil temperature, and established an empirical formula to compensate the oil temperature. However, nonlinear regression depends on the artificial selection of transfer function, which is not realistic for correlation analysis of multivariate parameters. Assian and Watson [6] promoted a similar empirical correlation coefficient of exhaust gas pressure and exhaust gas temperature, however, the correlation coefficient also has its limitations. The characteristics of peak and thick tail result in the difficulty of modelling with traditional multivariate statistical analysis method. Therefore, the new correlation measuring index shall be imported to implement the correlation analysis in allusion to the performance testing data of the diesel engine.

Copula function has been applied to the fields [7–9], such as finance, geology, wind power, etc. in recent years because of its great advantages in forming random marginal distribution function and random joint distribution function of relevant structural variables [10]. Further, the Copula theory has also been applied to study [3] for analysis in cold commissioning phase of the automobile engine. However, the advantages of different Copula function models in the Copula function family to fitting of relevant structure of the data could not be realized really because the single Copula function was applied to description of the correlation among the parameters, and the data could not be described fully. Therefore, the characteristics of the performance testing parameters of the diesel engine bench should be analyzed first in this text, relevant structure among the parameters should be formed by starting with the relevant structure and utilizing the advantages of the hybrid Copula function in description of data comprehensively, and the correlation of the parameters in the performance test of the diesel engine bench should be studied on the basis of the relevant structure.

2 Basic Theory of Mixed Copula Function

In 1959, Sklar [11] proposed the Copula function, which indicated that the Copula can be used to represent the n-dimensional joint distribution function with n edge distribution functions and a Copula function, which laid the foundation for the development of the method system. Nelsen [12] systematically summarizes the definition and construction of Copula function, and applies it to correlation analysis.

Definition of Copula Function 1 [13]: An n-dimensional Copula is a multivariate d.f., C, with uniform distributed margins in $[0, 1](U(0, 1))$ and the following properties: $C:[0, 1]^n \rightarrow [0, 1]$;$C$ is grounded and n-increasing; C has margins C_i, which satisfy $C_i(u) = C(1, \ldots, 1., u, 1, \ldots, 1) = u$ for all $u \in [0, 1]$.

Theorem 1 [14]: Let F be an n-dimensional d.f with continuous margins $F_1, \ldots, F_n$. Then it has the following unique copula representation:

$$F(x_1, \cdots, x_N) = C(F_1(x_1), \cdots, F_N(x_N)). \tag{1}$$

From Sklar's Theorem we see that, for continuous multivariate distribution functions, the univariate margins and the multivariate dependence structure can be separated. The dependence structure can be represented by a proper Copula function.

2.1 Single Copula Function

The common Copula function has the elliptic Copula function and the Archimedes Copula function.

The elliptic Copula function has a density function that obeys the elliptic distribution, which is commonly *Normal Copula*, *t-Copula*. The upper and lower tail correlation coefficient of the *Normal Copula* function is 0, so *Normal Copula* can`t capture the tail dependence of the variables. The thick tail of the *t-Copula* function is more obvious than the *Normal Copula* function, and has a symmetric tail dependence.

Archimedes Copula function common *Clayton Copula*, *Gumbel Copula* and *Frank Copula*. The *Clayton Copula* function is sensitive to the change of the lower tail; the *Gumbel Copula* function is sensitive to the variation of the upper tail; the density function of the *Frank Copula* function is symmetric, so it is not sensitive to the variation of the upper and lower tails.

2.2 Hybrid Copula Function

By the foregoing description, considering the respective advantages and disadvantages of different Copula function, it is easy to distortion with only one Copula function to fit the data. To avoid this problem, a hybrid of different Copula functions can be employed, then the hybrid Copula function is shown below:

$$C(\mu, v; \theta) = \sum_{i=1}^{n} \lambda_i C_i(\mu, v; \theta_i). \tag{2}$$

Which, $C_i(\mu, v, \theta_i)$ is the known single Copula function, θ_i is the relevant parameter, $0 \le \lambda_i \le 1$ is the weighting coefficient and $\sum_{i=1}^{n} \lambda_i = 1$.

By changing the weighting coefficients in Eq. (3), we can make the constructed hybrid Copula function include not only the properties of each Copula function, but also the mixing properties.

2.3   Correlation Index Based on Hybrid Copula Function

The correlation index based on Copula function is strictly monotonically increasing the correlation, that is, the strict monotone change of the variable, the Copula function derived from the correlation measure value will not change [15], and the tail correlation coefficient can be conveniently described with Copula function [16, 17].

Let $(X, Y)^T$ be a vector of continuous random variables with marginal distribution functions F and G. Trivially speaking, tail dependence expresses the probability of having a high (low) extreme value of Y given that a high (low) extreme value of X has occurred.

The analytic form for the coefficient of upper and lower tail dependence is:

$$\omega_u = \lim_{u^* \to 1} P(Y > G^{-1}(u^*) | X > F^{-1}(u^*)) = \lim_{u^* \to 1} \frac{C(1 - u^*, 1 - u^*)}{1 - u}.$$
$$\omega_l = \lim_{u^* \to 0} P(Y < G^{-1}(u^*) | X < F^{-1}(u^*)) = \lim_{u^* \to 0} \frac{C(u^*, u^*)}{u^*} \qquad (3)$$

provided that the limit $\omega \in (0, 1]$ exists. If $\omega \in (0, 1]$, the random variables X and Y are asymptotically dependent in the tail; on the contrary, if $\omega=0$, they are asymptotically independent in the tail.

So, the tail dependence of hybrid Copula function can be described as follows:

$$\omega_u = \sum_{i=1}^{n} \lambda_i \omega_{u,i} \quad \omega_l = \sum_{i=1}^{n} \lambda_i \omega_{l,i}. \qquad (4)$$

Which, λ_i is the weighting coefficient, $\omega_{u,i}$ and $\omega_{l,i}$ is the upper and lower tail dependence of $C_i(\mu, \nu, \theta_i)$.

3   Correlation Analysis of Diesel Engine Power

3.1   Modelling of Relevant Structure

The dependent structure and the marginal distribution of the variables could be divided into two independent parts on the basis of correlation analysis of Copula function.

The non-parametric kernel density estimation [18, 19] taken to set up the marginal distribution function of all parameters because the distribution types of all parameters could not be assumed.

Definition 3.1 [18]: As for variable x, kernel estimation should be implemented for the probability density function $f(x)$ of x when $n \to \infty$, $h_n \to \infty$ (h_n was related to n, and $h_n > 0$).

$$f_n(x) = \frac{1}{nh_n} \sum_{i=1}^{n} K(\frac{x - X_t}{h_n}). \qquad (5)$$

In the formula, $K(\bullet)$ is the kernel function; h_n is the width of Parzen window, and the size thereof is usually determined during actual application by using the rule of thumb.

Theorem 3.2: If the estimation of distribution function of (x, y) at (x_i, y_i) is

$$u_i = \int_{-\infty}^{x_i} f_X(x)dx \quad v_i = \int_{-\infty}^{y_i} f_Y(y)dy. \tag{6}$$

When the sample size is large, the choice of kernel function has limited effect on the normal distribution function, the most commonly used kernel function.

$$u_i = \frac{1}{T} \sum_{j=1}^{T} \phi(\frac{x_i - x_j}{h_X}) \quad v_i = \frac{1}{T} \sum_{j=1}^{T} \phi(\frac{y_i - y_j}{h_Y}). \tag{7}$$

Where $\phi(x) = \dfrac{1}{\sqrt{2\pi}} \int_{-\infty}^{x} e^{-\frac{t^2}{2}} dt$

3.2 Model Selection and Parameter Estimation of Hybrid Copula Function

The Copula function model was formed by using the Clayton-Copula function reflecting the characteristics of the lower tail, Gumbel-Copula function reflecting the characteristics of the upper tail and the t-Copula function reflecting the characteristics of the upper tail and the lower tail in order to describe the characteristics of the tail of the power of the diesel engine as follows:

$$C(u, v) = \lambda_1 C_1(u, v; \theta_1) + \lambda_2 C_2(u, v; \theta_2) + \lambda_3 C_3(u, v; \theta_3). \tag{8}$$

In the formula, C_1, C_2, C_3 are t-Copula, Clayton-Copula and Gumbel-Copula, respectively; $\lambda_1, \lambda_2, \lambda_3$ are weighting coefficients of corresponding Copula function, and meet the conditions: $0 \le \lambda_1, \lambda_2, \lambda_3 \le 1$, and $\lambda_1 + \lambda_2 + \lambda_3 = 1$.

Multiple unknown parameters exist between the weighting parameters λ_i and the relevant parameter Θ_i (i = 1, 2, 3) in the above formula, and the maximum likelihood estimation thereof is complicated; therefore, expectation maximization (EM) shall be taken to implement the parameter estimation [3].

Supposed that $y_i = (u_i, v_i)$ indicates one observation sample, the sample to be tested is indicated by $x_i = (y_i, z_i)$ after importing the hidden variable z_i; set $\varphi = (\lambda, \theta)$, the conditional probability of x_i can be expressed by:

$$P(x_i|\varphi) = \prod_{j=1}^{3} (\lambda_j c_j(y_i|\varphi))_{ij}^{z}. \tag{9}$$

Then the conditional probability of X can be expressed by:

$$P(X|\varphi) = \prod_{i=1}^{n} \prod_{j=1}^{3} (\lambda_j c_j(y_i|\varphi))_{ij}^{z}. \tag{10}$$

Step E of EM algorithm: The log-likelihood function anticipated of the above-mentioned formula is:

$$E(\ln P(X|\varphi^{(k)})) = \sum_{i=1}^{n} \sum_{j=1}^{3} z_{ij}^{(k)} \ln \lambda_j + \sum_{i=1}^{n} \sum_{j=1}^{3} z_{ij}^{(k)} \ln c_j(y_i|\varphi). \tag{11}$$

In the formula, k indicates iterative times.

Step M: The parameter φ to be acquired can be expressed as follows after obtaining the maximum conditional expectation:

$$\varphi^{(k+1)} = \arg \max E(\ln P(X|\varphi^{(k)})). \tag{12}$$

The estimated values of all parameters of the hybrid Copula function can be acquired by using iterative solution.

3.3 Evaluation of Model

Quality of the model could be judged by using the Euclidean distance as for the hybrid Copula function formed.

Definition 3.2: Supposed that the experiential Copula value [20, 21] of the observational data (X, Y) was $C_0(u, v)$, $C(u, v)$ can be acquired by using the Copula function, the average Euclidean distance could be expressed as

$$d_c^2 = \sum_{i=1}^{n} |C_0(u, v) - C(u, v)|^2. \tag{13}$$

The smaller the average Euclidean distance was, the better the degree of fitting would be, and the closer the distribution would be to the real distribution condition.

4 Experimental Study

Testing data of the diesel engine of certain model in a certain diesel engine manufacturer was selected to implement analysis and investigate the correlation among all parameters in order to verify the effectiveness of the method introduced. Moreover, the power and the temperature of the intake water of the diesel engine should be taken as the samples to illustrate in this paper because there are too many detection parameters.

The deviations further calculated of the two parameters including power and the temperature of the intake water were 0.3710 and −0.2718, respectively; the kurtosises at 3.2012 and 3.6915 reflected that the two random variables were distributed with characteristics of leptokurtic distribution and heavy-tailed distribution, and the test of normality were required. Jarque-Bera (JB) test, Kolmogorov-Smirnov (KS) test and Lilliefors (L) test [22], should be imported, wherein the test result is shown in Table 1.

Table 1. Normality test value

Parameters	JB test		KS test		L test	
	h	p	h	p	h	p
Power	0	0.098	0	0.912	0	0.500
Temperature of intake water	1	1.0e-03	1	2.8e-07	1	1.0e-03

Seen from the Table 1, h test values of the temperature of the intake water are 1, and p values are below 0.001, which indicate that the parameters are not in line with the normal distribution; therefore, the Gaussian kernel function shall be selected in the unified way; further, width of the window shall be determined by $h \approx 1.06\,\sigma n^{-1/5}$ as shown in Figs. 1 and 2. Moreover, the empirical distribution function of acquiring two variables by using the spline interpolation was given.

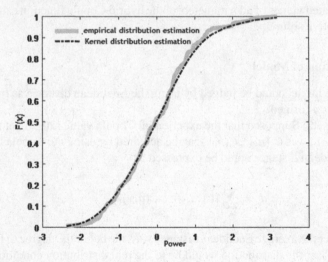

Fig. 1. Kernel distribution estimation and empirical distribution estimation of power

Seen from Figs. 1 and 2, the general distribution tended to be uniform with the empirical distribution if the kernel density function was taken to implement the non-parametric estimation; moreover, the kernel smoothing estimation was applied as for the general distribution, therefore the heterogeneous distribution of the original sample could be avoided and the condition were provided for selection and estimation of hybrid Copula in next step.

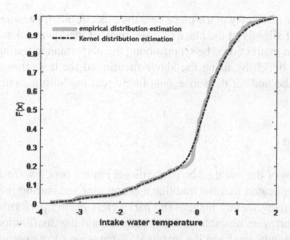

Fig. 2. Kernel distribution estimation and empirical distribution estimation of intake water temperature

The weighting parameters and the independent parameters estimated are shown in Table 2 in allusion to hybrid Copula model.

Table 2. The weight and structure parameters

Power	Weight			Structure parameters		
	λt	λc	λG	ρt	Θc	ΘG
Temperature of intake water	0.094	0.310	0.596	0.000	0.210	1.325

The average Euclidean distances of the Clayton-Copula function, the Gumbel-Copula function and the t-Copula function of the single parameter and the hybrid. Copula model were acquired respectively according to the Formula (13), with the results shown in Table 3.

Table 3. The average Euclidean distance

	t	Clayton	Gumbel	Mixed
Euclidean distance	0.447	1.598	1.598	0.170

Seen from Table 3, the hybrid Copula model had the minimum average Euclidean distance compared with the single Copula function, which indicated that the model fitted the original characteristics of the power and the temperature of the intake water more excellently; therefore, the correlation analysis to be implemented should be more accurate.

Therefore, the correlation coefficients of the upper tail and the lower tails among two parameters including the power and the temperature of the intake water were 0.2446 and 0.0172 calculated by using the hybrid Copula model. Seen from the point of the detection data, the analysis result indicated that the temperature of the intake water affected the power obviously at the place near the peak of the upper tolerance zone and affected the power slightly at the place near the peak of the lower tolerance zone.

The correlation matrix could be set up among the performance testing parameters of the diesel engine bench by using the above-mentioned method; thus, the theoretical foundation could be laid for the subsequent intelligent modelling, control and optimization.

5   Conclusion

The characteristics of the testing data of the diesel engine bench were analyzed in this text first, which indicated that the traditional correlation measuring method could not meet the need of analysis any longer. The parameters during performance test of the diesel engine bench were unstable and non-uniform and the distribution thereof could not be supposed in advance; and the marginal distribution of the parameters could be fitted excellently by using the nonparametric kernel density estimation method. Second, the joint distribution function of multiple random variables could be set up by using the Copula function which could not be limited by the distribution type of the variable and the relevant structure among the variables and were applicable for solving the problem related to correlation analysis of the parameters during performance test of the diesel engine bench. While the correlation among multiple random variables could be described more excellently by using the hybrid Copula model integrating advantages of all varieties of single Copula functions together as a whole and adjusting the contributions of all functions by using the weight; therefore, the theoretical foundation was laid for further intelligent control and optimization.

Acknowledgments.   The authors would like to acknowledge the financial support of the National Science Foundation of China (No. 51435009) and National Key Technology Research and Development Program of the Ministry of Science and Technology of China (No. 2015BAF 12B02).

References

1. Bo, G.: Research on improved design of T160 diesel engine. Wuhan University of Technology, Wuhan (2012)
2. Baolong, Z.: Internal Combustion Engine. China Machine Press, Beijing (1999)
3. Qingxu, J.: Research and Application of Multivariate Correlation and Data Processing Engine. Shanghai Jiao Tong University (2012)
4. Joachim, N., Song, P.: Cold test technology in the process of engine intake and exhaust. In: Measurement and Control Technology, pp. 16–18 (2004)
5. Su, P.: On line fault diagnosis of engine assembly based on cold test. HeFei University of Technology, Heifei (2011)
6. Assanis, D.N., Filipi, Z.S., Fiveland, S.B., Simiris, M.: A predictive ignition delay correlation under steady-state and transient operation of a direct injection diesel engine. J. Eng. Gas Turbines Power **125**, 450–457 (2003)

7. Yaoting, Z.: We should choose what kind of correlation index. In: Statistical Analysis, vol. 09, pp. 41–44 (2002)
8. Claudia, C., Rainer, K., Eike, C.B., Aleksey, M.: A mixed copula model for insurance claims and claim sizes. Scand. Actuarial J. **4**, 1–31 (2012)
9. Xiaosong, T., Dianqing, L., Chuangbing, Z., Guoguang, F.: Probabilistic analysis of load-displacement hyperbolic curves of single pile using copula. Rock and Soil Mechanics **33**(1), 171–178 (2012)
10. Romano, C.: Applying Copula function to risk management. http://www.icer.it/workshop/Romano.pdf
11. Sklar, A.: Function de Reparition an Dimensions et Leurs Marges. Publications del'Institute Statistique de I'Universite Paris, Pairs (1959)
12. Ding, A.A., Li, Y.: Copula correlation: an equitable dependence measure and extension of pearson's correlation. arXiv preprint arXiv:1312.7214 (2013)
13. Nelsen, R.B.: An Introduction to Copulas. Springer, New York (2006)
14. Wei, H.H., Zhang, S.Y.: Copula Theory and Its Application in Financial Analysis. Tsinghua University Press, Beijing
15. Pagaefthymiou, G.: Using copulas for modeling stochastic dependence in power system uncertainty analysis. IEEE Trans. Power Syst. **24**(1), 40–49 (2002)
16. Schwettzer, B., Wolff, E.: On nonparametric measures of dependence for random variables. Ann. Stat. **9**(4), 879–885 (1981)
17. Juri, A., Wutrich, M.V.: Copula convergence theorems for tail events. Insur. Math. Econ. **30**(3), 405–420 (2002)
18. Shunbo, H., Zhaolin, J., Xiangrong, Z.: Research on parzen window based on improved gaussian matrix in medical image registration. J. Comput. Inf. Syst. **8**(12), 5103–5110 (2012)
19. Kolee, K.K., Verbeek, M.: Selecting copulas for risk management. J. Bank. Finance **31**(8), 2405–2423 (2007)
20. Salim, B., Faouzi, E., Eddin, N., Tarek, Z.: On the multivariate two-sample problem using strong approximations of empirical copula processes. Commun. Stat. Theory Meth. **40**(8), 1490–1509 (2011)
21. Ju, Z., Liu, H.: Recognizing hand grasp and manipulation through empirical copula. Int. J. Soc. Robot. **2**(3), 321–328 (2010)
22. Berna, Y., Senay, Y.: A comparison of various tests of normality. J. Stat. Comput. Simul. **77**(2), 175–183 (2007)

Clustering

Comparative Study of Apache Spark MLlib Clustering Algorithms

Sasan Harifi[✉], Ebrahim Byagowi, and Madjid Khalilian

Department of Computer Engineering, Karaj Branch, Islamic Azad University, Karaj, Iran
{s.harifi,ebrahim.byagowi,khalilian}@kiau.ac.ir

Abstract. Clustering of big data has received much attention recently. Analytics algorithms on big datasets require tremendous computational capabilities. Apache Spark is a popular open-source platform for large-scale data processing that is well-suited for iterative machine learning tasks. This paper presents an overview of Apache Spark Machine Learning Library (Spark.MLlib) algorithms. The clustering methods consist of Gaussian Mixture Model (GMM), Power-Iteration Clustering method, Latent Dirichlet Allocation (LDA), and k-means are completely described. In this paper, three benchmark datasets include Forest Cover Type, KDD Cup 99 and Internet Advertisements used for experiments. The same algorithms that can be compared with each other, compared. For a better understanding of the results of the experiments, the algorithms are described with suitable tables and graphs.

Keywords: Clustering · k-means · Bisecting k-means · Spark MLlib · Big data · KDD cup 99 · Cover type · Train time · Cohesion

1 Introduction

Clustering is an unsupervised learning method which tries to find some distributions and patterns in unlabeled datasets. Usually, those points in the same cluster should have more similarity than other points in other clusters [1].

Considered that clustering pursues the arrangement of a family of items into homogeneous groups, taking into account inherent quantitative and qualitative information about them. However, the practical implementation of some clustering techniques requires certain mathematical assumptions that sometimes are difficult, if not impossible, to be checked. Some of these assumptions are quite often simply hidden in the interpreter's mind [2].

Also clustering items according to some notion of similarity is a major primitive in machine learning. Correlation clustering serves as a basic means to achieve this goal: given a similarity measure between items, the goal is to group similar items together and dissimilar items apart. In contrast to other clustering approaches, the number of clusters is not determined a priori, and good solutions aim to balance the tension between grouping all items together versus isolating them [3].

© Springer International Publishing AG 2017
Y. Tan et al. (Eds.): DMBD 2017, LNCS 10387, pp. 61–73, 2017.
DOI: 10.1007/978-3-319-61845-6_7

Clustering has been used in many areas such as machine learning, pattern recognition, image processing, marketing and customer analysis, agriculture, security and crime detection, information retrieval, and bioinformatics [1].

Clustering algorithms can be categorized into partitioning methods, hierarchical methods, density-based methods, grid-based methods, and model-based methods. Recently, quantum clustering, spectral clustering, and synchronization clustering have been presented and gained some attention.

In this paper MLlib clustering algorithms are described. Rest of the paper is structured as follows: Sect. 2 consists of related works, Sect. 3 describes Spark MLlib clustering algorithms. Section 4 includes comparison by experimental setup. Section 5 represents conclusions.

2 Related Works

In this section, we provide a brief description of the related works. Daniel Gómez et al. [2] introduced a hierarchical clustering algorithm in networks based upon a first divisive stage to break the graph and a second linking stage which is used to join nodes. They show that this algorithm is very flexible as well as quite competitive in relation with a set previous algorithms.

Madjid Khalilian et al. [4] proposed method Divide-and-Conquer Stream and compared it with Stream and incremental online Con-Stream for efficiency and accuracy in clustering results in data stream clustering. Stream utilizes Divide-and-Conquer method to overcome difficulties in data stream clustering and should be distinguished from the proposed method. Divide-and-Conquer Stream uses Divide-and-Conquer method based on length of vector as it is described later whereas Stream divides data by using sampling. In another study [5], he evaluate different aspects of existing obstacles in data stream clustering.

Renxia Wan et al. [6] extended Fuzzy C-Means and proposed a weighted fuzzy algorithm for clustering data stream. The algorithm tries to fuzzily cluster data stream. Their Experimental results on both standard datasets KDD-CUP'99 and synthetic datasets show its superiority over the traditional FCM algorithms.

Jingdong Wang et al. [7] proposed a novel approximate k-means algorithm to greatly reduce the computational complexity in the assignment step. Their approach is motivated by the observation that most active points changing their cluster assignments at each iteration are located on or near cluster boundaries.

Francesco Finazzi et al. [8] considered two approaches for clustering of time series. The first is a novel approach based on a modification of classic state-space modelling while the second is based on functional clustering. For the latter, both k-means and complete-linkage hierarchical clustering algorithms are adopted. The two approaches are compared using a simulation study. For more details see [8].

Matthias Brust et al. [9] proposed a 3-D clustering algorithm for autonomous positioning (virtual forces based clustering algorithm) of aerial drone networks based on virtual forces. These virtual forces induce interactions among drones and structure the system topology. According to their statements, the advantages of their approach are that virtual forces enable drones to self-organize the positioning process and virtual forces based clustering algorithm can be implemented entirely localized.

Celal Ozturk et al. [10] improved searching mechanism of the discrete binary artificial bee colony algorithm by the efficient genetic selection and they tested its performance on the dynamic clustering problem, in which the number of clusters is determined automatically. Moreover, they demonstrated the superiority of their proposed algorithm by comparing it with the discrete binary artificial bee colony, binary particle swarm optimization (BPSO), genetic algorithm (GA), Fuzzy C-means (FCM) and k-means algorithms on benchmark problems.

Shifei Ding et al. [11] reviewed the development and trend of data stream clustering and analyzes typical data stream clustering algorithms proposed in recent years, such as Birch algorithm, Local Search algorithm, Stream algorithm and CluStream algorithm. They also summarized the latest research achievements in this field and introduced some new strategies to deal with outliers and noise data.

Yan et al. [12] proposed a multitask clustering framework for activity of daily living analysis from visual data gathered from wearable cameras. Their intuition is that, even if the data are not annotated, it is possible to exploit the fact that the tasks of recognizing everyday activities of multiple individuals are related, since typically people perform the same actions in similar environments. For more details see [12].

3 Spark.MLlib Clustering Algorithms

Apache Spark is a cluster computing platform that is used for general purposes and designed to be fast [13, 14].

On the speed side, Spark expands MapReduce model to support more types of computing like interactive queries and stream processing. Speed is very important in processing large datasets, as it means the difference between exploring data interactively and waiting minutes or hours. One of the main features that Spark proposes for speed, is the ability to compute in memory. But this system is more efficient than the Hadoop MapReduce to run complex applications on disk, as well.

On the generality side, Spark is designed to cover a wide range of workloads that were previously required separate distributed systems, including batch applications, algorithms, iterative, interactive query and streaming. By supporting these workloads in the same engine, Spark makes easy and inexpensive, the combination of different types of processing are often required in production data analysis pipelines. Spark offers APIs in Java, Scala and Python. Spark components are shown in Fig. 1.

MLlib package includes common machine learning functionality that is the focus of this paper. MLlib includes several types of machine learning algorithms such as classification, regression, clustering and collaborative filtering and also includes model evaluation and data import. In the following MLlib clustering algorithms are described.

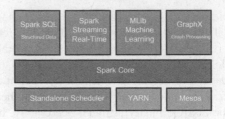

Fig. 1. Spark components [13].

3.1 Gaussian Mixture Model

Model-based clustering is assumed the data comes from a source with several subpopulations. Each subpopulations are modeled individually and the entire population is a mixture of these sub-populations. The final model is a finite mixture model [15]. When data are multivariate continuous observations, the component parameterized density is usually a multidimensional Gaussian density [15].

According to the model-based perspective, each cluster can be mathematically provided by a parametric distribution. All datasets can be modeled by a mixture of these distributions [16]. The model widely used is a mixture of Gaussians:

$$P(x|\Theta) = \sum\nolimits_{i=1}^{K} \alpha_i p_i(x|\theta_i). \tag{1}$$

where each p_i is a Gaussian density function parameterized by θ_i and $\Theta = (\alpha_1, \ldots, \alpha_K, \theta_1, \ldots, \theta_K)$ such that $\sum_{i=1}^{K} \alpha_i = 1$. Here it is assumed k component densities mixed together with k mixing coefficients α_i. In Eq. (1) $X = (x_1, \ldots, x_m)$ is a set of data points. It is expected to find Θ such that $p(X|\Theta)$ is a maximum. This is known as the Maximum Likelihood (ML) estimate for Θ. In order to estimate Θ, it is typical to introduce the log-likelihood function defined as follows:

$$\mathcal{L}(\Theta) = \log P(X|\Theta) = \log \prod\nolimits_{i=1}^{m} P(x_i|\Theta) = \sum\nolimits_{i=1}^{m} \log \left(\sum_{j=1}^{K} \alpha_i p_j(x_i|\theta_j) \right). \tag{2}$$

3.2 Latent Dirichlet Allocation

Latent Dirichlet Allocation (LDA) is a generative probabilistic model of a corpus. The basic idea is that documents are represented as random mixtures over latent topics, where each topic is characterized by a distribution over words [18]. More generally, LDA helps to explain the similarity of data by grouping features of this data into unobserved sets. A mixture of these sets then constitutes the observable data [19]. LDA was first introduced by Blei et al. [18]. The modeling process of LDA can be described as finding a mixture of topics for each resource, i.e., $P(z|d)$, with each topic described by terms following another probability distribution [19], i.e., $P(t|z)$. This can have such formula as:

$$P(t_i|d) = \sum_{j=1}^{z} P(t_i|z_i = j)P(z_i = j|d). \tag{3}$$

where $P(t_i|d)$ is the probability of the i th term for a given document d and z_i is the latent topic. $P(t_i|z_i = j)$ is the probability of t_i within topic j. $P(z_i = j|d)$ is the probability of picking a term from topic j in the document. The number of latent topics Z has to be defined in advance and allows to adjust the degree of specialization of the latent topics.

3.3 Power Iteration Clustering

For power iteration, Consider a set of data points: $\{x_1, x_2, \ldots, x_n\}$, where x is a d-dimensional vector, and some notion of similarity, for example:

$$s(x_i, x_j) = \exp\left(-\frac{\|x_i - x_j\|_2^2}{2\sigma^2}\right). \tag{4}$$

where σ is a scaling parameter that controls the kernel width. An affinity matrix A can built with $a_{ij} = s(x_i, x_j)$ if $i \neq j$ and $a_{ij} = 0$ if $i = j$. The degree matrix associated with A, denoted by D, is a diagonal matrix with the diagonal entries equal to the row sums of A, i.e., $D_{ii} = \sum_j A_{ij}$. A normalized random-walk Laplacian matrix L is defined as $L = \Delta - D^{-1}A$ [26], where Δ is the identity matrix. The intrinsic clustering structure is often revealed by representing the data in the basis composed of the smallest eigenvectors of L. The very smallest eigenvector is a constant vector that doesn't have discriminative power. In another matrix $W = D^{-1}A$ defining, the largest eigenvector is the smallest eigenvector of L. A well-known method for computing the largest eigenvector of a matrix is Power Iteration (PI), which randomly initializes an N-dimensional vector v^0 and iteratively updates the vector by multiplying it with W,

$$v^t = \gamma W v^{t-1}, \quad t = 1, 2, \ldots \tag{5}$$

where γ is a normalizing constant to keep v^t numerically stable.

An interesting property of the largest eigenvector of W discovered by Lin and Cohen [21]. They called their algorithm Power Iteration Clustering (PIC), which its details is in [21]. PIC is computationally efficient since it only involves iterative matrix–vector multiplications and clustering of the one dimensional embedding of the original data, which is relatively easy to do. However, similar to other spectral clustering algorithms, a bottleneck for PIC when applied to large datasets lies in the calculation and storage of big matrices [22].

The *spark.mllib* includes an implementation of PIC using GraphX as its backend. It takes a resilient distributed datasets of (*srcId, dstId, similarity*) tuples and outputs a model with the clustering assignments. The similarities must be nonnegative. PIC assumes that the similarity measure is symmetric. A pair (*srcId, dstId*) regardless of the ordering should appear at most once in the input data. If a pair is missing from input, their similarity is treated as zero.

3.4 k-means

k-means clustering is a method commonly used to automatically partition a dataset into k groups [23]. The number of clusters k is assumed to be fixed in k-means clustering. Let the k prototypes $(w_1, w_2, \ldots, w_k)$ be initialized to one of the n input patterns $(i_1, i_2, \ldots, i_n)$. Therefore, $w_j = i_l, j \in \{1, 2, \ldots, k\}, l \in \{1, 2, \ldots, n\}$. C_j is the jth cluster whose value is a disjoint subset of input patterns. The quality of the clustering is determined by the following error function [24]:

$$E = \sum_{j=1}^{k} \sum_{i_l \in C_j} \left| i_l - w_j \right|^2. \tag{6}$$

The number of iterations required can vary in a wide range from a few to several thousand depending on the number of patterns, number of clusters, and the input data distribution. Thus, a direct implementation of the k-means method can be computationally very intensive. This is especially true for typical data mining applications with large number of pattern vectors.

k-means‖. The *spark.mllib* implementation includes a parallelized variant of the k-means++ method called k-means‖ [17]. A parallel version of the k-means++ initialization algorithm is obtained and empirically demonstrate its practical effectiveness. The main idea is that instead of sampling a single point in each pass of the k-means++ algorithm, $O(k)$ points in each round is sampled and repeat the process for approximately $O(\log n)$ rounds [25]. At the end of the algorithm, $O(k \log n)$ points are left form a solution that is within a constant factor away from the optimum. These $O(k \log n)$ points into k initial centers for the Lloyd's iteration are clustered again. This initialization algorithm, which is called k-means‖, is quite simple and lends itself to easy parallel implementations. However, the analysis of the algorithm turns out to be highly non-trivial, requiring new insights, and is quite different from the analysis of k-means++.

Bisecting k-means. Bisecting k-means can often be much faster than regular k-means, but it will generally produce a different clustering [17]. Bisecting k-means is a kind of hierarchical clustering. Hierarchical clustering is one of the most commonly used method of cluster analysis which seeks to build a hierarchy of clusters. Also the bisecting k-means has a time complexity which is linear in the number of documents. If the number of clusters is large and if refinement is not used, then bisecting k-means is even more efficient than the regular k-means algorithm.

Streaming k-means. When data arrive in a stream, we may want to estimate clusters dynamically, updating them as new data arrive. The *spark.mllib* also provides support for streaming k-means clustering, with parameters to control the decay or forgetfulness of the estimates. The algorithm uses a generalization of the mini-batch k-means update rule. For each batch of data, all points to their nearest cluster are assigned, compute new cluster centers, then update each cluster.

4 Comparison by Experimental Setup

In this section at first the system configuration is introduced and then datasets used in this paper are described. Then the experiments along with their tables are explained and finally the graph of each experiment is shown.

4.1 Configuration

All evaluation experiments have been run on a Core™ processor Intel® CPU i7-4930MX 3.00 GHz with 32 GB RAM and GNU/Linux Ubuntu 16.04 operating system. Implementations have been run on Spark 1.6.1-Scala 2.10.5 for coding.

4.2 Data Sets

We use three benchmark datasets to evaluate the large scale clustering performance:

Forest Cover Type. The Forest Cover Type dataset [27] is composed of 581,012 data points from the US Geological Survey (USGS) and the US Forest Service (USFS). Each data point is represented by a vector of 54 dimensions and assigned to one of 7 classes, each class representing a Forest Cover Type. This is a challenging dataset for any clustering algorithm as it contains ten continuous features, and 44 binary features (four wilderness types and 40 soil types) [28].

KDD Cup 99. Since 1999, KDD'99 has been the most wildly used dataset for the evaluation of anomaly detection methods. This dataset is built based on the data captured in DARPA'98 IDS evaluation program. DARPA'98 is about four gigabytes of compressed raw (binary) tcpdump data of seven weeks of network traffic, which can be processed into about five million connection records, each with about 100 bytes. The two weeks of test data have around two million connection records. KDD training dataset consists of approximately 4,900,000 single connection vectors each of which contains 41 features and is labeled as either normal or an attack, with exactly one specific attack type [28].

Internet Advertisements. The Internet Advertisements dataset is available through the UCI Machine Learning Repository [30]. The 3279 instances of this dataset represent image advertisements and the rest do not. There are missing value in approximately 20 percent of the instances. The proposed task is to determine which instances contain advertisements based on 1557 other attributes related to image dimensions, phrases in the URL of the documents or the images, and text occurring in or near the image's anchor tag in the documents. The first three attributes encode the image's geometry. The binary local feature indicates whether the image URL points to a server in the same internet domain as the document URL. The remaining features are based on phrases in various parts of the documents [29].

4.3 Experiments

In this subsection, the experiments and their tables are explained. Please note that in all experiments, the unit of time is second and time refers to training time[1]. Also, unit of cohesion is Davies Bouldin Cohesion [20].

Table 1. Running k-means on KDD Cup 99 with/without Max Iterations

| K Value | A. k-means on KDD Cup 99 | | | | B. k-means on KDD Cup 99 (Max Iterations) | | | |
| | 1. KDD Cup 99 (10%) | | 2. KDD Cup 99 (100%) | | 1. KDD Cup 99 (10%) | | 2. KDD Cup 99 (100%) | |
	Cohesion	TT(S)	Cohesion	TT(S)	Cohesion	TT(S)	Cohesion	TT(S)
10	19.59062076	2.707	20.46709771	13.452	19.21086	2.472	20.15667	12.907
20	17.60970038	2.553	15.83935119	19.674	18.08498	3.773	19.62121	29.413
30	13.75523234	2.855	14.70325715	22.295	17.55708	5.778	19.4199	42.413
40	14.38443687	3.085	17.22224174	23.761	18.82096	3.213	18.88906	26.718
50	13.09798986	4.047	15.13062897	23.970	18.63252	3.701	19.83298	24.724
60	12.68632938	3.572	14.56952279	25.982	21.96627	3.250	19.8327	37.368
70	11.07109838	3.275	14.75204737	35.910	22.90242	4.655	19.00778	22.229
80	10.80793379	3.887	13.5979721	29.095	17.91366	3.836	22.73483	27.774
90	9.862917001	4.027	12.2244106	32.993	21.45907	6.883	19.41999	47.379
100	8.850165571	4.767	10.57837409	40.724	17.62087	3.485	19.18537	35.456

Experiment 1. In the first experiment k-means algorithm is applied on 10 percent KDD Cup 99 dataset and its iteration is considered 10. The results are shown in Table 1(A-1). As can be seen in the Table 1(A-1), if the lowest value (10) considered for k, In terms of time the best clustering time is achieved, but the cohesion of clusters are not suitable. If the value of k is increased, the cohesion is increased too, and therefore more time is spent on clustering. So the best time of clustering is when k value is low, and most coherent clusters are achieved when k value is high.

Experiment 2. In this experiment, experiment 1 is repeated with 100 percent data of KDD Cup 99 dataset. The expected results are achieved. The obtained time and cohesion in experiment 1 are repeated with higher scale in experiment 2. Table 1(A-2) shows k-means experiment with 100 percent of KDD Cup 99 data. As can be seen in Table 1(A-2), if k value is considered the highest value (100), the cohesion becomes very appropriate and the time becomes very inappropriate. The time is approximately 40 s.

Experiment 3. In this experiment, the previous experiments are done with max iterations. The results of experiment 3 are shown in Table 1(B). As can be seen in the Table 1(B-1), with max iterations the time is improved generally and clusters becomes more coherent as well. As can be seen in the Table 1(B-2), k-means algorithm run with

---

[1] TT(S) in all Tables describes Training Time (Second).

max iterations on full KDD Cup 99 dataset. If the iterations become more, the more coherent clusters with lower time of running are achieved.

Experiment 4. In this experiment, k-means algorithm is applied on the Forest Cover Type dataset. The iteration is considered 10 in this experiment. The results are shown in Table 2(C-1).

Table 2. Running k-means and Bisecting k-means on Forest Cover Type and Ads Dataset

K Value	C. Forest Cover Type dataset				D. Internet Advertisements dataset			
	1. k-means		2. Bisecting k-means		1. k-means		2. Bisecting k-means	
	Cohesion	TT(S)	Cohesion	TT(S)	Cohesion	TT(S)	Cohesion	TT(S)
10	43.0750461	4.034	44.0392431	11.906	1300.92123	1.340	1328.03257	2.107
20	32.5306901	9.409	38.6940653	17.051	1232.96979	1.725	1263.92542	1.927
30	27.413065	9.927	33.0507343	20.645	1153.17953	2.470	1213.8558	1.954
40	19.6669817	9.025	29.1864917	17.280	1066.70068	2.978	1141.27995	2.353
50	20.8943908	9.829	26.4425548	32.512	1035.76626	4.043	1051.90256	2.414
60	12.5754266	11.644	21.0348994	29.417	987.513667	4.500	1006.61273	2.994
70	14.8703284	13.782	20.0067657	25.880	873.877784	5.043	934.134936	3.113
80	10.0004226	15.651	18.7004353	26.935	816.97583	5.738	826.726687	3.107
90	9.5745834	17.078	17.4730561	28.264	784.340603	6.267	803.771187	3.787
100	10.9289989	16.593	17.0222951	26.964	727.975713	11.114	756.455351	3.847

Experiment 5. In this experiment, the Bisecting k-means algorithm is applied on the Forest Cover Type dataset and the iterations considered 10, the same as experiment 4. The results are shown in Table 2(C-2). Experiments 4 and 5 show that k-means algorithm, has better performance both in terms of time and in terms of cohesion in comparison with Bisecting k-means algorithm on the Forest Cover Type dataset.

Experiment 6. In this experiment k-means algorithm is applied on the Internet Advertisements dataset. The iteration is also considered 10 like the previous experiments. The results are shown in Table 2(D-1).

Experiment 7. In this experiment, Bisecting k-means algorithm is applied on the Internet Advertisements dataset and the iteration is considered 10 again and results are shown in Table 2(D-2). Experiments 6 and 7 show that in Internet Advertisements dataset, Bisecting k-means algorithm has better performance in terms of time. Both k-means and Bisecting k-means algorithms are approximately similar in terms of cohesion.

4.4 Comparisons

In this subsection for a better understanding of the results of the experiments, separately and based on dataset, time and cohesion, the algorithms are compared. Figures 2, 3, 4 and 5 show graphs related to each comparison.

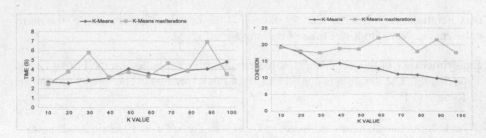

Fig. 2. Left graph: Comparison of k-means and k-means maxIterations training time on KDD Cup (10%). Right graph: Comparison of k-means and k-means maxIterations cohesion on KDD Cup (10%).

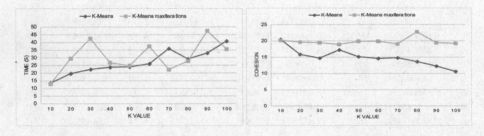

Fig. 3. Left graph: Comparison of k-means and k-means maxIterations training time on KDD Cup (100%). Right graph: Comparison of k-means and k-means maxIterations cohesion on KDD Cup (100%).

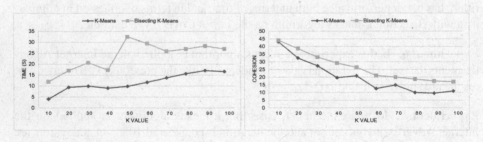

Fig. 4. Left graph: Comparison of k-means and Bisecting k-means training time on FCT dataset. Right graph: Comparison of k-means and Bisecting k-means cohesion on FCT dataset.

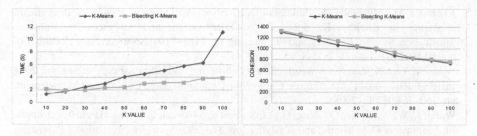

Fig. 5. Left graph: Comparison of k-means and Bisecting k-means training time on Ads dataset. Right graph: Comparison of k-means and Bisecting k-means cohesion on Ads dataset.

5   Conclusions

In this paper, at first MLlib library clustering algorithms are described in details. Being brief and usefulness of descriptions can help the researchers for future works. Then comparison of two algorithms, Bisecting k-means and k-means on three datasets, KDD Cup 99, forest cover type and was Internet Advertisements is done. Algorithms that compared with each other, were the same. In the experiments, Power-Iteration algorithm was not compared because it uses graph as an input. Low speed of GMM algorithm on the selected datasets was the reason of ignoring it in comparison with other algorithms. For example, for GMM algorithm in conditions similar to experiment 1, if the value of k equals 10, the amount of cohesion is 27.813 and train time is 973.058 s. So, this algorithm is non-optimal and slow and is not suitable for using in big data. LDA algorithm is also using with documents. The results of experiments and the comparisons show that depending on the circumstances, type and dimension of data, each algorithm can be best in clustering. Graphs of time and cohesion are applied in this paper to find a better view of the results of the experiments.

Future works can be the comparison of other MLlib library clustering algorithms. Even other parts of spark such as GraphX, Spark Streaming and Spark SQL can be compared. Being brief and usefulness of descriptions of other parts also can help to the quality of studies and reducing the time of research.

References

1. Chen, X.: A new clustering algorithm based on near neighbor influence. Expert Syst. Appl. **42**, 7746–7758 (2015)
2. Gómez, D., Zarrazola, E., Yáñez, J., Montero, J.: A Divide-and-Link algorithm for hierarchical clustering in networks. Inf. Sci. **316**, 308–328 (2015)
3. Pan, X., Papailiopoulos, D., Oymak, S., Recht, B., Ramchan-dran, K., I. Jordan, M.: Parallel correlation clustering on big graphs. In: Advances in Neural Information Processing Systems, pp. 82–90 (2015)
4. Khalilian, M., Mustapha, N., Sulaiman, N.: Data stream clustering by divide and conquer approach based on vector model. J. Big Data **3**, 1 (2016)

5. Khalilian, M., Mustapha, N., Sulaiman, N., Mamat, A.: Different aspects of data stream clustering. In: Elleithy, K., Sobh, T. (eds.) Innovations and Advances in Computer. Information, Systems Sciences, and Engineering, pp. 1181–1191. Springer, New York (2013). doi:10.1007/978-1-4614-3535-8_97
6. Wan, R., Yan, X., Su, X.: A weighted fuzzy clustering algorithm for data stream. In: 2008 ISECS International Colloquium on Computing, Communication, Control, and Management, pp. 360–364. IEEE (2008)
7. Wang, J., Wang, J., Ke, Q., Zeng, G., Li, S.: Fast approximate k-means via cluster closures. In: Multimedia Data Mining and Analytics, pp. 373–395. Springer International Publishing (2015)
8. Finazzi, F., Haggarty, R., Miller, C., Scott, M., Fassò, A.: A comparison of clustering approaches for the study of the temporal coherence of multiple time series. Stochast. Environ. Res. Risk Assess. **29**, 463–475 (2014)
9. Brust, M.R., Turgut, D.: VBCA: a virtual forces clustering algorithm for autonomous aerial drone systems. In: 2016 Annual IEEE Systems Conference (SysCon), pp. 1–6. IEEE (2016)
10. Ozturk, C., Hancer, E., Karaboga, D.: Dynamic clustering with improved binary artificial bee colony algorithm. Appl. Soft Comput. **28**, 69–80 (2015)
11. Ding, S., Wu, F., Qian, J., Jia, H., Jin, F.: Research on data stream clustering algorithms. Artif. Intell. Rev. **43**, 593–600 (2015)
12. Yan, Y., Ricci, E., Liu, G., Sebe, N.: Egocentric daily activity recognition via multitask clustering. IEEE Trans. Image Process. **24**, 2984–2995 (2015)
13. Karau, H., Konwinski, A., Wendell, P., Zaharia, M.: Learning Spark: Lightning-Fast Big Data Analysis. O'Reilly Media, Inc., (2015)
14. Meng, X., Bradley, J., Yuvaz, B., Sparks, E., Venkataraman, S., Liu, D., Freeman, J.: Mllib: machine learning in apache spark. JMLR **17**(34), 1–7 (2016)
15. Maugis, C., Celeux, G., Martin-Magniette, M.: Variable selection for clustering with gaussian mixture models. Biometrics **65**, 701–709 (2009)
16. He, X., Cai, D., Shao, Y., Bao, H., Han, J.: Laplacian regularized gaussian mixture model for data clustering. IEEE Trans. Knowl. Data Eng. **23**, 1406–1418 (2011)
17. Clustering - RDD-based API - Spark 2.1.0 Documentation. http://spark.apache.org/docs/latest/mllib-clustering.html
18. Blei, D.M., Ng, A.Y., Jordan, M.I.: Latent dirichlet allocation. J. Mach. Learn. Res. **3**, 993–1022 (2003)
19. Krestel, R., Fankhauser, P., Nejdl, W.: Latent dirichlet allocation for tag recommendation. In: Proceedings of the Third ACM Conference on Recommender Systems, pp. 61–68. ACM (2009)
20. Davies, D., Bouldin, D.: A cluster separation measure. IEEE Trans. Pattern Anal. Mach. Intell. **PAMI-1**, 224–227 (1979)
21. Lin, F., Cohen, W.: Power iteration clustering. In: Proceedings of the 27th International Conference on Machine Learning (ICML 2010), pp. 655–662 (2010)
22. Yan, W., Brahmakshatriya, U., Xue, Y., Gilder, M., Wise, B.: p-PIC: parallel power iteration clustering for big data. J. Parallel Distrib. Comput. **73**, 352–359 (2013)
23. Wagstaff, K., Cardie, C., Rogers, S., Schrödl, S.: Constrained k-means clustering with background knowledge. In: ICML, pp. 577–584 (2001)
24. Alsabti, K., Ranka, S., Singh, V.: An efficient k-means clustering algorithm. Electrical Engineering and Computer Science (1997)
25. Bahmani, B., Moseley, B., Vattani, A., Kumar, R., Vassil-vitskii, S.: Scalable k-means++. Proc. VLDB Endowment **5**, 622–633 (2012)
26. Meila, M., Shi, J.: A random walks view of spectral segmentation (2001)

27. Blackard, J., Dean, D.: Comparative accuracies of artificial neural networks and discriminant analysis in predicting forest cover types from cartographic variables. Comput. Electron. Agric. **24**, 131–151 (1999)
28. Kumar, D., Bezdek, J., Palaniswami, M., Rajasegarar, S., Leckie, C., Havens, T.: A hybrid approach to clustering in big data. IEEE Trans. Cybern. **46**, 2372–2385 (2016)
29. Alvarez, S.A., Kawato, T., Ruiz, C.: Mining over loosely coupled data sources using neural experts. In: International Workshop on Multimedia Data Mining. In Conjunction with the Ninth ACM SIGKDD International Conference on Knowledge Dis-cover and Data Mining (2003)
30. Lichman, M.: UCI Machine Learning Repository. University of California, School of Information and Computer Science, Irvine, CA (2013). http://archive.ics.uci.edu/ml

L-DP: A Hybrid Density Peaks Clustering Method

Mingjing Du[1] and Shifei Ding[1,2(✉)]

[1] School of Computer Science and Technology, China University of Mining and Technology,
Xuzhou 221116, China
`dingsf@cumt.edu.cn`, `dingshifei@sina.com`
[2] Key Laboratory of Intelligent Information Processing, Institute of Computing Technology,
Chinese Academy of Sciences, Beijing 100090, China

Abstract. Density peaks (DP) clustering is a new density-based clustering method. This algorithm can deal with some data sets having non-convex clusters. However, when the shape of clusters is very complicated, it cannot find the optimal structure of clusters. In other words, it cannot discover arbitrary shaped clusters. In order to solve this problem, a new hybrid clustering method, called L-DP, is proposed in this paper combines density peaks clustering with the leader clustering method. Experiments on synthetic datasets show L-DP could be a suitable one for arbitrary shaped clusters compared with the original DP clustering method. The experimental results on real-world data sets demonstrate that the proposed algorithm is competitive with the state-of-the-art clustering algorithms, such as DP, AP and DBSCAN.

Keywords: Clustering analysis · Density peaks clustering · Hybrid clustering method · Leader clustering method

1 Introduction

Clustering analysis has become one of the popular technologies of unsupervised learning, due to its usefulness in many applications, including pattern recognition, data mining, document retrieval, computer vision, and so forth [1–5].

Recently, Rodriguez and Laio [6] propose a density-based algorithm, called density peaks (DP) clustering. Unlike traditional density-based clustering methods, such as DBSCAN [7], the algorithm can be considered as a combination of the density-based and the centroid-based. DP clustering has wide applications in many fields, including image processing [8], text processing [9], community detection [10] and etc., due both to its low complexity and its ability to recognize non-spherical clusters. There still exist some shortages in the effectiveness and the efficiency of this algorithm. In the past years, researchers have made substantial contribution to improving density peaks clustering. For example, to eliminate the parameter, Mehmood et al. [11] propose a nonparametric method, called CFSFDP-HD. It proposes a nonparametric method for estimating the probability distribution of a given dataset. DP only has taken the global structure of data into account, which leads to missing many clusters. In order to overcome this problem,

© Springer International Publishing AG 2017
Y. Tan et al. (Eds.): DMBD 2017, LNCS 10387, pp. 74–80, 2017.
DOI: 10.1007/978-3-319-61845-6_8

Du et al. [12] propose a density peaks clustering based on k nearest neighbors (DPC-KNN) which introduces the idea of k nearest neighbors (KNN) into the original and has another option for the local density computation.

When the shape of clusters is very complicated, DP may not be able to generate the optimal structure of clusters. In order to solve this problem, we develop a new assignation strategy based on the leader clustering method [13]. We improve the leader clustering method to allow it to be incorporated into the density peaks clustering method. Then we introduce the assignation strategy based on the improved leader clustering into the density peaks method, and further propose a modified density peaks clustering algorithm, called L-DP.

The rest of this paper is organized as follows. Section 2 describes the related work. We make a detailed description of L-DP in Sect. 3. Experimental results are given in Sect. 4. Finally conclusions and future works appear in Sect. 5.

2 Related Work

The proposed algorithm is mainly based on the density peaks clustering and the leader clustering method, so this section provides a brief introduction to these two methods.

2.1 Original Density Peaks Clustering

We briefly introduce DP's idea and algorithm in this sub-section. The idea of the density peaks clustering is based on an assumption that the cluster centers generally can be regarded as a high-density point surrounded by neighbors with lower densities, and they are separated apart from each other.

For each point, DP needs to calculate two quantities, namely, its local density (denoted as ρ_i) and minimal distance to data points with higher density (denoted as δ_i).

A "cut off kernel" density estimator is adopted to estimate the local density ρ_i, as follow:

$$\rho_i = \sum_j \chi\left(dist(\mathbf{x}_i, \mathbf{x}_j) - d_c\right),$$

$$\chi(x) = \begin{cases} 1, x < 0 \\ 0, x \geq 0 \end{cases} \tag{1}$$

Where $dist(x_i, x_j)$ represents the Euclidean distance between objects $\mathbf{x}_i$ and $\mathbf{x}_j$, d_c represents a cutoff distance.

δ_i is calculated as:

$$\delta_i = \begin{cases} \max_j dist(\mathbf{x}_i, \mathbf{x}_j) \text{ if } \rho_i \text{ is the highest local density} \\ \min_{j:\rho_j > \rho_i} dist(\mathbf{x}_i, \mathbf{x}_j) \text{ otherwise,} \end{cases} \tag{2}$$

For the point $\mathbf{x}_i$ with the highest density, its distance $\mathbf{x}_i$ is defined as $max_j dist(x_i, x_j)$. In other cases, δ_i is the minimum distance between the point $\mathbf{x}_i$ and other points with higher density.

According to the local density and delta values for each point, DP draws a decision graph, where horizontal axis represents ρ_i and vertical axis represents δ_i. Based on the decision graph, this method identifies the cluster centers by searching anomalously large parameters ρ_i and δ_i. In fact, cluster centers always appear on the upper-right corner of the decision graph.

After determining cluster centers, DP adopts a simple and intuitive assignment strategy. Remaining points are sorted by local density, from largest to smallest. It assigns these points to the cluster to which its nearest neighbors with higher density belong, in that order.

2.2 Leader Clustering Method

The leader clustering method is an incremental algorithm that makes only a single pass through the data set and finds a set of leaders as the cluster representatives. Each leader representing a cluster is generated using a suitable threshold value τ.

For a user specified threshold distance τ, the leader clustering method works as follow. L denotes a set of leaders. In the beginning, L is an empty set and the method randomly selects any of the objects as the initial leader. For each object $\mathbf{x}_i$ in the data set, it finds the most similar cluster (the nearest leader $l_k \in L$). If $dist(\mathbf{x}_i, l_k) < \tau$, then $\mathbf{x}_i$ is assigned to the cluster represented by l_k and is called as a follower of the leader l_k, otherwise, the object $\mathbf{x}_i$ becomes a new leader and is added to L. Note that L is incrementally built, so before the method generates all leaders, L represents the set of all currently available leaders rather than the set of final leaders.

3 The Proposed Algorithm

Inspired by the leader clustering method, we propose a new assignment strategy for the remaining points in order to reduce the misclassification rate of DP. In other words, we propose a hybrid clustering method that combines density peaks clustering with the improved leader clustering method. The key idea of the proposed algorithm is based on adopting the improved leader clustering method to assign the remaining points to their correct clusters after cluster centers have been detected. This section makes a detailed description of L-DP and analyzes its complexity.

The hybrid clustering scheme can find cluster centers using the same way as DP, but it uses a new assignment strategy instead of the original strategy. In other words, the main difference between the hybrid scheme and DP is in using assignment strategy.

The proposed assignment strategy is inspired by the idea of the leader clustering method. However, the leader clustering method cannot be directly incorporated into the density peaks clustering method. Moreover, the leader clustering method is unsuitable for non-spherical distribution data due in part to its assignment based on centroids (leaders). Last but not least, the leader clustering method is highly sensitive to the initial leader and the access-order for points. To overcome these shortcomings, the leader clustering method is modified as enumerated below:

For each object x_i in the unvisited set, it finds the nearest visited point x_j (a leader or a follower). If $dist(x_i, x_j) < \tau$, then x_i is assigned to the leader l_k which the point x_j follows (that is, x_i is assigned to the cluster represented by l_k), and is marked as visited, otherwise, the object x_i is replaced in the unvisited set. Repeat the process, until the unvisited set is empty or the number of iterations exceeds the maximum limit users specify.

It is easy to find that the above method is sensitive to the order in which the data set is accessed. To avoid this issue, points in the data set are sorted by local density, from largest to smallest. This order is based on an assumptions that a cluster in a data space is a contiguous region of high point density. This means that cluster centers (or leaders) are closer to points with higher density than points with lower density.

Based on this assignment strategy, the L-DP algorithm is proposed. The L-DP algorithm can be summarized as follows:

Algorithm. L-DP.

Inputs:

The samples $\mathbf{X} \in \mathfrak{R}^{n \times m}$

The parameter d_c, τ

Outputs:

The label vector of cluster index: $\mathbf{y} \in \mathfrak{R}^{n \times 1}$

Method:

Step 1: Calculate distance matrix according to the distance measure, such as Euclidean measure

Step 2: Calculate ρ_i for point x_i according to Formula (1)

Step 3: Calculate δ_i for point x_i according to Formula (2)

Step 4: Plot decision graph and select cluster centers

Step 5: Assign other points according to the proposed assignment strategy

---

4 Experiments and Results

In order to test the power of our method, we select some synthetic data sets covering different separation and shape properties. To demonstrate the clustering performance of L-DP, we test it on some real-world data sets obtained from the UCI repository. The performance of the proposed method is compared with well-known clustering algorithms including AP [14], DBSCAN and DP. We select these algorithms as the comparison partner because they and L-DP have the common characteristic that the number of clusters does not need to be given in advance.

To test the clustering performance of L-DP, we select some synthetic data sets that cover the most important factors and properties influencing the methods' output. These properties include shape, separation and compactness. Moreover, the real world data sets used in the experiments are all from the UCI Machine Learning Repository. Brief information on these datasets is tabulated in Table 1.

Table 1. Synthetic and real world data sets

Data Sets	No. object	No. dimension	No. class
Twomoons	400	2	2
S3	5000	2	15
Iris	150	4	3
Wine	175	13	3
Waveform	5000	21	3
Penbased	10992	16	10

Table 2 shows comparisons against AP, DBSCAN and DP in terms of ACC and running time.

Table 2. The performance comparison of the proposed algorithm on the data sets

Twomoons	ACC	Time	Parameter	S3	ACC	Time	Parameter
L-DP	1	0.0629	$d_c = 2$	L-DP	0.8552	3.1681	$d_c = 6$
DP	0.86	**0.0575**	$d_c = 2$	DP	0.8552	**2.1609**	$d_c = 6$
DBSCAN	1	0.6381	$\varepsilon = 1.5$, $k = 10$	DBSCAN	*	*	*
AP	0.855	1.2222	$p = 8$	AP	**0.8556**	169.1537	$p = 2$
Iris	ACC	Time	Parameter	Wine	ACC	Time	Parameter
L-DP	**0.9667**	0.042	$d_c = 6$	L-DP	**0.9213**	0.0367	$d_c = 2$
DP	0.9	**0.0384**	$d_c = 6$	DP	0.8315	**0.0342**	$d_c = 2$
DBSCAN	0.8267	0.0536	$\varepsilon = 1.5$, $k = 10$	DBSCAN	0.5056	0.0552	$\varepsilon = 2.1$ $k = 4$
AP	0.9067	0.3411	$p = 2$	AP	0.8539	0.4084	$p = 6$
Waveform	ACC	Time	Parameter	Penbased	ACC	Time	Parameter
L-DP	**0.593**	27.3174	$d_c = 1$	L-DP	**0.8157**	27.3174	$d_c = 0.5$
DP	0.588	**16.435**	$d_c = 1$	DP	0.7611	**16.435**	$d_c = 0.5$
DBSCAN	*	*	*	DBSCAN	*	*	*
AP	0.5316	87.1189	$p = 100$	AP	0.6903	1033.1142	$p = 100$

The symbol "*" means that the corresponding algorithm cannot work on the corresponding data set (e.g., the corresponding algorithm generates only one cluster on the corresponding data set). Twomoons data set consists of two moon-shaped groups which are two non-spherical (or non-convex) clusters. Only DBSCAN and our methods achieve perfect clustering results on this data set. S3 has 15 overlapping clusters. On this data set, L-DP gives favorable performance compared to the state-of the-art clustering algorithms. Over all the real world data sets, results from Table 2 illustrate that the proposed clustering method holds against some of the state of the art methods, although most of the time slightly inferior to the DP clustering in terms of running time. From the above detailed analysis, we can conclude that L-DP has given an overall good performance in clustering.

5 Conclusions

In this paper, we propose a hybrid clustering algorithm, L-DP, to better find the true structure of clusters. Inspired by the leader clustering method, we introduce the idea of the leader clustering method to develop a new assignment strategy to improve clustering quality. That is, our method combines density peaks clustering with the leader clustering method. On a variety of synthetic data sets, our algorithm is very effective in grouping data with different separation and shape. Especially, some data sets with non-convex clusters, DP fails to find the optimal structure of clusters arbitrary, but L-DP obtains the best clustering results among four methods. We evaluate L-DP on a wide range of real world clustering tasks. Experimental results demonstrate that our proposed method gives favorable performance compared to other clustering algorithms in terms of both clustering quality and efficiency. However, L-DP has higher running cost compared with the original DP clustering method.

Acknowledgements. This work is supported by the National Natural Science Foundation of China (Nos.61672522,61379101), and the China Postdoctoral Science Foundation (No. 2016M601910).

References

1. Jain, A.K.: Data clustering: 50 years beyond k-means. Pattern Recogn. Lett. **31**(8), 651–666 (2010)
2. Li, X., Liang, Y., Cai, Y.: CC-K-means: a candidate centres-based k-means algorithm for text data. Int. J. Collaborative Intell. **1**(3), 189–204 (2016)
3. Kou, F., Du, J., He, Y.: Social network search based on semantic analysis and learning. CAAI Trans. Intell. Technol. **1**(2), 173–178 (2016)
4. Xu, R., Wunsch, D.: Survey of clustering algorithms. In: IEEE Trans. Neural Networks **16**(3), 645–678 (2005)
5. Ma, Z., Liu, Q., Sun, K.: A syncretic representation for image classification and face recognition. In: CAAI Trans. Intell. Technol. **1**(2) pp. 173–178 (2016)
6. Rodriguez, A., Laio, A.: Clustering by fast search and find of density peaks. Science **344**(6191), 1492–1496 (2014)
7. Ester, M., Kriegel, H.P., Sander, J., et al.: A density-based algorithm for discovering clusters in large spatial databases with noise. In: Proceedings of Second International Conference on Knowledge Discovery and Data Mining, vol. 96(34), pp. 226–231 (1996)
8. Chen, Y. W., Lai, D.H., Qi, H., et al.: A new method to estimate ages of facial image for large database. Multimedia Tools Appl. **75**(5), 2877–2895 (2016)
9. Wang, B., Zhang, J., Liu, Y., et al.: Density peaks clustering based integrate framework for multi-document summarization. CAAI Trans. Intell. Technol. (2017)

10. Bai, X., Yang, P., Shi, X.: An overlapping community detection algorithm based on DensityPeaks. Neurocomputing **226**, 7–15 (2016)
11. Mehmood, R., Zhang, G., Bie, R., et al.: Clustering by fast search and find of density peaks via heat diffusion. Neurocomputing **208**, 210–217 (2016)
12. Du, M., Ding, S., Jia, H.: Study on density peaks clustering based on k-nearest neighbors and principal component analysis. Knowl. Based Syst. **99**, 135–145 (2016)
13. Spath, H.: Cluster Analysis Algorithms for Data Reduction and Classification. Ellis Horwood, Chichester (1980)
14. Frey, B.J., Dueck, D.: Clustering by passing messages between data points. Science **315**(5814), 972–976 (2007)

Prediction

Incremental Adaptive Time Series Prediction for Power Demand Forecasting

Petra Vrablecová$^{(\boxtimes)}$, Viera Rozinajová, and Anna Bou Ezzeddine

Faculty of Informatics and Information Technologies, Slovak University
of Technology in Bratislava, Ilkovičova 2, 842 16 Bratislava, Slovakia
{petra.vrablecova,viera.rozinajova,
anna.bou.ezzeddine}@stuba.sk

Abstract. Accurate power demand forecasts can help power distributors to lower differences between contracted and demanded electricity and minimize the imbalance in grid and related costs. Our forecasting method is designed to process continuous stream of data from smart meters incrementally and to adapt the prediction model to concept drifts in power demand. It identifies drifts using a condition based on an acceptable distributor's daily imbalance. Using only the most recent data to adapt the model (in contrast to all historical data) and adapting the model only when the need for it is detected (in contrast to creating a whole new model every day) enables the method to handle stream data. The proposed model shows promising results.

Keywords: Power demand forecasting · Stream mining · Concept drift

1 Introduction

Processing of data streams coming from smart electricity meters is nowadays one of the biggest challenges in the data analytics area. The directive 2009/72/EC of the European parliament and of the Council states that EU members shall equip at least 80% of the consumers with intelligent metering systems by 2020. The smart meters send the measurements of consumption regularly, e.g., every 15 min. Power distributors collect this vast amount of measurements. The responsibility of the distributor is to effectively utilize information from these data to prevent imbalance in the grid (i.e. to minimize the online gap between the contracted energy supplies and the actual power demand) and to minimize regulation costs resulting from it. The accurate short-term forecast of the aggregated power demand of all distributor's customers for the next 24 h is important for trading energy on the liberated electricity market where the electricity can be contracted even 10 min ahead of its consumption.

The characteristics of this very specific type of data change over time are affected by many external factors. The power demand depends on the season, day of the week, time of the day, the type of the consumer (e.g., household, factory), weather, social activities, public events, and others [8]. Some of these factors are periodical, and one can predict them easily, e.g., seasonality, while others are quite unpredictable because there is often a lack of information about them, such as consumer behavior. Because of this, it is very important to consider sudden changes in power demand and to employ a

© Springer International Publishing AG 2017
Y. Tan et al. (Eds.): DMBD 2017, LNCS 10387, pp. 83–92, 2017.
DOI: 10.1007/978-3-319-61845-6_9

stream change detection mechanism that can quickly identify the ongoing changes in the monitored data – also called the concept drifts [5]. Drift detection can help to correct (or to adapt) the prediction model which can no longer (because of the drift) predict the values correctly. The challenge is to perform these model adaptations online. The parameters of the forecasting model are usually based on the data in a training set. But in the large online stream processing, there are usually no such data available nor there is enough time or memory to train new models from scratch. The aim of this paper is to propose a method for power demand forecasting with respect to the two main requirements for this type of predictions – *incremental computing* and *adaptivity to concept drifts*.

The rest of the paper is organized as follows: In Sect. 2 we present the related work concerning stream mining, power demand forecasting and concept drift. Sections 3 describes the proposed method. The performed experiments and their evaluation are outlined in Sect. 4. Section 5 is the conclusion and future work.

2 Related Work

Two basic properties (volume and velocity) of data stream, i.e. ordered sequence of samples transferred at the stable (high-speed) rate, determine the way it is processed. The third important property – variability – refers to the constantly changing data over time and the dynamic environment that affects it. The ideal stream mining method requires a short constant time per stream sample; uses a fixed amount of memory; builds a model by using a single scan over the data; can deal with concept drift; and its accuracy is comparable to a batch learner [6].

The concept drift can be formally defined as a change in joint probability distribution between a set of input variables and the target variable over time. The changes in data can be then seen as changes in the components of the relation, i.e. the input data characteristics or the relation between the inputs and the target variable [2]. There are different patterns of drifts in data (see Fig. 1). In power demand, the drift can be caused by replacement of a sensor with a differently calibrated one (abrupt), a broken sensor that loses its precision (incremental), or a change of tariff in a household (gradual). The reoccurring concepts represent the events that are repeated over time but in contrast to the seasonality they are not periodical. Therefore, we don't know when exactly they will happen, e.g., holidays, sport events, etc.

Fig. 1. Different patterns of concept drifts in power consumption (reoccurring, incremental, abrupt). The x-axis represents the time and the y-axis the consumption in kWh.

The prediction methods should therefore be able [5]: to detect the concept drift as soon as possible and adapt the prediction model to it if needed; to distinguish drifts from noise; and to operate in less than stream sample arrival time and use a fixed amount of memory. The procedure of changing the prediction model over time is referred to as the *adaptive learning* and consists of three steps: prediction; *diagnose* – after arrival of the next sample (the true value), the loss is estimated (usually as a difference between the predicted and the actual value); and update. By monitoring the losses, the changes in data can be detected. The update step of the procedure can be then carried out as an *informed* decision based on the detected changes (cf. [5, 14]).

The power demand forecasting methods can be divided into two groups – statistical and artificial intelligence (AI). The advantages of the statistical methods (e.g. linear regression, ARIMA [3], exponential smoothing [13]) are the solid theoretical foundations and simple, robust and explainable prediction models. On the other hand, they can usually model only linear relations between input variables and the target variable. In general, it is sufficient for the short term forecasting without weather variables (the relation between weather and power demand is usually nonlinear) [7]. AI methods (e.g. neural network, SVM, expert systems, fuzzy logic) can model the nonlinear relations and have low requirements for domain or statistics' knowledge.

The adaptive power demand forecasting methods we encountered in the literature usually use blind adaptation to cope with continually changing data. The most commonly utilized approach is the *regular batch learning*. The prediction model is created from scratch with the arrival of a new batch of data, e.g., once a day/hour/15 min. The new batch is appended at the end of the existing data and the new model is built from all available data. Sometimes only a part of data (i.e. a window) is used to speed up the training process, e.g. the last few (similar) days. Another popular approach is to create *a large group of context-specific models*, e.g. separate models for each day of the week, hour of the day, holiday. A proper model is chosen from this group for current circumstances [4]. *Periodical model parts' updates* assure the currency of the prediction model by updating its (all or some of) parameters regularly to minimize prediction error. The variations of the least squares' algorithm are employed for this purpose [9]. Some *adaptive models'* designs do not require the parameter updates. Instead, the parameters are defined as functions, e.g. STES [11].

Blind adaptive prediction methods are more common. They always update the prediction model on arrival of a new sample from a stream. The informed adaptive prediction methods utilize a concept drift detector and perform updates only when a drift is detected. This way, the possibility of overfitting is reduced and computing and memory resources can be spared. It has been shown that a concept drift detector can improve the prediction model accuracy [1]. In the next section, we propose a power demand forecasting method that is built on the incremental adaptive learning.

3 Incremental Adaptive Power Demand Forecasting

Our method predicts the aggregated power demand of a group of consumers for the next 24 h. The power demand is affected by multiple factors, such as seasonality (intra-day, intra-week and intra-year cycle), weather (temperature and humidity) and calendar or social events (holidays, summer leaves in factories, etc.). Figure 2 depicts the essence of our method based on the adaptive learning scheme. We assume that an initial prediction model exists. The first step – *prediction* – is executed on arrival of new data. *Diagnostics* happens after the arrival of the actual value of demand. The error is evaluated and sent to the change detection mechanism. If the prediction error increases over time, the alarm is set off and the third step – *update* – is triggered. The parameters of the prediction model are re-estimated from a sliding window of the most recent data. To complete our method, we had to *choose a proper prediction model* with a possible and effective parameter re-estimation in a stream environment, and *design a proper change detection mechanism* to timely spot demand changes.

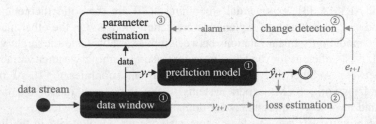

Fig. 2. The power forecasting method consists of three steps: predict (1), diagnose (2) and update (3), (y_t is the actual value of power consumption at time t, $\hat{y}_t$ is the predicted value of power consumption at time t, e_t is the error of prediction at time t).

3.1 Prediction Model and Parameter Estimation

The main criteria for the prediction model selection were simplicity, robustness, the ability to model the multiple seasonal cycles of power demand, the possibility to compute the predictions incrementally and easy parameter re-estimation. We decided to use the multiplicative variant of the double seasonal exponential smoothing (HWT) method [10], which conformed our criteria and was shown to have very good accuracy [12]. It calculates the predicted value recursively in a constant time utilizing the past results, so it is very suitable for incremental computing. The forecast for kth horizon from time t is defined as (1).

$$fcast_{t+k} = (level_t + k \cdot trend_t) \cdot dailyseas_{t-ds+k} \cdot weeklyseas_{t-ws+k} + \phi^k \cdot error_t \quad (1)$$

It is a combination of 4 time series' components; each is smoothed by its smoothing parameter: level (α), trend (β), and two seasonal components (daily – δ and weekly – ω). To improve accuracy, a simple adjustment for first-order autocorrelation (ϕ) is included. The parameters of this method come from the interval (0,1) and are estimated by the least squares method. ds and ws are lengths of daily and weekly cycles, e.g. 96 and 672 for 15-minute load measurements, $error_t$ is the difference between actual and the forecasted value at time t. We assume that the seasonal smoothing parameters (δ and ω) do not change significantly over time and the concept drifts affect only the level of the time series and can be managed by adjusting only the parameters α and ϕ. The parameters' values are re-estimated on drift detection by minimization of the sum of squared errors in a sliding window with the most recent stream data (2 weeks). The length of the window was determined experimentally, to be long enough to reflect the double seasonality and short enough to reflect the ongoing concept drift.

3.2 Change Detection

The timely warnings about the changes in the stream gives the distributor the opportunity for flexible reaction, e.g., to sell or to buy the electricity on short-term market. We propose a change detection mechanism that monitors the prediction errors, i.e. the differences between the predicted and the real values of power consumption ($e_t = y_t - \hat{y}_t$). It is based on the acceptable daily deviation of a power distributor. His difference between contracted and demanded electricity should not be higher than 5% a day. We evaluate the percentage absolute error (2) over the past day (e.g., in case of 15-min. measurements it is the last 96 observations).

$$pe_t = \frac{|e_{t-95}| + \ldots + |e_{t-1}| + |e_t|}{y_{t-95} + \ldots + y_{t-1} + y_t} \tag{2}$$

We use the absolute errors to penalize both higher and lower power supply. When the percentage error exceeds the defined threshold, we assume that the accuracy of the prediction model over time lowers and the concept drift is detected. The re-estimation of the model's parameters is triggered. We set the threshold to be 5% ($pe_t > 0.05$).

4 Evaluation

We tried to answer the following questions: (1) Can the informed adaptation be as accurate as blind adaptation? (2) Does informed adaptation require more time or memory? Is it suitable for stream processing? (3) Does the change detection improve the demand forecasting? If so, to what extent?

4.1 Data

Slovak smart meters measure in 15-min. intervals (i.e., 96 measurements per day). We used aggregated data from 6 regions in Slovakia – 12 weeks (8 for training, 4 for testing) with concept drifts (19 May to 10 August 2014) and 12 without drifts (13 January to 6 April 2014). As a second dataset, we selected 4 concept drift patterns from Slovak data (see Fig. 1). We found two test sets for each drift pattern.

4.2 Experiments

Comparison of blind and informed adaptation. Firstly, we applied our informed adaptive method and evaluated the prediction accuracy on data with and without concept drifts. We trained the initial double seasonal exponential smoothing model on the first 8 weeks of data. The remaining 4 weeks were used to sequentially analyze the residuals and improve the model's accuracy on the fly. To evaluate the accuracy, we used the *mean absolute percentage error (MAPE)*, which is the widely-used measure in power demand forecasting. Furthermore, we recorded how many times a drift was detected, i.e. the model was adapted (α and ϕ re-estimated from a sliding window).

Secondly, we computed the predictions on the same data with the use of the blind adaptive batch learning. Each day, the prediction model was created from scratch on all available data (the first 8 weeks of data plus the days that have passed until that moment) to predict the next day's demand. We used the 4 weeks long test set (28 models were created). In the end, we compared the blind and the informed adaptive approaches with respect to the accuracy and the needed computing resources.

Comparison with other prediction methods. We applied our method and seven incremental blind adaptive batch learners on 4 types of concept drift in power demand (see Fig. 1) and evaluated MAPE. The batch learners were re-trained every day from a sliding window to forecast the next 24 h (see Table 2). We used blind HWT model, seasonal random walk (RW), i.e. $ARIMA(0,0,0)(0,1,0)_{96}$, seasonal and trend decomposition using Loess with ARIMA (STL+ARIMA), multiple linear regression (MLR), artificial neural networks (ANN, feed-forward, one hidden layer), support vector regression (SVR, $\varepsilon = 0.06$, $C = 1$, RBF kernel) and random forest (RF).

4.3 Results

Comparison of blind and informed adaptation. The accuracy of both blind and informed adaptive approaches was comparable. We observed that their MAPE was slightly worse when the concept drift was present in data (cca by 1%, see Table 1). The informed adaptation had on average a 0.36% higher error than the blind approach. Wilcoxon signed rank test confirmed that this is not a significant difference. In several cases the accuracy of the informed approach was better (see green cells in Table 1). So, the informed adaptation can be as accurate as the blind one although it adapts only two model parameters (α and ϕ), only when the drift is detected and the learning is based only on the sliding window (14 days).

Secondly, we recorded the number of times the prediction model needed to be adapted during the test period. The informed adaptation required on average 55.35% less updates than the blind adaptation. With the respect to the memory usage, the informed adaptation needed only the last 14 days and the blind adaptation recomputed the model every day from all the available data. Since the accuracy results were very similar, we consider the informed adaptation more suitable for stream processing.

We managed to further improve the accuracy of the informed adaptive method at the cost of the needed resources. Figure 3 shows an example of the informed adaptation on data without (left) and with (right) drifts. To detect them, the error pe_t (2) of the last 96 observations is evaluated. So, after the prediction model is adapted, the next drift can be detected no sooner than after the next 96 observations arrive. We can't consider the prediction errors of the previous prediction model in the drift detection, because those were already too high to trigger the model adaptation. The mean value of those

Table 1. The MAPE results of blind and adaptive approaches on 4-week Slovak test sets with and without concept drifts. Legend: region name (# of consumers), #r – # of model adaptations.

| region | with concept drift | | | | without concept drift | | | |
| | blind | | informed | | blind | | informed | |
	MAPE	#r	MAPE	#r	MAPE	#r	MAPE	#r
BA (1314)	4.439	28	3.659	12	5.203	28	3.051	9
ZH (773)	3.680	28	4.143	13	1.777	28	3.202	5
TT (733)	4.064	28	3.949	13	3.486	28	3.745	11
PN (706)	4.013	28	6.134	14	1.932	28	2.764	5
ZV (605)	4.086	28	5.302	16	3.814	28	3.788	10
PE (584)	3.980	28	3.231	7	2.879	28	3.832	11

errors with the error from the adapted model would instantly result in drift detection. Also, it would not reflect the accuracy of the current model. Sometimes it is too late to start monitoring the prediction errors again after the 96 observations. The percentage error can be too high then (e.g., in Fig. 3 right, the error after the adaptation at day 70). We added a condition to detect drifts also when the percentage error of the last 48 observation exceeds 7.5%. This modification helps to react to drift sooner, preventing the higher absolute errors, but also results in more frequent model adaptations. In this case, the informed adaptation had on average a smaller error than the blind adaptation by 0.16% and required on average 38% less updates than the blind adaptation.

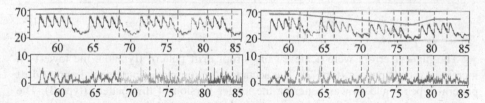

Fig. 3. Predictions (top) and prediction errors (bottom) for ZH region during days 59-84 (test set). Data without (left) and with concept drift (right). Measured in MW/15 min. The grey lines denote the times when a concept drift was detected and the prediction model was adapted.

To sum up, the concept drift detection significantly improved the power demand forecasting when considering time and memory resources. This feature is extremely useful in the stream environment. In return, the accuracy of the predictions did not significantly drop and we managed to maintain the daily 5% deviation that is acceptable in the power demand forecasting. Additional tests showed that if we would keep adapting the prediction model for another month (beyond the 4-week test set), the accuracy would drop on average by 0.86%. To maintain the accuracy, once in a longer period of time a new "initial" model should be created from the longer chunk of data, e.g. once in a few months. This model can be then maintained by the informed adaptation. We believe that even with this limitation, the informed adaptive forecasting is more suitable for the stream data than a blind adaptive approach.

Comparison with other prediction methods. The accuracy of our method (informed HWT) on datasets, which contained 4 concept drift patterns (see Fig. 1), was significantly higher than the accuracy of the HWT method with blind adaptation, three AI-based methods (artificial neural networks, random forest, support vector regression) and multiple linear regression (see Table 2). Two ARIMA-based models (RW and STL +ARIMA) had comparable accuracy. These models had lower number of parameters and lower training requirements.

Table 2. MAPE comparison of our method (informed HWT) with random walk (RW), seasonal and trend decomposition using Loess with ARIMA (STL+ARIMA), blind HWT, multiple linear regression (MLR), artificial neural networks (ANN), support vector regression (SVR) and random forest (RF) on 4 concept drift patterns (2 test sets for each pattern). The length of the sliding window was set experimentally for each prediction method.

	informed HWT	RW	STL+ARIMA	HWT	MLR	ANN	RF	SVR
1	4.607	4.179	3.991	10.804	5.768	4.633	5.769	6.171
	4.656	5.221	4.231	8.896	5.941	9.201	5.843	6.072
2	2.963	4.494	4.828	5.279	4.296	4.973	4.289	3.972
	3.699	4.701	5.132	5.515	5.152	4.849	5.154	5.511
3	3.574	3.764	3.963	5.220	4.846	3.876	4.845	4.865
	4.052	5.899	5.860	4.932	7.046	5.847	7.049	7.257
4	6.534	5.034	5.315	6.275	6.076	5.157	6.072	6.203
	3.450	5.227	4.803	4.060	6.032	11.035	5.900	6.228
sliding w. (days)	10	10	14	4	10	4	4

The highest errors were achieved on abrupt drift pattern (type 1), the lowest on incremental drift pattern (type 2). The accuracy on combined incremental-abrupt drift pattern (types 3 and 4) was better on the longer version of this pattern (type 3). The informed adaptation manifested better accuracy for most of the concept drift types.

5 Conclusion and Future Work

We introduced an informed adaptive approach for predicting the values of power demand time series. The precise short term forecasts are essential for minimizing the imbalance in power grid and resulting costs via operations on short term electricity market where electricity can be contracted even 10 min ahead of its consumption. It utilizes a simple, robust and easily interpretable statistical method – double seasonal exponential smoothing. A concept drift detector is employed to identify changes in power demand, particularly caused by holidays, summer leaves and weather. It monitors the daily percentage error and when it exceeds a defined acceptable threshold (e.g., 5%), the prediction model is adapted to the new concept in the data. The concept drifts manifest by time series level shifts, so the parameters concerning the level and residuals' correction are adapted based on a short window of recent values.

A further improvement might be achieved if the adaptations were based on data from a dynamic-length window. The shorter window is appropriate when drifts occur often, a longer one is suitable when there are no drifts for a longer period of time. In the future, external factor variables like temperature or humidity can be incorporated into the change detection. The indirect inclusion of the external factors makes the predictor more robust and functional even when the additional data are not available.

The proposed method is suitable especially for streams of numerical data with inherent seasonal dependencies. We made one-day ahead predictions of electricity demand and achieved similar error as the regular batch learning, which is a form of blind adaptation. The batch learner was trained every day and based on all historical data. With the informed adaptive method, the need of model adaptations decreased significantly, what can be considered a great benefit in the stream environment.

Evaluation of the method on datasets that contained four types of concept drift, which are typical for power demand time series, showed that the informed adaptive method had significantly better accuracy than several widely-used batch learners (e.g., multiple linear regression, artificial neural network, random forest). The method works best on incremental or partially incremental drifts.

Acknowledgements. This contribution was created with the support of the Research and Development Operational Programme for the project "International Centre of Excellence for Research of Intelligent and Secure Information-Communication Technologies and Systems", ITMS 26240120039, co-funded by the ERDF; the Scientific Grant Agency of the Slovak Republic, grants No. VG 1/0752/14 and VG 1/0646/15 and the STU Grant scheme for Support of Young Researchers.

References

1. Bosnić, Z. et al.: Enhancing data stream predictions with reliability estimators and explanation. Eng. Appl. Artif. Intell. **34**(C), 178–192 (2014)
2. Bouchachia, A.: Fuzzy classification in dynamic environments. Soft. Comput. **15**(5), 1009–1022 (2011)
3. Box, G.E.P., et al.: Time Series Analysis: Forecasting and Control, 4th edn. Wiley, New Jersey (2008)
4. Dannecker, L., Schulze, R., Böhm, M., Lehner, W., Hackenbroich, G.: Context-aware parameter estimation for forecast models in the energy domain. In: Bayard Cushing, J., French, J., Bowers, S. (eds.) SSDBM 2011. LNCS, vol. 6809, pp. 491–508. Springer, Heidelberg (2011). doi:10.1007/978-3-642-22351-8_33
5. Gama, J., et al.: A survey on concept drift adaptation. ACM Comput. Surv. **46**(4), 1–37 (2014)
6. Gama, J.: Data stream mining: the bounded rationality. Informatica **37**(1), 21–25 (2013)
7. Hong, T.: Energy forecasting: past, present and future. Foresight Int. J. Appl. Forecast. **32**, 43–48 (2014)
8. Hong, W.-C.: Intelligent Energy Demand Forecasting. Springer, London (2013)
9. Ma, S., et al.: The variable weight combination load forecasting based on grey model and semi-parametric regression model. In: 2013 IEEE International Conference IEEE Region 10 (TENCON 2013), pp. 1–4. IEEE, Xi'an (2013)

10. Taylor, J.W.: Short-term electricity demand forecasting using double seasonal exponential smoothing. J. Oper. Res. Soc. **54**(8), 799–805 (2003)
11. Taylor, J.W.: Smooth transition exponential smoothing. J. Forecast. **23**(6), 385–404 (2004)
12. Taylor, J.W., McSharry, P.E.: Short-term load forecasting methods: an evaluation based on european data. IEEE Trans. Power Syst. **22**(4), 2213–2219 (2007)
13. Winters, P.R.: Forecasting sales by exponentially weighted moving averages. Manag. Sci. **6** (3), 324–342 (1960)
14. Žliobaite, I.: Learning under Concept Drift: an Overview. Vilnius University (2010)

Food Sales Prediction with Meteorological Data — A Case Study of a Japanese Chain Supermarket

Xin Liu[✉] and Ryutaro Ichise

National Institute of Informatics, 2-1-2 Hitotsubashi, Chiyoda-ku,
Tokyo 101-8430, Japan
xin.liu@aist.go.jp, ichise@nii.ac.jp

Abstract. The weather has a strong influence on food retailers' sales, as it affects customers emotional state, drives their purchase decisions, and dictates how much they are willing to spend. In this paper, we introduce a deep learning based method which use meteorological data to predict sales of a Japanese chain supermarket. To be specific, our method contains a long short-term memory (LSTM) network and a stacked denoising autoencoder network, both of which are used to learn how sales changes with the weathers from a large amount of history data. We showed that our method gained initial success in predicting sales of some weather-sensitive products such as drinks. Particularly, our method outperforms traditional machine learning methods by 19.3%.

Keywords: Sales prediction · LSTM · Autoencoder · Meteorological data

1 Introduction

In Japan, each year as much as 6.42 million tons of food are thrown away and wasted due to expiration and left-over issues. Particularly, more than half of the waste which amounts to 169 billion JPY are occurred at the distribution stage of the delivery. To improve the situation, we should rely on forecasting as an important support of making decisions. Accurate forecasting of food sales can not only help in reducing waste but also in optimizing the supply chain system and enhancing efficiency of the whole society.

Meanwhile, the weather has a strong influence on food retailers' sales. First, it can affect customers' purchase method. For instance, during warm and sunny days customers often enjoy big, low-price, but relatively distant supermarkets, while during periods of inclement weather, customers often visit small, local, but relatively expensive convenience stores. Second, it can have impact on consumers emotional state and drives their demand. For example, a hot, humid summer can boost demand for fruits, vegetables, soft drinks while dampening demand for high calorie meats. Therefore, it is appropriate to connect weather to predicting food retailers' sales.

© Springer International Publishing AG 2017
Y. Tan et al. (Eds.): DMBD 2017, LNCS 10387, pp. 93–104, 2017.
DOI: 10.1007/978-3-319-61845-6_10

In this paper, we introduce a deep learning based method which use meteorological data to predict food sales. As a case study, we focus on the sales of a Japanese chain supermarket which has more than 200 shops[1]. Our method contains a long short-term memory (LSTM) network [20] followed by a stacked denoising autoencoder network [2,3,17,29]. The LSTM network is used for learning how the sales depends on the variation of weathers. The autoencoder network is used to reduce dimensions of the features which are learnt from the LSTM networks. Experiments showed that our method gained initial success in predicting sales of some weather-sensitive products such as drinks. In addition, our method outperforms traditional machine learning methods by 19.3% in accuracy.

The rest of the paper is organized as follows. Section 2 reviews related research. Section 3 gives a review of LSTM and autoencoder, which are used in our method. Section 4 formalize the problem of predicting food sales using meteorological data. Section 5 introduces our method in detail. Section 6 presents experimental results, followed by a conclusion in Sect. 7.

2   Related Works

Walmart has been basing decisions on meteorological data for years in obvious ways, such as putting up umbrella or snow-shovel displays in advance of rain or snow. In recent years, there has been a surge of interests in applying meteorological data to demand forecasting in various field, such as electricity demand [11], water and irrigation demand [6]. Particularly, NEC developed a commodity demand forecasting solution by applying the heterogeneous mixture learning technology [23]. However, their method needs various kinds of input data such as shipment data, warehouse-out data, advertising data. Such data are not always available. Besides, their forecasting models need to be reviewed and modified periodically to maintain the accuracy.

Our work is also closely related to deep learning for time-series analysis [14,25]. Recent work include the usage of Elman recurrent neural networks for chaotic time series prediction [7], employing neural network ensemble approach to forecast the amount of traffic in TCP/IP based networks [9], using multilayer perceptrons for modeling the amount of littering in the southern North Sea [26], using autoencoders to address the traffic flow prediction problem which considers the spatial and temporal correlations inherently [22], applying FPGA-Based Stochastic Echo State Networks for time-series forecasting [1], and comparing the performance of Convolutional Neural Network (CNN) with LSTM for the classification of visual and haptic data in robotics settings [15].

3   Preliminaries

Our proposed method builds upon LSTM and autoencoders. We first provide a brief description of them.

---

[1] Due to privacy issues, the supermarket's name has been omitted.

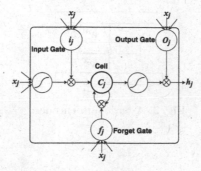

Fig. 1. A memory cell of LSTM.

3.1   LSTM

A recurrent neural network (RNN) is a class of artificial neural network where connections between units form a directed cycle and has the ability to process sequential data. Unfortunately, conventional RNNs are hard to be trained properly due to the vanishing gradient and exploding gradient problems [5]. To address these problems, Hochreiter et al. proposed LSTM [20], which uses *memory cells* to store information and is better at modeling long-range dependencies. Figure 1 illustrates a memory cell, which contains a few gate vectors to control the flow of information. Let us use $x = \langle x_1, \cdots, x_t \rangle$ and $y = \langle y_1, \cdots, y_t \rangle$ to denote input and output sequences, where $x_j \in R^l$ and $y_j \in R^s$, for $1 \le j \le t$. LSTM computes a mapping from x to y using the following equations iteratively from $j = 1$ to t:

$$i_j = \sigma(W^i x_j + U^i h_{j-1} + b^i), \tag{1}$$

$$f_j = \sigma(W^f x_j + U^f h_{j-1} + b^f), \tag{2}$$

$$o_j = \sigma(W^o x_j + U^o h_{j-1} + b^o), \tag{3}$$

$$c_j = f_j \odot c_{j-1} + i_j \odot tanh(W^c x_j + U^c h_{j-1} + b^c), \tag{4}$$

$$h_j = o_j \odot tanh(c_j), \tag{5}$$

$$y_j = W^y h_j + b^y, \tag{6}$$

where σ is the sigmoid function, $tanh$ is the hyperbolic tangent function, $\odot$ is the element-wise product of two vectors, i, f, o, c and h are respectively the input gate vector, forget gate vector, output gate vector, cell activation vector, and hidden state vector of cell j, $W^i, W^f, W^o, U^i, U^f, U^o \in R^{s \times l}$, $W^y \in R^{s \times s}$, $b^i, b^f, b^o \in R^s$ and $b^y \in R^s$ are weight matrices and bias vectors to be learned.

In addition, we can stack multiple LSTM layers on top of each other. Specifically, the hidden state h_j from the lower LSTM layer, is the input x_j of the upper LSTM layer. This stacked architecture, which can combine multiple levels or representations with flexible use of long range context, had a significant performance improvement compared with the shallow one [16].

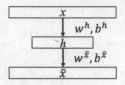

Fig. 2. A basic autoencoder.

3.2 Autoencoder

An autoencoder is an unsupervised neural network which attempts to reproduce its input [2,3,17,29]. Specifically, the basic autoencoder in Fig. 2 locally optimizes the hidden representation $h \in R^s$ of its input $x \in R^l$, such that h can be used to accurately reconstruct x

$$h = \rho_{en}(x) = \psi(W^h x + b^h) \tag{7}$$

$$\tilde{x} = \rho_{de}(h) = \psi(W^{\tilde{x}} x + b^{\tilde{x}}) \tag{8}$$

where $\tilde{x}$ is the reconstruction of x, the learning parameters $W^h \in R^{s \times l}$, and $W^{\tilde{x}} \in R^{l \times s}$ are weight matrices, $b^h \in R^s$ and $b^{\tilde{x}} \in R^l$ are bias vectors for the hidden and output layers respectively, and ψ is a nonlinear function such as the sigmoid function and the hyperbolic tangent function. Equation (7) encodes the input into an intermediate representation and Eq. (8) decodes the resulting representation.

The training objective is the determination of parameters $\theta = (W^h, W^{\tilde{x}}, b^h, b^{\tilde{x}})$ that minimize the average error between a set of inputs $\{x_1, \cdots, x_t\}$ and a set of reconstructions $\{\tilde{x}_1, \cdots, \tilde{x}_t\}$:

$$\hat{\theta} = \arg\min_\theta \frac{1}{t} \Sigma_{j=1}^t L(x_j, \tilde{x}_j) \tag{9}$$

$$= \arg\min_\theta \frac{1}{t} \Sigma_{j=1}^t L(x_j, \rho_{de}(\rho_{en}(x_j))) \tag{10}$$

where L is a loss function such as cross-entropy.

The training criterion with *denoising autoencoder* is the reconstruction of clean input x given a corrupted version $\tau(x)$ [27–29], where τ is a noise masking function that randomly sets a fraction of x's elements to 0. The underlying idea is that the learned latent representation is good if the autoencoder is capable of reconstructing the actual input from its corruption. Then, the average error becomes

$$\hat{\theta} = \arg\min_\theta \frac{1}{t} \Sigma_{j=1}^t L(x_j, \rho_{de}(\rho_{en}(\tau(x_j)))) \tag{11}$$

Multiple autoencoders can be used as building blocks to form a deep architecture, known as the *stacked autoencoder* [4,27,29]. For that purpose, the autoencoders are pre-trained layer by layer, with the current layer being fed the latent representation of the previous autoencoder as input. Using this unsupervised pre-training procedure, initial parameters are found which approximate a good solution. Subsequently, the original input layer and hidden representations of all the autoencoders are stacked and all network parameters are fine-tuned.

4   Problem Settings

In this section, we formalize the problem of food sales prediction with meteorological data. Suppose we focus on some product p. The task is to predict sales of p in shop s on day n, which is denoted by Sal_s^n. Since the absolute sale depends on shops, instead of simply taking the task as a regression problem we convert it to a classification problem. Specifically, we predict the label

$$
Label_s^n = \begin{cases}
0, & \text{if } Rate_s^n \in [0, 0.7) \\
1, & \text{if } Rate_s^n \in [0.7, 0.9) \\
2, & \text{if } Rate_s^n \in [0.9, 1.1) \\
3, & \text{if } Rate_s^n \in [1.1, 1.3) \\
4, & \text{if } Rate_s^n \in [1.3, \infty)
\end{cases}
\tag{12}
$$

where $Rate_s^n = Sal_s^n / Sal_s^{n-7}$. One truth is that sales fluctuate with day of the week cycle. For example, sales on Sunday is often higher than on Monday. Therefore, the essence of $Label_s^n$ is to take Sal_s^{n-7} which is the sales on the same day of the week as a reference, and predict how Sal_s^n depends on the weather.

To predict $Label_s^n$, the following information are provided.

- Sales of p in shop s on the previous d days: Sal_s^c for $c = n - 1, \cdots, n - d$.
- Forecasted weather condition in shop s on day n. Weather information includes 6 aspects:
  - Solar Radiation: Slr
  - Rainfall Precipitation: $Rain$
  - Relative Humidity: Hmd
  - Temperature: $Temp$
  - North Wind Velocity: Nwd
  - East Wind Velocity: Ewd

  That is, $Wea_s^n = (Slr_s^n, Rain_s^n, Hmd_s^n, Temp_s^n, Nwd_s^n, Ewd_s^n)$.
- Weather in shop s on the previous d days: Wea_s^c, for $c = n - 1, n - 2, \cdots, n - d$.
- Date information of day n and its previous d days. Date information includes 2 aspects:
  - Day of the week: $Dayw$
  - Month of the year: $Mthy$
  - Holiday or not: Hld

  That is, $Date^c = (Dayw^c, Mthy^c, Hld^c)$ for $c = n, n - 1, \cdots, n - d$.

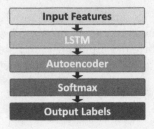

Fig. 3. The diagram of our method.

In other words, we use a bunch of features (the sales features, the weather features, and the date features) to predict $Label_s^n$. To be specific, the features are expressed as

$$Fea_s^n = (Sal_s^{n-1}, Sal_s^{n-2}, \cdots, Sal_s^{n-d}, Wea_s^n, Wea_s^{n-1}, \cdots$$
$$Wea_s^{n-d}, Date^n, Date^{n-1}, \cdots, Date^{n-d}) \tag{13}$$

In addition, we have a training dataset which is history data about the features and labels. The training dataset is $\{(Fea_i^m, Label_i^m)|i \in I, m \in M\}$, where $I = \{$All of the shops for which history data are available$\}$ and $M = \{$All of the days on which history data are available$\}$.

5   Our Method

Our basic strategy is to learn a prediction model based on the training dataset. It contains 2 steps to learn the model. In Step 1 we use a LSTM network, which learns how the sales depends on the variation of weathers. In Step 2 we use an autoencoder network, which helps to reduce dimensions of the features which are learnt from the LSTM network. A diagram is shown in Fig. 3. In the following, we explain each step in detail.

5.1   Step 1: LSTM

To predict $Label_i^m$, we use sales, weather, and date features of day m and the previous d days. Each day's features are not independent of each other, but are closely related and determines the label in a joint way. For example, in summer if the temperature of previous days is low and it suddenly rises by 10 degrees then the sales of drinks will probably increases dramatically. However, if the temperature has been keeping high for a long time then the sales will not change so much. In other words, it is not the high temperature but the relative variation of temperature that affects the sales. Therefore, one important point to the success of our prediction is to understand how the sales depends on the variation trend of previous days' features. For this purpose, we use LSTM.

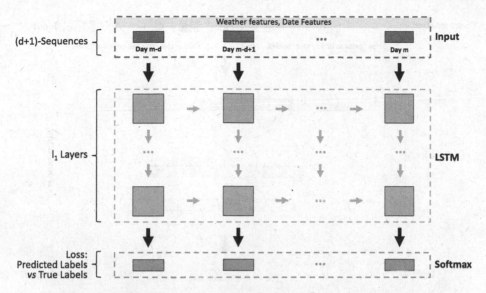

Fig. 4. The diagram of Step 1. A $(d+1)$-sequential data which includes the weather and date features are taken as input of the LSTM network. The LSTM network has l_1 layers which are stacked on top of each other. Each layer contains $d+1$ cells. Finally, a sequence of labels are calculated based on the softmax function on the output of each cell of the last layer. This sequence of labels is compared with a sequence of true labels to train the parameters of the LSTM network.

In step 1, from the training dataset we prepare a set of sequential data $\{(SeqX_i^m, SeqY_i^m)|i \in I, m \in M\}$ and feed it to a LSTM network. Specifically,

$$SeqX_i^m = \langle SeqX_1, SeqX_2, \cdots, SeqX_{d+1} \rangle \qquad (14)$$

$$SeqY_i^m = \langle SeqY_1, SeqY_2, \cdots, SeqY_{d+1} \rangle \qquad (15)$$

The sequence length of $SeqX_i^m$ and $SeqY_i^m$ is $d+1$. The j-th element of $SeqX_i^m$

$$SeqX_j = (Wea_i^{m-d-1+j}, Date^{m-d-1+j}), \text{for } j = 1, 2, \cdots, d+1 \qquad (16)$$

denotes the weather and date features of day $m - d - 1 + j$ in shop i. The j-th element of $SeqY_i^m$

$$SeqY_j = Label_i^{m-d-1+j}, \text{for } j = 1, 2, \cdots, d+1 \qquad (17)$$

denotes the label of day $m - d - 1 + j$ in shop i.

We take $\{SeqX_i^m | i \in I, m \in M\}$ as input to the LSTM network, which contains l_1 layers. Let the output of LSTM be

$$SeqO_i^m = \langle SeqO_1, SeqO_2, \cdots, SeqO_{d+1} \rangle \qquad (18)$$

where $SeqO_j$ has the same dimension as $SeqX_j$. Then we calculate the softmax outputs of $SeqO_j$, for $j = 1, 2, \cdots, d+1$, and suppose them as a sequence of

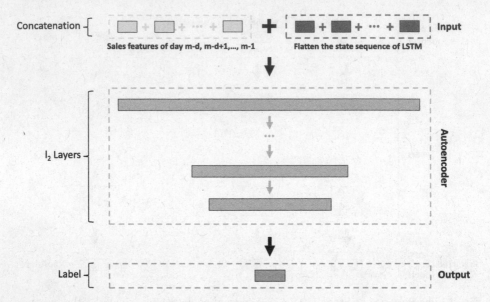

Fig. 5. The diagram of Step 2. The latent features learnt from step 1 in concatenation with the sales features are taken as input of the autoencoder network. The autoencoder network has l_2 layers which are stacked on top of each other and are trained independently. Finally, a label is calculated based on the softmax function on the output of the last layer. This label is compared with a true label to fine-tune the parameters of the whole network.

predicted labels for day $m - d$, $m - d + 1$, $\cdots$, m in shop i. So, the loss function which is used to train parameters of the LSTM network is calculated as the mean of the cross entropy of the sequence of predicted labels and true labels. A diagram of step 1 is shown in Fig. 4.

5.2   Step 2: Autoencoder

In step 1, we use sequential data which is composed of weather and date features to produce a sequence of hidden features. In this step, we use this hidden features plus sales features to predict $Label_i^m$. We use $Sales_i^m$ to denote the sales features of the previous d days in shop i

$$Sales_i^m = (Sal_i^{m-1}, Sal_i^{m-2}, \cdots, Sal_i^{m-d}) \tag{19}$$

We first flatten $SeqO_i^m$ and then concatenate it with $Sales_i^m$ to obtain a high dimensional vector $Hvec$. That is,

$$Hvec_i^m = (SeqO_i^m, Sales_i^m) \tag{20}$$

$Hvec_i^m$ represents latent features that can be used to predict $Label_i^m$. However, The dimension of $Hvec_i^m$ is so high that we cannot directly arrive at an accurate prediction based on it. Therefore, we feed $\{Hvec_i^m | i \in I, m \in M\}$ to a stacked denoising autoencoder to learn an efficient shortened features, and then use that shortened features to predict $Label_i^m$. A diagram of step 2 is shown in Fig. 5.

6   Experiments

In this section, we present experiment results of our method in comparison to other machine learning methods. We mainly studied the sales of the Japanese chain supermarket during Jan 2012 to Dec 2013. As an initial try toward predicting food sales with meteorological data, we set $p = drinks$, $d = 28$. As a result, our dataset contains 131,493 examples. That is

$$|\{(Fea_i^m, Label_i^m) | i \in I, m \in M\}| = 131493 \tag{21}$$

The percentage of the five labels are 9.32%, 25.50%, 29.85%, 19.03%, 16.30%, respectively. We split the dataset randomly with the ratios of 0.8 and 0.2 into the training and test datasets, respectively. We feed the training dataset to a method and evaluate its performance by the test dataset. We compare our method with five traditional machine learning methods.

- *SVM:* The support vector machine classifier [8] with radial basis function as kernels and based on one-vs-one scheme [21] for multi-class classification.
- *LR:* The regularized logistic regression classifier [10] based on one-vs-rest scheme for multi-class classification.
- *RandForest:* The random decision forests classifier [18,19] based on ensemble learning.
- *AdaBoost:* The adaptive boosting classifier [12] with decision tree classifier [24] as the base estimator.
- *GBDT:* The gradient boosting decision tree classifier [13].

We also consider using LSTM and autoencoder alone for comparison. For SVM, LR, RandForest, AdaBoost, GDBT, and autoencoder, we directly take Fea_i^m as input to learn the output label. For LSTM, we also prepare a set of sequential data $\{(SeqX_i^m, SeqY_i^m) | i \in I, m \in M\}$. In order to effectively utilize the sales feature, we incorporate it into $SeqX_i^m$. However, since we do not know Sal_i^m, we shift the sales feature by one day. That is,

$$SeqX_j = (Wea_i^{m-d-1+j}, Date_i^{m-d-1+j}, Sal_i^{m-d-2+j}), j = 1, 2, \cdots, d+1 \tag{22}$$

Besides, the label $Label_i^m$ is calculated as the softmax function on the last cell.

The average accuracy of different methods are shown in Table 1. We can find that our method outperforms traditional methods by a large margin. Particularly, our method achieved an average accuracy of 0.6194, 19.3% = $(0.6194/0.5193 - 1) * 100\%$ higher than SVM and GBDT which achieved the best result among the traditional methods. Our method which combines LSTM and autoencoder is also better than using alone. Table 2 shows a confusion matrix obtained by our method. We can find that most of the mistakes by our method

Table 1. The average accuracy of different methods achieved on the test dataset.

Method	Accuracy (%)
Our Method	0.6194
LSTM	0.5573
Autoencoder	0.5821
SVM	0.5193
LR	0.5145
RandForest	0.4963
AdaBoost	0.4317
GBDT	0.5193

Table 2. Confusion matrix of our method.

		Predicted				
		0	1	2	3	4
True	0	1573	835	42	2	0
	1	506	4266	1827	98	10
	2	13	1427	5102	1160	147
	3	0	125	1777	2257	846
	4	2	8	246	939	3091

Table 3. Accuracy of our method on each label.

Label	Precision	Recall	F1-Score
0	0.75	0.64	0.69
1	0.64	0.64	0.64
2	0.57	0.65	0.61
3	0.51	0.45	0.48
4	0.76	0.72	0.74

are not so serious, as the mistakes are just made for adjacent labels. For example, our prediction is label 2 for an "invariant" sales in the range of [0.9,1.1], while the true label is 1 for a slight drop in the range of [0.7,0.9). When look at performance for each label, as shown in Table 3, we can find that our method is good at predicting label 0 and 4 while struggling in label 1, 2, 3. As omitted here, this trend is evident for all of the methods we compared, meaning that it is easy to predict an abrupt increase/decrease in sales.

7   Conclusion

We introduced a deep learning based method which use meteorological data to predict food retailers' sales. Our method gained initial success in predicting the sales of drinks of a Japanese chain supermarket. We plan to use our method to predict sales of other whether-sensitive products, such as fruits, ice-cream.

However, food sales prediction is still difficult to achieve, because it is easily affected by other factors except for weathers, such as how many populations and how many competitive supermarkets, convenience stores, or food retailers in the same area, the price strategy, the advertising campaigns, and the sales area of a particular shop. As part of our future work, we plan to collect more data about such factors to enhance the accuracy of prediction.

Acknowledgment. We gratefully thank Japan Weather Association especially Mr.Tomohiro Yoshikai for supporting our work.

References

1. Alomar, M.L., Canals, V., Perez-Mora, N., Martínez-Moll, V., Rosselló, J.L.: Fpga-based stochastic echo state networks for time-series forecasting. Comput. Intell. Neurosci. **3917892**, 1–14 (2016)
2. Amiri, H., Resnik, P., Boyd-Graber, J., Daumé III, H.: Learning text pair similarity with context-sensitive autoencoders. In: Proceedings of the 54th Annual Meeting of the Association for Computational Linguistics, Berlin, Germany, pp. 1882–1892, August 2016
3. Bengio, Y.: Learning deep architectures for ai. Foundations and trends® in. Mach. Learn. **2**(1), 1–127 (2009)
4. Bengio, Y., Lamblin, P., Popovici, D., Larochelle, H.: Greedy layer-wise training of deep networks. In: Proceedings of the 20th Annual Conference on Neural Information Processing Systems, Vancouver, Canada, pp. 153–160, December 2006
5. Bengio, Y., Simard, P., Frasconi, P.: Learning long-term dependencies with gradient descent is difficult. IEEE Trans. Neural Networks **5**(2), 157–166 (1994)
6. Brentan, B.M., Luvizotto Jr., E., Herrera, M., Izquierdo, J., Pérez-García, R.: Hybrid regression model for near real-time urban water demand forecasting. J. Comput. Appl. Math. **309**, 532–541 (2017)
7. Chandra, R., Zhang, M.: Cooperative coevolution of elman recurrent neural networks for chaotic time series prediction. Neurocomputing **86**, 116–123 (2012)
8. Cortes, C., Vapnik, V.: Support-vector networks. Mach. Learn. **20**(3), 273–297 (1995)
9. Cortez, P., Rio, M., Rocha, M., Sousa, P.: Multi-scale internet traffic forecasting using neural networks and time series methods. Exp. Syst. **29**(2), 143–155 (2012)
10. Cox, D.R.: The regression analysis of binary sequences. J. Roy. Stat. Soc. Ser. B **20**(2), 215–242 (1958)
11. De Felice, M., Alessandri, A., Ruti, P.M.: Electricity demand forecasting over italy: Potential benefits using numerical weather prediction models. Electr. Power Syst. Res. **104**, 71–79 (2013)
12. Freund, Y., Schapire, R.E.: A short introduction to boosting. J. Japan. Soc. Artif. Intell. **14**(5), 771–780 (1999)
13. Friedman, J.H.: Greedy function approximation: a gradient boosting machine. Ann. Stat. **29**(5), 1189–1232 (2001)
14. Gamboa, J.C.B.: Deep learning for time-series analysis. arXiv preprint (2017). arxiv:1701.01887

15. Gao, Y., Hendricks, L.A., Kuchenbecker, K.J., Darrell, T.: Deep learning for tactile understanding from visual and haptic data. In: Proceedings of the 2016 IEEE International Conference on Robotics and Automation, Stockholm, Sweden, pp. 536–543, May 2016

16. Graves, A.: Mohamed, A.r., Hinton, G.: Speech recognition with deep recurrent neural networks. In: Proceedings of the 2013 IEEE International Conference on Acoustics. Speech and Signal Processing, Vancouver, Canada, pp. 6645–6649, May 2013

17. Hinton, G.E., Salakhutdinov, R.R.: Reducing the dimensionality of data with neural networks. Science **313**(5786), 504–507 (2006)

18. Ho, T.K.: Random decision forests. In: Proceedings of the 3rd International Conference on Document Analysis and Recognition, Montreal, Canada, pp. 278–282, August 1995

19. Ho, T.K.: The random subspace method for constructing decision forests. IEEE Trans. Pattern Anal. Mach. Intell. **20**(8), 832–844 (1998)

20. Hochreiter, S., Schmidhuber, J.: Long short-term memory. Neural Comput. **9**(8), 1735–1780 (1997)

21. Knerr, S., Personnaz, L., Dreyfus, G.: Single-layer learning revisited: a stepwise procedure for building and training a neural network. Neurocomputing **68**, 41–50 (1990)

22. Lv, Y., Duan, Y., Kang, W., Li, Z., Wang, F.Y.: Traffic flow prediction with big data: a deep learning approach. IEEE Trans. Intell. Transp. Syst. **16**(2), 865–873 (2015)

23. Nita, S.: Application of big data technology in support of food manufacturers commodity demand forecasting. NEC Tech. J. **10**(1), 90–93 (2015)

24. Quinlan, J.R.: Induction of decision trees. Mach. Learn. **1**(1), 81–106 (1986)

25. Schmidhuber, J.: Deep learning in neural networks: an overview. Neural Networks **61**, 85–117 (2015)

26. Schulz, M., Matthies, M.: Artificial neural networks for modeling time series of beach litter in the southern north sea. Marine Environ. Res. **98**, 14–20 (2014)

27. Silberer, C., Lapata, M.: Learning grounded meaning representations with autoencoders. In: Proceedings of the 52nd Annual Meeting of the Association for Computational Linguistics, Baltimore, MD, USA, pp. 721–732, June 2014

28. Vincent, P., Larochelle, H., Bengio, Y., Manzagol, P.A.: Extracting and composing robust features with denoising autoencoders. In: Proceedings of the 25th International Conference on Machine Learning, Helsinki, Finland, pp. 1096–1103, July 2008

29. Vincent, P., Larochelle, H., Lajoie, I., Bengio, Y., Manzagol, P.A.: Stacked denoising autoencoders: learning useful representations in a deep network with a local denoising criterion. J. Mach. Learn. Res. **11**, 3371–3408 (2010)

Cascade Spatial Autoregression for Air Pollution Prediction

Yangping Li, Xiaorui Wei, and Tianming Hu[(⊠)]

Dongguan University of Technology, Dongguan, China
tmhu@ieee.org

Abstract. Recent years have witnessed a growing interest in air quality prediction and a variety of predictions models have been applied for this task. However, all of these models only use local attributes of each site for prediction and neglect the spatial context. Indeed, the concentrations of air pollutants follow the first law of geography: everything is related to everything else, but nearby things are more related than distinct things. To that end, in this paper, we apply the spatial autoregression model (SAR) to air pollution prediction, which considers both local attributes and predictions from the neighborhoods. Specifically, as SAR can only handle a snapshot of spatial data but our input data are time series, we develop the cascade SAR, which is able to take care of both spatial and temporal dimensions without incurring extra computation. Finally, the effectiveness of the cascade SAR is validated on the dataset of the London Air Quality Network.

Keywords: Cascade spatial autoregression · Air pollution prediction

1 Introduction

It has been recognized long ago that suspended particle matter in the atmosphere lead to harmful health effects [2, 7]. Since accurate prediction of air pollution would help a lot in planning of more efficient actions to protect the people involved, recent years have witnessed a growing interest in air pollution prediction. So far, classical statistical methods and neural networks have been used by researchers for short term prediction of gas and particulate matter pollution. These prediction models use the pollution data from the past, some plus meteorological information, to predict average pollutant concentrations between one half and 24 h in advance [5, 6].

Although various prediction models have been employed, most of them only use local attributes and discard the spatial context. That is, to predict the pollutant concentrations of a site, they only use the past pollution data of that site, without touching the data of its neighbors. On the other hand, the concentrations of air pollutants are spatially correlated in that a site's air pollutant concentrations are more similar to those of its neighboring sites than the sites farther away. Normally, the concentration is highest at the site of the pollution source, and decreases gradually as the site's distance to the source increases. Indeed, the distribution of air pollutants also follows the first law of geography: everything is related to everything else, but nearby things are more

© Springer International Publishing AG 2017
Y. Tan et al. (Eds.): DMBD 2017, LNCS 10387, pp. 105–112, 2017.
DOI: 10.1007/978-3-319-61845-6_11

related than distinct things [3]. Hence, to better predict the air pollution, we need to make use of both local attributes and information from the neighbors.

To that end, in this paper, we apply the spatial autoregression model (SAR) to air pollution prediction. SAR is a linear autoregression model that incorporates the spatial context [1]. The standard input to SAR is a snapshot of data for a set of sites in a region at a time instant. For instance, if we predict hourly average concentrations of PM_{10} at any hour of the day with the 24 hourly average concentrations measured on the previous day, the input to SAR is the hourly average concentrations of PM_{10} at an hour for all of the sites. However, our history data record hourly average concentrations for a period, which are essentially time series. Apparently, it is impractical and unreasonable to construct an SAR model for each time instant. To make use of the whole time series with still one SAR model, we extend standard SAR by cascading it to suit time series. Finally, the effectiveness of the cascade SAR model is validated on the well-known London Air Quality Network.

The rest of the paper is organized as follows. Section 2 gives problem formulation and related work. Section 3 presents the cascade spatial autoregression model. Section 4 reports the experimental results. Finally, Sect. 5 concludes this paper.

2   Problem Formulation and Related Work

2.1   Problem Formulation

Here we forecast the air pollution concentration for a set of sites with countable indices. Let S denote the set of locations. S could be $\{(i, x_i, y_i)\}$, the set of triple (index, latitude, longitude). Let $s_i = (i, x_i, y_i) \in S$ denote a site. Then the problem of air pollution prediction can be formulated as follows.

- **Given**

1. A spatial framework of n sites, $S = \{s_i\}$ with a neighbor relation $N \subseteq S \times S$. Sites s_i and s_j are neighbors iff $(s_i, s_j) \in N$, $i \neq j$. Let $N(s_i) \equiv \{s_j: (s_i, s_j) \in N\}$ denote the neighborhood of s_i. We assume that neighbor relation N is given by a row-normalized contiguity matrix $\mathbf{W}$, where $\mathbf{W}(i, j) = 1/|N(s_i)|$ iff $(s_i, s_j) \in N$ and $\mathbf{W}(i, j) = 0$ otherwise.
2. Associated with each s_i, there is a d-dimensional feature vector of explanatory attributes $\mathbf{x}_i$ and a dependent variable y_i to be predicted. In our case, we make pollution prediction for n sites and m days, so the input data are time series for m days. Let y_i^t denote the hourly average concentration at a certain fixed hour of the day t, $\mathbf{x}_i^{t-1}$ the 24-dimensional vector of the 24 hourly average concentrations measured on the previous day $t - 1$. Let $\mathbf{y}^t \equiv [y_1^t, ..., y_n^t]^T$.

- **Find**

A function f that maps $\mathbf{x}_i^{t-1}$ to y_i^t. Let $\hat{y}_i^t \equiv f(\mathbf{x}_i^{t-1})$, $\hat{\mathbf{y}}^t \equiv [\hat{y}_1^t, ..., \hat{y}_n^t]^T$.

- **Objective**

Let **y** denote the *nm-dimensional* vector concatenating the *n-dimensional* **y**^{t}'s of *m* days and **ŷ** its estimator. Our objective is to maximize similarity between **y** and **ŷ**. We employ the commonly used regression criterion, mean squared error (MSE). Since we make prediction for *n* sites and *m* days, MSE is $(\|\mathbf{y}^1 - \hat{\mathbf{y}}^1\|^2 \ldots + \ldots \|\mathbf{y}^D - \hat{\mathbf{y}}^D\|^2)/nm$.

- **Constraint**

Spatial autocorrelation exists, that is, y_i is not only affected by its own $\mathbf{x}_i$, but also by $\mathbf{x}_j$ and y_j of its neighbors $s_j \in N(s_i)$.

2.2   Related Work

A wide variety of empirical, causal, statistical and hybrid models have been developed in order to forecasting air pollution [8]. Besides the conventional linear regression models, neural networks, multilayer perceptron particularly, have been used for the forecasting of a wide range of pollutants and their concentrations at various time, with good results. The findings of numerous research studies also exhibit that the performance of neural networks is generally superior in comparison to traditional statistical methods, such as multiple regression, classification and regression trees and autoregressive models [4].

The advantage of multilayer perceptron over traditional statistical models like linear regression is not surprising, for multilayer perceptron is essentially a two-layered composite function, which consists a nonlinear function at the hidden layer and a linear function at the output layer. In principle, given enough hidden nodes, multilayer perceptron can approximate any function. However, compared to the fast parameter estimation of linear regression, the training of multilayer perceptron is much more expensive. Furthermore, the learning of linear regression is deterministic and will output global optimal solutions, while the learning of multilayer perceptron is non-deterministic and will output local optimal solutions. Finally, during air pollution prediction, both standard linear regression and multilayer perceptrons do not consider spatial context and only use the local attributes of each site.

3   Cascade Spatial Autoregression

In this paper, we incorporate the spatial context into the linear regression model. Specifically, on one hand, we want to remedy the simple assumption of linear regression by making the pollutant concentration of each site depend on both its local attributes and its neighbors as well. On the other hand, we still want to retain its fast learning feature. Hence we introduce the SAR model to the air pollution prediction. SAR is a linear regression model, which ensures fast parameter estimation. Meanwhile, SAR assumes every site's pollutant concentration is affected not only by its local attributes, but also by its neighboring sites. Such incorporation of spatial context into the dependence assumption enables SAR to yield more accurate forecasting than conventional linear regression models.

3.1 Linear Regression

With standard linear regression modeling, the relationship between $\mathbf{x}_i^{t-1}$ to y_i^t can be described in matrix form in (1),

$$\mathbf{y} = \mathbf{Xw}. \tag{1}$$

where each row of the $nm \times 24$ matrix $\mathbf{X}$ is the 24-dimensional vector of the 24 hourly average concentrations measured on the previous day, $\mathbf{y}$ is the nm-*dimensional* vector concatenating the m $\mathbf{y}^t$ s, and $\mathbf{w} = [w_1, \ldots, w_{24}]^T$ is the 24-dimensional vector of weights to be estimated. Differentiating MSE with respect to $\mathbf{w}$ and solving the resulting normal equations for $\mathbf{w}$ yields the least square solution in (2), where pseudo-inverse $\mathbf{X}^+ = (\mathbf{X}^T\mathbf{X})^{-1}\mathbf{X}^T$

$$\begin{aligned} \hat{\mathbf{w}} &= (\mathbf{X}^T\mathbf{X})^{-1}\mathbf{X}^T\mathbf{y} \\ &= \mathbf{X}^+\mathbf{y} \end{aligned}. \tag{2}$$

3.2 Spatial Autoregression

The general model of spatial data, which is the underlying model for spatial statistics [3], is as follows.

$$\text{spatial data} = \text{trend} + \text{dependence} + \text{error} \tag{3}$$

According to this model, SAR takes the form

$$\mathbf{y}^t = \mathbf{X}^{t-1}\mathbf{w} + \rho\mathbf{W}\mathbf{y}^t. \tag{4}$$

where the i-th element of $\mathbf{y}^t$ is the concentration on day t for site s_i, the i-th row of $\mathbf{X}^{t-1}$ is the corresponding 24-dimensional vector on day $t - 1$, $\mathbf{W}$ is the contiguity matrix that encodes the neighborhood relationship. $\mathbf{w}$ is the weight vector to be estimated and $\mathbf{X}^{t-1}\mathbf{w}$ represents the spatial trend. ρ is the coefficient that weighs the spatial dependence $\mathbf{W}\mathbf{y}^t$. When it is set to zero, SAR is reduced to standard linear regression. It can be either fixed in advance or jointly estimated with $\mathbf{w}$. In the latter case, computationally expensive techniques such as Monte Carlo sampling may be applied. In the former case, with identity matrix $\mathbf{I}$, the MSE solution is

$$\hat{\mathbf{w}} = \left[(\mathbf{I} - \rho\mathbf{W})^{-1}\mathbf{X}^{t-1}\right]^+ \mathbf{y}^t. \tag{5}$$

One can see that SAR can only be directly applied to data at a time instance t, while our input is time series of length m. At first glance, we can construct one SAR model for each t and combine the results. In a way, this is related to multiple regression. However, on one hand, the extra learning for combination brings more computation. On the other hand, since test data may be time series of any length, it may make the combination inapplicable.

3.3 Cascade Spatial Autoregression

To make the SAR model applicable to time series, we extend it by cascading the contiguity matrix $\mathbf{W}$. In this way, we fold the time series into a "flat" time instant.

Specifically, we use the input matrix $\mathbf{X}$ and $\mathbf{y}$ for the whole time series, as in standard linear regression in (1). The new cascaded contiguity matrix $\mathbf{W}_c$ is a diagonal matrix, with m $\mathbf{W}$s on its diagonal.

$$
\mathbf{W}_c = \begin{pmatrix} \mathbf{W} & \mathbf{0} & \cdots & \mathbf{0} \\ \mathbf{0} & \mathbf{W} & \cdots & \mathbf{0} \\ \mathbf{0} & \mathbf{0} & \cdots & \mathbf{0} \\ \mathbf{0} & \mathbf{0} & \cdots & \mathbf{W} \end{pmatrix}.
\tag{6}
$$

Then the cascade SAR for the time series of length m becomes

$$
\mathbf{y} = \mathbf{Xw} + \rho \mathbf{W}_c \mathbf{y}.
\tag{7}
$$

One can see that for any dependent variable y_i^t, the dependence relationship is correctly retained during the multiplication of its row $\mathbf{r}$ in $\mathbf{W}_c$ with $\mathbf{y}$. In detail, if another dependent variable y is from the same day, its corresponding element in $\mathbf{r}$ must be in the t-th $\mathbf{W}$ of $\mathbf{W}_c$ and hence nonzero. If y is from a different day, its corresponding element in $\mathbf{r}$ cannot be in the t-th $\mathbf{W}$ of $\mathbf{W}_c$ and hence must be zero.

Therefore, the cascade SAR is not only flexible to cope with time series of any length, but also simple in its solution determination. With fixed ρ, the new solution is

$$
\dot{\mathbf{w}} = \left[(\mathbf{I} - \rho \mathbf{W}_c)^{-1} \mathbf{X} \right]^+ \mathbf{y}.
\tag{8}
$$

4 Experimental Evaluation

In this section, we evaluate cascade SAR on the London Air Quality Network.

4.1 Data

Due to the availability and comprehensiveness, we select the database of the hourly average concentrations of various pollutants, including $PM_{2.5}$, PM_{10}, NO_x, and O_3, from 2000 to 2006. There are a total of 2557 days from 2000 to 2006. Thus, for each pollutant, we have 61368 (2557×24) hourly concentration records at the monitoring site. Because not all pollutants are recorded at every site, due to lack of space, below we only report the prediction of PM_{10}. Figure 1 gives its hourly average concentration grouped by month. It is observed that the peak values happen around 8:00–10:00 and 20:00–21:00. In general, these peaks overlap with or lag a little behind the commuter time.

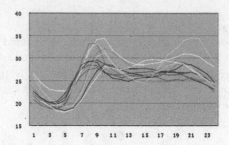

Fig. 1. The PM$_{10}$ hourly average concentration over 2000–2006 grouped by month.

4.2 Experimental Setting

We use the data of 2000 for model training and the data of 2001–2006 for testing. Before feeding the data into the models, they are normalized to [0, 1] with the conversion function, $x' = (x - x_{min})/(x_{max} - x_{min})$. After we get the prediction values, we convert them back to their original scale with unit μg/m^3.

As for the coefficient ρ, we choose to fix it in advance and then solve for the MSE weights analytically. After all, the major advantage of linear regression lies in its fast training, which is only feasible with a fixed ρ. We sample a wide range of [0, 2] for ρ and find the best results are obtained around $\rho = 1$. Hence, all results reported below are obtained at this value.

The results are evaluated with two measures. One is mean relative error (*mre*), $mre = \text{avg}_y |(\hat{y} - y)/y|$, where *avg* denotes the average function, y and $\hat{y}$ denote the true value and the predicted value, respectively. The other measure is squared error, $S = \text{avg}_y (\hat{y} - y)^2$, which is equivalent to MSE.

4.3 Empirical Results

Tables 1 and 2 give the test results of 24 h for the linear regression and cascade SAR, respectively. For each hour, we learn a linear model separately and use it for testing. One can see that cascade SAR beats the standard linear regression consistently, while their training costs are comparable.

Figure 2 plots the prediction results for cascade SAR. In terms of squared error, the error increases with the hour. However, in terms of *mre*, the peak values happen around 4:00 and 24:00. This can be explained by the denominator in *mre* definition, which takes on low values of true concentration, as illustrated in Fig. 3. For comparison, in Fig. 3, we also plot the average values of true concentration and prediction for each hour, which help explain again the peak values of *mre*.

Table 1. Prediction of PM$_{10}$ hourly average concentration by linear regression

hour	1:00	2:00	3:00	4:00	5:00	6:00	7:00	8:00
mre%	29.72	42.53	48.17	62.89	56.48	42.53	35.91	39.27
S	11.47	15.18	20.67	21.38	27.64	25.97	33.23	28.62
hour	9:00	10:00	11:00	12:00	13:00	14:00	15:00	16:00
mre%	39.94	40.46	30.70	38.61	41.32	36.91	41.51	33.02
S	31.82	32.54	34.59	40.667	41.97	41.48	47.18	42.99
hour	17:00	18:00	19:00	20:00	21:00	22:00	23:00	0:00
mre%	35.18	40.07	37.89	38.25	36.15	31.05	41.57	49.43
S	47.69	47.93	51.75	52.30	47.55	50.14	50.13	52.46

Table 2. Prediction of PM$_{10}$ hourly average concentration by cascade SAR

hour	1:00	2:00	3:00	4:00	5:00	6:00	7:00	8:00
mre%	17.5	28.95	46.27	49.19	47.00	41.07	31.74	31.07
S	4.69	7.61	13.24	17.46	21.09	24.26	26.17	28.31
hour	9:00	10:00	11:00	12:00	13:00	14:00	15:00	16:00
mre%	30.37	30.82	29.13	28.91	31.75	32.06	33.51	31.61
S	30.16	32.27	34.01	35.72	37.81	39.58	41.48	42.79
hour	17:00	18:00	19:00	20:00	21:00	22:00	23:00	0:00
mre%	30.97	30.92	29.97	28.66	29.6	30.7	33.08	40.1
S	44.18	44.88	45.63	45.94	46.06	46.23	46.57	47.29

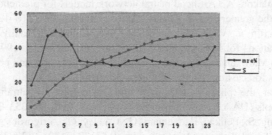

Fig. 2. The average prediction results of PM$_{10}$ hourly average concentration by cascade SAR.

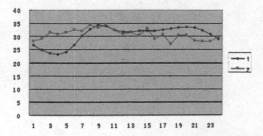

Fig. 3. The average of true (t) and predicted (y) PM$_{10}$ hourly average concentrations.

5 Concluding Remarks

In this paper, we applied the SAR model to air pollution prediction, which considers both local attributes and predictions from the neighborhoods. Specifically, we extended SAR by cascading the contiguity matrix, so that it can handle time series instead of a snapshot of spatial data. Finally, the experiments on the database of the London Air Quality Network validated the effectiveness of the cascade SAR model. For the future work, we plan to improve the learning process of SAR. For example, instead of using the data of the whole year for training, we can learn a model separately for each month or season. This is expected to yield more accurate prediction.

Acknowledgments. This work was supported by SDPD (No. 2014106101003 and 201510414000079).

References

1. Anselin, L.: Spatial regression. In: The Sage Handbook of Spatial Analysis, pp. 255–275. Sage, Thousand Oaks (2009)
2. Burnett, R.T., Smith-Doiron, M., Stieb, D., Cakmak, S., Brook, J.R.: Effects of particulate and gaseous air pollution on cardiorespiratory hospitalizations. Arch. Environ. Heal. **54**(2), 130–139 (1999)
3. Chun, Y., Griffith, D.: Spatial Statistics and Geostatistics: Theory and Applications for Geographic Information Science and Technology. SAGE, Los Angeles (2013)
4. Grivas, G., Chaloulakou, A.: Artificial neural network models for prediction of PM10 hourly concentrations in the greater area of Athens. Atmos. Environ. **40**, 1216–1229 (2006). Greece
5. Hernandez, E., Martin, F., Valero, F.: Statistical forecast models for daily air particulate iron and lead concentrations for Madrid. Atmos. Environ. **26B**, 107–116 (1992). Spain
6. Kolehmainen, M., Martikainen, H., Ruuskanen, J.: Neural networks and periodic components used in air quality forecasting. Atmos. Environ. **35**, 815–825 (2001)
7. Spurny, K.R.: On the physics, chemistry and toxicology of ultrafine anthropogenic, atmospheric aerosols (UAAA): new advances. Toxicolog. Lett. **96**, 253–261 (1998)
8. Schlink, U., Dorling, S., Pelikan, E., Nunnari, G., Cawley, G.: A rigorous inter-comparison of ground-level oregion predictions. Atmos. Environ. **37**, 3237–3253 (2003)

Machine Learning Techniques for Prediction of Pre-fetched Objects in Handling Big Data Storage

Nur Syahela Hussien, Sarina Sulaiman[(⊠)],
and Siti Mariyam Shamsuddin

UTM Big Data Centre, Ibnu Sina Institute for Scientific
and Industrial Research, Faculty of Computing, Universiti Teknologi Malaysia,
81310 Skudai, Johor, Malaysia
Nursyahela_90@yahoo.com, {sarina,mariyam}@utm.my

Abstract. Large data storage has to serve high volume transactions of data everyday when users request the data that can cause latency. Therefore, intelligent methods are required to solve the insufficient data storage experienced by some providers. Pre-fetching technique is one of the best techniques that enable assuming the data will be needed by the user in the near future. Consequently, users easily access their data at high speed to avoid latency. However, pre-fetch the wrong objects cause slow down the data management performance. In this context, this research proposes Machine Learning (ML) techniques to predicting the pre-fetched objects accurately. This paper also compares the Rough Decision Tree (RDT) with others ML techniques including J48 Decision Tree, Random Tree (RT), Naïve Bayes (NB), and Rough Set (RS). The experimental results reveal the propose RDT performs better compared with RS single-alone. However, J48 performs well in classifying the web objects for IrCache, UTM blog data, and Proxy Cloud Storage (CS) data sets. Hence, J48 was proposing to be implementing into the future work of mobile cloud storage services.

Keywords: Pre-fetching · Machine Learning · Prediction · Big data storage · Latency

1 Introduction

Pre-fetching is one of the most popular techniques that deals with the slow access speed of the World Wide Web (WWW), is commonly used and has quite effective algorithmic approach to reduce latency [1, 2]. However, overaggressive pre-fetching affects the performance because pre-fetching overhead occurs if the pre-fetching data is unused. The latency problem occurs when the big data not handle wisely. Hence, it needs the best pre-fetch algorithm to optimize the performance. Therefore, in this paper proposes a new work to determine and analyze the ML techniques that can be applied in optimizing the pre-fetching performance [2, 3]. There is no single technique that is better than others in solving all problems as discussed by several authors. Therefore, conducting a test of multiple algorithms and parameter settings is the best way to determine which algorithm is suitable in this research work.

© Springer International Publishing AG 2017
Y. Tan et al. (Eds.): DMBD 2017, LNCS 10387, pp. 113–124, 2017.
DOI: 10.1007/978-3-319-61845-6_12

In this study, we investigate the performance of J48, RT, NB and RS to test their ability to discover patterns and make accurate pre-fetching predictions. These techniques were mainly applied by earlier researchers [4, 5]. In addition, further analysis was made in this research by analyzing a hybrid method of Decision Tree and Rough Set, so called as a Rough Decision Tree (RDT). The hybrid method RDT analyzed the prediction of pre-fetched objects because in this study the researcher wanted to see the accuracy of hybrid method as DT is limited in reduction while RS in feature selection [6, 7]. Then, we reveal the technique with the highest accuracy, which can be applied in generating the rules for pre-fetching of future objects being requested. Thus, an ML technique is required to enhance the pre-fetching techniques because ML can learn and predict the right data requested by users even on big data. If only pre-fetching is used, it may pre-fetch the wrong data that will cause overhead and use a lot of memory that store useless data. This is important to ensure the techniques have efficient use of the limited amount of memory available in mobile devices for big data storage.

2 Proposed Methodology

The experimental framework of this research was illustrates in Fig. 1. Three log datasets were used, which are IrCache, UTM blog data and proxy CS. A pre-processing step was first necessary to clean and normalize the data as explained in Sect. 2.1. Then, the data was split using 10-fold cross validation into training and testing, and the appropriate ML algorithms were applied as discussed in Sect. 2.2.

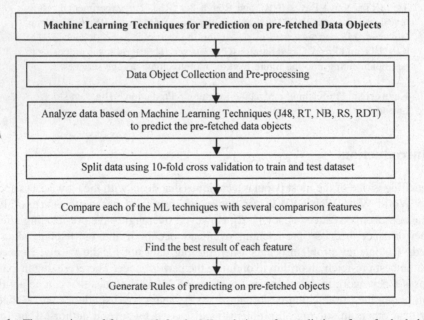

Fig. 1. The experimental framework for the ML techniques for prediction of pre-fetched objects

2.1 Datasets and Pre-processing

The process flow in identifying the most effective technique for this research included collecting data from the web log data. This research studies three datasets such as the IrCache, UTM blog and Proxy CS, from which the features were extracted. The first data from IrCache was collected from a proxy server installation ftp://ircache.net. IrCache is a National Laboratory of Applied Network Research project that encourages web caching and provides data for researchers. In this study we use 10591 files for IrCache log data. IrCache regularly makes traces available for academic researchers. Besides, IrCache data is often used by other researchers in their research studies [8, 9]. The second data is obtained from the UTM blog data consisting of 9103 requests for data transfer. The data was recorded on 13 January 2013, accessed by different client addresses. In addition, the third dataset that we use in this study is proxy CS dataset. Proxy CS is obtained by tracing the log data using Squid [10, 11]. All these three datasets were chosen because the log data field features were similar to the proposed work which is cloud computing data source. This research also conducted tests on cloud computing data, the Proxy CS. The data pre-processing was conducted by removing all irrelevant data. Then, RDT was used in the analysis by comparing them with other algorithms, J48, RT, NB, and RS. WEKA and ROSETTA tools were used to analyze the ML techniques that provided high accuracy in prediction. In this research used about 70% of each dataset for training and the remaining [4, 12, 13], were for testing purposed. The performance of trend prediction systems is evaluated using the cross validation method [14]. In cross-validation, a user needs to decide on a fixed number of folds or partitions of the data to find the best value. The data is randomly divided into 10 parts in which the class is represented in approximately the same proportions as in the full dataset [15].

2.2 Machine Learning Techniques

In this paper, J48, RT, NB, RS and RDT were analyzed to determine the best ML prediction.

A. J48

We use a simple C4.5 decision trees for prediction of the DT, the J48 classifier in this research. J48 is an open source java implementation of the C4.5 decision tree algorithm in the WEKA data mining tool that creates a binary tree. The DT approach is most useful in predicting problems [4, 16, 17]. We use this technique to construct a tree to model the prediction process. While building a tree, J48 ignores the missing values. For example, the value of the item can be predicted based on what is known about the attribute values for the other records. The basic idea is to divide the data into range based on the attribute values for the item found in the training sample. J48 allows classification via either decision trees or rules generated from them [16].

B. Random Tree

RT is built in two stages; the first is built independently from the training data, whereby a feature and cut value are randomly selected. Sometimes, the result structure is called a tree skeleton. It repeats until the tree achieves the specified depth [18]. The algorithm consists of two steps. The first step (BuildTreeStructure) generates the structure of each random tree. This is referred to as the skeleton since the leaves do not contain class distribution statistics. Building the skeleton does not require any training data, but instead only information about the features. In the second step called UpdateStatistics, the data training is used to compute class statistics for the leaf nodes.

C. Naïve Bayes

NB is based on Bayes' theorem and the theorem of total probability. The probability that a document d with vector $x = <x_1,...,x_n>$ belongs to hypothesis h as in Eq. (1) [16]:

$$P(h_1|x_i) = \frac{P.x_i.P.(h_1)}{P.x_i|h_1.P.h_1 + P.(x_i|h_2).P.(h_2)}. \tag{1}$$

Based on Eq. (1), $P(h_1|x_i)$ is posterior probability, while $P(h_1)$ is the prior probability associated with hypothesis h_1. Thus, Eq. (1) can be simplified as shown in Eq. (2),

$$P(h_1|x_i) = \frac{P.(x_i|h_1).P.(h_1)}{P(x_i)}. \tag{2}$$

D. Rough Set

RS, originally introduced by Pawlak in [19], is a valuable mathematical tool in dealing with imprecise or vague concepts. RS theory allows a concept to be described in terms of a pair of sets, lower approximation and upper approximation of the class. Let Y be a concept. The lower approximation $\underline{Y}$ and the upper approximation $\overline{Y}$ of Y are defined as in Eqs. (3) and (4) [20]:

$$\underline{Y} = \{e \in E| \ e \in \Omega_i \text{ and } X_i \subseteq Y\}. \tag{3}$$

$$\overline{Y} = \{e \in E| \ e \in \Omega_i \text{ and } X_i \cap Y = \emptyset\}. \tag{4}$$

Lower approximation is the intersection of all those elementary sets that are contained by Y, and upper approximation is the union of elementary sets that are contained by Y. In the proposed work, mathematical RS is applied to calculate the accuracy of classification. The method works as follows.

1. Calculate number of items pre-set in the dataset.
2. Discretization on data
3. Reduction based on Johnson's algorithm is implemented
4. Do cross validation to minimize the side effects of choosing a training set and a test set

Discretization is applied to training and testing set. Data reduction is an important step in knowledge discovery from data. This research use 10-fold cross validation method in proposed work, applying the ROSETTA, two algorithms namely, JohnsonReducer and BatchClassifierfor cross validation [21]. The first algorithm is the JohnsonReducer, which searches for reducts and produces rules. It is applied on the training set in each cross validation iteration. The second algorithm is the BatchClassifier, which uses the rules from the JohnsonReducer algorithm to predict the test set in that iteration.

E. Rough Decision Tree

The implementation of RDT is based on ROSETTA, which contains several rough set learning algorithms as well as discretization methods, but lacks algorithms for performing features selection. Hence, this research also includes hybrid method. These missing algorithms are used based on decision tree (J48) based on WEKA. The experimental results are revealed in Sect. 3. We analyze ML technique, which is effective and simple to use. The method to analyze the data based on RDT is as follows:

1. Select dataset
2. Extract features from the dataset
3. Select features using J48 algorithm decision tree
4. Apply rough set on trained data
   a. Calculate number of items pre-set in the dataset
   b. Discretization on data
   c. Reduction based on Johnson's algorithm is implemented
   d. Do cross validation to minimize the side effects of choosing a training set and a test set
5. Prediction is generated
6. Final results

3 Experimental Results

In order to evaluate the performance of ML techniques, it is important to measure its performance. Therefore, some common performance measures were used to evaluate the performance of particular ML techniques including J48, RT, NB, RS and RDT. Tables 1, 2 and 3 reveal the comparison of performance for J48, RT and NB used WEKA, then RS by using ROSETTA for three datasets. In addition, the hybrid method, based on RDT used J48 decision tree to select relevant features from the extracted feature set. The selected features were then applied onto a rough set based system to predict pre-fetched data objects. The results of features selection for the IrCache dataset were elapse time, size and hit; while for the UTM blog dataset were request method, size and hit. In addition, results for the Proxy CS dataset were elapse time, size, user agent and hit for. This evaluation helped to choose the most suitable ML algorithm to be implemented in the future proposed work of Mobile Intelligent Cloud Storage (MOBICS).

In this research, mean of accuracy and Standard Deviation (STD) were used as a statistical validation to verify the performance of the proposed algorithms. The accuracy was executed based on five types of ML algorithms namely, the J48, RT, NB, RS and RDT while the accuracy was measured as shown in Eq. (5).

Table 1. Mean and STD for accuracy of IrCache dataset based on fold 1 to 10 using five different algorithms.

Fold	Accuracy IrCache data set (%)									
	Training set					Test set				
	J48	RT	NB	RS	RDT	J48	RT	NB	RS	RDT
1	99.97	99.97	95.37	92.17	100	99.85	99.91	96.67	94.03	99.69
2	100	100	95.47	91.50	100	99.85	99.70	95.86	94.03	100
3	99.84	99.72	95.53	93.12	100	99.74	99.69	95.67	92.45	100
4	99.97	99.97	95.34	91.23	100	99.87	99.65	95.40	96.23	100
5	100	99.97	95.85	89.88	100	99.84	99.77	95.78	93.40	100
6	99.87	99.87	95.62	92.31	100	99.85	99.91	95.60	94.03	99.69
7	99.94	99.91	95.62	91.09	100	99.81	99.77	95.41	92.14	100
8	99.94	99.81	95.47	92.98	99.87	99.92	99.95	95.50	95.91	100
9	99.84	99.91	95.63	92.04	100	99.85	99.76	95.59	94.34	100
10	99.97	99.91	95.47	89.66	100	99.84	99.77	95.76	94.60	100
Mean	**99.93**	99.90	95.54	91.60	**99.99**	**99.84**	99.79	95.72	94.11	**99.94**
STD	0.057	0.081	0.140	0.112	0.004	0.042	0.095	0.347	0.113	0.001

Table 2. Mean and STD for accuracy of UTM Blog dataset based on fold 1 to 10 using five different algorithms.

Fold	Accuracy UTM blog data set (%)									
	Training set					Test set				
	J48	RT	NB	RS	RDT	J48	RT	NB	RS	RDT
1	97.18	95.57	91.21	75.98	90.42	98.57	96.19	87.38	65.93	82.78
2	97.36	97.44	91.17	73.78	88.07	98.48	93.53	88.02	66.67	79.12
3	97.00	97.00	91.21	74.25	90.89	98.41	95.68	87.00	63.00	80.95
4	96.52	95.02	90.70	74.73	87.13	98.95	96.50	88.89	65.57	79.85
5	96.96	97.03	91.54	75.67	84.93	98.46	93.64	89.45	65.20	78.75
6	96.78	95.13	91.25	74.10	89.80	98.63	95.64	88.29	65.57	71.79
7	97.22	95.61	91.58	72.21	88.07	98.65	94.19	89.30	69.60	78.02
8	97.62	96.08	91.43	77.55	90.89	98.85	96.17	88.60	61.90	77.29
9	97.03	95.31	90.11	74.57	87.60	98.76	93.80	86.99	65.20	75.09
10	97.33	97.47	91.79	75.20	88.21	98.85	96.19	88.38	72.63	82.42
Mean	**97.10**	**96.17**	91.20	74.80	88.60	**98.66**	**95.15**	88.23	66.13	78.61
STD	0.097	0.925	0.458	0.314	0.119	0.076	1.147	0.839	0.131	0.103

$$\text{Accuracy} = \frac{\text{Number of correct data}}{\text{Total data}} \times 100\%. \tag{5}$$

STD is a measure used to quantify the amount of variation of a set of data values [22]. STD close to 0 indicates that the data points tend to be very close to the mean which is also called the expected value of the set, while a high STD indicates that the data points

Table 3. Mean and STD for accuracy of proxy CS dataset based on fold 1 to 10 using five different algorithms.

Fold	Accuracy proxy CS data set (%)									
	Training set					Test set				
	J48	RT	NB	RS	RDT	J48	RT	NB	RS	RDT
1	100	99.34	94.08	95.77	100	98.59	95.07	93.80	74.19	96.77
2	99.67	98.36	92.11	90.14	100	98.59	96.91	93.67	80.65	96.77
3	99.34	96.38	93.42	90.14	98.59	93.39	95.92	93.53	83.87	90.32
4	100	99.67	93.75	92.96	98.59	94.80	96.62	94.37	80.65	96.77
5	99.34	99.01	93.09	90.14	100	97.61	95.36	93.39	70.97	93.55
6	99.34	98.36	91.80	91.55	100	96.20	96.48	93.39	80.65	96.77
7	97.70	98.03	91.15	90.14	98.59	90.58	95.78	94.09	77.42	100
8	99.67	98.36	92.76	95.77	100	99.30	99.86	93.53	80.65	93.55
9	98.68	96.38	91.45	91.55	100	99.30	96.48	93.66	77.42	96.77
10	99.01	97.70	93.75	94.44	100	95.36	95.78	93.11	73.08	100
Mean	**99.28**	98.16	92.74	92.26	**99.58**	**96.37**	**96.43**	93.66	77.95	96.13
STD	0.053	1.053	0.995	0.123	0.007	0.073	1.270	0.347	0.141	0.130

are spread out over a wider range of values [23]. The formula for calculation of STD is shown in Eq. (6) where $\sum$ is sum of, x is each values in the dataset, x is mean of all values in the dataset and n is number if value in the dataset.

$$\text{STD} = \sqrt{\frac{\sum(x - x)^2}{n - 1}}. \tag{6}$$

Tables 1, 2 and 3 show a comparison between the performance measures of J48, RT, NB, RS and RDT for the three datasets in both the training and testing datasets. In Tables 1, 2 and 3, the first best and second best values of the measures are highlighted with bold font. Based on Eq. 1, on average, J48 provided high accuracy compared with other ML. The J48 accuracy was 99.93%, 97.10% and 99.28% for IrCache, UTM blog data and proxy CS training datasets respectively. These were the best accuracy among the algorithms for all datasets. The full results are presented in Tables 1, 2 and 3. The results of data testing also showed that J48 had the best accuracy for UTM blog and proxy CS. Hybrid RDT was better than RS single-alone because DT overcame the limitation of RS in feature selection. In addition, RDT provided the best accuracy, but only for IrCache dataset. This is because the size of IrCache dataset was the biggest among the datasets and RS was not suitable to analyze small dataset. In this research, RS was not suitable for too small data analysis and less efficient in feature selection algorithm [24]. However, the hybrid RDT gave good result of accuracy compared with stand-alone RS based on trend prediction system, without any feature selection for all datasets. Nevertheless, RDT was not stable in this research as its performance was not highly accurate for all the UTM blog data compared with DT; besides, RDT was complex in processing the data. The results have determines J48 as the best ML technique in this research study and it was supported by the value of STD. Based on

Tables 1, 2 and 3, J48 had the averagely lowest STD, which means low error because the value was nearer to 0. Based on this result, J48 algorithm is proposed to be applied in the future work to generate the rules.

The accuracy of ML is used for evaluation the techniques as mentioned earlier. However, accuracy of prediction alone is insufficient to measure the performance of datasets. Therefore, like several previous researches works the precision, recall, error rate, sensitivity, specificity, area under ROC and F-Measure were also used in this study to evaluate the performance of ML techniques, as shown in Tables 4, 5 and 6. The confusion occurred only in the case of finding True positive (TP), False negative (FN), False positive (FP) and True negative (TN) [4, 16, 25].

Table 4. IrCache performance metric results

IrCache	J48	RT	NB	RS	RDT
Precision (%)	**99.96**	99.96	95.00	97.31	94.64
Recall (%)	**99.29**	98.72	5.19	84.90	85.23
Error Rate	**0.0037**	0.0066	0.4754	0.1662	0.1867
Sensitivity	**0.9929**	0.9872	0.0519	0.8490	0.8523
Specificity	**0.9996**	**0.9996**	0.9973	0.5298	0.0331
Area under ROC	**0.9963**	0.9934	0.8572	0.7512	0.4427
F- MEASURE (%)	**99.26**	98.89	90.60	90.68	89.69

Table 5. UTM blog performance metric results

UTM Blog	J48	RT	NB	RS	RDT
Precision (%)	92.34	90.01	95.20	93.92	**94.69**
Recall (%)	**99.38**	99.22	89.37	98.27	92.56
Error Rate	0.0444	0.0589	0.0757	0.0549	**0.0392**
Sensitivity	**0.9938**	0.9922	0.8937	0.9828	0.9256
Specificity	0.9175	0.8899	0.9549	0.8653	**0.9766**
Area under ROC	0.9794	0.9772	**0.9870**	0.9310	0.952
F- MEASURE (%)	**97.96**	97.32	93.44	96.04	93.61

Table 6. Proxy CS performance metric results

Proxy CS	J48	RT	NB	RS	RDT
Precision (%)	99.47	98.87	92.97	98.94	**99.98**
Recall (%)	**98.85**	96.56	92.08	94.44	97.47
Error Rate	**0.0084**	0.0227	0.0744	0.0426	0.0164
Sensitivity	**0.9885**	0.9656	0.9208	0.9444	0.9747
Specificity	0.9947	0.9890	0.9304	0.9813	**0.9989**
Area under ROC	**0.9944**	0.9825	0.9652	0.9626	0.4427
F- MEASURE (%)	**98.86**	97.08	88.90	96.64	98.72

Precision can be thought of as a model's ability to discern whether a document is relevant from a returned population and recall can be thought of as a model's ability to select relevant documents from the population at large [16]. Sensitivity is a probability that a test result will be positive when the disease is present (true positive rate, expressed as a percentage). Specificity is a probability that a test result will be negative when the disease is not present (true negative rate, expressed as a percentage) [16]. F-measure is a measure that combines precision and recall is the harmonic mean of precision and recall. The area under the ROC curve is called the Area Under the Curve (AUC), and takes on a value between 0.5 and 1.0 [21, 26]. It can also be said that the larger the value of the AUC, the better the classifier will be. Those performance metrics were evaluated with the results obtained from a sample data set collected from three datasets. The performance metric result values were evaluated and shown in Tables 4, 5 and 6 for IrCache, UTM blog and Proxy CS respectively. Based on the results output on Tables 4, 5 and 6, it is concluded that the most significant and best performance metric measurements averagely used J48 algorithm compared with other algorithms. The second best performance result was RDT. Although RDT was not the best performance, it proved that the hybrid of RDT was better than RS single-alone. Thus, J48 was selected as a propose algorithm to be used in propose MOBICS by implementing the rules generated by J48.

Previous results showed the proposed five ML techniques and J48 was one of the best algorithms for prediction of pre-fetched objects, especially for UTM blog and Proxy CS datasets while RDT produced better result compared to single RS. Due to the earlier analysis, the significant analyses of ML techniques were evaluated by using paired-samples t-test. The results of evaluation are tabulated in Tables 7, 8 and 9. The results of t-test in Tables 7, 8 and 9 show the accuracy of training and test set for IrCache, UTM blog and Proxy CS dataset respectively. The accuracy criterion for proposed RDT was significantly better than all other classifiers with a p-value <0.05 excluding J48 and RT for IrCache and Proxy CS dataset. This observation means that the performance of RDT and J48 was similar. However, a statistical significance was reported for all three datasets, as the p-values were always less than 0.05 for paired test of RDT - RS. It proved RDT is better than RS single-alone.

Table 7. Paired-samples t-test for accuracy of IrCache training and test set

Paired sample	Paired-samples T-test result of IrCache					
	Training set			Test set		
	t-value	*df*	*p-value*	*t-value*	*df*	*p-value*
RDT - J48	2.22188	9	0.053	2.13339	9	0.062
RDT - RT	3.29780	9	**0.009**	2.25225	9	0.051
RDT - NB	94.77863	9	**p < 0.001**	29.22477	9	**p < 0.001**
RDT - RS	22.29624	9	**p < 0.001**	14.15100	9	**p < 0.001**

Table 8. Paired-samples t-test for accuracy of UTM blog training and test set

Paired sample	Paired-samples T-test result of UTM blog					
	Training Set			Test Set		
	t-value	df	p-value	t-value	df	p-value
RDT - J48	−14.78692	9	**p < 0.001**	18.90532	9	**p < 0.001**
RDT - RT	−10.65016	9	**p < 0.001**	−16.38626	9	**p < 0.001**
RDT - NB	4.28344	9	**0.002**	8.64623	9	**p < 0.001**
RDT - RS	20.54985	9	**p < 0.001**	10.32427	9	**p < 0.001**

Table 9. Paired-samples t-test for accuracy of Proxy CS training and test set

Paired sample	Paired-samples T-test result of proxy CS					
	Training Set			Test Set		
	t-value	df	p-value	t-value	df	p-value
RDT - J48	1.14679	9	0.281	−0.17048	9	0.86840
RDT - RT	3.57798	9	**0.006**	−0.26770	9	0.79496
RDT - NB	17.11172	9	**p < 0.001**	2.67605	9	**0.02537**
RDT - RS	10.57826	9	**p < 0.001**	9.72821	9	**p < 0.001**

4   Discussions

There are many types of ML techniques used to optimize the current pre-fetching techniques. This research analyzed different ML techniques used to predict whether to pre-fetch web objects. The common ML techniques analyze the big data are J48, RT, NB, RS and RDT. The results reveal that RDT give better accuracy compared with RS single-alone. However, J48 was more precise with better accuracy in predicting web objects compared with other techniques of all three datasets with 99.84%, 98.66% and 96.37% for the IrCache, UTM blog data and Proxy CS respectively. In addition, it is supported by another performance metrics evaluation including precision, recall, error rate, sensitivity, specificity, area under ROC and F-measure which provided good results of evaluation. Hence, J48 algorithm was proposed to be implemented in future work of MOBICS to predict whether or not to pre-fetch the data objects so it would reduce the unused data web object that handles overhead of pre-fetching. Thus, the user can access the data more effectively and rapidly, even when there are big data on MOBICS. Besides, the use of J48 algorithm in predicting the cloud storage availability makes it easy for users to store their data without checking the availability of CS. However, in this research the RS was not suitable for small data because the results show slow accuracy for small datasets. Although the size of dataset influences the performance of pre-fetch object data, based on the result, J48 was a perfect algorithm in this research because high accuracy was maintained for both, small and big datasets. Hence, in this research propose to use J48 for generating rules on future work of MOBICS.

5 Conclusions

This research is an accomplishment of different ML techniques for prediction of pre-fetched objects mainly for IrCache, UTM blog data and Proxy CS datasets. Hence, this research determines the predictor that provides the highest accuracy with different datasets. Based on the result, the hybrid techniques RDT produced a better result compared to RS single-alone technique with any feature selection. However, the algorithm with the highest accuracy is J48. In addition, it is supported with the lowest value of STD which value is nearest to 0, indicating the lowest error occurrence among other ML techniques. Moreover, there are evaluations by another performance metrics to support the results of accuracy and averagely provide good result values. Hence, we propose J48 algorithm to be applied in Mobile Cloud Storage (MCS) as a future work. The result will be evaluated by comparing it with traditional MCS and MCS that applies rules based on J48algorithm called MOBICS. Due to this situation, we propose that the future work provides an intelligent MCS that produces more effective pre-fetch without useless data; hence, the work increases the performance in prediction of pre-fetched objects even on big data.

Acknowledgements. This research is supported by Ministry of Higher Education Malaysia (MOHE), Ministry of Science, Technology and Innovation Malaysia (MOSTI) and Universiti Teknologi Malaysia (UTM). This paper is financially supported by E-Science Fund, R. J130000.7928.4S117, PRGS Grant, R.J130000.7828.4L680, GUP Tier 1 UTM, Q.J130000. 2528.13H48, FRGS Grant, R.J130000.7828.4F634 and IDG Grant, R.J130000.7728.4J170. The authors would like to express their deepest gratitude to IrCache.net and CICT, UTM for their support in providing the datasets to ensure the success of this research, as well as Soft Computing Research Group (SCRG) for their continuous support and fondness in making this research possible.

References

1. Roudaki, A., Kong, J., Yu, N.: A classification of web browsing on mobile devices. J. Vis. Lang. Comput. **26**, 82–98 (2015)
2. Hussien, N.S., Sulaiman, S.: Mobile cloud computing architecture on data management for big data storage. Int. J. Adv. Soft Comput. Appl. **8**, 139–160 (2016)
3. Gao, J., Bai, X., Tsai, W.: Cloud testing- issues, challenges, needs and practice. Int. J. **1**, 9–23 (2011)
4. Kumar, P.N.V., Reddy, V.R.: Novel web proxy cache replacement algorithms using machine learning. Int. J. Eng. Sci. Res. Technol. **3**, 339–346 (2014)
5. Sulaiman, S., Shamsuddin, S.M., Abraham, A.: Meaningless to meaningful web log data for generation of web pre-caching decision rules using rough set. In: 4th Conference on Data Mining and Optimization, vol. 1, pp. 2–4 (2012)
6. Gupta, A.: A survey on stock market prediction using various algorithms. Int. J. Comput. Technol. Appl. **5**, 530–533 (2014)
7. Kim, Y., Enke, D.: Developing a rule change trading system for the futures market using rough set analysis. Expert Syst. Appl. **59**, 165–173 (2016)

8. Sathiyamoorthi, V., Bhaskaran, M.: Data preprocessing techniques for pre-fetching and caching of web data through proxy server. Int. J. Comput. Sci. Netw. Secur. **11**, 92–98 (2011)
9. Singh, N., Panwar, A., Raw, R.S.: Enhancing the performance of web proxy server through cluster based prefetching techniques. In: International Conference on Advances in Computing, Communications and Informatics, pp. 1158–1165 (2013)
10. Johann, M., Dom, J., Gil, A., Pont, A.: Exploring the benefits of caching and prefetching in the mobile web. In: 2nd IFIP International Symposium on Wireless Communications and Information Technology in Developing Countries, Pretoria, South Africa (2008)
11. Singh, A., Singh, A.K.: Web pre-fetching at proxy server using sequential data mining. In: 2012 Third International Conference on Computer Communication Technology, pp. 20–25 (2012)
12. Chang, J.-H., Lai, C.-F., Wang, M.-S., Wu, T.-Y.: A cloud-based intelligent TV program recommendation system. Int. J. Comput. Electr. Eng. **39**, 2379–2399 (2013)
13. Zissis, D., Xidias, E.K., Lekkas, D.: A cloud based architecture capable of perceiving and predicting multiple vessel behaviour. Appl. Soft Comput. **35**, 652–661 (2015)
14. Alemeye, F., Getahun, F.: Cloud readiness assessment framework and recommendation system. In: AFRICON 2015, Addis Ababa, pp. 1–5 (2015)
15. Witten, I.H., Frank, E.: Machine learning algorithms in java nuts and bolts: machine (2000)
16. Patil, T.R., Sherekar, S.S.: Performance analysis of naive bayes and j48 classification algorithm for data classification. Int. J. Comput. Sci. Appl. **6**, 256–261 (2013)
17. Suarez-Tangil, G., Tapiador, J.E., Peris-Lopez, P., Pastrana, S.: Power-aware anomaly detection in smartphones: an analysis of on-platform versus externalized operation. Pervasive Mob. Comput. **18**, 137–151 (2015)
18. Gao, W., Grossman, R., Gu, Y., Yu, P.S.: Why Naive ensembles do not work in cloud computing. In: IEEE International Computing Society, pp. 282–289 (2009)
19. Pawlak, Z.: Rough sets. Int. J. Comput. Inf. Sci. **11**, 341–356 (1982)
20. Sulaiman, S., Shamsuddin, S.M., Abraham, A., Sulaiman, S.: Rough set granularity in mobile web pre-caching. In: Eighth International Conference on Intelligent Systems Design and Applications, vol. 1, pp. 587–592 (2008)
21. Torgeir R. Hvidsten: A tutorial-based guide to the ROSETTA system: a rough set toolkit for analysis of data, pp. 1–44 (2013)
22. Sabitha, B., Amma, N.G.B., Annapoorani, G., Balasubramanian, P.: Implementation of data mining techniques to perform market analysis. Int. J. Innov. Res. Comput. Commun. Eng. **2**, 7003–7008 (2014)
23. Tiwari, S., Pandit, R., Richhariya, V.: Predicting future trends in stock market by decision tree rough-set based hybrid system with HHMM. Int. J. Electron. Comput. Sci. Eng. **3**, 1–10 (2010)
24. Voges, K.E., Pope, N.K.L.: Rough clustering using an evolutionary algorithm. In: 2012 45th Hawaii International Conference on System Science, pp. 1138–1145 (2012)
25. Moorthy, N.S.H.N., Poongavanam, V.: The KNIME based classification models for yellow fever virus inhibition. RSC Adv. R. Soc. Chem. **5**, 14663–14669 (2015)
26. Rojas, I., Work-conference, I., Hutchison, D.: Advances in Computational (2013)

Classification

Spectral-Spatial Mineral Classification of Lunar Surface Using Band Parameters with Active Learning

Sukanta Roy[1], Sujai Subbanna[2], Srinidhii Venkatesh Channagiri[2], Sharath R Raj[2], and Omkar S.N[1($\boxtimes$)]

[1] Indian Institute of Science, Bangalore, India
sukanta@aero.iisc.ernet.in.com
[2] BNM Institute of Technology, Bangalore, India
sujaisubbanna@gmail.com, srinidhii101@gmail.com

Abstract. In the field of remote sensing, the value of the large number of hyper spectral bands during classification is well documented. The collection of labeled samples is a costly affair and many semi-supervised classification methods are introduced that can make use of unlabeled samples for training. Due to the nature of these images, high dimensional spectral features must be distinctive with preservation of absorption band in mineral mapping. We propose the method in which we consider the band parameters of the spectral data combined with the neighborhood spatial information for mineral classification using Active Labeling to compensate for the lack of a large number of labeled samples. Here we demonstrate that by using these parameters for classification in conjunction with their spatial information, higher accuracies can be achieved during classification.

Keywords: SVM · RF · UV/VIS · MLR · RBM · EM · AL-MS · MLP

1 Introduction

In pattern recognition, classification of Hyperspectral images is used for discrimination and identification of minerals present on the surface. In this application, unique spectral signatures provide the recognizable feature corresponding minerals. Regarding this, a lot of research on machine learning comes to the light like SVM, RF, logistic regression etc. [1, 2, 3, 7]. These algorithms are under experimental observation as lunar data are collected at different instances and corruption level of space weathering is unpredictable in real scenario. The lunar surface is mainly made up of very few minerals including feldspar, pyroxene, olivine and ilmenite [14] which can be detected

S. Roy—Pursuing PhD in Department of Aerospace Engineering at IISc, Bangalore, Karnataka, India-560012.
S. Subbanna, S. Venkatesh Channagiri and S.R. Raj—Pursuing B.E in Computer Science & Engineering at BNMIT, Bangalore, Karnataka, India-560070.
S.N. Omkar—Chief Research Scientist in Department of Aerospace Engineering at IISc, Bangalore, Karnataka, India-560012.

© Springer International Publishing AG 2017
Y. Tan et al. (Eds.): DMBD 2017, LNCS 10387, pp. 127–136, 2017.
DOI: 10.1007/978-3-319-61845-6_13

with the use of hyperspectral images as they are sensitive to visible and infrared spectroscopy. Presence of silicate and ilmenite has been detected in Clementine data set obtained in 1994. The distribution of major lunar silicates was mapped based on a radiative transfer model. Once the composition of the minerals on the lunar surface was studied and documented, various methods were used to classify the spectral signatures and extraction of end members from hyper spectral data. In study of abundance mapping, several kinds of feature vectors are entertained according to their distinctive nature. In [5], several band parameters are introduced on the UV/VIS data processing. The parameters observed in the above study include, band strength values which are calculated from 1000/750 nm ratio. Band strength values are high for rocks with low ferrous absorption and high weathering. Band Curvature is measured along the 750 −900−1000 nm range and high curvature indicates presence of orthopyroxene. Band tilt find the presence of clinopyroxene when it is high.

However, these parametric studies suffer from impurity in pixels due to the limitation of our satellite configuration. The high spatial resolution can only assure of a single mineral at each pixel. Naturally, the task of labelling of mixed pixels become challenging during unmixing. As a consequence, the accuracy in proper labelling of minerals becomes an attentive research area. In this regard, the idea of active learning is very efficient to train a classifier on a small set of well-chosen samples. In this way, it can generalize the classification performance on the randomly chosen large data sets. In the literature, researchers mainly come up with three kinds of methodology: committee based, large margin based and posterior probability based active learning [7] for choosing the unlabelled data sets. In [8], Maximum Likelihood and Binary Hierarchical Classifier are implemented to get the classified hyper spectral data set using active learning approach. Unsupervised method of learning helps explore new findings on the investigation area. To acquire this achievement, a lot of semi supervised methods draw remarkable results in research. Jordi Munoz-Mari et al., introduces a semiautomatic procedure for selection of unlabeled samples that begins with the construction of a hierarchical clustering tree to find the most informative pixels with respect to the available classes [9].

In active learning, the cost of acquisition of labelling is not same for the data set at all times practically. Spatial dependency plays a key role in this context. In [10], a time bound regarding the cost of labelling of individual pixels is introduced. This paper outlines a method that relies on three variables: the cost of labelling a single point, the speed of the labeler's vehicle and the maximum time allowed per iteration. The spatial distribution of datasets also influences the classification accuracy in active learning. In [11], Mr Li et al., presents the effects of class imbalance in data sets to predict the unlabeled data and uses the Multinomial Logistic regression (MLR) via Logistic Regression with augmented Lagrangian in the learning stage. The data distribution should be learned from both spectral and spatial information to initialize the labelling phenomenon. Owing to this information, Jun Li et al., in [12] introduces loopy belief propagation for finding the marginal probability distribution. This work motivates authors to take into consideration the spatial information into the active labelling stage. The active labelling strategy is also studied with RBM in [4]. The randomness in selecting active samples, should also consider the spatial neighborhood information of the datasets.

This paper is organized as follows. Section 2 contains the information regarding the area under study. Section 3 states the basics and a description of our implementation strategy for active learning in our approach. In Sect. 4, the problem statement is defined. The following Sect. 5 shows the details for the implementation of our proposed approach. Section 6 discusses the results obtained during the experimentation done on the specified data sets. Finally, in Sect. 7 the conclusion with a future work is discussed.

2 Study Area

The Moon Mineralogy Mapper (M3) is reflectance spectrometer which is used for mapping and analyzing lunar surface [6]. It is configured with a spatial resolution of 140 to 280 m/pixel, across a 40 to 80 km swath. Spectral data were collected across 85 channels between 460 and 2980 nm at a resolution of 20 nm (from 750 to 1550 nm) and 40 nm (from 460 to 700 nm and 15802980 nm) and later all spectrums were smoothed [16]. This data set is available at http://pdsimaging.jpl.nasa.gov/data/m3/CH1M3_0004/DATA/20081118_20090214/200901/L2/. Photometrically and thermally corrected Level-2 data are used in this paper. A subset of image obtained from Planetary Data Archive (PDS) of size $1000 \times 300 \times 83$ is used.

3 Active Learning

Let $train\_X = \{x_i, y_i\}$ be the set of training samples where $x_i \in train\_X$ feature set and labels $y_i = \{1...train\_N\}$ where $train\_N$ is the length of the training set. Also $active\_X = \{ax_i\}$ be the set of active samples where $ax_i \in active\_X$ and length of $active\_X$ is equal to $train\_N$.

Let n_in be the number of features in the samples, $pretrain\_epochs$ be the number of pretraining epochs, n_epochs be the number of epochs for training the active samples and $pretrain\_lr$ and lr the learning rate for pretraining and active samples training respectively. n_add is the number of active samples that are chosen every epoch to be added to the training set.

Active Learning is a semi supervised method in which a selection of the most suitable samples from an unlabelled data set is added to the training data set. Availability of a large number of labelled samples for remote sensing compared to unlabelled samples is very low. In order to make use of these massive data sets of unlabelled samples in the training process, we can employ semi supervised learning methods that use these samples in order to fine-tune the weights of the network.

Active Labelling [4] specifies three methods in which the uncertainty of unlabelled samples can be calculated in order to select the most suitable samples. One of the three methods involves calculating the difference between the highest two classification scores for all the active samples and then choosing a specified number of samples with the least difference i.e. the highest uncertainty during classification. By choosing the most uncertain samples as predicted by the pre-trained network, we ensure that we are not choosing samples that only reinforce what the network has already learnt well during pre-training.

Algorithm. Active Labelling	
1:	Training Set $train\_X \leftarrow \{x_i, y_i\}$
2:	Active Samples Set $activeX \leftarrow \{ax_i\}$
3:	n_add and n_epochs
4:	repeat
5:	Pre-train the network with $train\_X$ with $pretrain\_lr$
6:	until $pretrain\_epochs$
7:	repeat
8:	for each candidate ax_i in $active\_X$ predict the output with trained network
9:	Calculate the difference between the maximum two classification probabilities
10:	Pick the first n_add samples based on minimal difference(signifies maximum uncertainty)
11:	Add these samples to $train\_X$
12:	Remove added samples from $active\_X$
13:	Train the network with new $train\_X$
14:	End for
15:	until n_epochs

By choosing the most uncertain samples, we make sure to better the generalization of the network during testing. The learning rate during the training of the active samples is kept higher in order to learn the features faster as the number of samples is lower compared to the number of original training samples. The number of epochs of training with the active samples is kept lower compared to pre training in order to reduce the impact of propagation of error from the originally trained network into the final network. The variation of these two parameters during training can result in an impact on the classification accuracy that depends on how low the error rate is of the original network that was pre-trained.

4   Problem Statement

In mineral mapping, machine learning algorithms must assign the class label corresponding to minerals with proper distinctive features. On the lunar surface, the compositions of rocks are most of the time influenced by the mixtures of feldspar, pyroxene (orthopyroxene and clinopyroxene), olivine, plagioclase and ilmenite [15]. Most of these minerals show specific absorption bands near 1000 nm and 2000 nm. These parametric values lose their uniqueness in data sets most of the time due to several kinds of noise. The label information also may be corrupted from the presence of mixed pixels in the investigation area. Because of these difficulties, it is very necessary to get the optimal training feature set for machine.

5 Proposed Methodology

As discussed in the previous section, we propose that the neighborhood weighted spatial mean features are computed and added to the band parameters feature vector for better classification accuracies. The active learning is carried out with these combined feature set.

5.1 Band Parameter Estimation

The entire M3 wavelength region is divided into two regions: namely the 1000 nm region and 2000 nm region. Around these area band parameters are calculated. Band Strength (BS) is defined as the area of highest reflectance value in a given region. Band Center (BC) is defined as the wavelength at which peak absorption occurs. Band Area(BA) is defined as the area under the curve represented by a given region. Area is obtained by performing trapezoidal integration by taking M3 wavelength on the x-axis and reflectance values in the y-axis. Band Area Ratio(BAR) is defined as the ratio of areas of 2000 nm to that of 1000 nm region. These parameters are calculated for both regions.

5.2 Initial Labeling Using EM Clustering Algorithm for Supervised Learning

Expectation–maximization (EM) algorithm is used iteratively method to find maximum likelihood or maximum a posteriori (MAP). It is a natural generalization of maximum likelihood estimation of the incomplete data case which is used for initial labeling in this case. A feature vector comprising of Band Strength, Band Center, Band Area for both the areas and Band Area Ratio as features is given as an input to the Expectation Maximization algorithm (Fig. 1).

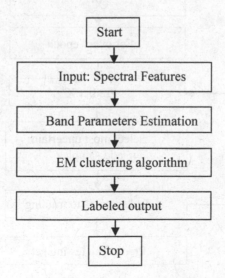

Fig. 1. Flowchart of initial labelling

5.3   Addition of Spatial Information to Band Parameters

The neighborhood spatial features for the calculate band parameters are computed for each pixel and added onto the feature vector in order to help increase classification accuracy. The inclusion of spatial features results in greater accuracy as seen in [12]. The spatial features are extracted using weighted mean of the surrounding pixels using the following formula:

$$pixel[x][y] = pixel[x][y] * \exp\left[-\left(i^2 + j^2\right)/2\sigma_S\right] * \exp\left[-(pixel[i] - pixel[j])^2/2\sigma_R\right]$$

(1)

where i and j move from 0 to window size and σ_S and σ_R are set to 0.1 and 0.8 respectively.

5.4   Algorithmic Flow of Active Learning

The band parameters augmented with the surrounding spatial features is first computed for both the training and the active sample set. The network is then pretrained with the provided training set in order to initialize the weights. Fine-tuning is then carried out using the active samples selected using AL-MS [4] selection criteria that adds n_{add} samples onto the training set every iteration. Using these added samples, the network is trained again with a lower learning rate in order to avail the information from the added samples while maintaining the accuracy of the underlying network (Fig. 2).

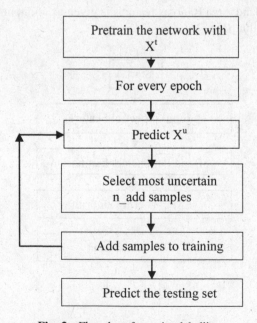

Fig. 2. Flowchart for active labelling

6   Experimental Study and Results

The investigation area is studied by Shivakumar et al. [13]. The result shows that Low Ca-Pyroxene, Plagioclase and the composition of (Olivine + Plagioclase) are present. The presence of Olivine composition indicates the hydrous phase in magmas.

For the experiment of active learning, 3 different sets of training and active sample sets are chosen in the following manner, as represented in the image below. This selection helps to generalize the learning in terms of both spectral and spatial information.

(1)   1200–1260 and 480–540
(2)   1380–1440 and 360–420
(3)   1140–1200 and 480–540

The above three sets are used to train the network and the entire set of 1500 X 300 samples is used for testing. Each of the training and active samples contain 60 X 300 samples each (Fig. 3).

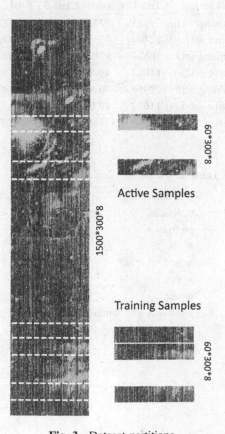

Fig. 3. Dataset partitions

6.1   Experimental Setup

In the experiment, the proposed methodology is implemented using a Multilayer Perceptron (MLP) with a single hidden layer with a multinomial logistic regression as the last layer for classification. The number of input features is 8 (band parameters) and the only hidden layer contains 7 nodes. C-support vector classification (gamma = 0.001 and C = 100) variant of SVM with an rbf kernel is also used to provide a comparison point for the accuracy of our network. The data set is classified using the above mentioned partitions of the data sets using SVM, MLP and MLP with the addition of spatial information along with Active Learning.

6.2   Results and Discussion

We can observe that the distribution of classes among each of the data sets is as follows (Table 1).

Table 1. Classwise distribution in the dataset

Data set	Class 1	Class 2	Class 3	Total
Training set1	9115	7722	1163	18000
Active set1	12404	4650	946	18000
Training set2	8677	8786	537	18000
Active set2	11662	4973	1365	18000
Training set3	9356	7030	1614	18000
Active set 3	11662	4973	1365	18000
Testing set	262243	153150	34607	450000

Table 2. Parameters during classification

pretrain_epochs	400
n_epochs	200
pretrain_lr	1
lr	4
n_add	45
n_in	8
n_out	3
windowX and windowY	3
σ_S	0.1
σ_R	0.8

Using the above selected samples (Table 1), we obtained the following results.

Table 3. Classification accuracy for each of the sets

Data set	Class 1	Class 2	Class 3	Overall
Spectral 1 -SVM	95.77	85.15	86.11	91.41
Spectral 1 - MLP	94.96	90.42	87.21	92.82
Spectral + Spatial 1 + AL - MLP	95.70	94.77	82.43	94.36
Spectral 2 - SVM	96.52	82.26	83.44	90.66
Spectral 2 - MLP	96.17	88.58	85.75	92.79
Spectral + Spatial 2 + AL - MLP	97.38	91.96	84.93	94.57
Spectral 3 - SVM	97.95	84.89	88.21	92.76
Spectral 3 - MLP	95.91	90.62	87.23	93.44
Spectral + Spatial 3 + AL - MLP	96.36	92.02	90.39	94.42

From Table 3, by adding spatial data along with the spectral information from the neighborhood pixels, classification accuracy increases. In further experiments, the authors found that subtle increases in classification accuracy can always be noticed whether in the overall or class wise prediction accuracy for other portions of the data set. Any increase in the active samples, epochs or the learning rate used for the same, caused the classification accuracy to decrease. With an increase in the number of pre training epochs, the classification accuracy rises across the board but the advantages of adding spatial information becomes less noticeable. Active Labeling is implemented for the spectral + spatial data sets where 45 X 200 samples are added to the training set by the end of the training. There is also a noticeable increase in accuracy by adding only a subset of the active samples when compared to adding all the samples.

7 Conclusion

In the present era of research in active labeling for classification of hyperspectral imagery, this paper shows that the addition of spatial neighborhood information to the spectral band parameters increases accuracy either class wise or overall. Active Labeling as implemented is successful in the usage of unlabeled samples for semi supervised classification of data using a pretrained network with a set of labeled samples. In future, different methods for selection and extraction of spatial information from hyper spectral data and their impact on the classification accuracy need to be studied.

References

1. Wu, H., Kuang, G., Yu, W.: Unsupervised classification method for hyperspectral image combining PCA and Gaussian mixture model. In: Third International Symposium on Multispectral Image Processing and Pattern Recognition, pp. 729–734. International Society for Optics and Photonics (2003)
2. Civco, D.L.: Artificial neural networks for land-cover classification and mapping. Int. J. Geograph. Inf. Sci. 7(2), 173–186 (1993)
3. Schölkopf, B., Smola, A.J.: Learning with kernels: support vector machines, regularization, optimization, and beyond. MIT press, Cambridge (2002)
4. Wang, D., Shang, Y.: A new active labeling method for deep learning. In: 2014 International Joint Conference on Neural Networks (IJCNN), pp. 112–119. IEEE (2014)
5. Borst, A.M., Foing, B.H., Davies, G.R., Van Westrenen, W.: Surface mineralogy and stratigraphy of the lunar south pole-aitken basin determined from clementine UV/VIS and NIR data. Planet. Space Sci. 68(1), 76–85 (2012)
6. Goswami, J.N., Annadurai, M.: Chandrayaan-1: India's first planetary science mission to the moon. Curr. Sci. 96(4), 486–491 (2009)
7. Tuia, D., Volpi, M., Copa, L., Kanevski, M., Munoz-Mari, J.: A survey of active learning algorithms for supervised remote sensing image classification. IEEE J. Sel. Top. Sign. Proces. 5(3), 606–617 (2011)
8. Rajan, S., Ghosh, J., Crawford, M.M.: An active learning approach to hyperspectral data classification. IEEE Trans. Geosci. Remote Sens. 46(4), 1231–1242 (2008)
9. Munoz-Mari, J., Tuia, D., Camps-Valls, G.: Semisupervised classification of remote sensing images with active queries. IEEE Trans. Geosci. Remote Sens. 50(10), 3751–3763 (2012)
10. Liu, A., Jun, G., Ghosh, J.: Active learning of hyperspectral data with spatially dependent label acquisition costs. In: 2009 IEEE International Geoscience and Remote Sensing Symposium, IGARSS 2009, vol. 5, pp. V–256. IEEE (2009)
11. Li, J., Bioucas-Dias, J.M., Plaza, A.: Hyperspectral image segmentation using a new Bayesian approach with active learning. IEEE Trans. Geosci. Remote Sens. 49(10), 3947–3960 (2011)
12. Li, J., Bioucas-Dias, J.M., Plaza, A.: Spectral–spatial classification of hyperspectral data using loopy belief propagation and active learning. IEEE Trans. Geosci. Remote Sens. 51(2), 844–856 (2013)
13. Sivakumar, V., Neelakantan, R., Santosh, M.: Lunar surface mineralogy using hyperspectral data: implications for primordial crust in the Earth-Moon system. Geosci. Frontiers 8(3), 457–465 (2017)
14. Lucey, P.G.: Mineral maps of the Moon. Geophys. Res. Lett. 31(8) (2004)
15. McKay, D.S., Heiken, G., Basu, A., Blanford, G., Simon, S., Reedy, R., French, B.M., Papike, J.: The lunar regolith. In: Lunar Sourcebook, pp. 285–356 (1991)
16. Clark, R.N., King, T.V.V.: Automatic continuum analysis of reflectance spectra (1987)

R²CEDM: A Rete-Based RFID Complex Event Detection Method

Xiaoli Peng[1,2,3], Linjiang Zheng[2(✉)], and Ting Liao[1,3]

[1] School of Intelligent Manufacturing,
Sichuan University of Arts and Science, Sichuan 635000, China
[2] College of Computer Science,
Chongqing University, Chongqing 400030, China
zlj_cqu@cqu.edu.cn
[3] Institute of Dazhou Intelligent Manufacturing Technology,
Sichuan 635000, China

Abstract. In various RFID application scenarios, RFID generates real-time and inherent unreliable raw data continually, which contain valuable enterprises business event. How to detect valuable event from RFID raw data has becoming a key issue. A Rete-based RFID complex event detection method called R²CEDM is proposed. Firstly, α_detecting net is established to detect all attribute of RFID primitive events in R²CEDM. Then β_detecting net is used to assemble RFID events into RFID complex events with the business rules, and concerned RFID complex events is obtained. The comparative experiments show the proposed R²CEDM can effectively improve the processing efficiency.

Keywords: Complex event detection · RFID · Rete · Rule match

1 Introduction

RFID system consists of tags, readers and management information systems. RFID is a non-contact automatic identification technology, which has some characteristic such as batch identification, fast mobile identification in comparison with the barcode. In recent years, RFID technology has been widely used in logistics, transportation, medical, agriculture, animal husbandry, special materials management, manufacturing lines and other fields [1]. In various application scenarios, RFID system will generate vast amounts of real-time raw data continually. Although these raw data contain valuable enterprises business event, they are difficult to be used directly by enterprises application system. There are also redundant, incorrect and other characteristics in the RFID raw data [1–3]. How to detect enterprises business logic events from RFID raw data, which are concerned by enterprises, has becoming a key issue that must be solved in RFID application system [4]. Event is a meaningful change in the system, or an occurrence of the interested content, such as a RFID reading process in the RFID system. In general, the RFID events can be classified into RFID primitive event and RFID complex event [5].

RFID event process is shown in Fig. 1 [6]. Through the data cleaning from event filter, the redundant and erroneous RFID primitive events have been removed. Then,

© Springer International Publishing AG 2017
Y. Tan et al. (Eds.): DMBD 2017, LNCS 10387, pp. 137–147, 2017.
DOI: 10.1007/978-3-319-61845-6_14

RFID complex event can be detected through RFID event aggregation according to the business logic rules, and actively inform to the business system such as WMS, ERP. RFID complex event detection is the key for RFID event processing, which directly affects the application of RFID system.

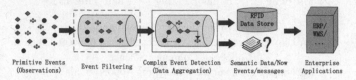

Primitive Events   Complex Event Detection   Semantic Data/Now   Enterprise
(Observations)   Event Filtering   (Data Aggregation)   Events/messages   Applications

Fig. 1. RFID event process

RFID complex event detection is an important topic in RFID application system. Now, there are some popular RFID complex event detection methods such as finite automata-based methods [7], matching tree-based methods [8], directed graph-based methods [9] and Petri net-based methods [10]. In the actual application scenarios such as manufacturing process, these objects such as materials, products, personnel and location are composed of many logical attributes. In addition, these objects always will extend some other attributes at any time as needed. When determining the concerned complex events, the user must decide several attributes and make their range as the detection rules. However, it is difficult that how to directly determine which kinds of primitive events are interested in, then we need to detect the attributes of each primitive event to determine whether it meets the rules.

Focusing on multi-attribute detection scenarios for RFID primitive events, this paper proposes a Rete-based RFID complex event detection method called R^2CEDM, which expands attributes of primitive event. The aim of this method is to do as follows:

(1) To extend object attributes for primitive events and do complex events detection. Extending all their attributes of the primitive events, and then transferring them to do complex event detection.
(2) To create an event detection network. Firstly, finding out the primary component events with detecting the attributes of input events. Secondly, generating parent events continually from Rete network. Finally, outputting the target complex events.

2   RFID Complex Event Detection Model

The primitive event is defined as PE = <OID, RID, T> [2] in traditional studies, but the definition is unsatisfied in actual application environment and cannot be used directly in complex event detection. Take a manufacturing workshop for example, managers are interested in all events which batch number is "M20160923". When the complex event detection system receives a primitive event <O1, R1, t>, it can't determine whether the object O1 belongs to the batch of "M20160923", unless we query all objects of the batch "M20160923" in the database.

In most scenarios, the complex event detection mainly focuses on those events, which are about a type of objects or happen in a region or period, rather than an individual, which is uniquely labeled by an OID. Moreover, the biggest drawback about focusing on the individual will generate a huge number of complex events, which are not all interested by the actual application. Therefore, in order to be more suitable for RFID complex event detection system in real scenario, this paper proposes expanding primitive event to expansion primitive event.

Definition 1: Expansion primitive event. EPE = <TypeID, O, R, L, T>, where TypeID is an identification of the expansion primitive event. O, R, L and T are defined as follows.

Definition 2: Monitor object. O = {OID, a1, a2, ..., an}, where O denotes a finite attribute set of RFID tag object, and its content can be increased or decreased on demand. OID denotes the object's unique identification (OID as a default tag data), and a1, a2, ..., an denote the other attributes of the monitor object. OID is the key of this data field, expressed as O.ID = OID.

Definition 3: RFID reader. R = {RID, ReadPointID, a1, a2, ..., an}, where R denotes a finite attribute set of RFID reader, and its content can be increased or decreased on demand, RID denotes a unique identification of the reader, ReadPointID is used to associate the incident locations of events, and a1, a2, ..., an express other attributes of the RFID reader. RID is the key of this data field, expressed as R.ID = RID.

Definition 4: Incident location of event. L = {ReadPointID, BusinessLocationID, a1, a2, ..., an}, where L denotes a finite attribute set of incident location of event, and its content can be increased or decreased on demand. ReadPointID denotes the identification of physical reading position. BusinessLocationID denotes the location identification specified by the upper system. And a1, a2, ..., an denotes other attributes to incident location of event. ReadPointID is the key to incident location of event, expressed as L.ID = ReadPointID.

Definition 5: Event time. T = {Timestamp, a1, a2, ..., an}, where T denotes a finite attribute set of event time, and its content can be increased or decreased on demand, Timestamp is defined in primitive event PE, and a1, a2, ..., an denotes other attributes of event time. Timestamp is the key to event time, expressed as T.ID = Timestamp.

In the definition of PE, the location of fixed RFID reader is treated as the incident location of event, but actually, this has the following three questions:

(1) There may have multiple readers at the same location.
(2) Without being specified explicitly, the reader located between two areas cannot judge where the object with tag enters into.
(3) The reader is moving in control scene, so it may read the tag in many locations not just one.

Therefore, we propose to define L separately. In practical applications, to solve the above problems is to read the "location tag". When the reader's location changes, the reader can read the location tag, which is previously fixed on a specified object, to update the associated relationship between RFID reader and incident location of event.

We can do complex event detection after expanding PE to EPE for the processing object. The R^2CEDM model is shown in Fig. 2.

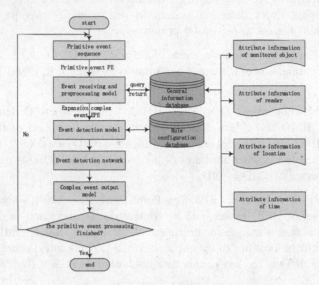

Fig. 2. RFID complex event detection model

3 R^2CEDM

The complex event consists of component events according to certain rules, so it is suitable to Rete algorithm, which is widely used for rule-based system. The main idea of Rete algorithm is as following: Firstly, to do routine detecting in α net. Secondly, to do detection and combination to these data arriving β net. Finally, to output rules which are triggered. EPE is also need to detect attributes and judge primary component event where it belongs at first, do combination to triggered primary component event with rules, and generate triggered target complex event at last. Consequently, we propose a Rete-based RFID complex event detection method.

3.1 R^2CEDM Network

R^2CEDM network consists of two stages, attribute detection and event logic relationship detection. Attribute detection, called α net, detects attributes of O, R, L and T in EPE to determine whether it matches the attribute constraint of the primary component events of target complex event. Event logic relationship detection, called β net, detects to determine whether it matches the logic relationship between primary component events of target complex event for the EPE from α net.

Let's take the event rules of SASE language [11] described as Table 1 for example to illustrate and show the structure of R²CEDM network shown in Fig. 3.

Table 1. Example of SASE described target complex event

	Event_Description	Remarks
Rule 1	**EVENT** ANY(2;A,B,C) **WHERE** (A.O.model = B. O.model) ∪ (B.O.model = C.O.model) ∪ (A.O.model = C.O.model)	ANY(2;A,B,C) represents the complex event that any two occur in events A,B and C.
Rule 2	**EVENT** OR(A,B,C) **WHERE** A.L.BLid='assembly line' ∪ B	
Rule 3	**EVENT** SEQ(AND(A,B),C) **WHERE** (A.O.size > B.size) ∩ C. O.model='TX125' **WITHIN** 10 min	
Rule 4	**EVENT** NObserve(AND(A,B);3)	
Rule 5	**EVENT** TNObserve(A;6;t_begin, t_end)	

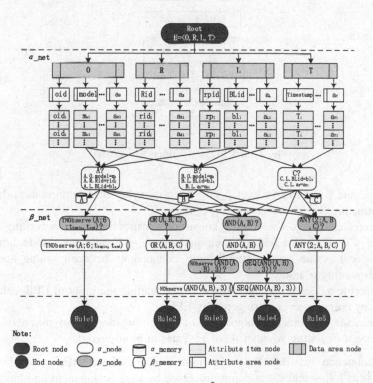

Fig. 3. The example of R²CEDM network

Because there need to record and transfer intermediate results of detection in β net, we define an event_token structure to illustrate R²CEDM algorithm.

Definition 6: Event_token. It is an event flag to record and transfer intermediate results in α_memory and β_memory. Event_token = (Component_event_List, Event_ Description), where Event_Description denotes complex event expression with SASE language [11], Component_event_List denotes recording the entering component events and their attributes in WHERE clause of Event_Description and discarding attribute values which are unconcerned.

3.2   The Steps of R²CEDM Algorithm

As shown in Fig. 4, the steps of R²CEDM algorithm are as following:

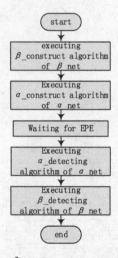

Fig. 4. R²CEDM algorithm flow chart

(1) To execute β net to construct β_construct: querying rule configuration database, constructing β net, and generating β_node.
(2) To execute α net to construct α_construct: constructing α net according to attributes of O, R, L and T in information database, generatingarea_node, attribute_node and value_node, establishing connection between value_node and corresponding α_node.
(3) To execute α_Detecting algorithm of α net: waiting the input of EPE, calculating primary complex events triggered by EPE.
(4) To execute β_Detecting algorithm of β net: waiting event_token output by α_node, calculating whether there are rules to be triggered.

The difference between Rete algorithm and traditional RFID complex event detection is as following: Every input processed by Rete is statement and has a single attribute that has a unique value. However, RFID events need to detect every attribute of each data area, compare every value to determine which type of event it belongs to. Based on the comparison above, it is important for R^2CEDM to optimize α_Detecting.

3.3 β_Construct Algorithm

β_construct algorithm works as follows: Taking out the rule from rule database one by one, constructing β net, α_node and α_memory according to EVENT clause, WHERE clause and WITHIN clause of the rules, and determining the event_token structures of α_memory and β_memory.

3.4 α_Construct Algorithm
3.4.1 Algorithm Description

Process description: Creating attribute_node and value_node according to the attribute which appears in the event_token's WHERE clause of α_node, and establishing a connection between α_node and value_node.

Initialization:

(1) Generating root_node and 4 area_node of O, R, L and T respectively.
(2) Constructing Hash mapping between root_node and area_node.
(3) Using constraint to represent a comparison condition of WHERE clause.

3.4.2 Algorithm Steps

Step 1:   To take out the next unprocessed _node $\alpha_i (1 \leq i \leq 4)$. If α_i = null, it represents all α_node have been disposed, then go to step 6. Else, go to step 2

Step 2:   To take out the next unprocessed constraint$_j (1 \leq j \leq \alpha_i$.compare_condition_ count) in α_i. If constraint$_j$ = null, it represents all attribute constraints of α_i have been disposed, i++, then go to step 1. Else, go to step 3

Step 3:   To take out an unprocessed attribute constraint constraint$_j$ in α_i. If the attribute of constraint$_j$ is not in α net which has constructed, we need to construct a new attribute_node required by constraint$_j$. Construct Hash mapping between attribute_node and corresponding area_node, and let attribute_node.α_relate_set = $\{\alpha_i\}$. Else, let attribute_node.α_relate_set = attribute_node.α_ relate_set $\cup \{\alpha_i\}$. Then, go to step 4

Step 4:   If the attribute value required by constraint$_j$ is not in α net, we need to construct a value_node for the attribute value, connect α_i, construct Hash mapping between attribute_node and its value_node, let value_node.α_ relate_set = $\{\alpha_i\}$. Else, let value_node.α_relate_set = value_node.α_ relate_set $\cup \{\alpha_i\}$. Then, go to step 5

Step 5:   j++, go to step 2

Step 6:   For all attribute_node, sorting the values of attribute_node.value_ node_count descending, if the value of attribute_node.value_node_count is equal to attribute_node, sorting descending according to the value |attribute_node.α_relate_set|, and outputting Sequence_attribute_node to end α_construct

3.5   α_Detecting Algorithm

α_Detecting in R^2CEDM algorithm is based on bidirectional reasoning of production system: Firstly, reasoning from the attribute values of EPE to α_node. Secondly, selecting appropriate time to do verification from the alternative targets α_node set (Set_target_α), whose range has been narrowed, to the attribute values of EPE to find the triggered primary complex event quickly. This α_Detecting algorithm avoids comparing each attribute of EPE or every rule.

So, there are two problems in α_Detecting: One is how to greatly reduce the number of target α_node by less comparison times. The other is when reverse reasoning starts.

For the first problem, we detect an attribute value of EPE in α net. To choose a corresponding attribut_node to do Hash mapping for an attribute value of EPE to see if it can hit the corresponding of the attribut_node. Assuming the hit rate of each value_node is roughly equal. If we compare the attribute value of this attribut_node in EPE at first, the greater the number of value_node, the greater the probability of being hit, and the more α_node related by value_node which is not hit being ruled out at the same time. Even if not being hit, it also can directly rule out all α_node related by this attribut_node. So the target scope is narrowed quickly. If we compare those attribut_node whose quantity is small in value_node at first, not only the smaller the hit rate, but also the less the irrelevant α_node being ruled out. The target scope is narrowed slowly. By the same token, it also can narrow the scope quickly to compare this attribute item at first whose |attribute_node.α_relate_set| is greater. So α_detecting compares those attribut_node that the quantity of αrewaa_node is greater and the value of |attribute_node.α_relate_set| is bigger at first. In addition, the ID value of a data area can uniquely determine all the other attribute values of this data area, therefore, if the ID attribute value of an EPE is hit, the EPE's other attributes can no longer be detected. This can greatly avoid unnecessary comparison and reduce detection times.

For the second problem, the paper proposes that we should set different inputs to different rules and amend the opportunity according to the actual situation. It is not always high efficiency to set a fixed bidirectional reasoning conversion opportunity. So we use configurable parameters δ in α_Detecting algorithm to set when the event detection process satisfies the Eq. (1), we can start backward reasoning. In Eq. (1), α_node.wait_fulfil_constraint denotes the number of constraint to be compared in α_node of detection process, and EPE.wait_check_attribute_node denotes the number of attributes to be compared in EPE of detection process. That is to say, when the number of attributes to be compared in default rules is less than the number of input attributes in EPE, we can start backward reasoning.

$$\sum_{m=1}^{|Set\_target\_\alpha|} \alpha\_node_m.wait\_fulfil\_constraint < \frac{1}{\delta} * EPE.wait\_check\_attribute\_node(\delta \geq 1) \quad (1)$$

$\alpha\_Detecting$ Algorithm description is as following.

Process description: according to the sequence of attribute nodes in Sequence_attribute_node ,detecting the attributes of EPE in the sequence to narrow the target scope, when the Eq. (1) is satisfied, reversely detecting attributes of EPE putting the target α_node as the starting point to make those α_node active.

Input: every attributes of EPE.

Output: Set_target_α, whose event_token is output by the α_node to corresponding β_node.

Initialization:

(1) To let $\alpha\_net.\alpha\_node\_count$ denote the total number of α_node in α net. EPE. wait_check_attribute_node $= |Sequence\_attribute\_node|$, where $|Sequence\_attribute\_node|$ denotes the number of elements in the attribute of the queue.

(2) To let $Set\_target\_\alpha = \sum_{j=1}^{\alpha\_net.\alpha\_node\_count} \{\alpha_j\}$ denote the set of alternative target α_node, where $|Set\_target\_\alpha|$ denotes the number of elements of alternative target α_node.

(3) To let $\alpha\_node.constraint\_count$ denote the number of attribute constraint condition of α_node.

(4) To let SEQ_wait_check_attribute_node denote the attribute_node queue to be compared and SEQ_wait_check_attribute_node = Sequence_attribute_node.

(5) To generate the ID attribute queue ID_node_list according to the sequence of the ID attribute in each data area of O, R, L, T in SEQ_wait_check_attribute_node.

3.6 β_Detecting Algorithm

Because $\beta\_Detecting$ algorithm is not the emphasis of R^{2}CEDM, we simply expound the work process as follows: To β_node whose input comes from α_node, to detect the constraints of EVENT clause, WHERE clause and WITHIN clause of β_node. To record the input of the satisfied constraint. If the event_token in β_node is activated, outputting down to trigger the father β_node to do the same detection, and traveling all β_node which has received input until the target complex event is output.

4 Experiment

R^{2}CEDM is for multi-attribute detection scenes of RFID primitive event, and optimize algorithm for event detection at the primary complex events stage of attribute detection. So one of the core of R^{2}CEDM is $\alpha\_detecting$ algorithm. Generally, the multi-attribute procession is not considered by the complex event processing mechanism. Therefore, we treat each EPE to be detected as a simple primitive event in the experiment. In this condition, we do the performance test for optimized α_detecting algorithm, and compare it with non- optimized α_detecting algorithm. Experimental platform are as follows: Inter (R) Core (TM) 2 T5500 1.66GHZ, 1.5 GB RAM. Windows XP. JRE1.6.0 compiler environment.

To set there are 200 attribute items at most in the data areas of O and R of EPE, 80 and 20 attribute items in the data areas of L and T of EPE respectively. 60 primitive complex events are averagely divided into 3 groups A, B and C, that each contains 4, 3, 2 attribute constraints and has 20 events respectively. Group A is divided equally into 5 parts, that each contains 4 to 0 attribute constraints and has 4 events respectively. Group B is divided equally into 4 parts, that each contains 3 to 0 attribute constraints and has 5 events respectively. Group C is divided into 3 parts, and each has 5, 5 and 10 events. To set each attribute has the range of 100 discrete attribute values, meanwhile to randomly set the constraint value of primary complex events.

As shown in Fig. 5, the optimized α_detecting algorithm has certain advantage comparing to non-optimized α_detecting algorithm, and with the increase in the number of expansion primitive events, the advantage of optimized α_detecting algorithm in efficiency increases gradually. The experiment result shows that α_detecting algorithm has the feasibility, and has improved the computation efficiency.

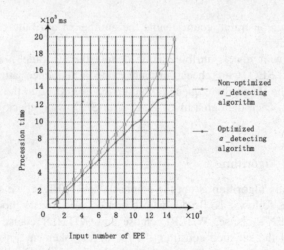

Fig. 5. The comparison of optimized and non-optimized α_detecting algorithm

5 Summary

With the wide application of the RFID technology in various industries, the efficiency of RFID complex event processing will be a key issue. Taking manufacturing enterprise environment as the background, this paper proposes a Rete-based RFID complex event detection method called R^2CEDM, using expansion primitive event and Rete algorithm to do complex detection. The core of the R^2CEDM is α_Detecting. On the attribute detection stage of R^2CEDM, we use the method of rapid narrowing the scope of the target nodes and bidirectional reasoning to improve the efficiency. Experiment shows that the optimized algorithm have improved the processing efficiency and reduced the processing time.

Acknowledgments. This work was supported by the National Science Foundation of China (61203135), National High-tech R&D Program of China (2015AA015308), China Post-doctoral Science Foundation (2014T70852), Chongqing Postdoctoral Science Foundation Project (Xm201305), Sichuan provincial education department (15ZB0327) and the general project of Sichuan University of Arts and Science (2015TP002Y).

References

1. Barenji, R.V., Barenji, A.V., Hashemipour, M.: A multi-agent RFID-enabled distributed control system for a flexible manufacturing shop. Int. J. Adv. Manuf. Technol. **71**, 1773–1791 (2014)
2. Yao, Z.L., Zhang, H., Yong, L.W.: RFID complex event processing: applications in real-time locating system. J. Int. J. Intell. Sci. **2**, 160–165 (2012)
3. Cugola, G., Margara, A.: Processing flows of information: from data stream to complex event processing. In: ACM International Conference on Distributed Event-Based Systems, pp. 359–360. ACM Press, New York (2011)
4. Yao, W., Chu, C.H., Li, Z.: Leveraging complex event processing for smart hospitals using RFID. J. Netw. Comput. Appl. **34**, 799–810 (2011)
5. Bok, K.S., Yeo, M.H., Lee, B.Y., et al.: Efficient complex event processing over RFID streams. Int. J. Distrib. Sens. Netw. **2012**(1550–1329), 53–62 (2012)
6. Wang, F.S., Liu, S.R., Liu, P.Y.: Complex RFID event processing. Int. J. Very Large Data Bases **18**, 913–931 (2009)
7. Nie, Y., Li, Z., Chen, Q.: Complex event processing over unreliable RFID data streams. In: Du, X., Fan, W., Wang, J., Peng, Z., Sharaf, Mohamed A. (eds.) APWeb 2011. LNCS, vol. 6612, pp. 278–289. Springer, Heidelberg (2011). doi:10.1007/978-3-642-20291-9_29
8. Yeh, M.K., Jiang, J.R., Huang, S.T.: Four-ary query tree splitting with parallel responses for RFID tag anti-collision. Int. J. AD Hoc Ubiquit. Comput. **16**, 193–205 (2014)
9. Lange, S., Donges, J.F., Volkholz, J., et al.: Local difference measures between complex networks for dynamical system model evaluation. Multi. Sci. **10**(6), e0129413–e0129413 (2015)
10. Wang, F., Liu, S., Liu, P., Bai, Y.: Bridging physical and virtual worlds: complex event processing for RFID data streams. In: Ioannidis, Y., Scholl, Marc H., Schmidt, Joachim W., Matthes, F., Hatzopoulos, M., Boehm, K., Kemper, A., Grust, T., Boehm, C. (eds.) EDBT 2006. LNCS, vol. 3896, pp. 588–607. Springer, Heidelberg (2006). doi:10.1007/11687238_36
11. Wu, E., Diao, Y., Rizvi, S.: High-performance complex event processing over streams. In: ACM SIGMOD, pp. 407–418. ACM Press, Chicago (2006)

Learning Analytics for SNS-Integrated Virtual Learning Environment

Fang-Fang Chua[✉], Chia-Ying Khor, and Su-Cheng Haw

Faculty of Computing and Informatics, Multimedia University, Cyberjaya,
Selangor, Malaysia
fang2x81@gmail.com, {cykhor,schaw}@mmu.edu.my

Abstract. With the increasing interest of social media usage among the students, we are motivated to integrate this informal mode of socialized learning
environment into the formal learning system to engage students for their
learning activities. The existing Learning Analytics (LA) focused only on
analyzing formal data obtained from controlled online learning environments
and the social connections and learning experience of students are not analyzed.
The expected output for this proposed work is a SNS-integrated Virtual
Learning Environment (VLE) named *Shelter* which provides formal and
informal learning for any subject domain area. User testing is conducted and
both informal and formal data are stored and elicited from *Shelter* to investigate
the impact of this data combination for more insightful LA results.

Keywords: Learning Analytics (LA) · SNS-integrated VLE · Formal ·
Informal

1   Introduction

In the traditional learning environment, there exists lacking of fair monitoring on all
levels of students by the educators due to time and learning technology constraint. To
overcome these limitations, there has been an increasing number of Virtual Learning
Environment (VLE) developed and used to facilitate the learning and teaching activities. However, the engagement of learners to use the learning system is always a
challenge as they are easily distracted, lack of motivation and interest. The factors
which contribute to the problems are mainly the availability of the learning resources in
the system and the lack of interactive components. With the increasing of social media
usage among the students, we are motivated to integrate this informal mode of
socialized learning environment into the formal learning system to engage students for
their learning activities. The existing Learning Analytics (LA) process has focused
mainly on analyzing formal data gathered from learning management systems,
face-to-face learning or controlled learning environments [1–4]. The social connections
and learning experience of students are not discovered and being analyzed. On the
other hand, the existing social networking sites are mainly focusing on communication
and socialization aspect and learners only utilize some of the features such as chatting
to facilitate their learning process. Also, not all students are active in social networking
sites to expose their feelings and thoughts. For the existing literature on mining social

© Springer International Publishing AG 2017
Y. Tan et al. (Eds.): DMBD 2017, LNCS 10387, pp. 148–155, 2017.
DOI: 10.1007/978-3-319-61845-6_15

media data (informal data) proposed by [5], LA is only being performed on informal learning environment with social networking sites (i.e. Twitter) whereby student-posted content is directly analyzed from uncontrolled spaces on the social web to understand students' learning experiences. No analysis has been done so far on the combination of both formal (student academic performance) and informal (social network services) data.

With the motivation as stated, we are developing a SNS-Integrated VLE, *Shelter*, which facilitates both formal and informal learning with the aim to increase learning interest and engagement. The impact of combining formal and informal data to be used for performing LA will be evaluated and analyzed. Few general goals are specified to derive the experiments design for performing LA: to predict student performance, to identify student which needs assistance or at risk of failing a particular course and thus affecting study duration, to understand learning experience/behavior and to discover social connections. The following sections describe our work as such: Sect. 2 explains the motivation of proposing a SNS-Integrated VLE and reviews on LA and educational data mining. Section 3 includes the literature on related works on LA. Section 4 reflects the stages of work and proposed solution. Finally, we conclude our proposed work in Sect. 5.

2   Learning Activities in SNS-Integrated VLE

The main limitations of VLE include lack of personalized control for learners over learning process and collaboration manner between learners and educators [6]. A VLE with Web 2.0 in a university environment which includes adaptation and adoption of few types of SNS was developed by [7]. [8] is a "revolutionary social learning platform that integrates learning content, assignment tracking facilities and social learning". [9] presented a socialized learning system whereby it allows educators to perform tasks such as tracking attendance, grades and performance. [10] proposed an e-learning application which is able to work with Google Applications. It has the quality factors of ease-of-use and fun to interactive with. Besides, [11] presented an online learning system and a social network environment that enables institutions to conduct an online academic program. There is a need to implement new assessments, faculty course and syllabus redesign and communication channel between educators and students. The data sources for all these activities can be gathered and analyzed from student learning activities [12].

LA and mining techniques analyze data generated in learning systems to understand students' learning behavior and consequently generating suggested decision making ideas [13]. Educational Data Mining focuses on applying "clustering, classification and association rule mining" to facilitate educators and students in analyzing learning activities [14]. According to [15], LA refers to "the analysis of a wide range of data produced by students in order to assess academic performance progress and predict future performance". [16] explained that LA uses intelligent and learner-produced data as well as analysis models to predict on learning and to discover information and social connections. LA is used for predicting and modelling learner activities in [17] and this facilitates adaptation or integration into existing learning content or services. LA is used to increase student retention, improve student success and identify the flows or learning process [18]. The data sources for LA consist of formal or informal [19] whereby in informal environment, data is being extracted from social networks. For our type of data sources, we are extracting from our proposed VLE, *Shelter*, which has both formal and informal data captured.

3   Related Works on Learning Analytics

A system named "Signals" that analyzed student performance data from Blackboard course system was created by a group of researchers at Purdue University [20]. LA support educators in the continuous process of improving quality of their teaching content or materials and to improve students' learning interest and performance [21]. [22] presented a LA framework for 3D educational virtual worlds that focuses on discovering learning flows and by using processing mining approach, the conformance can be checked and verified. [23] proposed different components for LA in order to help learners to improve their learning achievement continuously through an educational technology approach. [24] presented a cloud-based assessment tool called Learning Analytics Enriched Rubric (LAe-R) which is integrated into Moodle. [25] introduced time and students' perceptions as important predictors of actual performance in their study. Machine learning and statistical analysis techniques are applied to examine the effect of students' response time on the prediction of their learning performance [26].

Social Learning Analytics (SLA) gathers data from online learning activities and focuses on social activity engagement process. The interaction process includes messaging, friending, tagging or rating [27]. Interaction process can be reflected through our SNS-integrated VLE, *Shelter*, and it also better justifies our motivation to embed social communication components in our formal learning environment. SLA utilizes the data generated by learners' online activity to identify behaviors and patterns within the learning environment [28]. [29] proposed a e-learning analytics framework for asynchronous discussions where the interaction data is collected and to compute indicators for analyzing the teaching style and behavioral patterns of online asynchronous interaction. [30] proposed a set of services for the lecturer to evaluate the learners' progress and thus allowing the evaluator to easily track down the learners' online behavior. [5] extended the data scope of learning analytics to include informal twitter data. They also extended the understanding of students' experiences from students' academic performance to the social and emotional aspects based on their informal online conversations. However, not all students are active in Twitter to expose their real feelings and thoughts and the workflow proposed requires human effort and intervention for data analysis and interpretation.

4 Methodology and Proposed Solution

The proposed methodology consists of four main stages. For the **first stage** of the project, proposed SNS (Facebook service) is integrated with the designed learning services in the VLE implementation (Overall Architecture as shown in Fig. 1).

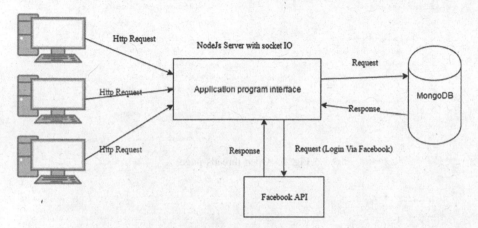

Fig. 1. Overall architecture

The proposed system named 'Shelter' (as shown in Fig. 2) includes discussion and comments posting in both open threads and group learning environment, materials management and collaboration portal (for group discussion and diagramming). Users are allowed to edit and draw three types of UML diagrams collaboratively by using the collaboration portal. The diagrams include Use case Diagram, Class Diagram and Sequence Diagram. Members are able to communicate and collaborate using real-time chatting component in the collaboration portal. Notepad++, node.js, jsUML2 library and MongoDB server are used for the development of Shelter.

For the **second stage** of the work, user testing is carried out for data collection and storage. 80 undergraduate students which taking Software Engineering subject were using Shelter to learn the subject for one semester. Shelter is deployed to a cloud server and learning interaction has been logged. Figure 3 shows the Collaboration Portal Page whereby a Use case Diagram is generated collaboratively.

The **third stage** of work includes data preprocessing. Table 1 presents the range of each particular group of data. The attributes have been discretized into 4 groups based on their quartile 1, median, and quartile 3 dividing points (sorted by data frequency). The 4th and 5th attributes are discretized to only 3 groups as they share similar quartile whereby 75% of the value is 0 and 1. Some of the attributes extracted (both formal and informal data) include the total time spent using Shelter (in minutes), frequency of posting comments, participation rate, number of likes on the threads, total materials downloaded and the total number of mentioned keywords in Shelter in matching with the keywords provided by the teaching lecturer. The discretized data will later be analyzed to find out the relation between the attributes and the grades performance.

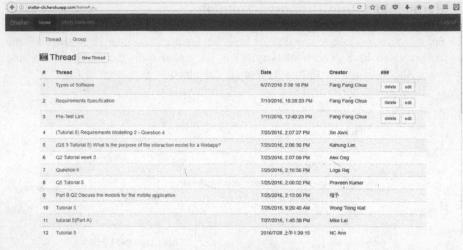

Fig. 2. Open threads page

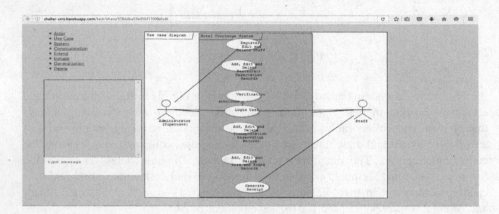

Fig. 3. Collaboration portal page

Finally during the **fourth stage** of work, we are performing LA on both formal and informal data as stated in Table 1. The collected formal data can be analyzed using descriptive statistics method, such as the mean and standard deviation of students' time-spent reading materials, the correlation between time-spent engaging in discussion forums and final exam grades. Once the collected formal data that contribute significantly towards learning performance are identified, it is interested to discover the strength of the correlation between the identified formal data and learning performance. By examining the strength level between the identified formal data and learning performance, it is able to predict student performance and further identify student who needs help and at the risk of failing.

To discover social connections and understand student learning behavior, social network analysis using informal data is performed by [5, 24]. In this proposed work,

Table 1. Data distribution

No.	Attributes	Group 1	Group 2	Group 3	Group 4
1.	TotalTimeSpend	[169–666]	[709–1042]	[1139–1852]	[2014–4456]
2.	Freq_of_post	[0]	[2–7]	[8–13]	[14–48]
3.	ParticipantRate	[0]	[0.0119–0.0595]	[0.0714–0.1429]	[0.1548–0.3452]
4.	NumOfLikeOnThread	[0]	[1]	[2–27]	[X]
5.	TotalDownload	[0]	[1]	[2–5]	[X]
6.	TotalMaterial	[1–3]	[4–5]	[6–7]	[8–21]
7.	NumGroupThread	[0]	[1]	[2–5]	[6–28]
8.	NumInitiateGT	[0]	[1]	[2–3]	[5–12]
9.	TotalMentionKeyWords	[0]	[13–43]	[44–98]	[104–458]

social network analysis is conducted using the collected informal data as mentioned in Table 1. A social network is quantified by transforming the social network into a "directed graph G = (V, E), where V is a set of nodes, indicating actors and E is a set of edges, indicating ties". The directed graph G is then transformed into square adjacency matrix [14]. Finally, we aim to analyze student performance and learning behavior based on the combination of formal and informal data. Association rule mining and classification with decision trees will be applied. For examples, are the time-spent (formal data) and the number of posted comments (informal data) significantly affecting student performance? Or, whether the percentage of material read (formal data) and the density level in the social network (informal data) has a significant impact on learning behavior? Figure 4 shows the overall architecture of the proposed LA framework. It consists of User Layer, UI Layer, Services Layer which includes VLE services (i.e. collaboration panel, open threads page, etc.) and SNS (Facebook services), Data Layer and finally proposed LA techniques to analyze the extracted data.

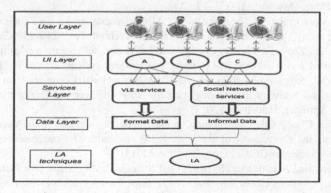

Fig. 4. Overall architecture of LA framework

5 Conclusion and Future Work

Student's learning behavior has evolved recently with the introduction of social media in our daily lives. They are showing lack of interest and easily distracted to learn in the traditional learning environment or either using the formal learning system. Students tend to spending time in using social network services for connectivity with peers. With this motivation, we are developing a learning environment which integrates both formal and informal learning approach to allow students to learn in a socialized environment for any subject domain. Both informal and formal data are stored and then to be elicited from the designed SNS-Integrated VLE, *Shelter*, to investigate the impact of this data combination for more insightful LA. With the proposed work, we aim to provide more insightful and useful information to students' learning performance prediction and to understand learning behavior. Future works include conducting in-depth LA evaluation based on the collected data from user testing and analysis results will be presented in a visualization tool.

Acknowledgments. This work is supported by funding of Fundamental Research Grant Scheme (FRGS), from the Ministry of Higher Learning Education (MOHE).

References

1. Ferguson, R.: The state of learning analytics in 2012: a review and future challenges. Technical report KMI-2012-01, Knowledge Media Inst. (2012)
2. Siemens, G., Long, P.: Penetrating the fog: analytics in learning and education. Educause Rev. **46**(5), 30–32 (2011)
3. Cetintas, S., Si, L., Aagard, H., Bowen, K., Cordova-Sanchez, M.: Microblogging in classroom: classifying students' relevant and irrelevant questions in a microblogging-supported classroom. IEEE Trans. Learn. Technol. **4**(4), 292–300 (2011)
4. Baker, R., Yacef, K.: The state of educational data mining in 2009: a review and future visions. J. Educ. Data Min. **1**(1), 3–17 (2009)
5. Chen, X., Vorvoreanu, M., Madhavan, K.: Mining social media data for understanding students' learning experiences. IEEE Trans. Learn. Technol. **7**(3), 246–259 (2014)
6. Du, Z., Fu, X., Zhao, C., Liu, Q., Liu, T.: Interactive and collaborative e-learning platform with integrated social software and learning management system. In: Lu, W., Cai, G., Liu, W., Xing, W. (eds.) Proceedings of the 2012 International Conference on Information Technology and Software Engineering. Lecture Notes in Electrical Engineering, vol. 212, pp. 11–18. Springer, Heidelberg (2013)
7. Chua, F.F., Choo, C.H.: Integrating social network services into virtual learning environment. In: Proceedings of the 13th IEEE International Conference on Advanced Learning Technologies (ICALT), pp. 264–266 (2013)
8. TOPYX. Learning Management System|TOPYX Social LMS (2012). http://interactyx.com/
9. Schoology. Schoology|Learn. Together (2012). https://www.schoology.com/home.php
10. Jotterlab. Collaborative classroom environment (2012). http://jotterlab.com/
11. Edvance360. Networking learning environment with integrated social network (2012). http://edvance360.com/

12. Baepler, P., Murdoch, C.J.: Academic analytics and data mining in higher education. Int. J. Scholarsh. Teach. Learn. **4**(2), 17 (2010)
13. Siemens, G., d Baker, R.S.: Learning analytics and educational data mining: towards communication and collaboration. In: Proceedings Second International Conference on Learning Analytics and Knowledge, pp. 252–254 (2012)
14. Chatti, M.A., Dyckhoff, A.L., Schroeder, U., Thüs, H.: A reference model for learning analytics. Int. J. Technol. Enhanced Learn. **4**(5), 318–331 (2012)
15. Johnson, L., Smith, R., Willis, H., Levine, A., Haywood, K.: The 2011 horizon report. The New Media Consortium, Austin, Texas (2011)
16. Siemens, G.: What are learning analytics? (2010). http://www.elearnspace.org/blog/2010/08/25/what-are-learning-analytics/
17. Greller, W., Drachsler, H.: Translating learning into numbers: a generic framework for learning analytics. Educ. Technol. Soc. **15**(3), 42–57 (2012)
18. Dietz-Uhler, B., Hurn, J.: Using learning analytics to predict (and improve) student success: a faculty perspective. J. Interact. Online Learn. **12**(1), 17–26 (2013)
19. Drachsler, H., et al.: Issues and considerations regarding sharable data sets for recommender systems in technology enhanced learning. Procedia Comput. Sci. **1**(2), 2849–2858 (2010)
20. Arnold, K.E., Pistilli, M.D.: Course signals at Purdue: using learning analytics to increase student success. In: Proceedings of the 2nd International Conference on Learning Analytics and Knowledge, pp. 267–270. ACM (2012)
21. Dyckhoff, A.L., Zielke, D., Bültmann, M., Chatti, M.A., Schroeder, U.: Design and implementation of a learning analytics toolkit for teachers. Educ. Technol. Soc. **15**(3), 58–76 (2012)
22. Fernández-Gallego, B., Lama, M., Vidal, J.C., Mucientes, M.: Learning analytics framework for educational virtual worlds. Procedia Comput. Sci. **25**, 443–447 (2013)
23. Yu, T., Jo, I.H.: Educational technology approach toward learning analytics: relationship between student online behavior and learning performance in higher education. In: Proceedings of the Fourth International Conference on Learning Analytics and Knowledge, pp. 269–270. ACM (2014)
24. Petropoulou, O., et al.: LAe-R: a new learning analytics tool in Moodle for assessing students' performance. Bull. IEEE Techn. Comm. Learn. Technol. **16**(1), 2–5 (2014)
25. Papamitsiou, Z.K., Terzis, V., Economides, A.A.: Temporal learning analytics for computer based testing. In: Proceedings of the Fourth International Conference on Learning Analytics and Knowledge, pp. 31–35. ACM (2014)
26. Xiong, X., Pardos, Z., Heffernan, N.: An analysis of response time data for improving student performance prediction. In: Proceedings of KDD 2011 Workshop: Knowledge Discovery in Educational Data, Held as part of 17th ACM SIGKDD Conference on Knowledge Discovery and Data Mining (2011)
27. Shum, S.B., Ferguson, R.: Social learning analytics. Educ. Technol. Soc. **15**(3), 3–26 (2012)
28. Ferguson, R., Shum, S.B.: Social learning analytics: five approaches. In: Proceedings of the 2nd International Conference on Learning Analytics and Knowledge. ACM (2012)
29. Cheng-Huang, Y.: A framework of e-learning analytics for asynchronous discussion forums. In: IEEE 13th International Conference on Advanced Learning Technologies (2013)
30. Retalis, S., et al.: Towards networked learning analytics – a concept and a tool. In: Proceedings of the Fifth International Conference on Networked Learning (2006)

LBP vs. LBP Variance for Texture Classification

Gerald Schaefer[1]([⊠]) and Niraj Doshi[2]

[1] Department of Computer Science, Loughborough University,
Loughborough, UK
[2] dMacVis Research Lab, India

Abstract. Texture classification algorithms are utilised in various image
analysis and medical imaging applications. A number of high performing texture algorithms are based on the concept of local binary patterns
(LBP) characterising the relationships of pixels to their local neighbourhood. LBP descriptors are simple to calculate, are invariant to intensity
changes and can be calculated in a rotation invariant manner as well as
at different scales. Incorporating variance information, leading to LBP
variance (LBPV) texture descriptors, has been claimed to lead to more
versatile and more effective texture features. In this paper, we investigate
this in more detail, benchmarking and contrasting the classification performance of several LBP and LBPV descriptors for generic image texture
classification as well as two medical tasks. We show that while LBPV-
based methods typically lead to improved classification performance this
is not always so and that thus the inclusion of variance information is
task dependent.

1   Introduction

Texture classification algorithms are employed in many computer vision and
biomedical image analysis applications and consequently much research has been
devoted to identifying powerful and efficient texture descriptors that can also
cope with challenges such as changes in orientation, scale, or illumination.

Local binary patterns (LBP), originally introduced in [14], provide a relatively simple yet powerful texture descriptor describing the relationships of a
pixel to its local neighbourhood. This approach is generalised in [15] where rotation invariance for LBP is derived, 'uniform' LBP patterns are proposed, and, by
calculating descriptors at different scales, multi-resolution LBP features are proposed. These latter multi-scale texture descriptors can also be more effectively
formulated in terms of multi-dimensional LBP histograms [17].

Local contrast information can be combined with LBP as demonstrated
in [10] where an LBP variance (LBPV) descriptor is proposed. LBPV is obtained
by using the local variance as adaptive weights to adjust the contribution of LBP
codes. In [4], it was found that multi-scale LBPV provides very good classification performance for textures captured under different orientations and different
illuminations.

In this paper, we compare the performance of LBP and LBPV-based descriptors for several texture classification tasks. In particular, we benchmark them

© Springer International Publishing AG 2017
Y. Tan et al. (Eds.): DMBD 2017, LNCS 10387, pp. 156–164, 2017.
DOI: 10.1007/978-3-319-61845-6_16

on a test suite of texture databases as well as two medical imaging applications. Our results show that while in general incorporating variance information leads to improved classification performance this is not always the case, and thus that the choice of texture features is typically task dependent.

2   Local Binary Patterns (LBP)

Local binary patterns (LBP) allow for simple yet effective texture analysis. The original LBP variant [14] operates on a per-pixel basis, and describes the 8-neighbourhood pattern of a pixel in binary form. If $\{g_1, g_2, \ldots, g_8\}$ is the set of 8-neighbourhood pixels of a centre pixel g_c, then the neighbouring pixels are set to 0 and 1 respectively by thresholding them with the centre pixel value. An LBP pattern is thus obtained by

$$\text{LBP} = \sum_{p=1}^{8} s(g_p - g_c)2^{p-1}, \tag{1}$$

where

$$s(x) = \begin{cases} 1 \text{ if } x \geq 0 \\ 0 \text{ if } x < 0 \end{cases}. \tag{2}$$

The 256 possible resulting patterns are then typically used to build a histogram, which serves as a texture feature for an image or an image region.

Alternatively, a circular neighbourhood can be employed [15] by R and P, where R defines the distance of the neighbours to the centre, and P is the number of samples at that distance that are employed as neighbours. Locations that do not fall exactly at the centre of a pixel are obtained through interpolation.

Rotation invariance can be easily achieved in LBP. If a texture is rotated, essentially the patterns (that is the 0s and 1s around the centre pixel) rotate with respect to the centre. Rotation invariant LBP codes can thus be obtained by grouping together corresponding rotated LBP patterns [15].

LBP patterns can also be grouped based on the number of spatial transitions from 0s to 1s and vice versa in the bit pattern. These patterns are called uniform [15], and are defined by a uniformity measure (typically set to 2) which corresponds to the maximum number of spatial transitions in the LBP code.

Clearly, rotation invariant and uniform patterns can be combined. For eight neighbours, there are nine rotation invariant uniform LBP codes, two without any 0–1 changes (i.e., one with all 0s and one with all 1s) and the remaining seven with $1, \ldots, 7$ ones in sequence. It has been shown [15] that focussing on these uniform patterns while aggregating all other (i.e., non-uniform) patterns into one group leads to improved texture descriptors. While LBP generates 256 patterns for an 8-neighbourhood, rotation invariant LBP^{ri} generates 36 patterns, while rotation invariant uniform LBP^{riu2} results in 10 pattern classes for the same neighbourhood.

By defining several radii around a pixel multiple concentric neighbourhood LBP codes can be extracted and thus multi-resolution texture descriptors obtained [15]. While in principle any radius is feasible, attention is often restricted to the sets $r = \{1,3\}$ and $r = \{1,3,5\}$. Also, while in general any number of neighbours could be defined, we found in our experiments that choosing 8 neighbours at all distances does not compromise accuracy while also corresponding to those directions (horizontal, vertical, and plus/minus $45°$) to which the human visual system is most sensitive to.

When recording multi-resolution texture information using LBP, a histogram is generated for each scale/radius, while the histograms are concatenated to form a one-dimensional feature vector. In [17], it was shown that storing multi-scale LBP features in such a fashion leads to a loss of information between the different scales and added ambiguity. The joint distribution of LBP codes at different scales can be preserved by building a multi-dimensional LBP (MD-LBP) histogram [17]. To do so, LBP codes are calculated at different scales while the combination of the codes identifies the histogram bin that is incremented.

3   LBP Variance (LBPV)

The contrast in an image

$$\text{VAR}_{P,R} = \frac{1}{P} \sum_{p=0}^{p-1} (g_p - \mu)^2 \tag{3}$$

with $\mu = \frac{1}{P} \sum_{p=0}^{P-1} g_p$ can be combined with $\text{LBP}_{P,R}$ to generate a joint distribution of $\text{LBP}_{P,R}/\text{VAR}_{P,R}$ which gives a powerful texture descriptor as it contains both local pattern and local contrast information. An alternative is the use of a hybrid scheme, LBP variance (LBPV) [10], which also captures joint LBP and contrast information but where the variance $\text{VAR}_{P,R}$ is used as an adaptive weight to adjust the contribution of the LBP code in histogram calculation.

LBPV histograms are calculated as

$$\text{LBPV}_{P,R}(k) = \sum_{i=1}^{N} \sum_{j=1}^{M} \omega(\text{LBP}_{P,R}(i,j), k), \tag{4}$$

with

$$\omega(\text{LBP}_{P,R}(i,j), k) = \begin{cases} \text{VAR}_{P,R}(i,j) & \text{if } \text{LBP}_{P,R}(i,j) = k \\ 0 & \text{otherwise} \end{cases}, \tag{5}$$

and $k \in [0, K]$ defining the various LBP codes.

An approach similar to MD-LBP can also be devised to obtain multi-dimensional LBPV (MD-LBPV) texture descriptors, which incorporate image variance information as adaptive weights to build multi-dimensional LBP histograms [5]. MD-LBPV histograms are calculated as

$$\text{MD-LBPV}_{P,R=\{r_1,r_2,\ldots,r_n\}}(k_1, k_2, \ldots, k_R) \tag{6}$$

$$= \sum_{i=1}^{N} \sum_{j=1}^{M} \omega(\text{LBP}_{P,R=\{r_1,r_2,\ldots,r_R\}}(i,j), k_1, k_2, \ldots, k_R),$$

with

$$\omega(\text{LBP}_{P,R=\{r_1,r_2,\ldots,r_R\}}(i,j), k_1, k_2, \ldots, k_R) \tag{7}$$
$$= \begin{cases} f(\mathcal{V}) & \text{if } \text{LBP}_{P,R=r_s}(i,j) = k_s \quad \forall s \in \{1, 2, \ldots, R\} \\ 0 & \text{otherwise} \end{cases},$$

and

$$\mathcal{V} = \{\text{VAR}_{P,r_1}(i,j), \text{VAR}_{P,r_2}(i,j), \ldots, \text{VAR}_{P,r_R}(i,j)\}. \tag{8}$$

MD-LBPV based on 2 radii will hence yield a 2-dimensional histogram, MD-LBPV based on 3 radii a 3-dimensional one and so on.

While in MD-LBP the histogram is always incremented in unit values and local contrast information is not utilised, in MD-LBPV local contrast is integrated into the way multi-dimensional texture histograms are generated. Of the various ways of how variance information can be incorporated into MD-LBPV histograms [5], we chose the maximum variance method which uses the maximum value of variance over all scales

$$f(\mathcal{V}) = \max\{\text{VAR}_{P,r_1}(i,j), \text{VAR}_{P,r_2}(i,j), \ldots, \text{VAR}_{P,r_R}(i,j)\} \tag{9}$$

and typically gives the best results.

4   LBP vs. LBPV Descriptors for Texture Classification

In the first of the three experiments we conduct, we evaluate the performance of LBP and LBPV-based texture descriptors on a set of benchmark texture classification tasks. For this, we select the Outex texture benchmarking suite [13], which provides both the diversity as well as the size required for an appropriate evaluation of texture descriptors. In total, we use eight Outex datasets which provide variations in terms of scale, rotation, etc. in order to evaluate texture classification performance.

For classification, we employ – throughout the paper – a standard approach based on support vector machines (SVMs). SVMs achieve classification of two separable classes by maximising the margin between the two classes [18]. Since we have more than two classes, we employ a one-against-one multi-class SVM [11] where for each SVM, we use a linear kernel, and optimise the cost parameter $C \in [-1.1; 3.1]$ using a cross validation approach [1]. The results, on all Outex datasets, and for both LBP and LBPV-based descriptors are given in Table 1.

As we can see from Table 1, all LBP features yield relatively good texture classification performance. We can however also notice that LBPV-based descriptors consistently give higher classification accuracies and that thus incorporating variance information is beneficial here.

Table 1. Texture classification results on Outex datasets.

	TC_00	TC_01	TC_10	TC_11	TC_12	TC_13	TC_14	TC_15	Average
LBP^{riu2}	93.80	85.98	82.76	76.88	63.47	86.03	31.18	73.56	74.21
$LBP^{riu2}_{R=1,3}$	98.87	95.72	95.81	80.83	85.35	90.59	48.75	82.18	84.76
$LBP^{riu2}_{R=1,3,5}$	99.73	97.65	94.61	87.29	86.18	91.32	54.34	83.33	86.81
$MD\text{-}LBP^{riu2}_{R=1,3}$	99.15	97.19	97.58	83.75	90.76	90.29	53.24	83.46	86.93
$MD\text{-}LBP^{riu2}_{R=1,3,5}$	99.60	97.95	95.34	86.04	91.96	89.41	60.96	84.29	88.19
$LBPV^{riu2}$	95.52	93.58	88.41	87.50	73.13	87.50	43.24	76.01	80.61
$LBPV^{riu2}_{R=1,3}$	97.90	96.52	97.42	91.46	92.44	87.21	54.34	80.50	87.22
$LBPV^{riu2}_{R=1,3,5}$	98.65	97.66	98.13	92.71	93.81	88.09	58.31	81.83	88.65
$MD\text{-}LBPV^{riu2}_{R=1,3}$	98.66	97.33	97.81	92.50	94.29	87.65	57.57	82.64	88.56
$MD\text{-}LBPV^{riu2}_{R=1,3,5}$	99.00	97.68	98.59	88.13	96.00	87.35	59.85	82.95	88.69

5   LBP vs. LBPV Descriptors for HEp-2 Cell Classification

Indirect immunofluorescence (IIF) imaging is routinely used for the screening of antinuclear antibodies (ANAs) based on HEp-2 cells. Analysis of ANAs is important and employed in the diagnosis of systemic rheumatic disease, systemic sclerosis and mellitus (type-I) diabetes [9]. Typically, an expert observes cultured HEp-2 cells under a fluorescence microscope and categorises the cells, based on fluorescence intensity and the type of staining pattern, into a number of groups (often six: homogeneous, fine speckled, coarse speckled, nucleolar, cytoplasmic, and centromere cells, examples of which are shown in Fig. 1). This visual classification is challenging as it is not only subjective and relies on the expertise of the specialist, but is also a laborious and time consuming process. Recently, there is hence significant interest in computer-aided approaches to perform this analysis to both speed up the task and provide objective, reproducible results.

In our experiments, we use the ICPR 2012 HEp-2 classification contest dataset [16] which is based on 28 HEp-2 images stored in 24-bit true-colour format with a resolution of 1388×1038 pixels. Cells were manually segmented and annotated by a specialist to obtain a ground truth.

The training dataset provided to contestants comprises 721 samples of individual cells, extracted from part of the captured images. There are 150 homo-

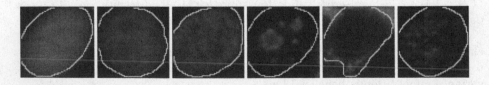

Fig. 1. Sample HEp-2 cell images (with manually defined borders in white) from the ICPR 2012 contest dataset. From left to right: homogeneous, fine speckled, coarse speckled, nucleolar, cytoplasmic, centromere.

geneous, 94 fine speckled, 109 coarse speckled, 102 nucleolar, 58 cytoplasmic, and 208 centromere cells (an example for each is given in Fig. 1). The testing dataset provided comprises 734 samples of individual cells obtained from different images. There are 172 homogeneous, 114 fine speckled, 101 coarse speckled, 139 nucleolar, 51 cytoplasmic, and 149 centromere cells to classify. The results obtained on the testing dataset are given in Table 2.

Table 2. Classification accuracy of HEp-2 cell images from the ICPR 2012 contest dataset.

	Accuracy(%)
$\text{LBP}_{R=1,3}$	64.99
$\text{LBP}_{R=1,3,5}$	67.03
$\text{MD-LBP}^{riu2}_{R=1,3}$	66.62
$\text{MD-LBP}^{riu2}_{R=1,3,5}$	70.30
$\text{LBPV}^{riu2}_{R=1,3}$	53.30
$\text{LBPV}^{riu2}_{R=1,3,5}$	56.27
$\text{MD-LBPV}^{riu2}_{R=1,3}$	54.77
$\text{MD-LBPV}^{riu2}_{R=1,3,5}$	58.58

It is evident from Table 2 that while conventional LBP methods provide very good classification performance (for comparison, the winner of the ICPR competition achieved a classification accuracy of about 68%), LBPV-based descriptors give relatively poor results. For this application, incorporating variance information thus not only does not help, it actually leads to significantly deteriorated classification performance.

6   LBP vs. LBPV Descriptors for Nailfold Capillary Image Analysis

Nailfold capillaroscopy (NC) is a non-invasive imaging technique employed to assess the condition of blood capillaries in the nailfold. It is particularly useful for early detection of scleroderma spectrum disorders and evaluation of Raynaud's phenomenon. Diagnosis using NC images involves the classification into Early, Active and Late groups, also known as NC patterns or scleroderma (SD) patterns [2] (see Fig. 2 for example images) based on the identification of enlarged or giant capillaries, haemorrhages, loss of capillaries, disorganisation of the vascular array, and ramified/bushy capillaries in the images [3].

The degree of these abnormalities indicates the severity and progression of the disease. Three NC patterns can be defined and characterised by [2]:

– **Early:** few giant capillaries, few capillary haemorrhages, relatively well preserved capillary distribution, no evident loss of capillaries.

Fig. 2. NC image samples, from left to right: healthy subject, early SD pattern, active SD pattern, late SD pattern.

- **Active:** frequent giant capillaries, frequent capillary haemorrhages, moderate loss of capillaries with some avascular areas, mild disorganisation of the capillary architecture, absent or some ramified capillaries.
- **Late:** irregular enlargement of the capillaries, few or absent giant capillaries, absence of haemorrhages, severe loss of capillaries with large avascular areas, severe disorganisation of the normal capillary array, frequent ramified/bushy capillaries.

While diagnosis based on NC is typically performed by manual inspection, computerised nailfold capillaroscopy can help to reduce the inherent ambiguity in human judgement while greatly reducing the time for diagnosis [6]. This is typically approached by trying to segment the capillaries and analysing the extracted structures [8,12,19]. In contrast, in [7] a novel holistic approach for analysing NC images was proposed based on texture analysis using multi-scale LBP information.

Since LBP features were used in [7], in here we inspect the performance of LBP and LBPV as well as their multi-scale extensions for the purpose of NC image analysis on a dataset of 12 subjects with NC images captured for three to four fingers for each patient and three patients for each class (i.e. control, early, active, and late). The outcome of this is presented in Table 3, which gives results both based on classification of individual finger images as well as 'patient classifications' based on a simple majority rule.

From Table 3 we can see that texture descriptors based on conventional LBP do give relatively poor results. On the other hand, LBPV-based features

Table 3. Classification accuracy of nailfold capillary classification experiment (reject refers to cases where the majority vote did not yield a single winner).

	Finger classification [%]	Patient classification [%] (reject)
$LBP^{riu2}_{R=1,3}$	46.34	33.33 (33.33)
$LBP^{riu2}_{R=1,3,5}$	51.22	50.00 (16.67)
$MD\text{-}LBP^{riu2}_{R=1,3}$	43.90	41.67 (8.33)
$MD\text{-}LBP^{riu2}_{R=1,3,5}$	43.90	33.33 (25.0)
$LBPV^{riu2}_{R=1,3}$	70.73	66.67 (8.33)
$LBPV^{riu2}_{R=1,3,5}$	70.73	75.00 (8.33)
$MD\text{-}LBPV^{riu2}_{R=1,3}$	73.17	83.33 (0.00)
$MD\text{-}LBPV^{riu2}_{R=1,3,5}$	65.85	83.33 (0.00)

yield significantly better performance and make useful descriptors for identifying Scleroderma patterns in NC images.

7   Conclusions

In this paper we have investigated the performance of LBP and LBP variance texture features for several texture classification problems. In particular we benchmarked them on a set of generic texture databases and on two medical image analysis tasks. We have shown that while in general incorporating variance information gives improved classification performance this is not always the case and thus that the choice of appropriate texture features is task dependent.

References

1. Chang, C.C., Lin, C.J.: libSVM: a library for support vector machines. ACM Trans. Intell. Syst. Technol. **2**, 27:1–27:27 (2011)
2. Cutolo, M., Sulli, A., Pizzorni, C., Accardo, S.: Nailfold videocapillaroscopy assessment of microvascular damage in systemic sclerosis. J. Rheumatol. **27**, 155–160 (2000)
3. Cutolo, M., Pizzorni, C., Sulli, A.: Capillaroscopy. Best Pract. Res. Clin. Rheumatol. **19**(3), 437–452 (2005)
4. Doshi, N.P., Schaefer, G.: A comparative analysis of local binary pattern texture classification. In: Visual Communications and Image Processing (2012)
5. Doshi, N.P., Schaefer, G.: Texture classification using multi-dimensional LBP variance. In: 2nd IAPR Asian Conference on Pattern Recognition (2013)
6. Doshi, N.P., Schaefer, G., Howell, K.: A review of computerised nailfold capillaroscopy. In: 17th Annual Conference in Medical Image Understanding and Analysis (2013)
7. Doshi, N.P., Schaefer, G., Merla, A.: Nailfold capillaroscopy pattern recognition using texture analysis. In: IEEE-EMBS International Conference on Biomedical and Health Informatics (2012)
8. Doshi, N.P., Schaefer, G., Zhu, S.Y.: An improved binarisation algorithm for nailfold capillary skeleton extraction. In: IEEE International Conference on Systems, Man, and Cybernetics (2013)
9. Egerer, K., Roggenbuck, D., Hiemann, R., Weyer, M.G., Büttner, T., Radau, B., Krause, R., Lehmann, B., Feist, E., Burmester, G.R.: Automated evaluation of autoantibodies on human epithelial-2 cells as an approach to standardize cell-based immunofluorescence tests. Arthritis Res. Ther. **12**(2), R40 (2010)
10. Guo, Z., Zhang, L., Zhang, D.: Rotation invariant texture classification using LBP variance (LBPV) with global matching. Pattern Recogn. **43**(3), 706–719 (2010)
11. Hsu, C.W., Lin, C.J.: A comparison of methods for multiclass support vector machines. IEEE Trans. Neural Netw. **13**(2), 415–425 (2002)
12. Kwasnicka, H., Paradowski, M., Borysewicz, K.: Capillaroscopy image analysis as an automatic image annotation problem. In: 6th International Conference on Computer Information Systems and Industrial Management Applications (2007)
13. Ojala, T., Maenpää, T., Pietikäinen, M., Viertola, J., Kyllonen, J., Huovinen, S.: Outex - new framework for empirical evaluation of texture analysis algorithms. In: 16th International Conference on Pattern Recognition, pp. 1:701–1:706 (2002)

14. Ojala, T., Pietikäinen, M., Harwood, D.: A comparative study for texture measures with classification based on feature distributions. Pattern Recogn. **29**, 51–59 (1996)
15. Ojala, T., Pietikäinen, M., Maenpää, T.: Multiresolution gray-scale and rotation invariant texture classification with local binary patterns. IEEE Trans. Pattern Anal. Mach. Intell. **24**, 971–987 (2002)
16. Percannella, G., Foggia, P., Soda, P.: 1st International Contest on HEp-2 Cells Classification (2012). http://mivia.unisa.it/hep2contest/HEp2-Contest_Report.pdf
17. Schaefer, G., Doshi, N.P.: Multi-dimensional local binary pattern descriptors for improved texture analysis. In: 21st International Conference on Pattern Recognition, pp. 2500–2503 (2012)
18. Vapnik, V.N.: Statistical Learning Theory. Wiley, New York (1998)
19. Wen, C.H., Hsieh, T.Y., Liao, W.D., Lan, J.L., Chen, D.Y., Li, K.C., Tsai, Y.T.: A novel method for classification of high-resolution nailfold capillary microscopy images. In: 1st IEEE International Conference on Ubi-Media Computing, pp. 513–518 (2008)

A Decision Tree of Ignition Point for Simple Inflammable Chemical Compounds

Ryoko Hayashi[✉]

Kanazawa Institute of Technology, 3-1 Yatsukaho, Hakusan,
Ishikawa 924-0838, Japan
ryoko@neptune.kanazawa-it.ac.jp

Abstract. Ignition point, the temperature at which a chemical compound begins to burn naturally, is one of the important values from the viewpoint of industry and safety. This manuscript addresses a trial prediction of the ignition point for relatively simple chemical compounds including carbon, oxygen and hydrogen via data mining such as decision tree and random forest. I used fundamental material values and the number of characteristic structures as descriptors for chemical compounds. Our input data file includes 240 kinds of chemical compounds and we prepared other 10 as the test data. At first, I used "rpart" package of the "R", one of the statistical programming language, in order to process decision tree. Furthermore I used "randomForest" with more data and more number of descriptors and I got better estimation of ignition point.

Keywords: Ignition point · Decision tree · Random forest · Chemical compound · Molecule

1 Introduction

Recently, data mining technology has became well-matured so that anyone can use data mining without difficulty. This paper reports a trial result of data mining techniques applied to material science field. Chemical compounds already have an enormous varieties over 20 million and continue increasing everyday. When we use a new, unknown chemical compound, the prediction of its characteristics by data mining will be helpful. Thus this paper reports the result of a trial prediction about the ignition point for simple chemical compounds using well-known data mining technique such as decision tree and random forest.

Prediction of ignition point is mainly studied in chemoinformatics [1] field, especially using the Quantitative Structure-Activity Relationship (QSAR) approach. With the QSAR approach, the estimated ignition point for a chemical compound is based on the linear equation of,

$$T_e = \Sigma_{i=1}^{n} C_i D_i, \tag{1}$$

where T_e represents the estimated ignition point, n is the number of descriptors, D_i is the value of the i-th descriptor and C_i is the coefficient for the i-th

© Springer International Publishing AG 2017
Y. Tan et al. (Eds.): DMBD 2017, LNCS 10387, pp. 165–172, 2017.
DOI: 10.1007/978-3-319-61845-6_17

descriptor. The natural complexities of chemical compounds are included on the descriptors or on the coefficients. The method to select descriptors is still a difficult problem in QSAR field [2] so that there are many works about it.

I will introduce a few recent works about the prediction of ignition point with QSAR approach. Tsai, Chen, Liaw predicted ignition point using 820 observation and 4 descriptors [3]. They got predictions with $89K$ of maximum error and $36K$ of average error. Their descriptors are based on numerical simulation result of molecular structure optimization so that the descriptors are notional. Shi, Chen, Chen used 265 observations and descriptors based on the fragments of molecules and they got the prediction with the error between $50K$ and $90K$ [4]. In the QSAR approach with linear combination formula, primitive material data cannot predict ignition point directly so that the descriptors have a tendency to be far apart from primitive material characteristics.

This manuscript uses data mining approach [5] with descriptors based on primitive material characteristics so that the natural complexities such as non-linearity will be included in data mining processes. Here I use decision tree [6] at first with the anticipation for automatic extracting important descriptors, then I use random forest [7] for comparing with decision tree. Our target chemical compounds are simple molecules [8], because they are not too difficult and are studied very well.

This manuscript is organized as follows: Sect. 2 shows the detail of data and descriptors; In Sect. 3, both of the results using decision tree and random forest are explained; Sect. 4 compares the two methods' estimations of ignition point for test data; Sect. 5 concludes the remarks.

2   Data and Descriptors

In this manuscript, I selected the observations of the chemical compounds include carbon atoms and hydrogen atoms mainly. Oxygen atoms are also available, but other elements are excluded in order to limit the variety of chemical characteristics. I used 240 observations for learning data to make decision tree and I prepared 10 independent observations as the test data. Furthermore, I added 44 more observations for learning data to execute random forest. The observations are obtained from ICSC database [8], a database containing ignition point if available, molecular formula, molecular weight, CAS number, and other information about safety such as toxicity.

Figure 1 shows the histogram for ignition point in the learning data. I used 240 observations for making decision tree at first, then I added 44 observations for random forest. Therefore Fig. 1 has two parts: bars filled with dot pattern for original 240 observations and bars filled with a pattern of slanted lines for additional 44 observations. In Fig. 1, many observations have ignition point between 200 °C and 600 °C. Additional 44 observations seem to have almost same distribution with original 240 observations.

Table 1 shows examples of an observation called "diketene", one of the cyclic ester chemical compound. In this manuscript, descriptors include the fundamental material data such as molecular weight, boiling point and melting point.

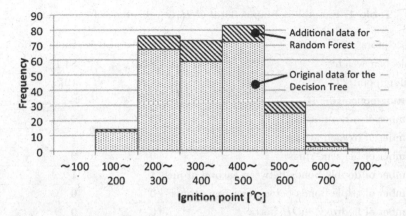

Fig. 1. Histogram for ignition point in the learning data.

Furthermore, the number of carbon atoms and oxygen atoms, the number of other chemical structures such as benzene rings are also used as descriptors. After the analysis using decision tree, more detailed descriptors seemed to be suitable so that I added descriptors represent several chemical structures for random forest as shown in Table 1.

Some of the descriptors have inclusion relationships. For example, an ester includes an ether and a ketone. The descriptors for decision tree are decided exclusively and the highest level of structure has the priority so that "the number of ester" descriptor for diketene is "1" in Table 1, but do not count the number of ether and the number of ketone. In order to deal with chemical substructures as same as superstructure, the descriptors for random forest are decided inclusively: in Table 1, diketene is cyclic ester so that many descriptors such as "number of cyclic", "number of cyclic ester", "number of ester", "number of cyclic ether", "number of ether" and "number of ketone" are "1" for random forest.

3   Decision Tree and Random Forest Execution

3.1   Decision Tree for Ignition Point

The decision tree in this manuscript is processed by "R" [5], one of the statistical programming environment, R is open-source and there are many data mining packages now so that we can execute various data mining technique with R. Here I use "rpart" package [6] to process decision tree.

Using "rpart" package, we can get a decision tree, but the tree often includes over-fitting so that we should discuss pruning at first. Figure 2 shows the relationships between the size of original decision tree or complexity parameter with relative cross validation error. The dotted horizontal line in Fig. 2 means the summation of the average cross validation error and a standard deviation; the line of the guide for over-fitting. In Fig. 2, the first point under the line from the left side indicates the number of terminal node often selected to avoid the effect

Table 1. Examples of input data: $C_4H_4O_2$, diketene of cyclic ester.

Descriptors	Decision tree	Random forest
Molecular weight	84.1	84.1
Boiling point [°C]	127	127
Melting point [°C]	−7	−7
Number of carbon atoms	4	4
Number of oxygen atoms	2	2
Number of benzene rings	0	0
Number of double bonds between carbon atoms	1	1
Number of triple bonds between carbon atoms	0	0
Number of hydroxy, $-OH$ parts	0	0
Number of aldehyde, $-CHO$ parts	(undefined)	0
Number of ketone, $-C(=O)-$	0	1
Number of ether, $-O-$	0	1
Number of cyclic ether	(undefined)	1
Number of carboxy $-COOH$	0	0
Number of ester $-COO-$	1	1
Number of cyclic ester	(undefined)	1
Number of ring structure	(undefined)	1
Straight-chain	(undefined)	0
Ignition point [°C]	275	275

of over-fitting. From the Fig. 2, the number of effective terminal node seems to be 2, but it is few so that here I adopt 4 as the number of terminal node of the tree because of almost same cross validation error.

Figure 3 shows the pruned decision tree with 4 terminal nodes. From Fig. 3, we can mention top three important rules:

Rule 1 (on the root node of the tree): If a compound has no benzene ring, the average ignition point is 343 °C, otherwise, a compound has at least one benzene ring, the average ignition point is 494 °C.

Rule 2: Among the compounds with no benzene ring, if a compound has two or more ether linkage, the average ignition point is 246 °C. Otherwise, the average ignition point is 352 °C.

Rule 3: Among the compounds have no benzene ring and one or no ether linkage, if a compound has seven or more carbon atoms, the average ignition point is 306 °C. Otherwise, the average ignition point is 369 °C.

The rule 1 is already known empirically so that it seems to be a valid rule. Ether is known as one of the relatively stable chemical structure so that the rule 2 also seems to be a valid rule. The rule 3 requires more detailed discussion because there are some relationships about the size of molecule and vaporization and they will have some influence on ignition point.

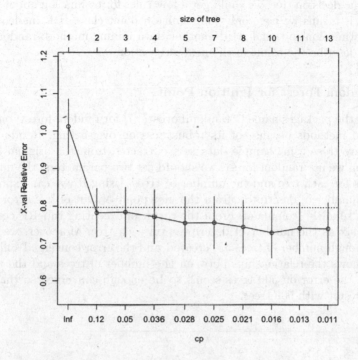

Fig. 2. The relationships between complexity parameter between cross validation error on original decision tree.

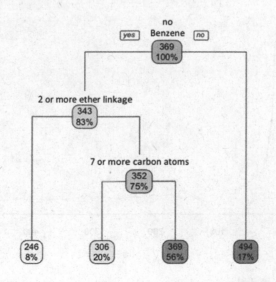

Fig. 3. Pruned decision tree with 4 terminal nodes.

From the decision tree, we could get a few rules for estimation about ignition point, but it means we have only four ignition point class. It is inadequate for effective estimation so that I tried another data mining method: random forest and I will describe the result in the next subsection.

3.2 Random Forest for Ignition Point

R also has the packages named "randomForest" [7] for random forest, one of the well-known methods because of its robustness on over-fitting. Random forest makes many trees using sample data sets extracted from the original learning data. When we use random forest, we should set two parameters: the number of descriptors for each tree and the number of trees. Using R, we can examine the optimum number of descriptor, from the viewpoint of Out-of-bag error estimation with "tuneRF" command. From the result of executing tuneRF command, I selected six as the number of descriptor on each tree. Moreover, we have to confirm enough number of trees are created and the error converged sufficiently. Figure 4 shows the relationships between the number of trees and the error. In the Fig. 4, the error on 500 trees seems to be enough converged so that here I show the result with 500 trees.

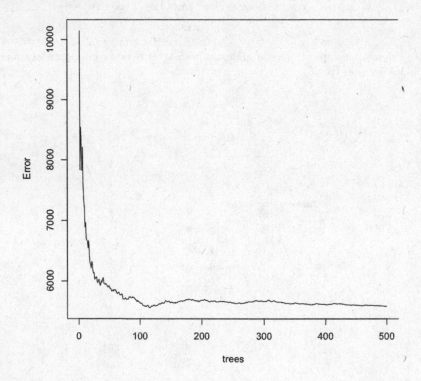

Fig. 4. The relationships between the number of trees and the error.

Unlike decision tree, random forest does not show rules to decide ignition point, but shows the importance of the descriptors. Figure 5 shows the relationships between the decrease of node impurities using Geni index and the descriptors. In Fig. 5, upper descriptors are more important and the expressions for descriptors omitted the words "number of". From the Fig. 5, the most important descriptor is the number of benzene, as same as the result of decision tree. Melting point, molecular weight, boiling point and the number of carbon atoms are also important descriptors, but they have some correlation each other actually.

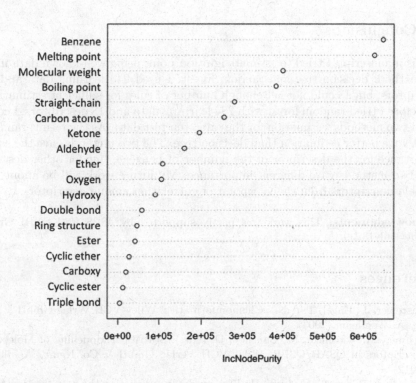

Fig. 5. The relationships between importance and descriptors.

4   Prediction Quality Evaluation

Table 2 show the result of estimation for same ten test data using decision tree and random forest described in Sect. 3. In Table 2, decision tree's result show that the difference from the actual ignition point is over 150 [°C] in the worst case. On the other hand, random forest gave better result and the difference from the actual value is smaller than 110 [°C] in the worst case. RMSE values from the ten estimation results also show the random forest gave better estimation than decision tree.

Table 2. Summary of estimation results for ignition point.

The worst case	Actual ignition point [°C]	Decision tree		Random forest	
		Estimation	difference	Estimation	difference
(decision tree's)	202	369	167	251	49
(random forest's)	238	369	131	347	109
RMSE			79		53

5   Conclusions

In this manuscript, I tried to estimate ignition point using well-known data mining method: decision tree and random forest. I used 240 learning data and 13 descriptors, but I could not get enough number of rules for sufficient estimation. Therefore I tried random forest with 284 learning data and 18 descriptors. I compared two methods' estimation for the same ten test data. As the result, random forest gave better estimation than decision tree. The two methods gave the same importance for the descriptor of the number of benzene, but for other descriptors, two methods gave different importance. My future work will be about the suitable descriptors from the viewpoint of correlation among descriptors.

Acknowledgments. This work was partly supported by JSPS KAKENHI Grant Number 16K13739.

References

1. Gasteiger, J., Engel, T. (eds.): Chemoinformatics. Wiley-VCH Verlag GmbH & Co. KgaA, Weinheim (2003)
2. Dehmer, M., Varmuza, K., Bonchev, D. (eds.): Statistical Modelling of Molecular Descriptors in QSAR/QSPR. Wiley-VCH Verlag GmbH & Co. KgaA, Weinheim (2012)
3. Tsai, F.-Y., Chen, C.-C., Liaw, H.-J.: A model for predicting the auto-ignition temperature using quantitative structure property relationship approach. Procedia Eng. **45**, 512–517 (2012)
4. Shi, J., Chen, L., Chen, W.: Prediction on the auto-ignition temperature using substructural molecular fragments. Procedia Eng. **84**, 879–886 (2014)
5. R Core Team: R: A language and environment for statistical computing. R Foundation for Statistical Computing, Vienna, Austria (2015). http://www.R-project.org/
6. Package "rpart". https://cran.r-project.org/web/packages/rpart/rpart.pdf
7. Package "randomForest". https://cran.r-project.org/web/packages/randomForest/randomForest.pdf
8. ICSC database. http://www.ilo.org/dyn/icsc/showcard.home

Analyzing Consumption Behaviors of Pet Owners in a Veterinary Hospital

Jo Ting Wei[1], Shih-Yen Lin[2], You-Zhen Yang[3],
and Hsin-Hung Wu[3(✉)]

[1] International Business Department, Providence University, Taichung City,
Taiwan
[2] Department of Tourism, Leisure and Hospitality Management, National Chi
Nan University, Nantou, Taiwan
[3] Department of Business Administration, National Changhua University of
Education, Changhua, Taiwan
hhwu@cc.ncue.edu.tw

Abstract. The purpose of this study is to identify different consumption behaviors of pet owners in a veterinary hospital so as to provide proper marketing strategies. A case study was conducted by combining data mining techniques and RFM model for a veterinary hospital located in Taichung City, Taiwan by examining its transactions data focusing on pet mice in 2014. The development of marketing strategies for the veterinary hospital is important to improve its service quality and strengthen the positive relationship between the pet owners and the case veterinary hospital.

Keywords: Data mining · Customer relationship management · Veterinary · Marketing strategies

1 Introduction

Retaining the existing customers costs less than attracting new customers [1]. Customer relationship management (CRM) has been viewed as the core marketing activity for firms facing increasingly complex transaction types and service processes that can provide an effective way to achieve relationship marketing by building good interactions with customers [2]. CRM can effectively record customer behaviors in transactions and enhance customer profitability such as customer retention and loyalty by retaining existing customers and finding new customers [3]. Many industries rely on CRM to maintain a long-term and stable relationship with their customers.

The pet markets have grown rapidly in Taiwan. Based on the survey of a recent census in Taiwan, pet owners have increasingly spent much on pet-related products and services. The increasing expenditure on pets shows the rapid growth of the pet industry, reflecting the pet industry has become more and more competitive [4, 5]. Hence, the veterinary hospital faces fierce competitions due to the great demand for veterinary services from pet owners [6].

Data mining techniques are an essential tool to achieve CRM, particularly the cluster analysis for market segmentation. This study intends to use cluster analysis

© Springer International Publishing AG 2017
Y. Tan et al. (Eds.): DMBD 2017, LNCS 10387, pp. 173–180, 2017.
DOI: 10.1007/978-3-319-61845-6_18

along with RFM (recency, frequency, and monetary) model to examine customer behaviors of a veterinary hospital located in Taichung City, Taiwan by analyzing the transactions data focusing on pet mice in 2014 as well as developing its marketing strategies.

2 Literature Review

2.1 Data Mining

Data mining (DM) is defined as the process adopting computer learning techniques to automatically search abundant data to analyze and summarize them into useful information [7]. DM plays an important role in CRM and has been widely applied in different areas such as medicine, tourism, telecommunications, banking, and retailing [8]. Clustering analysis, one of the DM techniques, is to discover knowledge based on the data similarity by minimizing variance within groups and maximizing variance among groups [2]. Self-organizing maps (SOM) and K-means methods are the most commonly seen approaches in cluster analysis.

SOM is an unsupervised learning algorithm that uses a visualization method to discover statistical insights and models from big data by a two-dimensional discrete map to seek the underlying hidden patterns [9–11]. K-means method which is a non-hierarchical approach has been widely applied in practice due to its ease of use [12, 13]. K-means method has the following steps: (1) beginning with choosing the number of clusters, k, and calculating the centroid of these clusters; (2) assigning each point to the nearest cluster centroid; (3) after reallocating all data, computing the new centroid of the clusters; and (4) repeating the two prior steps until centroids do not change any more at each step [2].

Kuo et al. [14] proposed a two-stage clustering approach to improve the weaknesses of SOM and K-means by adopting SOM followed by the K-means method. Prior findings show that the combination of these two methods outperforms either SOM or K-means method [14, 15]. This study adopts the two-stage approach for cluster analysis. That is, SOM is applied to identify the number of the clusters. Later, the determined number of the clusters is adopted for K-means method to partition the data set.

2.2 RFM Model

RFM model consists of recency (R), frequency (F), and monetary (M) [15]. Recency is calculated according to the number of periods since the last purchase. Frequency is computed in accordance with the number of purchase made in a given time period. Monetary is to sum the dollar value that customers spent in a specified time period. To sum up, RFM model measures when customers buy, how often they buy, and how much they buy [8, 16].

Traditionally, RFM model is implemented by sorting the customer database and then dividing the data into five equal segments for each measure. The top 20% segment is coded as a value of 5; the next 20% segment is coded as a value of 4, and so forth [17, 18]. Wei et al. [8], on the other hand, used the original data to perform RFM model rather than the coded numbers, and their definitions of RFM model is as follows. Recency refers to the number of days since the most recent purchase; frequency is the number of purchase in a specified time period; and monetary is the dollar value spent on all purchases in the same time period. This study follows Wei et al. [8] to use the original data to implement RFM model.

2.3 The Classification of Customer Types

Ha and Park [19] used average RFM values of each cluster to compare the total average RFM values of all clusters to segment customers. An upward arrow ($\uparrow$) is given to a particular symbol if the average value is greater than the total average, whereas a downward arrow ($\downarrow$) is denoted if the average value is less than the total average. That is, there are eight combinations of RFM model. Following Wei et al. [8], R is defined as the number of days since the last visit when the first day of the specified time period is set to one. Larger R value indicates that the customer has visited the organization more recent. Four major types pf customers based on Ha and Park [19] include lost customers ($R\downarrow F\downarrow M\downarrow$ and $R\downarrow F\uparrow M\uparrow$), new customers ($R\uparrow F\downarrow M\downarrow$), loyal customers ($R\uparrow F\uparrow M\uparrow$), and promising customers ($R\uparrow F\downarrow M\uparrow$). In addition, the customer value matrix using F and M variables can group the customers into best customers ($F\uparrow M\uparrow$), spender customers ($F\downarrow M\uparrow$), uncertain customers ($F\downarrow M\downarrow$), and frequent customers ($F\uparrow M\downarrow$) as shown in Fig. 1 [20]. With different RFM combinations, customer value can be evaluated and, more importantly, different marketing strategies can be designed to meet different customer needs.

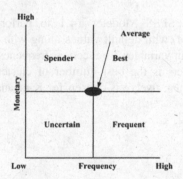

Fig. 1. Customer value matrix

3 A Case Study of a Veterinary Hospital in Taiwan

The purpose of this study is to identify valuable customers in a veterinary hospital in Taiwan. The transactions data set which focuses solely on pet mice in 2014 from a veterinary hospital located in Taichung City, Taiwan is adopted. The data set has the basic information with 175 customers, including membership numbers, owners' genders, pets' birthdays, visiting dates, monetary spent each time, and examination records. Data integration, cleaning, and transferring were implemented. Those consumptions of zero dollar records were deleted. To transform the data set into frequency, the number of purchase for a specific customer in 2014 was counted. Furthermore, if a customer visits more than once in a specific day, this study regards that the customer only has one visit. The definitions of the RFM model are as follows. Recency (R) is defined as the number of days since the last visit in 2014. This paper coded a value of one to January 1, 2014, a value of two to January 2, 2014, and a value of 365 to December 31, 2014. Accordingly, the values of recency are between 1 and 365. Frequency (F) is defined as the number of visit in 2014. Monetary (M) is defined as the total money spent in 2014.

Table 1 reports the descriptive statistics of recency, frequency, and monetary and the symbol of R, F, and M above the average. Larger variables (R, F, and M) represent the customer visits more recent, visits more frequent, and spends more money, respectively. The maximum and minimum values of recency are 363 and 6, respectively. The maximum and minimum frequency values are 9 and 1, respectively. The respective maximum and minimum monetary values are $4,714 and $60, respectively (reported in New Taiwan Dollars).

Table 1. The descriptions of recency, frequency, and monetary

	Maximum	Minimum	Average	Standard deviation
Recency	363	6	189.51	102.79
Frequency	9	1	1.71	1.31
Monetary	4,714	60	768.07	756.49

This study uses IBM SPSS Modeler 14.1 to perform cluster analysis. The "Kohonen node" (i.e., SOM) with default values along with the Kohonen mode set to "simple" are used. The input variables are recency, frequency, and monetary. Figure 2 indicates that twelve clusters is the best number of clusters among 175 customers recommended by SOM. Then, twelve is chosen for K-means method.

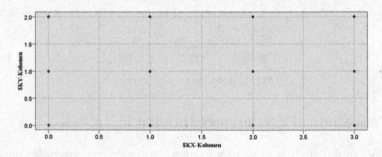

Fig. 2. Twelve clusters generated by SOM technique

4 Marketing Implications

Table 2 shows the descriptive statistics of twelve clusters based on sample size, average numbers of R, F, and M of the customers, and the symbol(s) of R, F, and M greater than the averages of R, F, and M. These twelve clusters can be further classified into four types including RFM, FM, R, and none. Most of the customers are in Clusters 2, 5, 8, 9, and 12. Five clusters (Clusters 1, 3, 7, 10, and 11) with all R, F, and M values greater than the average values are the customers providing high profitability to this veterinary hospital, regarded as loyal customers. Clusters 8 and 12 with the symbol of R↑F↓M↓ are new customers in this veterinary hospital. In contrast, Clusters 4, 5, and 6 have higher F and M values but lower R values (R↓F↑M↑) and Clusters 2 and 9 with the symbol of R↓F↓M↓ are all regarded as lost customers.

Table 2. Descriptive statistics of twelve clusters

Cluster	Sample size	Average R	Average F	Average M	Item(s) above average
1	1	347.00	9.00	4,714.00	RFM
2	28	39.71	1.18	478.29	–
3	8	325.38	2.75	918.50	RFM
4	1	184.00	2.00	4,658.00	FM
5	22	144.91	2.36	1,157.59	FM
6	2	37.50	3.00	3,132.00	FM
7	4	295.50	6.50	2,435.25	RFM
8	37	218.35	1.30	563.95	R
9	35	133.20	1.09	355.97	–
10	5	228.40	3.80	1,977.60	RFM
11	3	363.00	5.33	2,519.67	RFM
12	29	327.66	1.00	415.72	R
Average		189.51	1.71	768.07	

To simplify the RFM analysis, two types of customers based on the customer value matrix are identified, i.e., best customers (F↑M↑) and uncertain customers (F↓M↓). Clusters 1, 3, 4, 5, 6, 7, 10, and 11 are classified as best customers, whereas Clusters 2, 8, 9, and 12 are classified as uncertain customers (Fig. 3).

Clusters 1, 3, 7, 10, and 11 are defined as the best or loyal customers, indicating that these customers are core customers. This veterinary hospital can develop customized services and products to meet their unique needs via observing their transactions to understand their shopping habits. For example, the veterinary hospital can offer compassionate and personalized care and delicate food for the pets of the customers so that the customers can enjoy their visit here without worrying about their pets. It is suggested that enhancing customer perceptions of services and products can achieve CRM effectively [21]. Besides, the veterinary hospital can provide VIP discounts for the customers when they purchase particular product and service packages, such as dental service package (covering teeth cleaning and other medical

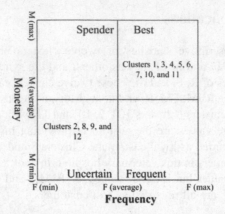

Fig. 3. Distributions of twelve clusters in customer value matrix

examination). Moreover, the veterinary hospital can extend its hospital operating time for VIP customers based on their demand.

In contrary to the similarities, Clusters 4, 5, and 6 have significant differences. Clusters 4, 5, and 6 are classified either the best or lost customers and Clusters 2 and 9 are classified either uncertain or lost customers. Uncertain customers include lost customers and new customers. Although Clusters 4, 5, and 6 are lost customers, they are also best customers due to the high visiting frequency and high monetary spent. This reveals that this veterinary hospital must target and retain the customers by adopting particular marketing strategies to mend fences with them. That is, the veterinary hospital can provide particular promotional channels for them such as holding particular activities to invite them to participate and sending mobile phone text messages regarding particular price discounts for limited time only. It is necessary for the veterinary hospital to examine the reasons why the customers are not interested in their services and products and further utilize the feedback information to restore their interests to regain them.

Clusters 2 and 9 are classified either uncertain customers or lost customers. The veterinary hospital can still further strengthen customer relationships with them because they are likely to be interested in the services and products in the future. As the customers only can provide limited contributions, the veterinary hospital should reduce marketing resource allocation on them. Cheap and convenient marketing channels become the first choice, such as informing the promotional news via instant messaging software. In addition, Clusters 8 and 12 are classified either uncertain customers or new customers. Despite of little visit and consuming amount, the customers visit recently and still have the potential to become the core customers in the future.

The veterinary hospital can motivate the lost and new customers to come often and consume more by offering preferential price discounts for bundled service and product packages such as dental services and geriatric pet care. In addition, the veterinary hospital can enhance its brand by actively participating in community events to enhance its awareness in the community. The brand enhancement can attract new customers and regain lost customers. Therefore, lost or new customers are likely to become other types of customers when the veterinary hospital aims to improve particular dimensions of RFM model by making particular marketing strategies. Obviously, a particular type of customers is likely to be upgraded to other types of customers when particular dimensions of RFM model are improved. For example, customers in Cluster 4, 5, and 6 can be upgraded to the loyal customers when R can be increased.

5 Conclusions

This study adopts a two-stage cluster analysis by combining SOM and K-means method to analyze the transactions data focusing on pet mice in 2014 from a veterinary hospital located in Taichung City, Taiwan. Twelve clusters have been found based on 175 customers. In order to develop marketing strategies, this study first compares average RFM values of each cluster to the total average RFM values of all clusters based on the study of Ha and Park [19] to segment customers. The customers can be grouped into loyal customers with the symbol of $R\uparrow F\uparrow M\uparrow$, lost customers with the symbol of $R\downarrow F\downarrow M\downarrow$ or $R\downarrow F\uparrow M\uparrow$, and new customers with the symbol of $R\uparrow F\downarrow M\downarrow$. Second, customers are further grouped based on the customer value matrix by developing a 2×2 matrix combining R and M. The customers can be classified into two types of customers, i.e., best customers with the symbol of $F\uparrow M\uparrow$ and uncertain customers with the symbol of $F\downarrow M\downarrow$.

Considering these two types of analyses for segmenting customers, this study develops different marketing strategies for the veterinary hospital so as to keep good relationships with the loyal customers, regain lost customers, and attract new customers. It is necessary to pay attention to the adoption of different RFM variables which leads to different viewpoints in examining customer value. Clusters 4, 5, and 6 are regarded as best customers based on the suggestions of Ha and Park [19] whereas these clusters are regarded as lost customers when Marcus's [20] suggestions are considered. Besides, Clusters 8 and 12 have relatively higher R value, revealing that the customers have visited this veterinary hospital more recent and thus can be regarded as new customers. However, from the viewpoint of Ha and Park [19], Clusters 8 and 12 are regarded as uncertain customers.

More insights from the customer types for this veterinary hospital can be obtained when different variables are included in the analysis. In order to offer further marketing implications, future studies can analyze more customer demographic features such as age and gender or even their real viewpoint of the veterinary hospital such as their satisfaction for the veterinary hospital. The implications of this study may be limited due to the examination on customer transactions data that are only related to recency, frequency, and monetary. Moreover, the results of this study are likely not to be generalized to the pet markets outside Taiwan because of the differences in consuming habits of customers and the culture.

References

1. Massey, A.P., Montoya-Weiss, M.M., Holcom, K.: Re-engineering the customer relationship: leveraging knowledge assets at IBM. Decis. Support Syst. **32**, 155–170 (2001)
2. Chang, E.C., Huang, H.C., Wu, H.H.: Using K-means method and spectral clustering technique in an outfitter's value analysis. Qual. Quant. **44**(4), 807–815 (2010)
3. Yang, Y.F.: Service capabilities and customer relationship management: an investigation of the banks in Taiwan. Serv. Ind. J. **32**(6), 1–24 (2011)
4. Mosteller, J.: Animal-companion extremes and underlying consumer themes. J. Bus. Res. **61**, 512–521 (2008)
5. Chen, A., Hung, K.P., Peng, N.: A cluster analysis examination of pet owners' consumption values and behavior – segmenting owners strategically. J. Target. Meas. Anal. Mark. **20**(2), 117–132 (2012)
6. Hsiao, T.C., Shieh, S.L., Chen, T.L., Liu, C.H., Yeh, Y.C.: Data analysis of medical records in veterinary hospital using clustering method and association rule. Appl. Math. Inform. Sci. **9**(6), 3319–3326 (2015)
7. Mckechnie, S.: Integrating intelligent systems into marketing to support market segmentation decisions. Intel. Syst. Account. Finance Manage. **14**, 117–127 (2006)
8. Wei, J.T., Lin, S.Y., Weng, C.C., Wu, H.H.: A case study of applying LRFM model in market segmentation of a children's dental clinic. Expert Syst. Appl. **39**(5), 5529–5533 (2012)
9. Wang, S.: Cluster analysis using a validated self-organizing method: cases of problem identification. Int. J. Intel. Syst. Account. Finance Manage. **10**(2), 127–138 (2001)
10. Fish, K., Ruby, P.: An artificial intelligence foreign market screening method for small businesses. Int. J. Entrep. **13**, 65–81 (2009)
11. Larose, D.T.: Discovering Knowledge in Data: An Introduction to Data Mining. Wiley, Hoboken (2005)
12. Cox, D.R.: Note on grouping. J. Am. Stat. Assoc. **52**, 543–547 (1957)
13. Fisher, W.D.: On gouping for maximum homogeneity. J. Am. Stat. Assoc. **53**(284), 789–798 (1958)
14. Kuo, R.J., Ho, L.M., Hu, C.M.: Integration of self-organizing feature map and K-means algorithm for market segmentation. Comput. Oper. Res. **29**(11), 1475–1493 (2002)
15. Chiu, C.Y., Chen, Y.F., Kuo, I.T., Ku, H.C.: An intelligent market segmentation system using K-means and particle swarm optimization. Expert Syst. Appl. **36**, 4558–4565 (2009)
16. Wu, H.H., Lin, S.Y., Liu, C.W.: Analyzing Patients' Values by Applying Cluster Analysis and LRFM Model in A Pediatric Dental Clinic in Taiwan. Sci. World J. **2014**, 685495 (2014)
17. Hughes, A.M.: Strategic Database Marketing. McGraw-Hill, New York (1994)
18. Wei, J.T., Lin, S.Y., Wu, H.H.: A review of the application of RFM model. Afr. J. Bus. Manage. **4**(19), 4199–4206 (2010)
19. Ha, S.H., Park, S.C.: Application of data mining tools to hotel data mart on the intranet for database marketing. Expert Syst. Appl. **15**(1), 1–31 (1998)
20. Marcus, C.: A practical yet meaningful approach to customer segmentation. J. Consum. Mark. **15**(5), 494–504 (1998)
21. Ince, T., Bowen, D.: Consumer satisfaction and services: insights from dive tourism. Serv. Ind. J. **31**(11), 1769–1792 (2011)

Schedule and Sequence Analysis

Modeling Inter-country Connection
from Geotagged News Reports:
A Time-Series Analysis

Yihong Yuan[(✉)]

Department of Geography, Texas State University, San Marcos, TX 78666, USA
yuan@txstate.edu

Abstract. The rapid development of big data techniques provides growing opportunities to investigate large-scale events that emerge over space and time. This research utilizes a unique open-access dataset, "The Global Data on Events, Location and Tone" (GDELT), to model how China has connected to the rest of the world, as well as predicting how this connection may evolve over time based on an autoregressive integrated moving average (ARIMA) model. Methodologically, we examined the effectiveness of traditional time series models in predicting trends in long-term mass media data. Empirically, we identified various types of ARIMA models to depict the connection patterns between China and its top 15 related countries. This study demonstrates the power of applying GDELT and big data analytics to investigate informative patterns for interdisciplinary researchers, as well as provides valuable references to interpret regional patterns and international relations in the age of instant access.

Keywords: Time series analysis · ARIMA · Inter-country relations · Mass media events · GDELT

1 Introduction

In recent decades, the rapid development of techniques and theories in the big data field has introduced new challenges and opportunities to analyze the large amount of information available online [1–3], including user-contributed (personalized) information such as social media data and traditional mass media that targets a larger audience. Social media are best characterized by a series of Social Network Sites (SNS) (such as Facebook and Twitter) that have attracted worldwide users to communicate, socialize, and share their daily lives, whereas mass media refers to various forms of media technologies that aim to reach a large audience via mass communication, including broadcast, print, film, and new channels developed with the growth of the world wide web (WWW), such as online news reports [4]. Although many studies have focused on how user-generated content has revolutionized the traditional media landscape, especially in the marketing field [5, 6], there has not been sufficient study on how these mass media datasets (e.g., massive online news archives) can be utilized to track, analyze, and model societal issues, such as the conflict and interaction between regions and countries. Realizing the necessity to explore the geographic component of

© Springer International Publishing AG 2017
Y. Tan et al. (Eds.): DMBD 2017, LNCS 10387, pp. 183–190, 2017.
DOI: 10.1007/978-3-319-61845-6_19

these geotagged news reports, this research utilizes an open-source dataset, "The Global Data on Events, Location and Tone" (GDELT), to analyze the time series of China's inter-country connections with respect to time. GDELT monitors print, broadcast, and web news media in over 100 languages worldwide and automatically encodes such data into a structured database. Although researchers in various fields such as sociology and communication have explored the potential of such data in analyzing societal events [7, 8], there is very limited research in utilizing these extracted mass media data in geography, such as analyzing the evolution of a geographic entity or the connection between geographic entities upon time [9, 10]. We adopt an autoregressive integrated moving average (ARIMA) model to analyze time series due to its capability of dealing with both Autoaggressive (AR) moving average (MA) and "Integrated" components. These models are appropriate for time series data either to better interpret the autocorrelation of the data or to forecast future points in the series [11]. Additionally, the ARIMA model is capable of dealing with non-stationary time series data, which is typically associated with long-term news events. This research concentrates on demonstrating the effectiveness of applying time series analysis to geotagged mass media data. We do not aim to interpret these patterns from a sociological or political perspective. The applied methodology can be further extended to other fields as a data pre-processing strategy, such as public relations, communications, and political geography.

2   Related Work

In the age of instant access, the wide spread usage of the Internet has introduced multiple new channels in the field of communication. On the one hand, researchers have investigated how "individual-oriented" SNS have revolutionized where, when, and how people communicate and share their daily life [12–14]. These social media platforms not only provide multiple avenues for communication, but they also generate rich data sources that allow researchers to analyze human behavior patterns from both individual and aggregated perspectives [2]. On the other hand, compared to individual-oriented social media data, traditional mass media channels, such as newspapers and TV programs, concentrate on delivering information to a larger audience [4, 15]. Researchers have realized the advantages of mass media content in its professorial nature: compared to social media, traditional mass media often addresses significant and aggregated events [16], thereby playing an important role in analyzing the social, economic, and cultural status of a society. Many researchers have applied mass media data to modeling short-term events [17], such as the response of the stock market corresponding to major social events [18]. In addition, since traditional mass media has evolved for decades (or even centuries), and the data are often collected over a longer time span, they are more appropriate for investigating long-term socio- economic trends and patterns, such as the evolution of an urban system over decades or the collective patterns of a society or the connection between societal systems [17]. For example, the machine-coded GDELT dataset [19] utilized in this research is updated daily and consists of over a quarter-billion news event records dating back to 1979. It captures what has happened/is happening worldwide [7, 17], and therefore has been utilized in many previous studies to analyze various long-term collective patterns [20]. One example study was conducted by Yonamine [8], in which the researchers constructed a predictive model to explore conflict levels in Afghanistan by incorporating various socio-economic indicators such as unemployment levels and ethnic diversity. Yuan, Liu, and Wei [17] also utilized GDELT to analyze how China was connected to other countries in the past few decades and how these patterns can be clustered into different categories. As mentioned in Sect. 1, this study focuses on constructing ARIMA time series models to predict the strength of international relations, which can be considered an extension of Yuan, Liu, and Wei [17] from a time series modeling perspective.

3 Methodology

The GDELT data used in this research include multiple columns such as the source, actors, time, and approximated location of recorded events. For instance, in a news report entitled "In Malaysia, Obama carefully calibrates message to Beijing," Actor 1 would be "United States government" and Actor 2 would be "Chinese government". The associated geographic locations of Actor 1, Actor 2, and the actual action are "Beijing, China", "Washington DC, United States", and "Kuala Lumpur, Malaysia".

As discussed in Sect. 2, this research concentrates on the inter-country relatedness between China and foreign countries. The analyses will be conducted using the following two steps:

- **Data Preprocessing**

First, we extract all news records involving China and another country as two parties. Note that the location of "action" is not a substantial factor here since an event related to a certain country can happen inside or outside of that country. Based on the pre-processed data, we calculate descriptive statistics to provide a general interpretation of the trend at various spatio-temporal scales. For each year and each country, we calculate the frequencies of "co-occurrence" with China (donated as C) in the dataset. The frequencies are noted as $F_y(i,c)$, which stands for the "co-occurrence" frequency between China and country i in year y. Here we first define connection strength as follows:

$$Co_y(i, c) = \frac{F_y(i, c)}{\sum\limits_{j \neq c} F_y(j, c)}. \tag{1}$$

where $\sum\limits_{j \neq c} F_y(j, c)$ is the total number of records that involve China and another country as two actors. Note that the connection strength is not normalized by the total occurrence of country i.

To explore the changing dynamics of this pattern, we compute the yearly connection strength between China and the top 15 countries, represented as time series data.

The following series provides an example series between United States and China, which indicates that the connection strength is 0.162 in the year 1979 and 0.179 in 2013:

US [0.162, 0.174, 0.191, 0.193, 0.189, 0.189, 0.181, 0.177, 0.174, 0.169, 0.17, 0.165, 0.162, 0.157, 0.157, 0.161, 0.17, 0.165, 0.164, 0.165, 0.166, 0.16, 0.164, 0.16, 0.159, 0.155, 0.153, 0.151, 0.153, 0.156, 0.162, 0.169, 0.175, 0.178, 0.179]

- **Modeling and interpreting time series data**

As discussed in Sect. 1, ARIMA models can be applied to both stationary and non-stationary time series data. Due to its flexibility in data processing, this research constructed ARIMA models to better interpret the summarized time series. ARIMA model is generally referred to as an ARIMA(p,d,q) model where three parameters p, d, and q are non-negative integers. They refer to the autoregressive, integrated, and moving average parts of the model respectively, and are interpreted as follows:

- p: the autoregressive parameter indicates how much the output variable depends linearly on its own previous values (e.g., how much the value in 2010 depends on the years 2009, 2008, etc.).
- d: the integrated parameter is the number of non-seasonal differences and long term trend. For instance, the random walk model $Y(t) - Y(t-1) = \mu$ (where the average difference in Y over time t is a constant, denoted by μ), since it includes (only) a non-seasonal difference and a constant term, is classified as an "ARIMA(0,1,0) model with constant."
- q: the order of lagged forecast errors in the prediction. For instance, if series μ_t can be represented by the weighted average of q white noise patterns (Eq. 1, where ε_t are white noise series, $\theta_1 \ldots \theta_q$ are constants), then μ_t corresponds to ARIMA (0,0,q). q can be interpreted as a level of uncertainty in time series analysis:

$$\mu_t = \varepsilon_t + \theta_1 \varepsilon_{t-1} + \cdots \theta_q \varepsilon_{t-q}. \tag{2}$$

The construction of ARIMA models provides quantitative evidence of how the inter-nation connection of China has changed upon time, and the fitted parameters can be applied for predictions and estimates of future patterns.

4 Results and Discussion

The ARIMA models are constructed based on the yearly connection strength defined in Sect. 3 step 1. Table 1 presents the models and fitted results. To test the effectiveness of the models, we utilized data from 1979–2010 as a training set and the years 2011, 2012, and 2013 as a testing set for model validation.

Table 1. ARIMA models and predicted results ('Obs.' indicates observed data)

Country		ARIMA model	Fitted 2011	Obs. 2011	Fitted 2012	Obs. 2012	Fitted 2013	Obs. 2013
United States	US	(1,1,0)	0.172	0.175	0.1732	0.1781	0.1737	0.1792
Japan	JA	(1,0,0)	0.0991	0.0946	0.0984	0.0927	0.0977	0.0929
Russia	RS	(1,0,0)	0.0858	0.0806	0.087	0.0806	0.0881	0.0808
South Korea	KS	(0,1,1)	0.0525	0.0493	0.054	0.0467	0.0555	0.0465
North Korea	KN	(0,1,1)	0.0488	0.0455	0.0503	0.0423	0.0517	0.0424
United Kingdom	UK	(1,0,2)	0.0446	0.0409	0.0458	0.0413	0.0483	0.0414
France	FR	(0,1,0)	0.0295	0.0292	0.0289	0.029	0.0285	0.029
Iran	IR	(0,1,0)	0.0225	0.0215	0.0232	0.0235	0.0239	0.0238
Pakistan	PK	(2,0,0)	0.0242	0.0249	0.0234	0.0236	0.022	0.0236
India	IN	(1,1,0)	0.0236	0.0226	0.0245	0.0226	0.0252	0.0227
Australia	AS	(1,1,0)	0.022	0.022	0.0226	0.022	0.0232	0.0219
Vietnam	VM	(1,2,0)	0.0191	0.0208	0.0192	0.0195	0.0198	0.0193
Germany	GM	(0,1,1)	0.0183	0.0186	0.0175	0.0184	0.017	0.0184
Philippines	RP	(0,1,1)	0.0109	0.0127	0.0139	0.0151	0.0149	0.0152
Canada	CA	(2,1,0)	0.0128	0.0124	0.0133	0.0138	0.0135	0.014

The fitted ARIMA models in Table 1 show interesting patterns. The non-zero d value (integrated parameter) for most countries indicates that a non-stationary long term trend exists in the connection between China and these countries. Figure 1 shows an example time series in South Korea showing a clear increasing trend ($d = 1$). This reflects the rapidly growing connection between China and South Korea since the 1970s.

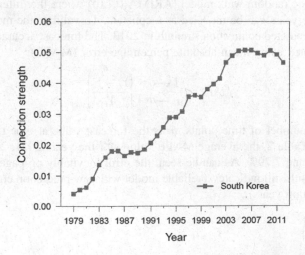

Fig. 1. Yearly connection strength between China and South Korea

Moreover, Table 2 indicates that the 15 countries can be characterized into the following categories (Table 2):

Table 2. Categorized ARIMA models and countries

Heading level	Characteristics	Countries
Autoregressive models $(p > 0, d = 0, q = 0)$	The output variable depends linearly on its own previous values	JA, RS, PK
Autoregressive integrated models $(p > 0, d > 0, q = 0)$	Autoregressive models with non-stationary behavior (e.g., long-term trend)	US, IN, AS, VM, CA
Integrated moving average models $(p = 0, d > 0, q > 0)$	For moving average models, the output variable is conceptually a linear regression of the linear combination of $q + 1$ white noise variables. Integrated moving average models is MA model with non-stationary behavior	KS, KN, GM, RP
Autoregressive moving average models $(p > 0, d = 0, q > 0)$	A combination of MA and AR models without a non-stationary component	UK
General integrated models $(p = 0, d > 0, q = 0)$	The output variable depends only on the orders of a non-stationary component	FR, IR

Table 2 indicates varying patterns between different countries and China. For instance, the connection strength between China and Russia is fitted as a stationary process, in which the connection strength for a certain year auto-correlates with the value of the previous year. However, between China and France, the connection strength is a basic random walk model (ARIMA(0,1,0)) where the difference between two consecutive years can be modeled as a constant. To validate the models, we also computed the predicted connection strength in 2013. The forecast accuracy level of the model is evaluated using mean absolute percentage error (MAPE):

$$MAPE = \frac{1}{n} \sum_{t=1}^{n} \left| \frac{Y_t - F_t}{Y_t} \right|. \tag{3}$$

where n is the number of time points, F_t is the forecast value at time t, and Y_t is the actual data. In Table 2, the average MAPE values for the years 2011, 2012, 2013 are 4.20%, 6.26%, and 7.79%. As can be seen, the error inevitably propagates over time; however, the result still indicates a reliable model with low prediction error rates (<8% for all three testing years).

5 Conclusion

This paper applied the GDELT dataset to examine the connection between China and foreign countries based on time series analysis. We examined the effectiveness of ARIMA models in predicting trends in long-term mass media data. Although ARIMA has been previously applied in fields such as political geography and communication, its utility for determining inter-country relations in the big data era is limited. We also demonstrated the power of applying GDELT and big data techniques to investigate informative patterns for interdisciplinary researchers. This research does not aim to provide in-depth interpretation of the causes and consequences of these international events from a political perspective; instead, we proposed a method to discover the patterns that can provide insights in different research fields.

Potential future directions include extending this method to other countries to test its robustness. GDELT provides a rich data source to analyze inter-region relations at various spatial scales, such as investigating the connection between different provinces in China. Another valuable direction is to compare the performance of mass media and social media in characterizing urban-level patterns. Future study can also look into the correlation between connection strength and various demographic variables such as population, economic status, and the tone of each event record.

References

1. Eagle, N., Pentland, A., Lazer, D.: Inferring friendship network structure by using mobile phone data. Proc. Natl. Acad. Sci. USA **106**, 15274–15278 (2009)
2. Liben-Nowell, D., Novak, J., Kumar, R., Raghavan, P., Tomkins, A.: Geographic routing in social networks. Proc. Natl. Acad. Sci. USA **102**, 11623–11628 (2005)
3. Yuan, Y., Liu, Y.: Exploring inter-country connection in mass media: a case study of China. In: International Conference on Location-based Social Media, Athens, Georgia (2015)
4. Mazzitello, K.I., Candia, J., Dossetti, V.: Effects of mass media and cultural drift in a model for social influence. Int. J. Mod. Phys. C **18**, 1475–1482 (2007)
5. Stephen, A., Galak, J.: The effects of traditional and social earned media on sales: a study of a microlending marketplace. J. Mark. Res. **49**, 624–639 (2012)
6. Meraz, S.: Is there an elite hold? Traditional media to social media agenda setting influence in blogs networks. J. Comput. Mediated Commun. **14**, 682–707 (2009)
7. Leetaru, K., Schrodt, P.: GDELT: global data on events, language, and tone, 1979–2012. In: International Studies Association Annual Conference, San Diego, CA (2013)
8. Yonamine, J.E.: Predicting future levels of violence in Afghanistan district using GDELT. UT Dallas (2013)
9. Cohen, S.B., Cohen, S.B.: Geopolitics: The Geography of International Relations. Rowman & Littlefield, Lanham (2009)
10. Liu, Y., Wang, F.H., Kang, C.G., Gao, Y., Lu, Y.M.: Analyzing relatedness by toponym co-occurrences on web pages. Trans. GIS **18**, 89–107 (2014)
11. Wilde, G.J.S.: Effects of mass-media communications on health and safety habits - an overview of issues and evidence. Addiction **88**, 983–996 (1993)
12. Gao, H., Liu, H.: Mining Human Mobility in Location-Based Social Networks. Morgan & Claypool Publisher (2015)

13. Memon, I., Chen, L., Majid, A., Lv, M.Q., Hussain, I., Chen, G.C.: Travel recommendation using geo-tagged photos in social media for tourist. Wireless Pers. Commun. **80**, 1347–1362 (2015)

14. Wu, L., Zhi, Y., Sui, Z.W., Liu, Y.: Intra-urban human mobility and activity transition: evidence from social media check-in data. PLoS ONE **9**, e97010 (2014)

15. McQuail, D.: The influence and effects of mass media. In: Graber, D.A. (ed.) Media Power in Politics. CQ Press, Washington, D.C. (1979)

16. Liebert, R.M., Schwartzberg, N.S.: Effects of mass-media. Annu. Rev. Psychol. **28**, 141–173 (1977)

17. Yuan, Y., Liu, Y., Wei, G.: Exploring inter-country connection in mass media: a case study of China. Comput. Environ. Urban Syst. **62**, 86–96 (2017)

18. Yu, T., Jan, T., Debenham, J., Simoff, S.: Classify unexpected news impacts to stock price by incorporating time series analysis into support vector machine. In: 2006 IEEE International Joint Conference on Neural Network Proceedings, vols. 1–10, pp. 2993–2998 (2006)

19. Schrodt, P.: Conflict and Mediation Event Observations Event and Actor Codebook V.1.1b3. (2012)

20. Jiang, L., Mai, F.: Discovering bilateral and multilateral causal events in GDELT. In: International Conference on Social Computing, Behavioral-Cultural Modeling, & Prediction (2014)

Mining Sequential Patterns of Students' Access on Learning Management System

Leonard K.M. Poon$^{(\boxtimes)}$, Siu-Cheung Kong, Michael Y.W. Wong, and Thomas S.H. Yau

Department of Mathematics and Information Technology,
The Education University of Hong Kong, Hong Kong SAR, China
{kmpoon,sckong,mywwong,shyau}@eduhk.hk

Abstract. Novel pedagogical approaches supported by digital technologies such as blended learning and flipped classroom are prevalent in recent years. To implement such learning strategies, learning resources are often put online on learning management systems. The log data on those systems provide an excellent opportunity for getting more understanding about the students through data mining techniques. In this paper, we propose to use sequential pattern mining (SPM) to discover navigational patterns on a learning platform. We attempt to address the lack of literature support about conducting SPM on Moodle. We propose a method to apply SPM that is more appropriate for mining user navigational patterns. We further propose three sequence modeling strategies for mining patterns with educational implications. Results of a study on a statistics course show the effectiveness of the proposed method and the proposed sequence modeling strategies.

Keywords: Sequential pattern mining · Educational data mining · Learning management systems · Moodle · Navigational patterns

1 Introduction

Novel pedagogical approaches supported by digital technologies such as blended learning and flipped classroom are prevalent in recent years. To implement such learning strategies, learning resources such as lecture notes, simulations, videos, and quizzes are often put online on learning management systems (LMSs). Since the LMSs usually store information about students and log their access, they provide an excellent opportunity for getting more understanding about the students through data mining techniques.

Sequential pattern mining (SPM) [6,14] is a data mining technique for finding patterns among sequences of ordered items. It was first proposed for studying the customer purchase sequence for pattern discovery [1]. With the growing interest in e-learning and educational data mining [2,9], SPM has also been applied in education to further facilitate teaching and learning with technology.

© Springer International Publishing AG 2017
Y. Tan et al. (Eds.): DMBD 2017, LNCS 10387, pp. 191–198, 2017.
DOI: 10.1007/978-3-319-61845-6_20

Despite the report of many successful cases showing that the implementation of approaches such as blended learning is effective in enhancing students' learning, few studies have explored the ways that students navigate and interact with the learning resources in the online learning environment and the possible pedagogical implications derived from the activities of students online. One possible reason is that the volume of data acquired for the users' activities can be large. Although current LMSs such as Moodle provide different reports with user statistics, such information might not be sufficient for instructors to draw meaningful conclusion regarding the whole course or behaviors of users online [10, 12, 13]. In view of this, this study proposes to use SPM to discover sequential patterns, or navigational patterns, on LMSs.

Previous studies have attempted to apply SPM to analyze the usage patterns on various e-learning systems. Surprisingly, few studies have attempted to apply this technique to LMSs for the analysis of learning behavior of students. [15] use both association rule mining and sequential pattern mining to look for resource access patterns of students on Moodle for exam preparation. However, their study does not discuss the potential issues in the implementation of SPM on Moodle. Romero et al. [11] introduce the theoretical and practical way to apply various data mining techniques on Moodle, but the coverage on SPM is limited due to the wide scope of their paper.

This paper attempts to address the lack of literature support about conducting SPM on Moodle. We aim to discover the navigational patterns of students. We propose a method to apply SPM that is more appropriate for mining navigational patterns on LMSs. We further propose three sequence modeling strategies with reference to Zhou et al. [14] for mining patterns with educational implications. We show some results using the proposed method and strategies in a statistics course.

2   Sequential Pattern Mining

Consider a set of items such as resources on a learning management system. A *sequence* is an ordered list of items. It may indicate the order of which resources are accessed by a student. An item can occur multiple times in a sequence. The number of items in a sequence is known as its *length*. A sequence $\alpha = \langle a_1 a_2 \ldots a_n \rangle$ is called a *subsequence* of another sequence $\beta = \langle b_1 b_2 \ldots b_m \rangle$, or α is *contained* in β, if there exists integers $1 \leq j_1 \leq j_2 \leq \cdots \leq j_n \leq m$ such that $a_1 = b_{j_1}, a_2 = b_{j_2}, \ldots, a_n = b_{j_n}$. In other words, the subsequence α can be formed by removing some items from β with the order of the remaining items preserved. For example, the sequence $\langle a, c, d \rangle$ is a subsequence of the sequence $\langle a, b, c, d, e \rangle$, where a, b, c, d, and e are items. Due to brevity, we do not consider the general case where an element of a sequence can be a set of items.

Let S be a set of sequences. We refer to the number or proportion of sequences in S that contain a sequence α is known as the *support* of α in S. We denote the support as $\text{support}_S(\alpha)$ and omit the subscript when it is clear from context.

Sequential pattern mining (SPM) aims to discover frequent subsequences among a set of sequences. The frequent subsequences are also called *sequential patterns.* Specifically, given a set of sequences S and a real number $\xi \in [0,1]$ as threshold, the problem of SPM is to find all the sequences α such that $\text{support}_S(\alpha) \geq \xi$. The threshold ξ is also known as the *minimum support*.

Often we are interested in only the set of closed sequential patterns. A sequential pattern α is *closed* if it is not contained in another sequential pattern that has the same support. More formally, α is closed if there exists no other sequential pattern α' where α is a subsequence of α' and $\text{support}(\alpha) = \text{support}(\alpha')$.

As an example, consider a database with four sequences $\langle a,d,c \rangle$, $\langle a,c,d,c,b,a \rangle$, $\langle b,a,d,c \rangle$, and $\langle a,c \rangle$. The items a, b, c, and d may refer to four different resources or pages on Moodle. If the minimum support is set to be 1.0, the set of sequential patterns resulting from SPM should comprise the three sequences $\langle a \rangle$, $\langle c \rangle$, and $\langle a,c \rangle$. We can check that each of them is contained in all the sequences in the database and hence their supports are all equal to 4. As a counter example, the sequence $\langle a,d \rangle$ has a support of 3 and thus is not included in the result. The result set will be reduced to $\{\langle a,c \rangle\}$ if only closed sequence is considered.

Many algorithms have been developed to mine sequential patterns efficiently. Some of the earliest attempts, e.g. [1], are based on the well-known APRIORI algorithm for association rule mining. However, Apriori-like algorithms need to consider an exponential number of candidate sequences in the worst case and they require repeated scanning of the data set to check the support of candidate sequences [6]. Those problems make Apriori-like algorithms infeasible for large data sets or when the sequential patterns are expected to be long.

PREFIXSPAN [7], on the other hand, avoids those problems by taking another approach. Its main idea is to project the database of sequences into a set of smaller databases based on a set of frequent subsequences. The frequent subsequences are then grown and checked in the smaller projected databases separately. The projection step and the growth step are done recursively until no more longer frequent sequences are found. PREFIXSPAN has been shown empirically to be considerably faster than another Apriori-like algorithm.

SPM may result in a large number of sequential patterns. This may lead to a long processing time and make the mining results hard to understand and use. Therefore, constraints have been imposed to limit the mining results to those more interesting to users [8]. Among other constraints that have been used, three kinds of constraints are related to our work. The first kind is item constraint, which specifics a subset of items that should or should not be present in the sequential patterns. The second kind is duration constraint, which requires the duration of sequential patterns to be shorter or longer than a given period. The third kind is gap constraint, which requires the time difference between two items in sequential patterns to be shorter or longer than a given gap.

To support the duration and gap constraints, a sequence database has to been extended to contain time information. A time-extended database is defined to be a set of time-extended sequences $\alpha = \langle (t_1, a_1), (t_2, a_2), \ldots, (t_n, a_n) \rangle$, where each item a_i is annotated with a timestamp t_i. The gap between two items with consecutive indices i and j is then defined as $|t_j - t_i|$ and the duration of a sequence is defined as $|t_n - t_1|$.

Algorithms have to been adapted to respect the constraints specified. Hirate and Yamana [5] extend PREFIXSPAN to support mining sequential patterns with constraints on minimum and maximum gaps and minimum and maximum durations in a time-extended database. Their algorithm was further extended by Fournier-Viger et al. [4] to give only closed sequential patterns.

3   Mining Navigational Patterns on Moodle

Potential Issues. There are some potential issues when conducting sequential pattern analysis based on the log from LMSs. Zhou et al. [14] suggest three main challenges. The primary challenge concerns the granularity level of log data. If fine-grained events such as a single key-stroke or mouse click are recorded by the system, it may obscure the mining of patterns with coarse-grained events. The second challenge concerns the mined results from SPM algorithms. Most existing algorithms are not designed to be applied in educational context, so excessive patterns with limited relevancy and value could be generated, which also cost additional processing time. The last concerns the identification of learning strategies in the pattern analysis process. Domain knowledge would be needed during the mining and post-hoc stage in order to deduce the educational implications from the discovered patterns. In the following, we describe our proposed approach and explain how we handle those three challenges.

Context of Study. We have chosen Moodle[1] to conduct our study for two reasons. First, Moodle is a free and open-source LMS widely used in educational institutions. Second, unlike other cloud-based platforms, Moodle can be deployed on a private server allowing access to the data for data mining.

Our study was conducted in a statistics course. The main types of digital statistics resources provided on Moodle were simulations, online videos and online quizzes in three selected topics: sampling distribution, central limit theorem, and confidence interval. Students could access the learning platform for pre-learning before class. Teachers also offered some learning activities for student to conduct online after the face-to-face lessons. A total of 123 students participated in the course.

Data Extraction. In Moodle, the log data contain fine-grained information about the interaction behaviors of students on the system. Consider the quiz activity on Moodle as an example. Moodle has different descriptions in the log to record every action of students when answering the quizzes (e.g. "view the quiz", attempt the quiz, "submit the quiz", "review the quiz", etc.). If we include all those actions, we may find many trivial patterns showing sequences of work on a single quiz. Therefore, we aggregate those actions by combining consecutive

---

[1] https://moodle.org/.

Table 1. Targeted actions under study.

Behavior	Description from Moodle as in the pattern
Visited simulation	Viewed the url "[name of the simulation]" e.g. Viewed the url "[Simulation] What is sampling distribution"
Watched online videos	Viewed the page "[name of the video]" e.g. Viewed the page "[Video] What is sampling distribution"
Answered questions within the quizzes	Has viewed the attempt for the quiz "[name of the quiz]" e.g. Has viewed the quiz "[Quiz] Definition of sampling distribution"

quiz-related actions into a single quiz action. We originally recorded the access to individual questions in a quiz, but it also turned out to be excessive.

The actions (or items) we included in the sequential pattern analysis are listed in Table 1. The time at which an action is taken is used as the timestamp in the input to a SPM algorithm if it is needed.

SPM Algorithms. We used a Java library called SPMF [3] for the implementation of SPM algorithms. We consider three algorithms in our study. PREFIXSPAN is used due to its efficiency. To allow constraints, we use also the algorithm by [4], called FOURNIER08 below following the convention in SPMF. We use the maximum gap constraint to restrict the time interval between two actions in a sequential pattern to be shorter than 30 min.

An issue was discovered with FOURNIER08 during our trial. The algorithm considers the time gap as part of an item in a sequence. For example, the sequence $\langle (0, a), (1, b) \rangle$ is contained in $\langle (10, a), (11, b) \rangle$, but not in $\langle (10, a), (20, b) \rangle$ due to difference in time gap. This restriction is unrealistic in our study because it means that the actions of students must be carried out with exactly the same amount of time elapsed for that sequence to be considered as frequent.

To avoid the above issue, we propose to preprocess the timestamps as follows before we use FOURNIER08. If the time gap of an action with its previous action is smaller than 30 min, we change the timestamp to 0. Otherwise, we keep the original timestamp. We further specify the maximum duration constraint to 30 min. As a result, the proposed method ignores the time gap difference between two sequences and restricts the time gap to be less than 30 min. Note that the change voids the maximum duration constraint and the resulting sequential patterns could last longer than 30 min in the actual sequences.

Sequence Modeling Strategies. The introduction of time constraint in the SPM algorithm could effectively filter patterns which are not educationally meaningful for the interpretation of students' learning behavior online. However, the pattern discovery process could be further stratified for the exploration of specific patterns under different research contexts. Zhou et al. [14] consider this step in processing the log files before pattern discovery as *sequence modeling*. In the current study, three different sequence modeling strategies are proposed. They aim to yield results with various implications in different perspectives.

Quiz-performance sequence modeling. The students in the course are divided into different groups based on their performance in the online quizzes available in the system. The sequential patterns obtained from each group could tell the common patterns leading to good results in the course or the patterns which indicate students are not performing well in the course.

Resource oriented sequence modeling. Sometimes we are interested about the access situation of certain resources. For example the access patterns before or after watching a video. Therefore, we can include only students that have used certain resources in the analysis. The support of a sequential pattern would then refer to the proportion of students following the pattern among those who have used certain resources. An additional advantage of this filtering is the significant reduction of processing time.

Evaluation oriented sequence modeling. The integration of evaluation results based on online questionnaires and log data could provide more information for us to investigate the behavior of students on the platform. Results from questionnaires, such as those adopting a Likert-scale, could only provide an average numerical rating for reference and hence offering limited insights. However, if students are grouped based on the evaluation results (e.g. high and low ratings) before aggregating their log data for pattern discovery, the obtained patterns could further reveal how the ratings from students reflect their navigational behaviors. Teachers could be better informed about the needs of students for possible adjustments on the platform and resources.

4   Empirical Results

In this section we present some results from the study described in the previous section. We first compare the three SPM algorithms described in Fig. 1. As shown in the left chart, PREFIXSPAN is drastically slower than the other two methods with time constraints. The proposed method is slower than FOURNIER08 when the minimum support is low. The reason can be explained by the right chart. We see that the proposed method can discover much more sequential patterns with low minimum support. Besides, our experimental results show that when minimum support is set to 0.2, the length of sequential patterns found by FOURNIER08 is at most one, whereas the length of those found by the proposed method can be seven. This shows that our proposed method can successfully relax the unnecessarily restrictive constraints imposed by FOURNIER08.

We now look at the sequential patterns returned by the proposed method when minimum support was set to 0.2. Some generated patterns indicate that students often attempted a quiz after watching a video on the same topic. For example, 40% of the students attempted the quiz "Definition of sampling distribution" after watching the video "What is sampling distribution". On the other hand, there are some patterns containing only quiz activities. For example, there is a pattern with 0.2 support showing the access solely to seven different quizzes,

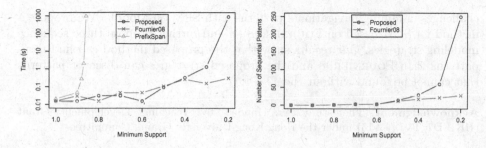

Fig. 1. Comparing the proposed method with FOURNIER08 and PREFIXSPAN.

which implies 20% of students attempted all the quizzes without taking any break longer than 30 min. This might suggest students with such pattern were unengaged in learning online and finished the quizzes in a casual manner, rather than using the quizzes and feedbacks for reflection and knowledge consolidation. These students hence deserve more attention from teachers.

We also tested two sequence modeling strategies in the study. The results show that these strategies can yield some interesting patterns.

For quiz-performance sequence modeling, the overall mean score of all quizzes were calculated and students were divided into two groups, one group with students scored above the mean and another below the mean. SPM was ran separately on both groups' data and the generated patterns were compared. The patterns show that those students scored above the mean demonstrate a better utilization of the learning resources. For example, there are two patterns "viewing a video → attempting a quiz → using statistical simulations → attempting a quiz" and "viewing a video → viewing a video → using statistical simulation" for the topic of sampling distribution with a support value of approximately 0.4. In contrast, the patterns obtained from the group of below mean score mostly consist of one or two actions only with limited implications.

Among the three types of resources, students were found to be less familiar with the simulations. We used resource oriented sequence modeling to focus on students who had used any one of the three available simulations and obtained three sets of sequential patterns. In general, the results show that students who have accessed the simulations would also access other kinds of resource. Among the 28 students who used the simulation for confidence interval, some exhibited navigational behaviors in accessing other resources within the same topic such as "viewing a video → attempting a quiz → viewing a video → attempting a quiz → using the statistical simulation". Such behaviors indicate an effective use of online resources for learning.

5   Conclusion

This paper attempts to provide some insights for researchers who are interested to investigate the navigational patterns of students on Moodle. We discuss some challenges in mining navigational patterns with SPM on Moodle. We propose a method for SPM based on FOURNIER08 [4] and further propose three sequence modeling strategies. Our results show that the proposed method can find more patterns than FOURNIER08 and the proposed strategies can discover patterns that cannot be found without them.

Acknowledgment. The study was funded by Teaching Development Grant (HKIED7/T&L/12-15) under the Hong Kong University Grants Committee.

References

1. Agrawal, R., Srikant, R.: Mining sequential patterns. In: ICDE (1995)
2. ElAtia, S., Ipperciel, D., Zaïane, O.R. (eds.): Data Mining and Learning Analytics: Applications in Educational Research. Wiley, Hoboken (2016)
3. Fournier-Viger, P., Gomariz, A., Gueniche, T., Soltani, A., Wu, C.W., Tseng, V.S.: SPMF: a java open-source pattern mining library. J. Mach. Learn. Res. **15**, 3569–3573 (2014)
4. Fournier-Viger, P., Nkambou, R., Nguifo, E.M.: A knowledge discovery framework for learning task models from user interactions in intelligent tutoring systems. In: Gelbukh, A., Morales, E.F. (eds.) MICAI 2008. LNCS (LNAI), vol. 5317, pp. 765–778. Springer, Heidelberg (2008). doi:10.1007/978-3-540-88636-5_72
5. Hirate, Y., Yamana, H.: Generalized sequential pattern mining with item intervals. J. Comput. **1**(3), 51–60 (2006)
6. Mooney, C.H., Roddick, J.F.: Sequential pattern mining - approaches and algorithms. ACM Comput. Surv. **45**(2), 19:1–19:39 (2013)
7. Pei, J., Han, J., Mortazavi-Asl, B., Pinto, H.: PrefixSpan: mining sequential patterns efficiently by prefix-projected pattern growth. In: ICDE (2001)
8. Pei, J., Han, J., Wang, W.: Constraint-based sequential pattern mining: the pattern-growth methods. J. Intell. Inf. Syst. **28**, 133–160 (2007)
9. Peña-Ayala, A. (ed.): Educational Data Mining: Applications and Trends. Springer, Cham (2014)
10. Psaromiligkos, Y., Orfanidou, M., Kytagias, C., Zafiri, E.: Mining log data for the analysis of learners' behaviour in web-based learning management systems. Oper. Res. Int. J. **11**(2), 187–200 (2011)
11. Romero, C., Ventura, S., García, E.: Data mining in course management systems: Moodle case study and tutorial. Comput. Educ. **51**(1), 368–384 (2008)
12. Valsamidis, S., Kontogiannis, S., Kazanidis, I., Karakos, A.: E-learning platform usage analysis. Interdiscip. J. E-Learn. Learn. Objects **7**(1), 185–204 (2011)
13. Zaiane, O.R., Luo, J.: Towards evaluating learners' behaviour in a web-based distance learning environment. In: Proceedings of IEEE International Conference on Advanced Learning Technologies, pp. 357–360 (2001)
14. Zhou, M., Xu, Y., Nesbit, J.C., Winne, P.H.: Sequential pattern analysis of learning logs: methodology and applications. In: Handbook of Educational Data Mining, pp. 107–121. CRC Press (2010)
15. Ziebarth, S., Chounta, I.-A., Hoppe, H.U.: Resource access patterns in exam preparation activities. In: Conole, G., Klobučar, T., Rensing, C., Konert, J., Lavoué, É. (eds.) EC-TEL 2015. LNCS, vol. 9307, pp. 497–502. Springer, Cham (2015). doi:10.1007/978-3-319-24258-3_46

Design of a Dynamic Electricity Trade Scheduler Based on Genetic Algorithms

Junghoon Lee and Gyung-Leen Park[✉]

Department of Computer Science and Statistics, Jeju National University,
Jeju City, Republic of Korea
{jhlee,glpark}@jejunu.ac.kr

Abstract. This paper presents a design and measures the performance of a dynamic electricity trade scheduler employing genetic algorithms for the convenient application of vehicle-to-grid services. Arriving at and being plugged-in to a microgrid, each electric vehicle specifies its stay time and sales amount, while the scheduler, invoked before each time slot, creates a connection schedule considering the microgrid-side demand and available electricity from vehicles for the given scheduling window. For the application of genetic operations, each schedule is encoded to an integer-valued vector with the complementary definition of C-space, which orderly lists all combinatory allocation maps for a task. Then, each integer element indexes a map entry in its C-space. The performance measurement result, obtained from a prototype implementation, reveals that our scheduler can stably work even when the number of sellers exceeds 100 as well as improves demand meet ratio by up to 6.3% compared with the conventional scheduler for the given parameter set.

Keywords: Vehicle-to-grid · Dynamic schedule · Genetic algorithm · Encoding scheme · Demand meet ratio

1 Introduction

Electric vehicles, or EVs in short, make the transportation system as a part of the power network, as they obtain energy for driving from the grid [1]. The energy comes not from gasoline or diesel engines but from nuclear, thermal, or other power plants. Moreover, EV batteries can alleviate the time disparity problem between generation and consumption of renewable energies. In the meantime, EVs can send electricity stored in their batteries, back to the grid, enabled by two-way communication and bidirectional energy flow. Called V2G (Vehicle-to-Grid), this technology can shave the peak load over the grid. Here, EVs can be charged during the low-load interval and inject electricity to the grid during the peak-load interval. As power plants are built to catch up with the peak energy

This research was supported by Korea Electric Power Corporation through Korea Electrical Engineering & Science Research Institute. (Grant number: R15XA03-62).

© Springer International Publishing AG 2017
Y. Tan et al. (Eds.): DMBD 2017, LNCS 10387, pp. 199–208, 2017.
DOI: 10.1007/978-3-319-61845-6_21

demand across the community, such V2G capability can suppress the construction of new power generation facilities, bringing many benefits in environmental and economic aspects [2].

However, as indicated in [3], V2G poses several critical problems stemmed from increased uncertainty in day-to-day system operations and extended potential disturbance in the power network. Here, uncontrolled charging or discharging can possibly lead to voltage instability and other malicious situations. Hence, the penetration of V2G strongly requires sophisticated control and coordination strategies for the safe operation of the power system. V2G coordination must put emphasis on power balancing, that is, power generation and consumption must be always equal, for EVs to be invited in energy markets [4]. Here, it is important to make EVs send electricity to the grid at an optimal location at an optimal time interval [5]. The coordination can take fully advantage of real-time interaction mechanisms supported by modern information and communication technologies just like other smart grid services [6].

In V2G electricity trades, microgrids having their own battery systems will be active purchasers in practice. They can buy electricity by autonomously selecting line frequencies to control the direction of electricity flow as well as making an appropriate reward plan to EV owners [7]. Microgrids essentially comprise distributed generators, storage devices, and smart meters, possibly making a decision on whether they will buy electricity from the main grid, EVs, or other renewable energy sources [8]. Here, the energy supply strategy can be selected upon the demand forecast created from the past data analysis. From the viewpoint of a microgrid, it wants to take EV-stored electricity, especially when it is forced by a demand response plan which imposes a high rate during peak hours. It must be mentioned that when there are many EVs to sell electricity at a specific time interval, the microgrid must decide from which EVs it will buy now and from which next according to the demand forecast.

In the meantime, the number of EVs will get larger. This situation is easily expected considering the ever-growing penetration of EVs into our daily lives and the well-available web interface allowing massive participation in Internet-based trade [9]. Then, brokering between the two parties will be a problem of severe time complexity. It is not possible to find an optimal solution for a specific goal such as minimization of the demand-supply mismatch within an acceptable response time. To solve this problem, this paper designs a genetic algorithm for scheduling V2G electricity trade. Here, it is necessary to encode a V2G trade schedule into an inter-valued vector to apply genetic operations such as crossover, selection, mutation, and fitness evaluation [10]. The encoding process may define an additional data structure to efficiently represent a trade schedule and integrate given constraints.

This paper is organized as follows: After issuing the problem in Sects. 1 and 2 reviews related work. Section 3 explains the system model and designs a genetic scheduler, focusing on the encoding scheme. Section 4 measures the performance of the proposed scheme in terms of demand meet ratio. Finally, Sect. 5 summarizes and concludes this paper with a brief introduction of future work.

2 Related Work

To begin with, as an example of energy source selection in a microgrid, M. Lopez et al.'s scheme exploits an auction to get a better price for both buyers and sellers [8]. The auction is carried out for each round until no available generation remains or no demand can be satisfied. It does not consider a scheduling mechanism integrating the next round allocation. Next, in S. Xie et al.'s work, a V2G control scheme is considered targeting at a distribution network serving a number of houses equipped with EV chargers [11]. Assuming that the limited number of EVs are plugged-in for quite a long time, for example, 20 h a day, the scheduler selects EVs to discharge or to be charged, mainly using a fleet of EVs as static spinning reserve equipment. In this approach, EVs with high SoC (State-of-Charge) will be picked to discharge and obtain a profit, by which the EV will be prioritized in charging phases.

In a survey on energy trading in smart grids, different objectives for the optimization problems are categorized into cost minimization, cost-emission minimization, power-loss minimization, and peak-to-average minimization [5]. In addition, one more important aspect is whether the schedule is carried out dynamically or a priori. On cyber-physical modeling for smart grids [12], our previous work has designed a one-day ahead electricity trade scheduler [7]. First of all, it takes the demand forecast on each time slot and the feasible arrival time provided from each EV. Then, it builds a whole day trade schedule pursuing the goal of minimizing the demand-supply mismatch. The schedule can arrange the arrival time of each EV at the microgrid. To reduce the enormous search space brought by a large number of EVs participating in the trade process, this scheme develops a heuristic which iteratively pairs the slot having the smallest number of available EVs and the EV having the least slack. Dynamic schedulers are also needed for those EVs which cannot reserve time slots for energy trading in advance.

3 Scheduling Scheme Design

3.1 System Model

Figure 1 depicts our system architecture. Upon arrival, an EV specifies its stay time and amount to sell via any available mobile terminal such as a cell phone or a tablet PC. The EV is also plugged-in to the microgrid, but the electricity does not flow yet. This sales request is stored in the predefined database table. The scheduler periodically runs once a fixed size time slot. The scheduler retrieves all requests from the table. We assume that a demand forecast model of suitable accuracy is already available. By the demand forecast, it knows the number of EVs to select for each slot and decides the EVs to sell electricity for each slot during the given window. Then, the trade coordinator allows those EVs to inject electricity to the microgrid by connecting the switch between them. Here, the EV which leaves the parking lot soon will be more prioritized to better match the demand. At the end of each slot, or just before the beginning of a new scheduler invocation, the stay time of each EV is decreased by one, while the amount-to-sell of an EV which has sold its electricity during the last slot will be updated.

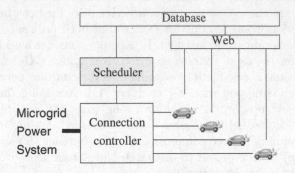

Fig. 1. Data processing framework

To make the scheduling time manageable, this paper assumes the fixed size time slot, and the slot length corresponds to the price signal change interval [13]. Each parameter is aligned with the time slot length. The stay time of an EV can be represented by the number of slots without any problem. In addition, the amount of electricity to sell can be also expressed by the number of slots, considering most power lines allow about 3 kw electricity flow per hour. If the slot length is 20 min, 1 kwh can flow during a single slot. Hence, in case an EV wants to sell 2 kwh, it is equivalent to 2 slots. We also assume that each EV decides the amount of electricity it will sell based on the current SoC and daily driving plan, as there are quite many SoC consumption models available [14]. The per-slot demand is also represented by the number of slots and it can be forecasted either for short or long term basis, mainly exploiting past consumption records.

In addition, it is not necessary to make a whole day plan as no prior knowledge is available to the scheduler on how many EVs will arrive and how much they want to sell. The schedule is required to embrace the average stay time. We set the scheduling window to 6 to 8 slots, as this paper mainly targets at microgrids such as shopping malls. The stay time or sales amount exceeding the scheduling window will be trimmed to the window size. For a slot, there can be insufficient number of EVs, while others have more than their demand. Hence, it is important to select EVs to discharge considering the demand on subsequent slots and the stay time of EVs. For an EV staying m slots and expecting to sell e slots, the scheduler has $_mC_e$ options. If the number of EVs grows, the search space will explode soon. For a shopping mall microgrid, tens or hundreds of EVs can possibly gather in a small number of time slots. It is not possible to traverse the whole search space to find the optimal solution.

3.2   Genetic Algorithm Design

Even if the time scale in the physical world is much larger than in cyberspace, the execution time approaching several minutes cannot be tolerable. Hence, a suboptimal scheme, such as genetic algorithms, is preferred to generate a reasonable quality schedule within an acceptable time bound. The genetic algorithm is one of the most widely-used suboptimal search techniques, built upon the principle of natural selection and evolution [10]. Beginning from the initial population consisting of a given number of feasible solutions, genetic iterations improve the fitness of chromosomes generation by generation. In our scheme, a feasible solution corresponds to a V2G trade schedule. Each generation is created from its parent generation by applying genetic operators such as selection, crossover, and mutation. As details on genetic operations are quite well known, we just mention that our implementation takes the roulette-wheel selection scheme and random initial population, as they empirically show a better performance.

To apply genetic operators, each trade schedule must be mapped to an integer vector, and how to encode is the core of genetic algorithm designs in a specific problem. Our encoding scheme is explained by an example shown in Fig. 2. To begin with, suppose that 4 EVs are plugged-in to the microgrid and ready to be scheduled at a specific scheduling time. They are denoted as tasks from T_0 to T_3, and the stay time and amount-to-sell of each task is listed in Fig. 2(a). Our scheduler adds one more column, named C_{num}, which is the number of combinations of the previous two column values. For example, the C_{num} field of T_0 will be $_4C_2$, namely, 6. C_{num} will be the number of feasible allocations for each task. For T_0 again, the EV will stay 4 more slots and wants to sell for 2 slots. Then, to meet its request, the scheduler needs to pick 2 slots out of 4, and there are 6 ways for this selection.

If we restrict the scheduling window to a fixed value, say 6, the number of feasible pairs of a stay time and an amount-to-sell is calculated. They will be $_1C_1, _2C_1, _2C_2, ..., _6C_5, _6C_6$. We can build C-space as a 2-dimensional matrix as depicted in Fig. 2(b). Here, only $_3C_1, _3C_2, _4C_2$, are shown due to space limitation, and each column lists all feasible combinations. As can be inferred intuitively, the number of rows is equal to C_{num}. The maximum number of rows will be 20 ($_6C_3$) when the scheduling window is 6 slots, and 70 ($_8C_4$) when 8 slots. Now, a chromosome (4, 2, 2, 1) is given in this figure and each location is associated with a task sequentially. Obviously, the vector length is equal to the number of EV tasks. The first element is an allocation for T_0 and as C_{num} of T_0 is 6, those numbers from 0 to 5 can appear at this location. This number is an index within C-space. As the number for T_0 is 4, the allocation vector in C-space is 1010. It means that electricity will flow from T_0 to the microgrid at the first and third slots. Likewise, the index of T_1 is 2, and we can find 110 at the indexed C-space of $_3C_2$.

Figure 2(c) shows the electricity trade schedule mapped from (4, 2, 2, 1). Hence, any valid trade schedule can be encoded by an integer-valued vector with an additional definition of C-space, which is built just once before starting the genetic iterations. Now, it is necessary to evaluate a trade schedule. Our approach aims at meeting the demand from the microgrid as much as possible. For a given slot by slot demand and a trade schedule, the fitness function first calculates the available electricity for each slot by summing up all rows for each

task	stay	sales	C_{num}
T_0	4	2	6
T_1	3	2	3
T_2	4	2	6
T_3	3	1	3

(a) Task set

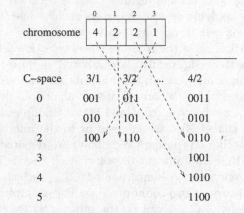

(b) Encoding scheme

slot	0	1	2	3	4	5
T_0	1	0	1	0	-	-
T_1	0	1	1	-	-	-
T_2	0	1	1	0	-	-
T_3	0	1	0	-	-	-

(c) Electricity trade schedule

Fig. 2. Example of trade schedule encoding

column. In the example of Fig. 2(c), the per-slot supply will be $(1, 3, 3, 0, 0, 0)$. If the demand is $(2, 0, 2, 3, 1, 1)$, 3 slots can be met, namely, 1 at the first slot and 2 at the third. In the second slot, the demand is 0 but the supply is uselessly 3. It is impossible to meet the demand in the 5-th and 6-th slots, as none of 4 EVs will stay at those slots.

4   Performance Measurement

This section implements a prototype of the proposed allocation scheme using a C program on an average-performance desktop computer to assess its performance. For a task, the stay time randomly distributes from 1 to the scheduling window size, which is set to 6. The amount to sell distributes exponentially with the given average. For performance comparison, Random allocation generates random schedules and evaluates to find the best one during the approximately same time interval needed to execute the proposed scheme. This selection works quite well as the candidate allocation is selected only within the valid range for each set. Particularly, if the stay time is not so long for some tasks and the number of tasks is small, this scheme can be sometimes efficient. After all, for fair comparison, the same task sets are given to both schemes in each parameter configuration.

The experiment measures how much microgrid-side demand is covered by EV-supplying energy, according to the number of EVs, population size, average amount to sell, and demand-supply ratio. For the performance metric, we define demand meet ratio, which is the ratio between the amount of bought electricity and total demand. In addition, with the given demand-supply ratio, the average of the exponential distribution of per-EV sales amount will be adjusted. Hence, even if the number of EVs increases, the total supply will remain almost constant. In the subsequent experiments, the demand meet ratio is measured changing one performance parameter with the others fixed to their default values. By default, the number of EVs is 50, the population size is 100, the per-slot demand is 1.5, and the demand-supply ratio is set to 1.0, respectively. For each parameter setting, 10 sets are generated and their results are averaged.

The first experiment measures the effect of the number of EVs ranging from 20 to 100. Here, with more EVs, the amount to sell for an EV is adjusted to make the demand-supply ratio close to 1.0. Each addition of an EV task extends the search space size by $\sqrt{3}^6$ times for the scheduling window size of 6 [15]. However, both schemes show quite stable performance behavior. It is true that they may fail in finding a reasonable solution within the limited number of iterations and random generations, but such a case is not observed. For the case of 20 EVs, both schemes find the same solution for all generated task sets, possibly equal to the optimal. However, the gap gets enlarged on a large number of EVs. The genetic scheduler shows just 4.2% degradation in the demand meet ratio even if the number of EVs changes from 20 to 100. For 100 EVs, the performance gap reaches 6.3% and more than 81.7% of demand is covered by the electricity purchase from EVs for the whole range (Fig. 3).

Next, Fig. 4 shows the effect of population size to the demand meet ratio. A large population can accommodate more candidate solutions, improving the diversity of chromosomes to mate. However, as each genetic loop necessarily includes a sorting procedure to select the solutions of better fitness, the execution time increases. In the experiment, population size is changed from 30 to 120. When population size is 120, the execution time becomes tens of seconds, and the number of random generations is made to increase to approximate the equal execution time. For each population size, different 10 sets are generated, yet the demand meet ratio for the random scheduling scheme is almost same. However, in the genetic scheduler, demand meet ratio is improved by 3.2%. Actually, a large population hardly leads to performance improvement beyond a certain point, it is necessary to find an optimal size considering the trade-off between execution time and performance.

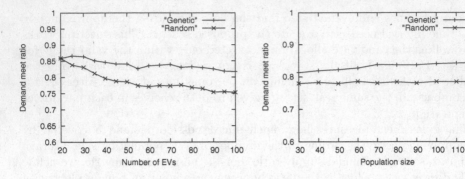

Fig. 3. Effect of the number of EVs   **Fig. 4.** Effect of population size

Figure 5 plots the result of the experiment on the effect of the sales amount to demand meet ratio. In this experiment, the average sales amount for an EV is changed from 0.5 to 3.0 slots, actually, 0.5 to 3.0 kwh. Those values are fed to the exponential distribution generator. When this average increases, the supply will increase. However, we made the demand also increase to maintain the demand-supply ratio close to 1.0, as more supply definitely leads to a better demand meet ratio. As shown in this figure, when the average sales amount is 0.5, all electricity can be purchased by the microgrid with both schemes. It indicates that both schemes work efficiently in relatively small search space. However, according to the increase in the sales amount, the demand meet ratio decreases. The genetic scheduler meets demand better, the performance gap reaching 6.2% when the average is 2 slots. Meanwhile, when the average becomes 3.0 slots, the improvement is cut down to just 1.4%.

Finally, Fig. 6 plots the result of the experiment on the effect of the sales amount to demand meet ratio. If the demand-supply ratio is 0.5, supply is twice as much as demand. In this case, both schemes meet the given demand by up to 97.6% and 96.6%, respectively, enjoying abundant supply. The demand ratio decreases as expected as supply gets smaller than demand. The genetic scheduler seems to be less affected by supply insufficiency, showing the maximum

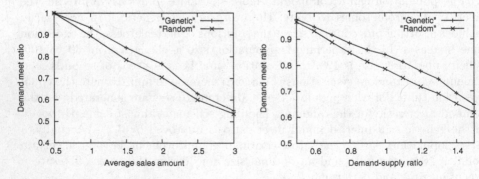

Fig. 5. Effect of average sales amount   **Fig. 6.** Effect of demand-supply ratio

performance gap of 6.2% when the demand-supply ratio is 1.1. Beyond this point, the improvement also gets smaller. In the random scheme, demand meet ratio is almost linearly cut down according to the increase in the demand-supply ratio.

5 Conclusions

Sophisticated computer algorithms can make V2G services smarter and more convenient to microgrids and EVs. In this paper, we have designed a dynamic energy trade scheduler which decides EVs to inject electricity to the microgrid on each time slot during the upcoming time interval. EV arrival patterns are not known in advance, and an EV specifies its stay time and amount to sell just when it is plugged-in to the microgrid. To build a reasonable quality schedule within an acceptable time bound, genetic algorithms are exploited and C-space is defined to encode a trade schedule. Each vector element is the index to the allocation map entry. Judging from the performance measurement result obtained from a prototype implementation, we can find out that the genetic scheduler can stably work even when the number of sellers exceeds 100, improving demand meet ratio by up to 6.3% compared with the random scheduling scheme for the given parameter set.

As future work, we are planning to conduct the performance analysis with the actual demand pattern of a microgrid and its price plan. In addition, as our region is accumulating an enormous amount of monitoring data stream from a variety of smart grid entities, a research on how to orchestrate them to find valuable information is scheduled.

References

1. Chakraborty, C., Ho-Ching Iu, H., Dah-Chuan Lu, D.: Power converters, control, and energy management for distributed generation. IEEE Trans. Ind. Electron. **62**(7), 4466–4470 (2015)
2. Bhattarai, B., Levesque, M., Maier, M., Bak-Jensen, B., Pllai, J.: Optimizing electric vehicle coordination over a heterogeneous mesh network in scaled-down smart grid testbed. IEEE Trans. Smart Grid **6**(2), 784–794 (2015)
3. Chukwu, U., Mahajan, S.: Real-time management of power systems with V2G facility for smart-grid applications. IEEE Trans. Sustain. Energy **5**(2), 558–566 (2014)
4. Kumar, L., Sivaneasan, B., Cheah, P., So, P., Wang, D.: V2G capacity estimation using dynamic EV scheduling. IEEE Trans. Smart Grid **5**(2), 1051–1060 (2014)
5. Bayram, I., Shakir, M., Abdallah, M., Qaraqe, K.: A survey on energy trading in smart grid. In: Signal and Information Processing for Energy Exchange and Intelligent Trading, pp. 258–262 (2014)
6. Ramchurn, S., Vytelingum, R., Rogers, A., Jennings, N.: Putting the 'Smarts' into the smart grid: a grand challenge for artificial intelligence. Commun. ACM **55**(4), 89–97 (2012)
7. Lee, J., Park, G.: A heuristic-based electricity trade coordination for microgrid-level V2G services. Int. J. Veh. Des. **69**(1–4), 208–223 (2015)

8. Lopez, M., Martin, S., Aguado, J., de la Torre, S.: V2G strategies for congestion management in microgrids with high penetration of electric vehicles. Electr. Power Syst. Res. **104**, 28–34 (2014)
9. Ansari, M., Al-Awami, A., Sortmme, E., Abido, M.: Coordinated bidding of ancillary services for vehicle-to-grid using fuzzy optimization. IEEE Trans. Smart Grid **6**(1), 261–270 (2015)
10. Sivanandam, S., Deepa, S.: Introduction to Genetic Algorithms. Springer, Heidelberg (2008)
11. Xie, S., Zhong, W., Xie, K., Yu, R., Zhang, Y.: Fair Energy Scheduling for Vehicle-to-Grid Networks using adaptive dynamic programming. IEEE Trans. Neural Netw. Learn. Syst. **27**, 1697–1707 (2016)
12. Liu, R., Vellaithurai, C., Biswas, S., Gamage, T., Srivasta, A.: Analyzing the cyber-physical impact of cyber events on the power grid. IEEE Trans. Smart Grid **6**(5), 2444–2453 (2015)
13. Facchinetti, T., Bibi, E., Bertogna, M.: Reducing the peak power through real-time scheduling techniques in cyber-physical energy systems. In: 1st International Workshop on Energy Aware Design and Analysis of Cyber Physical Systems (2010)
14. Kim, E., Lee, J., Shin, K.: Modeling and real-time scheduling of large-scale batteries for maximizing performance. In: IEEE Real-Time Systems Symposium, pp. 33–42 (2015)
15. Lee, J., Kim, H., Park, G., Kang, M.: Energy consumption scheduler for demand response systems in the smart grid. J. Inf. Sci. Eng. **28**(5), 955–969 (2012)

Increasing Coverage of Information Spreading in Social Networks with Supporting Seeding

Jarosław Jankowski[1]([⊠]) and Radosław Michalski[2]

[1] Faculty of Computer Science and Information Technology, West Pomeranian University of Technology, ul. Żołnierska 49, 71-210 Szczecin, Poland
jjankowski@wi.zut.edu.pl
[2] Department of Computational Intelligence, Wrocław University of Science and Technology, Wrocław, Poland
radoslaw.michalski@pwr.edu.pl

Abstract. Campaigns based on information spreading processes within online networks have become a key feature of marketing landscapes. Most research in the field has concentrated on propagation models and improving seeding strategies as a way to increase coverage. Proponents of such research usually assume selection of seed set and the initialization of the process without any additional support in following stages. The approach presented in this paper shows how initiation by seed set process can be supported by selection and activation of additional nodes within network. The relationship between the number of additional activations and the size of initial seed set is dependent on network structures and propagation parameters with the highest performance observed for networks with low average degree and smallest propagation probability in a chosen model.

Keywords: Information spreading · Supporting seeding · Viral marketing · Word of mouth · Complex networks

1 Introduction

The development of electronic platforms and social big data [1] has led to a situation in which the most crucial point to understand the patterns, behaviours, and predispositions of millions of users online [5]. Online marketers use new opportunities and put increasing effort to engage consumers in propagation of information about the products and services. In many cases, viral marketing results in better outcomes than conventional advertising campaigns due to the presence of social influence and the spreading of information among close friends [29]. Such socially-oriented recommendations have a higher influence on target consumers than conventional commercial messages owing to higher trust in such a communication received from strong ties within social network [9]. Early research in the field applied a diffusion of innovations framework and offered a macroscopic view on the quantity of acquired customers [24]. Online social networking websites allowed more detailed microscopic monitoring of the process of commercial information spreading by identifying and assessing message senders and recipients [3]. Research related to viral marketing and diffusion of

© Springer International Publishing AG 2017
Y. Tan et al. (Eds.): DMBD 2017, LNCS 10387, pp. 209–218, 2017.
DOI: 10.1007/978-3-319-61845-6_22

marketing content within complex networks has considered identification of factors affecting successful campaigns [2, 8], modelling diffusion processes with the use of epidemic models and extensions [13], selection of initial seed sets for campaign initialization [7]. More recent research has been related to spreading processes within temporal networks [20, 28] and multilayer structures [25]. While most of the earlier research concentrated on initial seed set selection [7], the research presented in this paper focuses on supporting seeding during the process and improvement of coverage. Our study assumes differing intensity of support during the process and comparison of results with reference based on primary seeding only. The rest of the paper is organized as follows: Sect. 2 provides a review of literature,

Section 3 presents the conceptual framework and assumptions for the proposed approach. Next, in experimental part, empirical results are presented, and Sect. 5 concludes the work.

2 Related Work

The growing number of users of social networking platforms has increased the interest of marketers in campaigns targeting users as well as in social big data analysis [1]. Differing from typical forms of contextual advertising or the use of demographics and behavioural targeting, marketers try to motivate their users to spread information about products among friends within social connections. Research in this field follows an interdisciplinary approach and attracts marketers, computer scientists, physicists and sociologists with a broad scope of research aims and approaches [7, 9, 13]. For modelling the spread of information models from epidemic research such as SIR or SIS are implemented [13], as well more dedicated solutions such as independent cascades model [14] or linear threshold [24]. Macroscopic analysis of total number of activations within network is used as well as microscopic view for detailed monitoring of processes at the social network and its participants level [23].

Several directions of current research in the field can be identified. A bulk of studies have been related to selection of network nodes in a form of seed set to initiate the spreading processes. Such approaches are often based on structural network measures such as degree centrality, closeness or betweenness centrality for creation of rankings of nodes and selection of top nodes with assumed high potential of spreading information due to important role within network structures [7]. Due to limited computational resources required, such approaches are often used albeit failing to deliver optimal seed set. More sophisticated solutions such as greedy based selection and its extensions deliver better results but their usage is associated with substantial computational costs and very limited ability of implementation on real complex networks [14]. Other research further optimizes the usage of structural measures to refrain from selecting nodes in the same segments of network for better allocation of seeds. Solutions of this type are based on sequential seeding for better usage of natural diffusion processes [10], sequential seeding with dynamic rankings [11], targeting communities to avoid seeding of nodes within same communities with close intra connections [6] a usage of voting mechanisms with decreased weighs after detection of already activated nodes [30]. In other studies, a k-shell based approach was implemented to detect central nodes within the networks [15].

Most of earlier approaches are based on static networks and modelling within them dynamic processes while recent studies take into account dynamic networks with used temporal characteristics and closer to reality specifics than static snapshots [20]. Other directions are based on multilayer networks and intra layer information spreading processes [25]. While most of solutions are targeted to processes initiated by seed set, attempts are observed to use additional knowledge about the ongoing process to improve results in a form of adaptive seeding [26]. Other approaches take into account multiple ongoing campaigns and relations between them [4].

Solutions based on single stage seeding are simplified representation of real world marketing campaigns where marketers divide marketing budgets into several stages for better allocations based on campaign monitoring. Then more natural and close to reality solutions should assume usage of initial budgets and then implementing supporting activities during campaign.

3 Conceptual Framework

The review of literature presented shows that most of research concentrates on seed set selection on the beginning of the process with further goal to maximize final coverage at the end of the process. Real viral marketing campaigns started by initial seeding are subject of monitoring and companies are using various mechanisms to improve the results including teasers, incentives during campaign and other techniques to attract more users within networks. Techniques of this type can be based on selection of additional seeds within network apart from initial seed set. The main goal of research presented was to evaluate results with the use of additional seeding during information spreading process. The main assumptions for research are presented in Fig. 1 with distinguished primary seeding (PD) denoted by red line and secondary seeding (SS) with three variants of intensity denoted with green, light blue and blue.

Conventional information spreading process initiated by seed set is illustrated in Fig. 1A. Primary seeding PS is based on p selected nodes from the network with high potential to influence others. Various strategies can be used to select initial seed set discussed in the literature based on centrality measures or more sophisticated approaches like k-shell, VoteRank or community based seeding. Primary seeding initiates information spreading process and it continues until process terminates after S_{PS} steps with achieved coverage C_{PS}. Results from process based on primary seeding are treated as reference for further research in terms of coverage and number of steps till process end. The goal of research is analyse effect of additional seeding on the process with various intensities. It is assumed additional seeding is based on $SF * p$ seeds (SF denotes support intensity factor as a percentage of number of seeds used in primary seeding) divided by the number of steps S_{PS} in which process terminates. Additional seeds are selected with the use of same seed selection strategy as in the primary seeding. If primary seeding was based on $p = 100$ seeds and process terminates in $S_{PS} = 5$ steps and support intensity factor $SF_1 = 50\%$ then supporting seeding SS takes place in 5 steps and in each $s_1 = 10$ seeds are used as is illustrated in Fig. 1B.

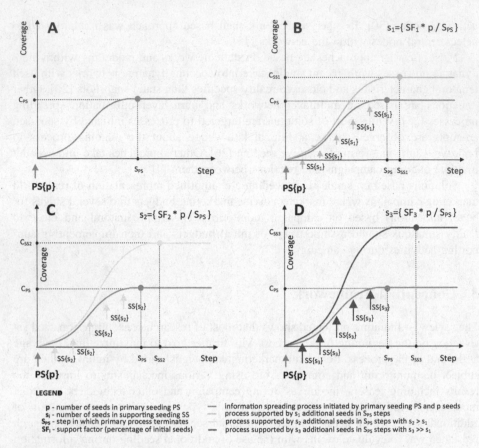

Fig. 1. Information spreading processes based on primary and secondary seeding (Color figure online)

Due to additional support process reaches coverage $C_{SS1} > C_{PS}$ and the duration of the process $S_{SS1} > S_{PS}$. Higher intensity of support is illustrated in the Fig. 1C where $SF_2 = 75\%$ and s_2 seeds are used in each step. Higher coverage and the duration of the process is reached. If the intensity is increased the further growth of coverage and duration is assumed as illustrated in Fig. 1D. Following the above assumptions, the main goal of research is to evaluate effect of additions seeds on the ongoing information spreading processes with the use of real social networks, various characteristics of information spreading processes and seed selection strategies used for seed selection.

4   The Empirical Study

In the study, experiments were conducted with the use of agent based simulations with the main goal to analyse the effect of number of additional activations within network on final coverage represented by a number of affected nodes. As a methodological basement for simulations, the independent cascades model (IC) was used with five propagation probabilities PP of activation of contacted neighbours during the process with assigned values 0.05, 0.1, 0.15, 0.2 and 0.25. All parameters used in simulations related to information propagation process, networks, and strategies for seeds selection are presented in Table 1.

Table 1. Parameters used for simulations

Symbol	Parameter	Variants	Values
N	Network	8	Real networks N1–N8 from various areas
SP	Initial seeding percentage	1	1%
A	Percent of additional activations	11	10%, 20%, 30%, 40%, 50%, 60%, 70%, 80%, 90%, 100% and 0% as reference
PP	Propagation probability	5	0.05, 0.1, 0.15, 0.2, 0.25
R	Seed ranking method	5	R - random, D - degree, P - PageRank E - Eigenvector, B - Betweenness

Parameters of simulations create experimental space N × SP × A × P × R with 2200 simulation configurations with results averaged from 100 runs on each configuration. Simulations seed selection approaches are based on typical heuristics for the selection of nodes within network with high potential to influence others such as degree based selection (D), PageRank based selection (P), eigenvector based selection (E), betweenness (B) and for comparison random selection (R). Seeding percentage SP represents the initial number of seed selected at the beginning to initiate the information spreading process. For the experiments, eight networks were used with the number of nodes in the range 4039–16264 with specification presented in Table 2. For each network average values of man parameters such as degree (D), second level degree (D2), closeness (CL), PageRank (PR), eigenvector (EV), clustering coefficient (CC) affecting the dynamics of information spreading processes are presented. The number of nodes in each network was enough to initiate diffusion processes with used seeding percentage and the number of additional activations assigned to each step of simulations.

In the first stage, simulations were performed to detect the processes duration S_{PS} represented by the number of simulations steps for each configuration when process dies out. It was used to establish the number of steps in which additional activations take place for each network N, seeding percentage SP, propagation probability PP and seed selection strategy R. Number of additional activations was starting from 10% to 100% of initial seeds and was divided by a number of steps after which the process finished S_{PS}. Results from additional activations were compared with the process without any additional support. Improvement of coverage after each increase of additional activations was dependent on overall process performance within network as is visible in Fig. 2. For total coverage below 20%, differences each additional package

Table 2. Characteristics of real networks used in experiment

Network	Nodes	Mean values of main network measures						Reference
		D	D2	CL	PR	EV	CC	
N1	16,726	5.85	40.68	0.000363	0.000061	0.005384	0.637985	[21]
N2	4,941	2.67	10.16	0.053679	0.000202	0.004790	0.080104	[22]
N3	4,039	43.69	717.17	0.276168	0.000248	0.040473	0.605547	[17]
N4	5,242	11.05	30.85	0.000769	0.000191	0.010351	0.106230	[18]
N5	12,591	7.90	248.11	0.009886	0.000079	0.009543	0.116553	[19]
N6	6,474	3.88	567.14	0.276570	0.000154	0.010303	0.252222	[16]
N7	6,120	45.33	5449.57	0.477727	0.000163	0.007144	0.259767	[27]
N8	6,327	46.93	504.37	0.003694	0.000161	0.036107	0.597632	[12]

of activations was improving total results, while for processes with coverage above 50% additional activations did not affect the process. Example results from network N1, propagation probability PP = 0.05 and betweenness based seeds selection are presented in Fig. 3 with showed the distance between results for 20, 40, 60, 80 and 100 percentage of additional activations.

The results of each additional seeding percentage were compared with approach based on initial seed set only with the use of Wilcoxon test for dependent group. Comparison showed the measure of difference represented by Hodges-Lehmann estimator $\Delta = 0.21$ for 10% of additional seeds, $\Delta = 0.38$ for 20%, $\Delta = 0.55$ for 30%, $\Delta = 0.71$ for 40%, $\Delta = 0.88$ for 50%, $\Delta = 1.03$ for 60%, $\Delta = 1.19$ for 70%, $\Delta = 1.35$ for 80%, $\Delta = 1.51$ for 90% and $\Delta = 1.68$ for 100% with p-value $< 2.2e\text{-}16$ for all additional seeds percentages. Results averaged for each network N1-N8, propagation probability PP and seed selection strategy R are presented in the Table 3. The results show that using additional initial seeds for additional activations resulted increase of coverage by 2.94% up to 22.84% for the 10% and up to 100% of initial seed set, respectively.

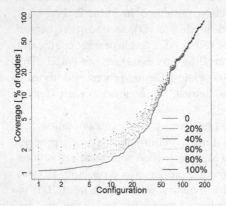

Fig. 2. Coverage improvement for each additional 10% of activations used.

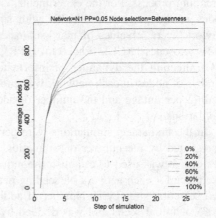

Fig. 3. Example diffusion process for mixtures with different proportion of seeds with high degree.

Table 3. Coverage increase for additional activations

Activations		10%	20%	30%	40%	50%	60%	70%	80%	90%	100%
All results		2.94	5.29	7.56	9.94	12.27	14.31	16.35	18.66	20.73	22.84
Network	N1	3.17	5.74	8.38	10.93	13.21	15.6	17.68	20.01	22.23	24.52
	N2	9.17	19.09	29.06	38.87	48.89	57.47	66.34	75.93	84.47	93.62
	N3	0.96	1.21	1.43	1.54	1.79	1.86	1.99	2.18	2.29	2.51
	N4	7.49	11.36	14.6	19.1	23.04	26.4	29.71	34.3	37.88	41.2
	N5	0.72	1.35	1.9	2.56	3.16	3.73	4.26	4.73	5.26	5.79
	N6	1.49	2.61	3.76	4.75	5.91	6.84	7.96	8.86	9.98	10.88
	N7	0.3	0.52	0.75	0.99	1.21	1.45	1.67	1.93	2.18	2.4
	N8	0.27	0.43	0.61	0.78	0.97	1.09	1.22	1.36	1.53	1.77
Probability	PP 0.05	5.03	8.95	12.63	16.21	19.83	23.21	26.81	30.33	33.67	37.57
	PP 0.10	3.89	6.54	9.02	12.12	14.85	17.26	19.4	22.36	24.99	27.31
	PP 0.15	2.65	4.52	6.58	8.58	10.74	12.47	14.25	16.02	17.66	19.32
	PP 0.20	1.69	3.41	5.02	6.7	8.4	9.73	11.1	12.86	14.28	15.58
	PP 0.25	1.47	3.03	4.56	6.09	7.54	8.86	10.21	11.75	13.04	14.41
Selection	Random	2.52	4.28	5.91	7.83	9.79	12.04	14.1	16.25	18.31	20.3
	Degree	3.02	5.68	8.38	11.23	13.32	14.94	16.72	18.95	21.09	23.2
	PageRank	2.99	5.24	7.3	9.54	11.71	13.62	15.54	17.68	19.39	21.47
	Eigenvector	2.75	4.94	7.09	9.4	11.78	13.84	16.29	19.37	21.6	23.9
	Betweenness	3.44	6.31	9.12	11.7	14.76	17.09	19.12	21.08	23.26	25.31

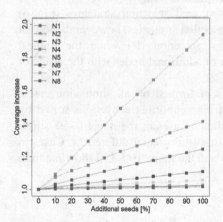

Fig. 4. Coverage increase for each used network and additional seeds percentage

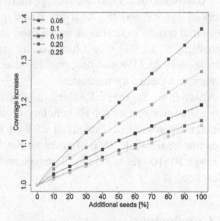

Fig. 5. Coverage increase for each used propagation probability and additional seeds percentage

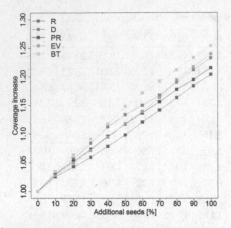
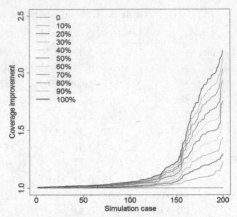

Fig. 6. Coverage increase for each used seed selection strategy and additional seeds percentage

Fig. 7. Results from all simulation cases sorted by coverage improvement

The results for each network are presented in Fig. 4. The highest increase after 100% more activations than in initial seed set was observed for network N2 with coverage higher by 93.62%. The lowest performance was observed for network N7 with only 1.77% increase. Results were dependent on propagation probability PP as is visible in Fig. 5. The best results were achieved for the lowest propagation probability PP = 0.05 with total 37.57% coverage increase while lowest improvement 14.41% was observed for PP = 0.25. Analysis based on used seed selection strategies (Fig. 6) shows that results were similar for all used seed selection strategies; however the lowest improvement 20.3% was observed for random selection (R) while the highest improvement 25.31% was observed for selection of additional nodes with the usage of betweenness centrality measure.

Apart from aggregated values, analysis was performed on all simulation cases. Figure 7 shows results for all simulation cases from supporting activations sorted by coverage improvement. For about 25% of simulation cases in the range 0–50 differences are small while 25% of cases above 150 show substantial differences. Cases in the range 50–100 show small differences and cases in the range 100–150 show medium differences.

5   Conclusions

The presented study focused on the influence on ongoing information spreading processes by activation of additional nodes during the process launched by initial seed set. While most of earlier research concentrated on the initial seed set, the present research shows how additional seeds used during the process affected number of activated nodes within the network. Results showed that using the same quantity of additional seeds such as were used to initiate the diffusion process increase the coverage within the network by 22.84% while small additional package of seeds equal to 10% of initial seed set has the potential of coverage increase by 2.94%. Results were highly dependent on networks and their characteristics. The lowest improvement was observed for networks N3, N7 and N8. Those networks had a high average degree with values and the final coverage was very high regardless of whether additional seeds were used. Similar conclusions come from influence of propagation probabilities on final results. The best coverage increase was observed for the lowest propagation probability while the highest propagation probability delivered much lower improvement. Results were dependent on the seed selection strategies used. Heuristics based on degree, betweenness of eigenvector delivered up to 20% better results than random selection of additional seeds. The presented research opens several questions for future research. Apart from flat number of additional seeds in simulation stages other distributions can be used to minimize number of additional seeds at the beginning and maximizing when the dynamics of the process drops. Another direction can be based on wider range of used additional seeds exceeding initial seed set.

Acknowledgements. This work was partially supported by the National Science Centre, Poland, grant no. 2016/21/B/HS4/01562.

References

1. Bello-Orgaz, G., Jung, J.J., Camacho, D.: Social big data: recent achievements and new challenges. Inf. Fusion **28**, 45–59 (2016)
2. Berger, J., Milkman, K.L.: What makes online content viral? J. Mark. Res. **49**(2), 192–205 (2012)
3. Chen, W., Wang, C., Wang, Y.: Scalable influence maximization for prevalent viral marketing in large-scale social networks. In: Proceedings of the 16th ACM SIGKDD International Conference on Knowledge Discovery and Data Mining, pp. 1029–1038. ACM (2010)
4. Granell, C., Gómez, S., Arenas, A.: Competing spreading processes on multiplex networks: awareness and epidemics. Phys. Rev. E **90**(1), 012808 (2014)
5. Hanna, R., Rohm, A., Crittenden, V.L.: We're all connected: the power of the social media ecosystem. Bus. Horiz. **54**(3), 265–273 (2011)
6. He, J.-L., Fu, Y., Chen, D.-B.: A novel top-k strategy for influence maximization in complex networks with community structure. PLoS ONE **10**, e0145283 (2015)
7. Hinz, O., Skiera, B., Barrot, C., Becker, J.U.: Seeding strategies for viral marketing: an empirical comparison. J. Mark. **75**(6), 55–71 (2011)
8. Ho, J.Y., Dempsey, M.: Viral marketing: motivations to forward online content. J. Bus. Res. **63**(9), 1000–1006 (2010)
9. Iribarren, J.L., Moro, E.: Impact of human activity patterns on the dynamics of information diffusion. Phys. Rev. Lett. **103**(3), 038702 (2009)
10. Jankowski, J., Bródka, P., Kazienko, P., Szymanski, B.K., Michalski, R., Kajdanowicz, T.: Balancing speed and coverage by sequential seeding in complex networks. Sci. Rep. **7**(1), 891 (2017)
11. Jankowski, J.: Dynamic rankings for seed selection in complex networks: balancing costs and coverage. Entropy **19**(4), 170 (2017)

12. Joshi-Tope, G., et al.: Reactome: a knowledgebase of biological pathways. Nucleic Acids Res. **33**, D428–D432 (2005)
13. Kandhway, K., Kuri, J.: How to run a campaign: optimal control of SIS and SIR information epidemics. Appl. Math. Comput. **231**, 79–92 (2014)
14. Kempe, D., Kleinberg, J., Tardos, É.: Maximizing the spread of influence through a social network. In: Proceedings of the Ninth ACM SIGKDD International Conference on Knowledge Discovery and Data Mining, pp. 137–146. ACM (2003)
15. Kitsak, M., Gallos, L.K., Havlin, S., Liljeros, F., Muchnik, L., Stanley, H.E., Makse, H.A.: Identification of influential spreaders in complex networks. Nat. Phys. **6**(11), 888–893 (2010)
16. Leskovec J., Kleinberg J., Faloutsos C.: Graphs over time: densification laws, shrinking diameters and possible explanations. In: Proceedings of the Eleventh ACM SIGKDD International Conference on Knowledge Discovery in Data Mining, pp. 177–187. ACM (2005)
17. Leskovec, J., Mcauley, J.J.: Learning to discover social circles in ego networks. In Advances in Neural Information Processing Systems, pp. 539–547 (2012)
18. Leskovec, J., Kleinberg, J., Faloutsos, C.: Graph evolution: densification and shrinking diameters. ACM Trans. Knowl. Discov. Data **1**(1), 2 (2007)
19. Ley, M.: The DBLP computer science bibliography: evolution, research issues, perspectives. In: International Symposium on String Processing and Information Retrieval, pp. 1–10 (2002)
20. Michalski, R., Kajdanowicz, T., Bródka, P., Kazienko, P.: Seed selection for spread of influence in social networks: temporal vs. static approach. New Gener. Comput. **32**(3–4), 213–235 (2014)
21. Newman, M.E.: Scientific collaboration networks. I. Network construction and fundamental results. Phys. Rev. E **64**, 016131 (2001)
22. Opsahl, T., Panzarasa, P.: Clustering in weighted networks. Soc. Netw. **31**, 155–163 (2009)
23. Pfitzner, R., Garas, A., Schweitzer, F.: Emotional divergence influences information spreading in twitter. In: Proceedings of Sixth International Conference on Weblogs and Social Media, pp. 2–5 (2012)
24. Rogers, E.M.: Diffusion of Innovations. Simon and Schuster, New York (2010)
25. Salehi, M., Sharma, R., Marzolla, M., Magnani, M., Siyari, P., Montesi, D.: Spreading processes in multilayer networks. IEEE Trans. Netw. Sci. Eng. **2**(2), 65–83 (2015)
26. Seeman, L., Singer, Y.: Adaptive seeding in social networks. In Foundations of Computer Science (FOCS), IEEE 54th Annual Symposium, pp. 459–468. IEEE (2013)
27. Subelj, L. Bajec, M.: Software systems through complex networks science: Review, analysis and applications. In: Proceedings of the First International Workshop on Software Mining, pp. 9–16. ACM (2012)
28. Tang, J., Musolesi, M., Mascolo, C., Latora, V., Nicosia, V.: Analysing information flows and key mediators through temporal centrality metrics. In: Proceedings of the 3rd Workshop on Social Network Systems, p. 3. ACM (2010)
29. Watts, D.J., Peretti, J., Frumin, M.: Viral Marketing for the Real World. Harvard Business School Pub, Boston (2007)
30. Zhang, J.-X., Duan-Bing Chen, Q.D., Zhao, Z.-D.: Identifying a set of influential spreaders in complex networks. Sci. Rep. **6** (2016)

Big Data

B-Learning and Big Data: Use in Training an Engineering Course

Leonardo Emiro Contreras Bravo[(⊠)],
Jose Ignacio Rodriguez Molano,
and Giovanny Mauricio Tarazona Bermudez

Universidad Distrital Francisco José de Caldas, Bogotá, Colombia
{lecontrerasb, jirodriguez,
gtarazona}@udistrital.edu.co

Abstract. Is presents a case study in a descriptive work of qualitative Court that seeks to evaluate the advantages of the deployment and the use of the b-learning methodology and big data in pedagogical processes. There is the need for evolution of the type of traditional education currently practised in the University by a methodology that allows for greater participation and respon-sibility on the part of the student and which present an opportunity for devel-opment of independent learning skills. Initially develops a theoretical reference framework associated with the traditional teaching, B-learning and Big data with its approach to the field of education. Subsequently, is an approach to the existing problems in a case study employing the use of descriptive records, participant observation and interviews not structured to analyze and compare the academic performance of students in a course implementing b-learning vs. a course with traditional methods.

Keywords: B-learning · Big data · Virtual engineering environments education

1 Introduction

Technological progress, and especially information and communication technologies have influenced many fields one of these education sector, in which it is possible to appreciate some changes such as cases of Board and marker, which has been replaced by multimedia (video beam) projectors; billboards and the centers of copied, replaced by virtual space or emails that make it possible to maintain communication in real time; Therefore the current cultural settings are crossed and stained by these new tech-nologies. In this direction, [1] says that "the market of the future and the labour demands will rotate around information and information management" and "media transformed the world and transforming teaching".

This work is carried out in the field mechanical processes, belonging to industrial engineering of the Universidad Distrital Francisco José de Caldas (Colombia). In order to anticipate changes in ICT is used the platform Moodle from the second half of 2010. the use of a Virtual Classroom, which includes the development of a series of activities to distance was introduced as an additional and complementary educational resource.

© Springer International Publishing AG 2017
Y. Tan et al. (Eds.): DMBD 2017, LNCS 10387, pp. 221–233, 2017.
DOI: 10.1007/978-3-319-61845-6_23

The design is oriented to a b-learning (blended or mixed learning) system, combining distance learning and blended since the formation of engineers demanding the permanent pursuit of pedagogical alternatives, as well as the continuous use of technological resources offered by today's world, to achieve higher levels of training [2].

2   General Information

Teaching through a traditional method as it is usually done in different engineering subjects, is characterized by a work in which the teacher teaches a series of concepts, the student tries to assimilate them and at the same time they are evaluated by quizzes and partial. In this, the student is a passive actor, who only uses group work in some cases at the time of study for evaluations written from textbooks, and the notes taken in lectures. Of course, many leaders, academics, and practical to believe that traditional approaches in education, as a dependency in the textbooks, mass instruction, lectures and multiple-choice tests, are obsolete in the age of information [3]. It is necessary to highlight the differences showing with respect to the traditional ICT-based education: concerning the modification of spaces, roles and methods characteristic of virtual education versus the traditional or face-to-face, paradigm according to media and handlings that were given by the institution, and its interrelation with contemporary paradigms, between them it was digital and the globalization as epicenter political, social and economic [4].

2.1   B-Learning

The simplest definition and also the most precise b-learning describes it as that mode of learning that it combines teaching through traditional classroom activities with non-contact technology: "which combines face-to-face and virtual teaching" [5]. Therefore, the blended learning, approaching more a hybrid training model that is able to collect the best of teaching at a distance and the best of classroom teaching; i.e., use correctly the electronic resources and infrastructure available digital and use appropriate methods of active participation in class in order to facilitate learning.

A drawback that the incursion of ICT in different fields of society, is the resistance to change and especially in the field of education, where some professors disagree of any change in an educational system that has worked for many years. Centuries. On the subject [7] justifies the "blended learning" as a "soft" option to enter the information technology between a reluctant Faculty, and especially the technologies of information and communication, it has often been acclaimed as a catalyst for change, but this change need not be radical. "Some useful ICT through well-planned easy ways can be, I suggest to use widely available technologies combined with more familiar approaches to teaching and learning".

2.2   Platform LMS

Authors as [8] give us an overview of the development of the e-learning over time, explaining that applied to the process of teaching-learning ICT can be classified as:

training-based computer (CBT), Web (WBT) or management platforms (LMS) learning based training. Within the platforms commercial more known are WebCT and Blackboard, which possess few institutions Latin American due to its high cost and difficulty to manage and maintain. In contrast are free software (available for free on the Internet)-based LMS platforms such as Claroline, Dokeos and Moodle (Modular Object-Oriented Dynamic Learning Environment) among others. The latter can be copied and modified and allow you to create a virtual learning environment with ease, without having to be expert programmers favoring the social construction of knowledge through a process of gradual transfer of control, as points out it [9]. Moodle is a virtual learning environment based on internet. It is a project in development designed to support a social constructionist education framework, therefore, is a system of free course management that helps the learning process-online collaborative learning. Some of the modules of academic activities integrated into the platform that can be used through Moodle are: chat, forums, quizzes, blogs and wikis. Additionally allows the link to Web pages, files and videos in order to make more interactive virtual education.

2.3 Big Data

The big data can be defined as a concept that refers to the collection, storage and processing of a set of data that is too large for the purpose of finding repetitive patterns within them, that it is impossible to manage them with databases and conventional analytical tools. The sector of information and communication technologies has been the leader in the trend to manipulate huge amounts of data that are generated in real time and that come from different media such as social networks, sensors, platforms and devices audio and video requiring to use such information in reporting statistical and predictive models that can be used in many areas of human endeavor. These volumes of data collected are usually classified in three main groups [10].

- Structured data: data that are formatted in order to clearly define aspects such as its length. Examples of this group are: dates, numbers or character strings. Typically, databases and spreadsheets contain structured data.
- Semi-structured data: data that have been processed to some extent, i.e., data which do not correspond to a specific format, but its elements are separated by markers [10]. An example of this type of data are those who come from web pages traditional (HTML)
- Unstructured data: are data that do not have any format, since they are as they were collected. Some examples of this type of data are PDFs, multimedia documents and e-mail [11].

Could mention are some of the possible applications of the big data in different fields. One of these medicine, where more clearly you are drawing these benefits, treatments and personalized medicine are certainly a desirable future for all. In the trade sector, allow greater visibility of the needs of customers, and the role of the customer in his social circle could be determined through an analysis of social networks. Similarly, there are other areas where being built the foundations to harness the potential of Big data, such as nutrition, sport, advertising and transport, among many others [12].

2.4   Exploration of Big Data in Education

Big technology data has dabbled in different fields, but the field of education as it has happened frequently, has never been a field pioneered the implementation of innovations. This has been characterized by changes perhaps late in terms of technological tools, something when other sectors did successfully [13]. The concepts of big data and data analysis can be applied to a variety of fields of higher education; in the administrative and instructional field, process admissions, financial planning, monitoring of teacher and student performance. This article also aims to identify the Big data and its analytical can support decision-making in environment b-learning. Thus, students of the platforms using educational LMS generate an amount of data that need to be analyzed. According to [14], these systems of registration of information on the results of the students, could be used by institutions to study the patterns of student performance over time. And collected data for each entry of a student in an online course. That is, input the Forum of discussion, blog or wiki activity input, completion of examinations, among other data [15]. This amount of data and transactions in real time of the students not only a course but an entire semester requires data mining Software to establish processes of conversion of data into useful information, in order to discover patterns and take corrective actions to the learning process of the students in real time. Although the majority of current data mining tools are too complex for teachers and their functions go far beyond the scope of which could require a teacher [16].

According to [17] you can set the Analytics of learning as the use of smart data; data produced by the student, and the analysis of models to discover the information and social relations, and thus predict and advise on their learning process. This can be supported by teachers and students to take action based on the evaluation of educational data. However, the technology to deliver this potential is still very young and research in the understanding of the pedagogical usefulness of learning Analytics is still in its infancy [18]. According to [19], there are eight categories of potential practical applications in which it is possible to use the Analytics of learning, of which the most common are the monitoring and early identification of weaknesses to an intervention of the teaching-learning process and student: monitoring of the student's individual performance; breakdown of the performance of students in different ways, such as by age, year of study, ethnicity, etc.; identification of outliers for early intervention in the learning process; to promote the potential of the students so that they develop an optimal learning process; avoid the wear and tear of a course or program; identify and develop techniques for effective teaching; analyze instruments and methods of evaluation; and evaluate resumes or curricula.

On the contrary Miguel Caulfield, director of learning combined and network of the University of the State of Washington, in Vancouver, cautioned that the early research on the effectiveness of learning and retention Analytics need additional verification and review [20]. Thus, the analysis of learning as a retention tool is still in its nascent stage. So much so, that society of researchers for the Analytics of the learning (SOLAR), which is an interdisciplinary network of leading international researchers are exploring the role and impact of analytics in the process of teaching and learning, training and development [20].

In accordance with [18], for the application of data mining to the b-learning it must collect data from course management system (for example: amount of approved not approved vs; required courses elective vs) system of qualifications and learning management system - LMS (e.g.: information of all members who have access to the course, frequency of use of each log in the course and data related to the activities of teaching and learning of) each Member of the specific course). The second step called mapping, consists of the combination of data for an ID key and before carrying out an analysis by tools of big data, it is recommended to make an analysis of descriptive statistics to identify the application of b-learning and identify if data, cleaning of existing data and removal of the values are missing. Then it will be possible to classify them by subgroups or cluster depending on the identified activities that must be analyzed by data mining for results about online activity, identify clear policies for the reduction of working hours face to face, among others. Something similar appears [21]; for the implementation of a process of learning Analytics should be started with data collection. Data collecting different student activities when they interact within a LMS platform or a personal learning (PLE) environment, (are examples of activities for data collection: the participation in collaborative, write in the Forum or reading a document). The second stage of the analytical process involves data mining based on different techniques, such as grouping, sorting, Association rule and social network analysis. Subsequently, the results of the mining process can be displayed as an application (widget), that can be integrated to PLE or LMS providing representations charts that facilitate decision making by the teacher as soon as the impact of his teaching method, the form for learning, student achievement, and effectiveness of his teaching. It is worth mentioning that it should be privacy in the collection and analysis of data from the students. Although many questions about education are not intended to examine the records of individual students. Rather, the data of all students or subgroups with specific characteristics, is the important thing.

2.5   Tools Big Data

He field of it education, especially in Latin America is of limited resources economic, being this main limitations for the adoption of the technology big data in any institution added to the of infrastructure and of capital human experienced that is require for its instrumentation. However, below, are a number of tools that are currently in storage, management and analysis of large amounts of data. This, with the purpose of having a clearer perspective when choosing the right tool that will allow better use of resources and data.

2.5.1   Hadoop
Hadoop (HDFS) is a distributed file system designed to be executed from the hardware of a computer or information system [22]. Among its characteristics: Hadoop is made up of hundred or thousands of connected servers that store and execute the user's tasks; the possibility of failure is high, since if only one of the servers fails, the whole system fails. Therefore, the Hadoop platform always has a certain percentage of inactivity [23]; the applications executed by means of Hadoop are not for general use, since it

processes sets of data without contact with the user; an advantage Hadoop has regarding the other systems is that the processing of information takes place in the same place where it is store, which does not overcrowd the network since it does not have to transmit the data elsewhere to be processed.

2.5.2 MongoDB

MongoDB is a distributed NoSQL data manager of a documental sort, which means that it is a non-relational database. Data exchange in this manager is done by means of BSON, which is a text that uses a binary representation for data structuring and mapping. This manager is written in C ++, may be executed from different operational systems; it is open-source code [24]. This data manager has the following characteristics: flexible storage, since it is sustained by JSON and does not need to define prior schemes; multiple indexes may be created starting on any attribute, which facilitates its use, since it is not necessary to define MapReduce or parallel processes.

3 . Materials and Methods

Matter has groups that have on average 40 students of both sexes, with hourly intensity of 4 h per week to master classes of exhibition type and 4 laboratory practice. It is worth mentioning that many of the students of the University as this mostly with population of layer 1, 2 and 3 possess features such as lack of own computers, limitation of 24-hour internet access and do not have a computer culture to become self-sufficient in their learning. Traditional teaching methodology has been based on a transfer of knowledge from the teacher to a group of students who listen carefully, trying to assimilate the ideas transmitted (expository teaching methodology).

3.1 Academic and Methodology Aspects

In order to advance a process of integration of the computer, educational, and other tools available on the internet with the classes, was used as a support a virtual course developed in the Moodle platform. The General characteristics of the phases of learning of the development of the course with the support of the virtual tool are performed in the following way (Table 1):

- Recognition: Design of activities which allow students to move from the preemptions to the notions. I.e., motivate him engage in the initial processes of learning and triggered their cognitive structures. Consists of: the review of Pre-sabers.
- Deepening: Design situations and activities in a didactic manner, leading to the appropriation of concepts, theories, and procedures, according to the purposes, objectives and competencies established in the course. In each of the units is developed through them lessons evaluative, of them quizzes on line and face-to-face, and partial.
- Transfer: Design of activities to be added reconceptualization and productivity values derived competences and knowledge that is learned. The transfer is promoted through face-to-face and collaborative group work

Table 1. Activities of the course of mechanical processes

Stage	Activity	Evaluation form	Learning phase
Activity unit 1	Recognition of prescriptions	Individual - on line	Recognition
	Quiz 1	Individual - on line	Deepening
	First Test	Single - face to face	Deepening
	Collaborative group work 1 CAD/CAM	Online - face to face	Transfer
Activity unit 2	Quiz 2	Individual - on line	Deepening
	Second Test	Single-face to face	Deepening
	Group collaborative work 2 CAD/CAM	On line-face to face	Transfer
Activity unit 3	Quiz 3	Individual - on line	Deepening
	Laboratory-Workshop	Single-face to face	Transfer
	Exposition	Single-face to face	Deepening
Final work	Collaborative group work 3 CAD/CAM	On line-face to face	Transfer
End test	Final exam	Single-face to face	Deepening

The Fig. 1, shows how to integrate the two methodologies (traditional and e-learning). The student has a series of technological resources to be described later, from a recipient subject of information only in master class to one that has access 24 h a day to different elements for the development of their learning. The student within this scheme raised, must develop in the course some hours of work autonomous (HTA), which are the Foundation of the training and of the learning. Develops through the personal work and group work. There will be hours of work collaborative (HTC) dedicated to orientation possibly requiring each of the preset student groups, aims to learning through a task force, involving the socialization of the results of personal work, development of activities in team and reporting according to scheduled activities. And finally you will have hours of direct labor (HTD) or hours of master classes, which are hours that the student must attend in person; in this space the teacher exercises accompanying the set of students in charge where relevant, systematized information is analyzed and deepens the key concepts of the course.

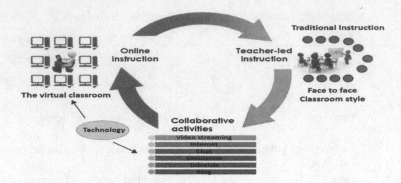

Fig. 1. Conceptual design of the propone academic work methodology

3.2 Technological Aspects

The methodology is generally designed in such a way that the student be reflective about the why of the activities it carries out, so you have a better understanding of the process and the outcome. She sets an interactive teaching leveraging resources in new technologies, with content more attractive to the student and where it has a high degree of participation. Then briefly describes some of the technological tools.

3.2.1 Online Moodle Course
This was created through the willing tool University (Fig. 2a), holds information such as the agenda of the course, quizzes and pre knowledge of each one of the issues that divides the syllabus, download link of the software CAD-CAM (computer - manufacturing assisted by computer-aided design), group collaborative work, etc.

3.2.2 Supporting Material
It includes the elaboration of transparencies-summaries in PowerPoint for each thematic block. As well as books and articles related, available in the Department, library or via the Internet, relating to the subject matter. This serves as half of share information to those students to facilitate their learning. Each of the thematic units has various videos (50 links), about manufacturing processes explained in theoretical matera and calculation of its variables in the traditional kind.

3.2.3 CAD/CAM Software
This educational resource helps the student to complement the laboratory practice which consists in the management of various machine tools with a CAD/CAM software called SpectraCAM Turning; likewise the virtual course has information such as learning the same lessons (Fig. 2b), so that the student can download it and with this make your reading online and offline. This is a help to develop some collaborative works.

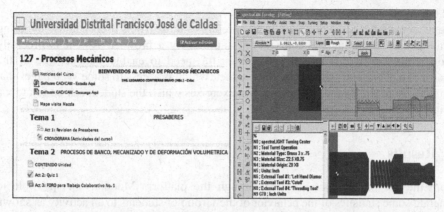

Fig. 2. (a) Overview of the virtual course, (b) image of the.pdf study file for students

3.2.4 Class Workshops

It involves the planning, development and solving different "base" of each of the thematic units. There are also various exercises so that students will develop them in their hours of autonomous work and with which doubts are resolved in the subsequent to the issue presented master classes.

3.2.5 Displays Graphic Interactive Environment

As previously mentioned, one of the two objectives pursued this work is the create an interactive database of inquiry with the aim that the student can establish key concepts needed to pursue the successful course that sometimes occurs that many basic concepts that the Professor is known, are not clear enough when the student begins with the development of exercises. In this type of subject (theoretical-practical) is important to the two-dimensional and three-dimensional graphics display. For this reason was created an interactive database of inquiry with the aim that the student can establish key concepts needed to pursue the successful course, hence is of great help show the systems that are being studied in a graphical environment and with the possibility to interact with them. Some used simulations were AutoCAD and Solidworks software product (Fig. 3).

Fig. 3. Prototypes. Result final work (Model 3D)

3.2.6   Web Applications

The resource educational web application, refers to those Tools users can use by going to a web server through Internet or one intranet by a browser. And it is precisely through the use of this tool where is born the need to enable students one of the resources that currently offered by the Internet, which is access to some Web pages that contain virtual simulations and interactive exercises which the student can be found via a simple 'click'.

4   Results

From the use of the course online through the platform Moodle, it is possible to generate some statistics of the behavior of the Group of students to an activity as shown in the Fig. 4a, or perhaps a comparative analysis of each student throughout the semester in terms of a specific activity as shown in Fig. 4b.

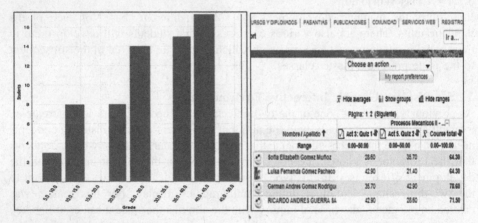

Fig. 4. (a) Group statistics of an activity: Quiz 1 (b) example of statistics of an online course activity

The average scores obtained through face-to-face system - traditional (31.1) are slightly lower than those obtained by students through the application of education B-learning (34.5). In terms of the standard deviation the group that received traditional teaching had a value of 5.49 while that of B-learning teaching had a value of 4.23. In this case, it can be concluded that the application of the methodology gave good results, although it is expected that in the second stage of application (extension to other related subjects), the results are better. Through a survey to students valued aspects of the implemented methodology. With regard to"the ease of using the Moodle platform and services that comprise", 97% said to agree that it is easy to use in varying degrees. The reading of these results, confirms the technological resource course virtual on line and platform Moodle does not present major difficulties in its use, even for students who have weaknesses when knowledge and access to computing resources. Observed the good rating obtained for the teaching of CAD/CAM software, through the virtual course created on Moodle. Is of highlight the 48% and 36% for them answers "enough" and "completely" concerning the question "the teaching of the software CAD/CAM by means of the course virtual you seemed successful".

As regards big application data to the teaching-learning process, some & that can be implemented after reviewing some papers were identified: is advisable to implement selected courses teachers initially learning Analytics involved in the development of the project already in this way, is could get feedback and immediate comments on the analytical data processed in real time; through semi-structured interviews to identify a set of graphical indicators and needs of users that will be displayed in the software design. The interviews are oriented through prepared questions, but it is also possible to spontaneously ask questions to investigate interesting details [26]; the application to create requires indicators such as: use of documents, evaluation/performance, activity of user and communication, among others. So therefore arises as a next stage, which is shown in Fig. 5, in which it is required to acquire data through the interaction of the student (producer of data) with different activities established in the Moodle platform. Using a tool of data due to the large volume of data mining is to achieve the transformation of the same as with the help of data analytics and applying algorithms and mathematical models, It is possible view in a platform some initial graphical results. Last planned future use of Machine Learning or machine learning which is one of the most important applications of artificial intelligence that evolved from the study of the recognition of patterns and computational learning theory. The aim is to create and study algorithms that are capable of supporting data and make predictions about its (academic activities data) base.

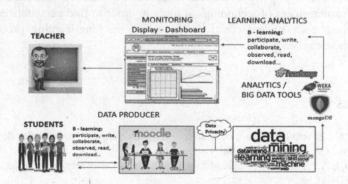

Fig. 5. The learning analytics process

5 Conclusions

The raid of B-learning was a methodology which provided educational experiences of quality in innovative formats (internet and computing), which encouraged the participation and integration in the areas of communication and collaboration among stakeholders in the education process and constitutes a first step to the decrease of resistance to change not only teachers and students, but executives.

Although it is possible to obtain some results through traditional means of data analysis by applying the methodology b-learning, the big Analytics data is the current key resource to understand it and improve it, since there are some constraints that could be solved. With the help of big software data, better monitoring could be made to students in the sense of improving tutorials and activities, objectives of the evaluation, which was seen issues and not an overall assessment of all students taking the subject information, predict academic risks or simply to understand the behavior of school groups.

The project is in the stage of maturity in the use of analysis of big data. Below briefly describes the stages to develop in order to achieve the specific objectives of the implementation and the use of the b-learning methodology and big data in pedagogical processes: generation of a development environment on BigData (Hadoop, MongoDB and analysis platform) technology, study of current models and development of specific models, enabling to identify students at risk in such a way that allow decisions to intervene in order to reduce the drop-out and increase yields, study of risks and ethical issues. [25]

Learning Analytics, basically consists of the measurement, collection, analysis and presentation of data on students, their contexts and interactions occurring there, in order to understand the process of learning that is developing and optimize the environments in which it is produced. Today, it is used to: detect the strengths and weaknesses of the educational systems, improve courses that offer educational institutions, reflect on the achievements and the patterns of behavior of each student regarding their colleagues or used educational curriculum and help teachers and support staff in its interventions. It is important to foster the analytical introduction of big data at universities in Colombia, train human resources to perform the analytical with big data and introduce them even more to the management of technological tools in order to find personalized learning contexts, where each student could carry out the teaching-learning process really adapted to their characteristics.

References

1. Garrido, M.: El reto del cambio educativo: nuevos escenarios y modalidades de formación, Educar, vol. 38, pp. 243–258 (2006)
2. Londoño, E.A.: Ambientes de aprendizaje para la educación en tecnología, Educ. en Tecnol. I, p. 11 (1996)
3. Cuban, L.: Oversold and Underused: Computers in the Classroom, vol. 9, no. 2 (2001)
4. Henao, O.: El aula escolar del futuro," Educ. y Pedagog. Exp. e Investig., no. 1, pp. 1–12 (1993)
5. Coaten, N.: Educaweb. Suplemento del boletín de educaweb, Barcelona (2003). ISSN:1578-5793
6. Contreras, L.: Use of ICT and especially of blended learning in higher education, Rev. Educ. Y Desarro. Soc., 151–160 (2011)
7. Dipro, E., Almenara, J.C., Díaz, M.: ICT training of university teachers in a Personal Learning. J. New Approaches Educ. Res. 1(1), 2 (2012)

8. García, P., Lacleta, M.: Moodle: Difusión y funcionalidades, I Jornadas Innovación Docente, Tecnol. la Inf. y la Comun. e Investig. Educ. en la Univ. Zaragoza (2006)
9. Rogoff, B.: Aprendices Del Pensamiento: El Desarrollo Cognitivo en el Contexto Social, p. 301 (1993)
10. Purcell, B.: The emergence of 'big data' technology and analytics. J. Technol. Res. **4**, 1–7 (2013)
11. Database SystemS. http://bit.sparcs.org/~dinggul/tools/1423902017.pdf
12. Carrillo, J., et al.: Big Data en los entornos de Defensa y Seguridad. Inst. Español Estud. Estratégicos **1**, 124 (2013)
13. B. I. G. Data and E. N. La, "Big data," vol. 17, pp. 1–16 (2016)
14. Picciano, A.: Big data and learning analytics in blended learning environments: benefits and concerns. Int. J. Interact. Multimed. Artif. Intell. **2**(7), 35 (2014)
15. Shvachko, K., Kuang, H., Radia, S., Chansler, R.: The hadoop distributed file system. In: 2010 IEEE 26th Symposium on Mass Storage Systems and Technologies, MSST 2010 (2010)
16. Alcalá-Fdez, A., et al.: KEEL: a software tool to assess evolutionary algorithms for data mining problems. Soft. Comput. **13**(3), 307–318 (2009)
17. Elearnspace: What are Learning Analytics?. http://www.elcarnspace.org/blog/2010/08/25/what-are-learning-analytics
18. Park, Y., Yu, J.H., Jo, I.-H.: Clustering blended learning courses by online behavior data case study in a Korean higher education institute. Internet High. Educ. **29**, 1–11 (2016)
19. Zhu, W.-D.J.: International Technical Support Organization. IBM Watson Content Analytics discovering actionable insight from your content (2014)
20. Iten, L., Arnold, K., Pistilli, M.: Mining Real-Time Data to Improve Student Success in a Gateway Course, Elev. Annu. TLT Conf
21. Dyckhoff, A.L., Zielke, D., Bültmann, M.: Design and implementation of a learning analytics toolkit for teachers. Educ. Technol. Soc. **15**(3), 58–76 (2012)
22. Dittrich, J., Quian, J.: Efficient big data processing in hadoop mapreduce. Proc. VLDB Endowment **5**(12), 2014–2015 (2012)
23. MongoDB Inc. 2008–2016. https://docs.mongodb.org/manual/introduction/
24. Hall, M., Frank, E., Holmes, G., Pfahringer, B., Reutemann, P., Witten, I.: The WEKA data mining software. ACM SIGKDD Explor. Newsl. **11**(1), 10 (2009)
25. Garner, S.: WEKA: the waikato environment for knowledge analysis. In: Proceedings of New Zealand Computer Science, pp. 57–64 (1995)
26. Lindlof, T., Taylor, B.: Qualitative Communication Research Methods, second edn., no. 1985, p. 195. Sage Publications, Thousand Oaks (2002)

Mapping Knowledge Domain Research in Big Data: From 2006 to 2016

Li Zeng$^{(\boxtimes)}$, Zili Li, Tong Wu, and Lixin Yang

Center for National Security and Strategic Studies (CNSS),
National University of Defense Technology Changsha, 410073 Changsha, China
crack521@163.com, {zlli,tongwu}@nudt.edu.cn,
717112894@qq.com

Abstract. This paper was explore a scientometric analysis of the research work in the emerging field of "Big Data" in recent years. Research on "Big Data" in the past few years, and in a short time has gained tremendous momentum. It is now considered one of the most important emerging research areas in computational science and related disciplines. By using the related literature in the Science Citation Index (SCI) database from 2006 to 2016, a scientometric approach was used to quantitatively assessing current research hotspots and trends. It shows that "Big Data" is a new emerging field with rapid development, the total of 2076 articles covered 131 countries (regions) and Top 3 countries (regions) were USA (731, 38.86%), China (373, 19.83%), England (93, 4.94%). In addition, Top 10 keywords are found to have citation bursts: epidemiology, scalability, social media, genomics, visualization, sequencing data, integration, intelligence, association, behavior. The results provided a dynamic view of the evolution of "Big Data" research hotpots and trends from various perspectives which may serve as a potential guide for future research.

Keywords: Big data · Scientometric analysis · Mapping knowledge domain

1 Introduction

With the rapid development of the Internet, cloud computing, mobile and Internet of Things, mobile devices, RFID, wireless sensors have been are producing huge amount data all the time, human society has entered the era of "Big Data". The term "Big Data" has become so important during last few years that Nature and Science have published special issues dedicated to discuss the opportunities and challenges brought by "Big Data" [1, 2]. Compared to traditional data, the features of 'Big Data' are characterized by 5 V, namely, huge Volume, high Velocity, high Variety, low Veracity, and high Value [3].

Cloud computing is a powerful technology to perform massive-scale and complex computing. The recent explosive publications of big data studies have well documented the rise of big data and its ongoing prevalence. The interest in Big Data has generated a broad range of new academic, corporate, and policy practices along with an evolving debate among its proponents, detractors, and skeptics. Ekbia, H et al. (2015) [4] provide a synthesis by drawing on relevant writings in the sciences, humanities, policy,

© Springer International Publishing AG 2017
Y. Tan et al. (Eds.): DMBD 2017, LNCS 10387, pp. 234–246, 2017.
DOI: 10.1007/978-3-319-61845-6_24

and trade literature. Al-Jarrah, OY (2015) [5] reviews the theoretical and experimental data-modeling literature, in large-scale data- intensive fields, relating to model efficiency and new algorithmic approaches. Hashem, IAT (2015) [6] has been observed that massive growth in the scale of data or big data generated through cloud computing. Hilbert, M (2016) [7] uses a conceptual framework to review empirical evidence and some 180 articles related to the opportunities and threats of Big Data Analytics for international development. Liu, JZ (2016) [8] indicated and summarized the problems faced by current big data studies with regard to data collection, processing and analysis: inauthentic data collection, information incompleteness and noise of big data, unrepresentativeness, consistency and reliability.

In this paper, the Bibliometric analysis tools Tviz and Citespace [9] was performed by investigating annual scientific outputs, distribution of countries, institutions, journals, research performances by individuals and subject categories to offer another perspective on the development of research in the field of "Big Data". Moreover, innovative methods such as Keyword co-citation analysis, semantic clustering and Keyword Frequent Burst Detection were applied to provide insights into the global research hotpots and trends from various perspectives which may serve as a potential guide for future research.

2 Data and Methods

We collected the bibliographic records from the Web of Science (WoS) of Thomson Reuters on January 31, 2017 and determined the time frame of this analysis to 2006 and 2016. The ultimate query string about "Big Data" looked like this: TOPIC: ("Big Data") Indexes = SCI-EXPANDED Timespan = 2006–2016. The query resulted in 2076 bibliographic records. The whole bibliographic records were then downloaded for subsequent analysis. Then we used a Bibliometric approach to quantitatively assessing current research hotspots and trends on "Big Data".

3 Result and Discussion

3.1 Characteristics of Article Outputs

Figure 1 shows the variation of article numbers of the research about Big Data between 2006 and 2016. Black curve stands for the annual number of publications about "Big Data". From the curve, we found that a substantial interest in "Big Data" research did not emerge until 2012, although a few articles related to "Big Data" were published previously. And the highest annual number occurred in the years of 2016. The red curve stands for the cumulative number of publication in the field of "Big Data". According to the theory of technology maturity, the cumulative number of the publication will be presented as an S-curve in general [10]. But the cumulative number cannot fit the S-curve which may implies that "Big Data" is a new emerging field with rapid development.

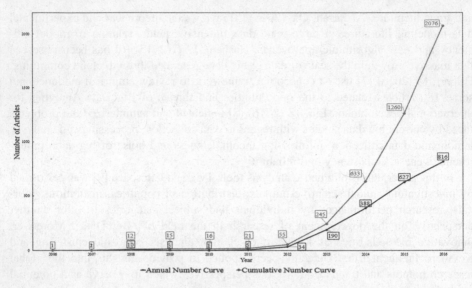

Fig. 1. Variation of article numbers

3.2   Characteristics of Document Type

The distribution of the document type was displayed in Fig. 2. Overall, the papers about "Big Data" involving a total of six document types, and the details are as follows: Article (1069, 51%), Editorial Material (574, 28%), Meeting Abstract review (149, 7%), Review (119, 6%). The distribution of document type suggested the high priority of Article in "Big Data" research.

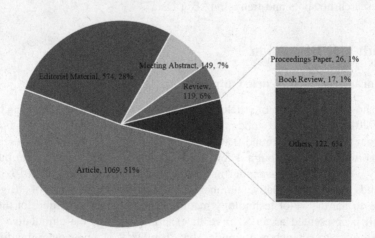

Fig. 2. Distribution of document type

3.3   Subject Categories Distribution and Co-occurring Network

The distribution of the subject categories identified by the Institute for Scientific Information (ISI) was analyzed and the result was displayed in Fig. 3.The total of 2076 articles covered 50 ISI identified subject categories in the SCI databases. The annual articles of the top ten productive subject categories were analyzed. The top ten categories were Computer Science (817, 28.05%), Engineering (390, 13.39%), Telecommunications (202, 6.93%), Science & Technology-Other Topics (136, 4.67%), Health Care Sciences & Services (77, 2.64%), Mathematics (71, 2.44%), Pharmacology & Pharmacy (70, 2.40%), Neurosciences & Neurology (59, 2.03%), General & Internal Medicine (54, 1.85%) and Biochemistry & Molecular Biology (52, 1.79%). We noticed that 41% of all articles were mostly related to Computer Science and Engineering, and the distribution of subject categories also suggested the high priority of Telecommunications, Health Care Sciences & Services, and Mathematics issues in the research fields of "Big Data".

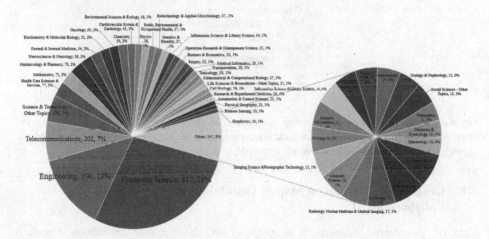

Fig. 3. Distribution of subject categories

We visualize a subject categories co-occurring network applying a threshold to the network between centrality in the network of subject categories. Network centrality measures the relative importance of nodes within networks and could be used as an indicator of a subject category's position within the network [11]. We can find that the Computer Science and Engineering took part in more co- occurring relationship, and SOFTWARE ENGINEERING took the central position in the co-occurring network, followed by MULTIDISCIPLINARY SCIENCES, HARDWARE & ARCHITECTURE and so on (Fig. 4).

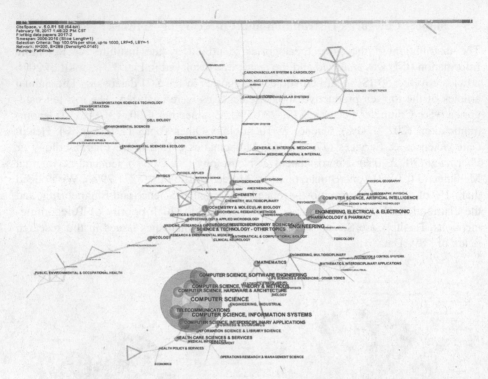

Fig. 4. Subject categories co-occurring network (the thickness of each link represents the intensity of co-occurring, the size of each node represents the number of total articles, and the purple color denotes the high betweenness centrality node)

3.4   Geographic Distribution Map of Countries and International Collaboration

Data on geographic information were generated from author affiliations. Figure 5 shows the geographic distribution of countries (regions) in the field of "Big Data". Overall, The total of 2076 articles covered 131 countries (regions) and ten most common countries (regions) were USA (731, 38.86%), China (373, 19.83%), England (93, 4.94%), South Korea (68, 3.62%), Germany (59, 3.14%), Spain (57, 3.03%), Australia (55, 2.92%), Canada (53, 2.82%), Italy (40, 2.13%), France (38, 2.02%).

Figure 6 depicts a network consisting of 82 nodes and 111 links on behalf of the collaborating countries between 2006 and 2016. As can be seen, the major contribution of the total output mainly came from two countries, namely, USA and China. Clearly, USA is the largest contributor publishing 731 papers. In other words, USA and China has a dominant status in the field of "Big Data", which produced about which produced about a half of world's total during this period. An interesting observation is that there are certain countries which have relatively low frequency but have high value of centrality among all other countries. JAPAN leads other countries, which are shown as node rings in purple in Fig. 6. This is followed by papers originating from FINLAND, USA, HUNGARY, MEXICO, FRANCE, SPAIN, SAUDI ARABIA, MALAYSIA,

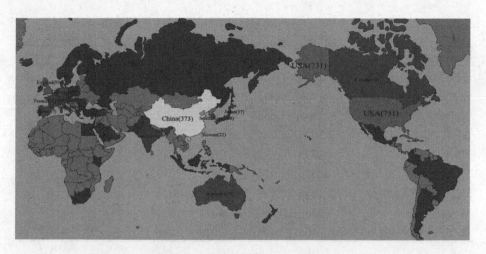

Fig. 5. Geographic distribution map of Countries (regions)

PAKISTAN, SOUTH AFRICA, GERMANY and so on. In other words, they are pivotal nodes in the network with the highest betweenness centrality. In addition, ten countries are found to have citation bursts: USA (15.93), SLOVENIA (1.9), JAPAN (1.65), DENMARK (1.34), NETHERLANDS (1.23), GERMANY (1.13), IRELAND (1.08), WALES (0.86), SOUTH AFRICA (0.84) and INDONESIA (0.83),suggesting that they have abrupt increases of citations, and the details are listed in the Table 1.

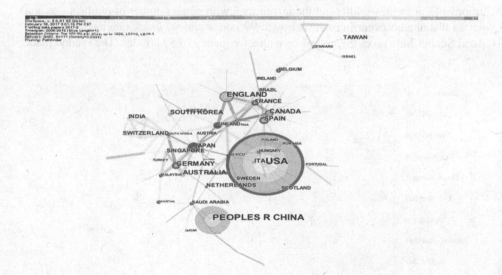

Fig. 6. Cooperation network of Countries on "Big Data" research papers

Table 1. Top 10 countries with the strongest citation bursts.

Countries	Year	Strength	Begin	End	2006 - 2016
GERMANY	2006	1.1298	2006	2011	▄▄▄▄▄▄▄ ▄ ▄ ▄ ▄
SOUTH AFRICA	2006	0.842	2006	2013	▄▄▄▄▄▄▄▄▄ ▄ ▄
JAPAN	2006	1.6547	2006	2012	▄▄▄▄▄▄▄▄ ▄ ▄ ▄
NETHERLANDS	2006	1.234	2011	2011	▄ ▄ ▄ ▄ ▄ ▄ ▄ ▄
USA	2006	15.925	2012	2013	▄ ▄ ▄ ▄ ▄ ▄ ▄ ▄
IRELAND	2006	1.0791	2013	2013	▄ ▄ ▄ ▄ ▄ ▄ ▄ ▄
INDONESIA	2006	0.835	2013	2014	▄ ▄ ▄ ▄ ▄ ▄ ▄ ▄
WALES	2006	0.8557	2014	2014	▄ ▄ ▄ ▄ ▄ ▄ ▄ ▄
CZECH REPUBLIC	2006	0.8243	2015	2016	▄ ▄ ▄ ▄ ▄ ▄ ▄ ▄▄

3.5   Institutions Distribution and Co-occurring Network

Overall, a total of 2163 research institutes in the world were engaged in "Big Data" during 2006 and 2016. Figure 7 lists the top ten of them: Chinese Acad Sci (57 papers), Harvard Univ (40 papers), Tsinghua Univ (34 papers), Univ Minnesota (23 papers), Univ Washington (22 papers), Univ Maryland (20 papers), Johns Hopkins Univ (19 papers), Univ Michigan (19 papers), NYU(17 papers) and Stanford Univ (17 papers).

Figure 8 shows Institutes co-occurring network. In order to show the core institutions of this field, we filter out the institutions with small number of publications and get an institute co-occurring network with 99 nodes and 90 links. Obviously, Chinese Acad Sci in China takes the first place with a frequency of 57 articles. The second place

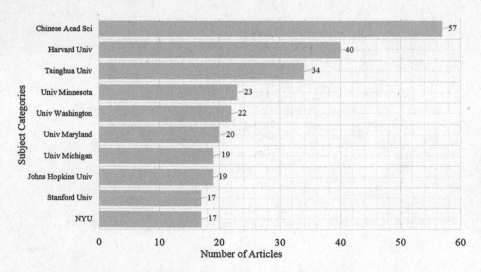

Fig. 7. Distribution of institutes

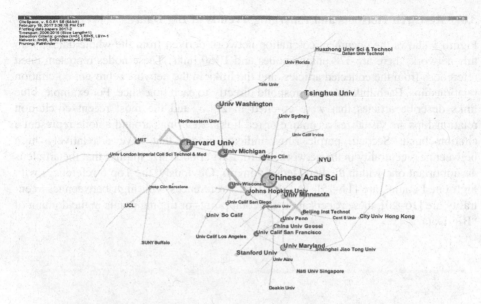

Fig. 8. Institutes co-occurring network with 99 Nodes and 90 Links

is Harvard Univ with a frequency of 40 articles. We also notice that China' institutes such as Chinese Acad and Tsinghua Univ were on the top of the list. In addition, ten institutes are found to have citation bursts: MIT (5.12), NYU (4.23), Natl Univ Singapore (3.98), Univ Calif San Diego (3.74), SUNY Buffalo (3.17), Univ Calif Los Angeles (2.87), Univ Roma La Sapienza (2.77), Univ Calif Berkeley (2.77), Univ Arizona (2.77) and Univ So Calif (2.63), suggesting that they have abrupt increases of citations detected, and we listed the details in the Table 2.

Table 2. Top 10 institutes with high citation bursts

Countries	Year	Strength	Begin	End	2006 - 2016
MIT	2006	5.124	2012	2014	▫▫▫▫▫▫◻◻▫▫▫
NYU	2006	4.2313	2014	2014	▫▫▫▫▫▫▫▫◻▫▫
Natl Univ Singapore	2006	3.9787	2015	2016	▫▫▫▫▫▫▫▫▫◻◻
Univ Calif San Diego	2006	3.7416	2014	2014	▫▫▫▫▫▫▫▫◻▫▫
SUNY Buffalo	2006	3.1713	2015	2016	▫▫▫▫▫▫▫▫▫◻◻
Univ Calif Los Angeles	2006	2.8693	2015	2016	▫▫▫▫▫▫▫▫▫◻◻
Univ Calif Berkeley	2006	2.7698	2015	2016	▫▫▫▫▫▫▫▫▫◻◻
Univ Roma La Sapienza	2006	2.7698	2015	2016	▫▫▫▫▫▫▫▫▫◻◻
Univ So Calif	2006	2.6298	2014	2016	▫▫▫▫▫▫▫▫◻◻◻
MIT	2006	5.124	2012	2014	▫▫▫▫▫▫◻◻▫▫▫

3.6   Research Hotspots and Emerging Trends of Big Data

Figure 9 shows the document co-citation network derived from the whole datasets. In this network, there are 378 unique nodes and 1360 links. These nodes represent cited references from the collected articles, and the links in the network represent co-citation relationships. Each link colors correspond directly to each time slice. For example, blue links describe articles that were co-cited in 2006, and the most recent co-citation relationships are visualized as orange or red links. Red rings around a node represent a citation burst. Second, purple rings indicate nodes that have a relatively high betweenness centrality in the network. Third, Larger node size implies that the article is an important one within the knowledge domain. Obviously, the Top 5 references with high cited counts are [11–15], and the Top 5 references with high betweenness centrality are [16–20], these papers are the pivotal points or tipping points in the domain of "Big Data".

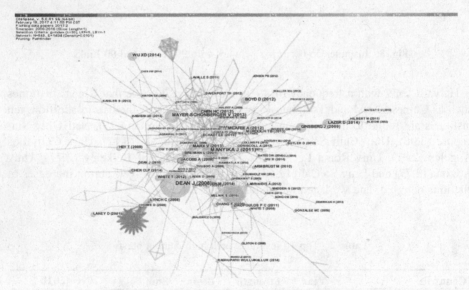

Fig. 9. References co-citation network of "Big Data"

Table 3 presents the Top 10 articles with highly co-cited counts which can represented the research hotspots of "Big Data". Dean, Jeffrey and Sanjay Ghemawat (2004) [21] proposed the algorithm of MapReduce: Simplified Data Processing on Large Clusters. Manyika, James, et al. (2011) [22] discussed the Big Data: The Next Frontier For Innovation, Competition, And Productivity. Mayer-Schnberger, Viktor (2013) [23] regarded Big Data as a revolution that will transform how we Live, work and think. Lazer, D, et al. (2014) [24] discussed the problem of the parable of google flu: traps in big data analysis. Ginsberg, J, et al. (2008) [25] detected influenza epidemics using search engine query data. Wu, Xindong, et al. (2014) [26] discussed the problem of "data mining with big data". danah boyd, and Kate Crawford (2012) [27]

Table 3. Top 10 articles with highly co-cited counts.

No	Title	First author	Co-cited counts	Year
1	MapReduce: Simplified Data Processing on Large Clusters	Dean, Jeffrey	123	2004
2	Big Data: The Next Frontier For Innovation, Competition, And Productivity	Manyika, James	71	2011
3	Big Data: A Revolution That Will Transform How We Live, Work and Think	Mayer-Schnberger, Viktor	60	2013
4	Big data. The parable of Google Flu: traps in big data analysis	Lazer, D	47	2014
5	Detecting influenza epidemics using search engine query data	Ginsberg, J	46	2008
6	Data Mining with Big Data	Wu, Xindong	42	2014
7	CRITICAL QUESTIONS FOR BIG DATA	danah boyd	36	2012
8	The big challenges of big data	Marx, Vivien	35	2013
9	The Inevitable Application of Big Data to Health Care	Murdoch, Travis B.	33	2013
10	Big Data: A Survey	Chen, Min	32	2014

discussed the critical questions for big data. Marx, Vivien (2013) [28] talked about the big challenges of big data. Murdoch, Travis B., and A. S. Detsky (2013) [29] discussed the inevitable application of big data to health care. Chen, Min, S. Mao, and Y. Liu (2014) [30] review the background and state-of-the-art of big data. (Fig. 10)

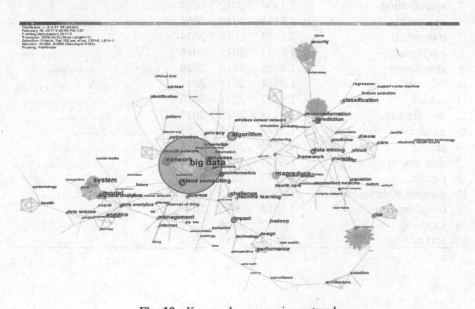

Fig. 10. Keyword co-occurring network

In order to find the research hotspots and frontiers about "Big Data" in detail, A keyword co-occurring method were used, and Fig. 8 show the result of such method. There are 286 keyword nodes, 659 vertex in the network and the keywords with high betweenness centrality are bioinformatics (0.54), prediction (0.43), cloud computing (0.42),alignment (0.42), artificial intelligence (0.41), data analytics (0.32), genomics (0.32), standard (0.32), data mining (0.31), database (0.3), big data (0.28), biodiversity (0.24), algorithm (0.22), evolution (0.22), impact (0.21), data compression (0.2) and so on.

In addition, twenty keywords are found to have citation bursts: epidemiology (4.3748), scalability (2.7789, social media (2.5466), genomics (2.5052), visualization (2.3594), sequencing data (2.1153), integration (2.1153), intelligence (1.9831), association (1.9385), behavior (1.8561), cohort (1.8497), data storage (1.8153), genome (1.814), cluster (1.6885), data quality (1.6177), parallel (1.6177), genome wide association (1.5858), image retrieval (1.5858), sampling (1.5857) and text mining (1.5857), suggesting that they have abrupt increases of citations detected, and we listed the details in the Table 4.

Table 4. Top twenty keywords with high citation burst

ID	Keywords	Burst Strength	Start	End	2006-2016
1	epidemiology	4.3748	2015	2016	
2	scalability	2.7789	2015	2016	
3	social media	2.5466	2014	2016	
4	genomics	2.5052	2012	2016	
5	visualization	2.3594	2012	2013	
6	sequencing data	2.1153	2014	2014	
7	integration	2.1153	2014	2014	
8	intelligence	1.9831	2015	2016	
9	association	1.9385	2013	2014	
10	behavior	1.8561	2014	2014	
11	cohort	1.8497	2014	2016	
12	data storage	1.8153	2012	2014	
13	genome	1.814	2011	2014	
14	cluster	1.6885	2013	2014	
15	data quality	1.6177	2014	2016	
16	parallel	1.6177	2014	2016	
17	genome wide association	1.5858	2015	2016	
18	image retrieval	1.5858	2015	2016	
19	sampling	1.5857	2014	2014	
20	text mining	1.5857	2014	2014	

4   Conclusion

In this paper, we used a scientometric method to quantitatively assessing current research hotspots and trends on "Big Data", using the related literature in the Science Citation Index (SCI) database from 2006 to 2016. Articles referring to "Big Data" were concentrated on the analysis of scientific outputs, distribution of countries, institutions, periodicals, subject categories and research performances by individuals. Moreover, in-innovative methods such as Keyword co-citation analysis, semantic clustering and Keyword Frequent Burst Detection were applied to provide a dynamic view of the evolution of "Big Data" research hotpots and trends from various perspectives which may serve as a potential guide for future research.

References

1. Graham-Rowe, D., Goldston, D., Doctorow, C., Waldrop, M., Lynch, C., Frankel, F., Reid, R., Nelson, S., Howe, D., Rhee, S.Y.: Big data: science in the petabyte era. Nature **455** (7209), 1–136 (2008)
2. Dealing with data. Science **331**(6018), 639–806 (2011
3. Jin, X., et al.: Significance and challenges of big data research. Big Data Res. **2**, 59–64 (2015)
4. Ekbia, H., Mattioli, M., Kouper, I., Arave, G., Ghazinejad, A., Bowman, T., Suri, V.R., Tsou, A., Weingart, S., Sugimoto, C.R.: Big data, bigger dilemmas: a critical review. J. Assoc. Inf. Sci. Technol. **66**(8), 1523–1545 (2015)
5. Al-Jarrah, O.Y., Yoo, P.D., Muhaidat, S., Karagiannidis, G.K., Taha, K.: Efficient machine learning for big data: a review. Big Data Res. **2**(3), 87–93 (2015)
6. Hashem, I.A.T., Yaqoob, I., Anuar, N.B., Mokhtar, S., Gani, A., Khan, S.U.: The rise of "Big Data" on cloud computing: review and open research issues. Inf. Syst. **47**, 98–115 (2015)
7. Hilbert, M.: Big data for development: a review of promises and challenges. Dev. Policy Rev. **34**(1), 135–174 (2016)
8. Liu, J.Z., Li, J., Li, W.F., Wu, J.Z.: Rethinking big data: a review on the data quality and usage issues. ISPRS J. Photogram. Remote Sens. **115**, 134–142 (2016)
9. Chen, C.: CiteSpace II: detecting and visualizing emerging trends and transient patterns in scientific literature. J. Am. Soc. Inf. Sci. Technol. **57**(3), 359–377 (2006)
10. Rogosa, D., Brandt, D., Zimowski, M.: A growth curve approach to the measurement of change. Psychol. Bull. **92**(3), 726 (1982)
11. Freeman, L.C.: Centrality in social networks conceptual clarification. Soc. Netw. **1**(3), 215–239 (1979)
12. Schadt, E.E.: Computational solutions to large-scale data management and analysis. Nat. Rev. Genet. **11**(9), 647–657 (2010)
13. Manyika, J.: Big data: the next frontier for innovation, competition, and productivity. Analytics (2011)
14. Schadt, E.E.: Computational solutions to large-scale data management and analysis. Nat. Rev. Genet. **11**(9), 647–657 (2010)
15. Ranger, C.: Evaluating MapReduce for multi-core and multiprocessor systems. In: HPCA (2007)

16. Schatz, M.C.: Highly sensitive read mapping with MapReduce. Bioinformatics **25**, 1363–1369 (2009)
17. Bell, G., Hey, T., Szalay, A.: Beyond the data deluge. Science **323**(5919), 1297–1298 (2009)
18. Jacobs, A.: The pathologies of big data. Queue **7**(6), 10 (2009)
19. Howe, D.: The future of biocuration. Nature **455**(7209), 47–50 (2008)
20. Bengio, Y.: Learning deep architectures for AI. Found. Trends® Mach. Learn. **2**, 11–55 (2009)
21. Kambatla, K.: Trends in big data analytics. J. Parallel Distrib. Comput. **74**(7), 2561–2573 (2014)
22. Dean, J., Ghemawat, S.: MapReduce: simplified data processing on large clusters. In: OSDI (2004)
23. Mayer-Schnberger, V., Cukier, K.: Big Data: A Revolution That Will Transform How We Live, Work and Think. Houghton Mifflin Harcourt, Boston (2013)
24. Lazer, D.: Big data. the parable of google flu: rraps in big data analysis. Science **343**(6176), 1203 (2014)
25. Ginsberg, J.: detecting influenza epidemics using search engine query data. Nature **457** (7232), 1012–1014 (2008)
26. Wu, X.: Data mining with big data. IEEE Trans. Knowl. Data Eng. **26**(1), 97–107 (2014)
27. Boyd, D., Crawford, K.: Critical questions for big data. Inf. Commun. Soc. **15**(5), 1–18 (2012)
28. Marx, V.: The big challenges of big data. Nature **498**(7453), 255–260 (2013)
29. Murdoch, T.B., Detsky, A.S.: The inevitable application of big data to health care. JAMA, J. Am. Med. Assoc. **309**(13), 1351–1352 (2013)
30. Chen, M., Mao, S., Liu, Y.: Big data: a survey. Mobile Netw. Appl. **19**(2), 171–209 (2014)

Template Based Industrial Big Data Information Extraction and Query System

Jie Wang$^{(\boxtimes)}$, Yan Peng, Yun Lin, and Kang Wang

School of Management, Capital Normal University,
105 North road of the Western 3rd-Ringroad, Beijing, China
cnu_wangjie@126.com

Abstract. Currently, with the rapid development of industry, the amount of data generated by industrial enterprises and industrial business website is exponential growth, and the big data has different types. In this paper, we design and implement an industrial big data information acquisition and query system. The system is based on big data acquisition and analysis of industry news data and industrial products data. We use a template based information acquisition method to crawl data from industry related news data and industry products data. We also discuss the query performance of text industry data with text index and only by SQL without index. The system is useful for analysis the hot news in industrial field and industry public opinion, and it is also useful for providing reference and rapid search and comparison of the relevant industrial products price, inventory and other information.

Keywords: Industry big data · Information extraction · Query system · Template based

1 Introduction

In recent years, with the rapid development of industrial technology, the promotion of industrial news and products information technology continue to enter the public view. Since the era of social media, all enterprises and individuals can become the publisher of information [1].

The system mainly faced to the industry of Internet information of collection and analysis, and is based on the technologies including the Chinese information processing technology, data mining technology as the core, mainly including industrial data acquisition module, data pre-processing module, industrial data analysis processing module and display module.

In this paper, we design and implement an industrial big data information acquisition and analysis system. The system is based on large data acquisition and analysis of Internet industry news data and industrial products information data, and it is useful for analysis the hot news in industrial field and industry public opinion, and it is also useful for providing reference and rapid search and comparison of the relevant industrial products price, inventory and other information.

In this paper, we introduce the acquisition algorithm of industrial big data and information extraction strategy based on template, and system structure, design of each

© Springer International Publishing AG 2017
Y. Tan et al. (Eds.): DMBD 2017, LNCS 10387, pp. 247–254, 2017.
DOI: 10.1007/978-3-319-61845-6_25

function module and the concrete realization method of the system are also introduced. We also discuss the performance of query text industry data with index and only by SQL without index. The system is useful for analysis the hot news in industrial field and industry public opinion, and it is also useful for providing reference and rapid search and comparison of the relevant industrial products price, inventory and other information.

2   The Overall Design of the System

2.1   System Architecture

The system architecture of the industrial big data information acquisition and analysis system is as follows (Fig. 1).

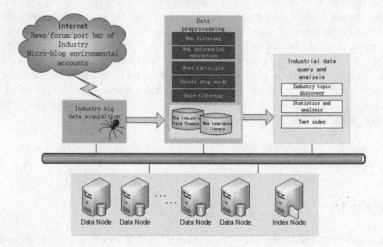

Fig. 1. System architecture

The core part of the system has four modules: industrial big data acquisition module, data preprocessing module, industrial data storage and data retrieval module and industrial big data analysis module.

This module is based on Jsoup [2], which is an open source tools to design the industrial data acquisition system, which analysis the specific targets in the Internet resources. Firstly, it obtains a relative URL, processes the URLs according to the preset crawling parameters (such as URL suffix, spider's climb take depth, etc.) and forms the crawl queue, and crawl web source resources and store them in the database [3].

The main task of data preprocessing module includes pre-built the DOM tree of the web page, web information extraction [4], web keyword extraction, generating page feature vector. Information extraction from web page by page purification technology on Internet information collecting module to the page source for cleaning, extract pages of useful information, such as news headlines, news time, news content, news writers.

In our system, the template based information extraction method is used and a new K-VSM model representation is proposed to extract the news Web text. In this way, the processing of text will be converted into a form of computer processing, for the next step of research and analysis of data preparation.

Industrial data storage module is the foundation of industrial data query, data analysis and mining, and which provides user defined keywords or time to extract the needed data, or view of all data, all data will be in reverse chronological order arrangement [5].

The industrial big data query module completes the index function of big volume text data, and 20 thousand industry news data retrieval from different sites are used to test the query performance with Lucene and only use SQL without text index.

In the next research plan, we will use the data to do emotional analysis of the data, the hot topic discovery, news clustering. Basic data processing consists of word processing, and feature vector representation [6].

3 Template Based Industrial Big Data Information Extraction

3.1 The Process of Data Acquisition

In the information acquisition system, take the Webpage as a node in the graph and the links in it are edges points to the other Webpages. Then the Internet Webpages are modeled as a directed graph [5]. By accessing specific areas of the page URL collection and getting Internet resources, we realized the bulk collection of related pages.

3.2 The Realization of Template Based Data Extraction

In this paper, the open source Java package Jsoup is used to complete data acquisition and information extraction.

In Web information acquisition, a web-based analytical method based on DOM is often used. DOM provides a structured way to describe documents, each element in the document, and the properties are in the tree node. At first, it uses the HTTP function to get the DOM of the whole target Web pages, and then uses the REMOVE function to go through all the DOM nodes in order to remove the CSS code and JavaScript which can reduce the workload later. According to statistical data show that noise pages often contain concentrated in the following tag: <script> </ script>, <object> </ object>, <input> </ input> tags in a class, so our de-noising work mainly on the Tag is removed.

In our industry information extraction and query system, we used a template based method to get information from industry news sites and products sites. And the whole process of the realization of template based data crawling is described as below (Fig. 2):

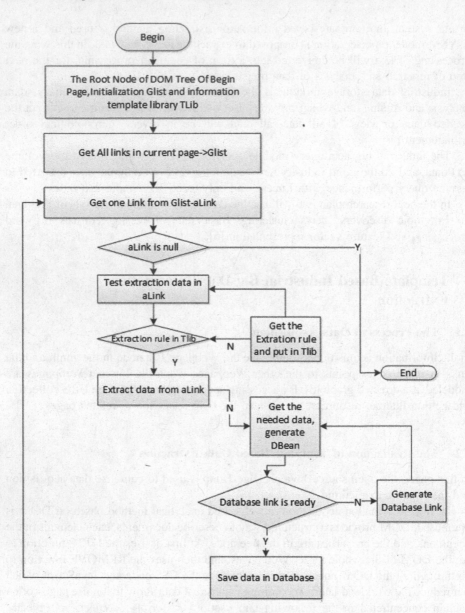

Fig. 2. The process of the of template based data acquisition

3.3   Information Extraction and Data Storage

Jsoup provides very convenient APIs, with the DOM, CSS and other methods similar to the jQuery, which retrieves data and finishes the preliminary data cleaning. For the same web site, the pages often written based on similar template, so when extract information from web pages from the same site, we can define the extraction rules by testing several pages from the site and put it into the extraction template in the template library. When a new page comes, we firstly test whether it can be extracted by the template library. If it's extraction rule is new, we test to get the new rule and define it as a new template and put it in the library.

For example, after analyzing the format of the first 100 data pages from one industry product site, we found that data in the <td> tag was ranged like "Products - Prices - market information - date -null-null". This structure was dealt as a prototype in the process to design the specific program. Each item corresponds to a link, we had to traverse and store all the URLs. After entering the search for how many goods in the sale, the results of the total number of pages is calculated. So we can define a template for the site and put it in the library. As to get available and complete data. To avoid data duplication, a unique code was generated valued as a primary key with the treated data in the database.

The algorithm of information extraction is as below:

```
Input: Url Queue
Output: Info Bean List extracted from Pages
For each Url in Url Queue
Get doc from Jsoup.connect (Url);
 If doc is not null do
   If Test info extraction template Ttemp in Tlib; //
extraction rule is in template library
       If info is not null do
Put info select by rule defined in Ttemp into in-
foBeans;
If infos is not null do
      For each Url
          Get each field of info from page in plain
text; //such as production, price, place, time fields
      End for
End for
```

We can distinguish different website login IP address and get the active query information that the server received from the browsers, in order to use CGI program to push accurate result or product data to consumers with the realization of user interaction.

4   Performance Analysis of Data Query

Industrial data storage and retrieval is the basis of the analysis and mining of large industry data. The storage and retrieval of industry products big data is relatively simple, just storage them by field in database and use SQL to retrieval can work well. The storage and retrieval of industrial news data is complicated. For news headlines or news content, which query has the key words or the time limit, when the query is doing in big text data and the results include news headlines, keywords, abstract, release time, the original link, etc. the query performance is a big data processing system.

We use Lucene which is an open source tools to retrieve industry news big data in our system. Lucene establishes inverted index for all data, and the query results will return to complete the sorting result set. Lucene turns fuzzy queries into multiple processes, which can use the index query and logical combination, and the retrieval efficiency is greatly improved when the big data.

We selected 20 thousand industry news data retrieval from different sites to test the query performance. Industry news headlines and news content were tested every 5000 data for a retrieval experiment, and a total of 16 sets of data having different data volume and different data size are tested by two different retrieval methods.

The experimental results are shown in Table 1:

As shown in Fig. 4, in the search of title, the retrieval methods with Lucene index use less time than the SQL query methods, but the gap is not large, until to the 35000 data volume, Lucene query has been significantly improved. As shown in Fig. 3, when the amount of data increases, the time of the two retrieval methods has been improved to some extent, but the time of Lucene retrieval on any data set is significantly lower than that of SQL fuzzy query.

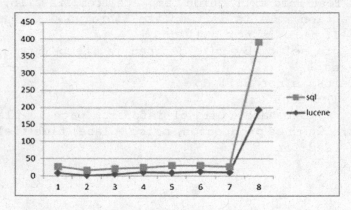

Fig. 3. SQL and Lucene query time comparison in content field

Therefore, it can be seen from the above experiments, in the large text data retrieval, the text index can effectively improve the query performance, especially for the content of the text data set.

Table 1. Comparison of like query and Lucene retrieval time

Heading level data volume	The query time of content with Lucene index	The time for SQL to query content without Lucene index	The time for query title with Lucene index	The time for query title without Lucene index
5000	5	316	8	19
10000	124	328	1	15
15000	140	312	5	16
20000	192	320	10	14
25000	169	373	9	21
30000	142	471	12	18
35000	176	415	10	17
40000	192	383	19	20

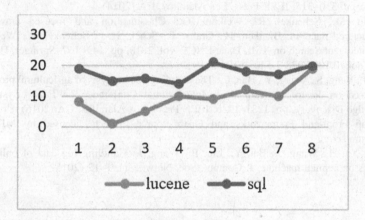

Fig. 4. SQL and Lucene query time comparison in title field

5 Conclusion

Currently, with the rapid development of industrial 4.0 and industrial products of electronic commerce, the amount of data generated by industrial enterprises and industrial business website is exponential growing, and the big data has the character of big amount, unstructured, professional growth trend. It is difficult for people to quickly retrieve and dig the industrial news from the mass of information and obtain the information of industrial products [8].

In this paper, we design and implement a template based industrial big data information extraction and query system. The system is based on big data acquisition and analysis of Internet industry news data and industrial products of data, and it is useful for analysis the hot news in industrial field and industry public opinion, and it is also useful for providing reference and rapid search and comparison of the relevant industrial products price, inventory and other information.

Acknowledgments. This work was supported by the Key Laboratory of machine intelligence and advanced computing (MSC-201707A); Capital Normal University interdisciplinary research project; Capital Normal University science and technology innovation platform project.

References

1. Viktor, M.S., Kenneth, C.: Big Data: A Revolution That Will Trans-Form How We Live, Work, and Think. Houghton Mifflin Harcourt, Boston (2013)
2. Jsoup Open Source Project Distributed under the Liberal MIT License. http://jsoup.org/
3. Wang, J., Wu, J., Zhang, Y., He, G.: Content information extraction of theme web pages based on tag information. In: 7th IEEE International Symposium on Computational Intelligence and Design, pp. 501–504. IEEE Press, Los Alamitos, CA (2015)
4. He, G., Wang, J., Zhang, Y., Peng, Y.: Keyword extraction of web pages based on domain thesaurus. In: 3th IEEE International Conference on Cloud Computing and Intelligence Systems, pp. 310–315. IEEE Press, Los Alamitos, CA (2014)
5. Theobald, M., Schenkel, R., Weikum, G.: Classification and focused crawling for semistructured data. In: Blanken, H., Grabs, T., Schek, H.-J., Schenkel, R., Weikum, G. (eds.) Intelligent Search on XML Data. LNCS, vol. 2818, pp. 145–157. Springer, Heidelberg (2003). doi:10.1007/978-3-540-45194-5_10
6. Wang, J., Yang, S., Wang, Y., Han, C.: The crawling and analysis of agricultural products big data based on Jsoup. In: 12th IEEE International Conference on Fuzzy Systems and Knowledge Discovery, pp. 1231–1236. IEEE Press, Los Alamitos, CA (2016)
7. Bootstrap Front-end Frameworks and Open Source Projects licensed by MIT. http://getbootstrap.com/
8. Jia, M., Xu, H., Wang, J., Bai, Y., Liu, B., Wang, J.: Handling big data of online social networks on a small machine. J. Comput. Soc. Netw. 2(1), 1–12 (2015)

Finding the Typical Communication Black Hole in Big Data Environment

Jinbo Zhang[1,2], Shenda Hong[1,2], Yiyong Lin[1,2], Yanchao Mou[1,2], and Hongyan Li[1,2(✉)]

[1] Key Laboratory of Machine Perception, Peking University, Ministry of Education, Beijing, China
ihy@cis.pku.edu.cn
[2] School of Electronics Engineering and Computer Science, Peking University, Beijing, China

Abstract. "Black hole" are widely spread in the mobile communication data, which will highly downgrade the mobile service quality. OLAP tools are extensively used for the decision-support application in the multidimensional data model, which just like the mobile communication case. As different dimensions of the mobile data are incomparable and, thus, can hardly generate one unique final value that satisfies all dimensions. We exploit the skyline operator as the postoperation while building data cubes, named as data cube of skyline. As the skyline of a cuboid is not derivable from another cuboid and the skyline operation is holistic, which makes this problem even challeging. In this paper, we propose a method in materializing the cube of skyline in the big communication data and proof its effectiveness and efficiency by extensive experiments.

1 Introduction

"Black hole", which is defined as an area with a weak quality of the wireless signal, paging channel congestion, or inability to establish a control channel, are widely spread in the mobile communication data. In the "black hole" areas, mobile users will face with resetting up connection, sending messages errors or communication interrupted problems, which will highly downgrade the mobile service quality.

Communication data is usually stored in the multidimensional data model, which includes delay time of different reasons, location, occur time and communication data type information. OLAP tools are extensively used for the decision-support application in the multidimensional data model, which just like the mobile communication case. The traditional OLAP aggregation function on data cubes generally takes a set of measure values as input content and, after computing according to pre-defined criteria, returns a unique numeric value. As different dimensions of the mobile data, however, are incomparable[1] and, thus, can hardly generate one unique final value that satisfies all dimensions.

---

[1] For example, it's meaningless to compare "year" dimension to "region" dimension.

© Springer International Publishing AG 2017
Y. Tan et al. (Eds.): DMBD 2017, LNCS 10387, pp. 255–262, 2017.
DOI: 10.1007/978-3-319-61845-6_26

We, instead, resort to the skyline operator. By extending data cube with a new perspective, this paper exploits the skyline operator as the postoperation while building data cubes, named as data cube of skyline.

For example, Table 1 shows the China Mobile communication session data. Each tuple represents a communication session record, with the record "Region", "Year", as well as its evaluated scores "DelayAuthReq", "DelayAuthResp", and "DelaySetup". Suppose that lower delay is preferable over higher ones. If we want to find the outstanding session of each year, the table in Table 3 can answer the query directly. That table is called a skyline data cuboid with grouping dimension set $G = \{$"Type"$\}$ and skyline dimension set $S = \{$"DelayAuthReq", "DelayAuthResp", "DelaySetup"$\}$. That essentially means that the communication sessions in D are first grouped by regions and then the skyline operation conduct on each group to find those which are not dominated by the others on all three delays. On the other hand, If we want to find the outstanding communication of each region in parts of "DelayAuthReq" and "DelayAuthResp", then the table in Table 2 can be used to solve the query.

The key in exploiting the skyline operator on data cubes is in efficient computation of the data cube.

Several methods have paved ways towards this goal. For example, working partitioning methods [3] assign the computation of different cuboids to different nodes. Data partitioning methods [5] partition the raw data set into p subsets and store each subset locally on one computer node. Each node computers all

Table 1. China mobile black holes database D.

ID	Region	Type	DelayAuthReq	DelayAuthResp	DelaySetup
51713	Haidian	4	5	5	3
51714	Haidian	4	4	8	7
51715	Haidian	3	3	3	4
...	...	...	...	...	...
51717	Haidian	3	6	6	6
51718	Fengtai	3	2	4	2
51719	Fengtai	3	2	5	3
51720	Fengtai	3	7	7	4

Table 2. $S = \{$"DelayAuth Req", "DelayAuthResp"$\}$

Region	ID	Req	Resp
Haidian	51715	3	3
Technical	51718	2	4
	51719	2	5

Table 3. $G = \{$"Type"$\}$, $S = \{$"DelayAuthReq", "DelayAuthResp", "DelaySetup"$\}$

Type	ID	DelayAuthReq	DelayAuthResp	DelaySetup
4	51713	5	5	3
	51712	4	8	7
3	51715	3	3	4
	51718	2	4	2
	51719	2	5	3

Table 4. $S = \{$"DelayAuth Req", "DelaySetup"$\}$

Region	ID	Req	DS
Haidian	51713	5	3
	51715	3	4
Technical	51718	2	2

cuboids with the subset of data which store locally. These methods aim at the algebraic aggregate function instead of skyline ones and can hardly adapt to big data. For the skyline ones,Index-based solutions, such as [7–9], operate on bitmap and B^+ tree indices on the skyline attributes, whereas the others operator on an R-tree that indexes the data set by skyline attributes. Moreover, the skycube [15–17], which focus on analysing subspace skyline, of a data set $\mathcal{D}$ is defined as the collection of skyline result set $\Psi(\mathcal{D}, \mathcal{S})$ for each nonempty subset S of S^* but without any grouping attributes.

What's more, as the skyline of a cuboid is not derivable from another cuboid, he skyline operation is, therefore, *holistic*, which makes this problem even challeging. For example, consider the cuboids in Tables 2 and 4, although they share the same set of grouping dimensions and their skyline results are not derivable from each other. Specifically, we can see that Table 2 is not derived from Table 4, because the skyline black hole 51719 does not exist in Table 4. Moreover, Table 4 is not derived from Table 2, because the skyline black hole 51713 does not exist in Table 2 as well.

Furthermore, the mobile communication dataset is usually very huge, and it's saved in distributed storage. However, The dimensions that define a data cube of skyline include not only the usual grouping dimensions but also the measure attributes, which makes the corresponding data size will be exponentially huge.

2 Implementation

Given a set S of skyline attributes, a tuple t is said to *dominate* another tuple t', denoted by $t \succ_S t'$, if $(\exists A_i \in S, t[A_i] < t') \land (\forall A_i \in S, t[A_i] \leq t'[A_i])$ assuming that smaller values are preferable over larger ones. Here, we use $t[A_i]$ to represent the value of the attribute A_i of the tuple t. Given a set $\mathcal{D}$ of tuples, the skyline operation ψ on $\mathcal{D}$ is defined as

$$\psi(\mathcal{D}, S) = \{t \in \mathcal{D} | \nexists t' \in \mathcal{D}, t' \succ_S t\}. \tag{1}$$

in other words, a tuple t belongs to the skyline result set if no other tuple dominates it.

Let R is a relation which has k attributes, and these attributes are $X = \{A_1, A_2, ..., A_i, B_1, B_2, ..., A_j\}$. And i + j = k. Given a set $G \subset A$ of grouping attributes, which can be non-numeric, a set $S \subset B$ of measure attributes, which are all numeric. Accordingly, a data cube of skyline query is defined as follows:

Definition 1 (Data cube of skyline query). *Given the sets G group attributes and S measure attributes, a data cube of skyline query $Q = (G, S)$ computes a skyline result set $\Psi(\mathcal{D}(g), S)$ for each group instance g defined on G.*

Considering the data set $\mathcal{D}$ in Table 1, the result of a data cube of skyline query Q with its grouping dimension set $G = \{\text{"Type"}\}$, and its skyline dimension set $S = \{\text{"DelayAuthReq"}, \text{"DelayAuthResp"}, \text{"DelaySetup"}\}$ is shown in Table 3. And example of a group instance g is ("4").

In order to evaluate various data cube of skyline queries efficiently, we introduce the concepts of data cube of skyline cuboid and group-by skyline cube.

Definition 2 *(Data cube of skyline). Let G^* be the set of all possible grouping dimensions and S be the set of skyline dimensions. Given a subset, a Two-part group-by skyline cuboid $C(G, S)$ is defined as a collections of cells. The Two-part group-by skyline cube is defined as the collection of $C(G, S)$, for any subset.*

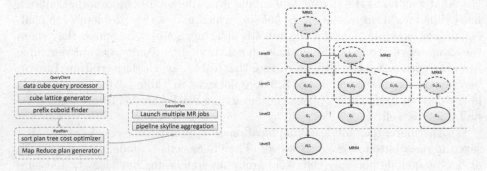

Fig. 1. Architecture overview   **Fig. 2.** Example of sort tree

The construction of data cube of skyline process is divided into three stages. The Fig. 1 shows the all stages: QueryClient, PipePlan and ExecutePlan. In the QueryClient stage, there are three steps. In these steps, Data cube query processor aim to processing the user query, in which the algorithm can get the skyline data cube query. Cube lattice generator can generate the proper cube lattice for the query. As soon as getting the proper cube lattice, the prefix cuboid finder will compute a set of cuboids sharing the same sort order together with one scan of cuboids lattice. Then, the algorithm will launch next step – PipePlan.

2.1   PipePlan Stage

The PipePlan is a stage which generates the minimum cost sort plan tree from a cube lattice. It uses a set of cuboids generated by prefix cuboid finder before, and builds sort trees and pipelines from a cube lattice structure to compute data cubes efficiently. The sort trees represent the cuboids which don't share the sort order of their parent cuboids thus have to sort these parents to compute them. For example, Fig. 2 includes three sort trees whose root nodes are represented as dotted circles(i.e. RAW, $G_1G_2G_3$ and G_1G_3 cuboids with dotted circles). The tree in the middle of the figure shows that the cuboid G_2G_3 is computed by sorting the cuboid $G_1G_2G_3$ in the order of G_1G_2 and the cuboid G_1G_3 by sorting the $G_1G_2G_3$ in the order of G_1G_3. In order to reduce the number of MapReduce phases and to maximize the degree of parallelism, the PipePlan stage processes each sort tree level by level which all the cuboids in the same level are sorted together from their parents by one MapReduce phase. Thus three

MapReduce phases plan are generated to process the sort trees in the example of Fig. 2. After planing the cuboids in sort trees, the Pipeplan execute next stage – ExecutePlan – to execute all the computation to output the data cube of skyline.

2.2   Execution Stage

From the PipePlan stage, as showed in Fig. 2, there are four MR jobs. MR #1, MR#2 and MR#3 are all from Algorithm 2. In execution stage, there are two techniques: MapReduce jobs and multipipe procedure. The details in MapReduce and MultiPipeMap are showed in Algorithm 2–4. And the most important challenge is how to aggregate the skyline points through the MapReduce framework. Since the pipe plan is built, the MapReduce procedure will be launched. Algorithm 2,3 describe the MapReduce procedure, which are referred as MR#1-3 in Fig. 2. In the MapReduce procedure, The Map procedure loads local data according to sort tree, and emit key-value pair, in which cuboid as key and LSP as value. The LSP is produced by The skylineAggregation procedure, whose details are explained below. The Reduce procedure combines the skyline points in the same cuboid, and outputs the final skyline points set.

Algorithm 1. Pipeplan

```
Input: CL: coboids lattice
Output: R: Construction Plan of Data Cube of Skyline
1  forall level k in CL do
2      Pipeline P < −P ∪ FindPrefixCuboid (k + 1− > k) ;
3      SortTree S < −P ∪ MinimumCostMatching (k + 1− > k);
4  end
5  forall SortSubTree s_k in S do
6      R < −R ∪ Map (s_k) & Reduce () ,
7      R < −R ∪ MultiPipemap (P);
8  end
9  return R
```

Algorithm 2. Reduce

```
Input: (cuboid, LSP): key-value
    pair from the map procedure
Output: (cuboid, FSP): the final
    skyline points in the exact
    cuboid
1  finalSkylinePoints < − SkylineCombination
    (cuboids, LSP) ;
2  emit (coboids, FSP);
```

Algorithm 3. MultiPipeMap

```
Input: P: pipelines for pipeplan stage
Output: (cuboid, FSP): the final skyline
    points in the exact cuboid
1  forall cuboids C in P do
2      skylinePoints <- skylineAggregation (PDataset,
        C) ;
3      emit (coboids, FSP);
4  end
```

Algorithm 4. Map

```
Input: S: sort tree
Output: (cuboid, LSP): for each cuboid, get
    the local skyline points set(LSP)
1  foreach parent cuboid P in S do
2      foreach child cuboid C in S do
3          SetofCuboid < − C ∪ SetofCuboid;
4      end cuboid, LSP
5      < − skylineAggregation
        (pDataset,SetofCuboid);
6      emit (cell, LSP);
7  end
```

Algorithm 5. SkylineAggregation

```
Input: PDataset: Parent cuboid P dataset
    SetofCuboid: all child cuboids for cuboid P
Output: SL: LCRS Skyline Tree
1  Sort PDataset by a topological monotone function F ;
2  SkylineFirst <- a skyline object in PDataset;
3  SL.insert (SkylineFirst) ;
4  for o ∈ PDataset do
5      PreOrderDominate (SL, o);
6  end
7  return SL
```

In order to maximize the degree of parallelism, the FSP produced from MapReduce procedure can be used to produce the children cuboids in parallel as showed in MR#4 in Fig. 2. In this procedure, because of the skyline points are all sorted by grouping attributes, the children cuboids' skyline points can be computed by skylineAggregation procedure. The SkylineAggregation procedure is adopted from OSP as the skyline computation in the solution. Thus,

This paper describe the method in detail below. The skylineAggregation procedure first sort the data set by a monotone score function $\mathcal{F}$. $\mathcal{F}$ is simply the SUM function in skyineAggregation procedure.

3   Experiments

In this section, we experimentally evaluate the efficiency and effectiveness of the proposed technique.

3.1   Experimental Setup

In the experiments, we use 1 NameNode, 20 DataNodes, and total 20 PCs in a cluster. All PCs are with 2.4 GHz Quad-Core AMD OpteronTM Processor 2378, 16 GB RAM, and a 500 GB HDD. All the PCs are installed with Ubuntu Linux 12.04, the Java version is JDK 1.6, and the MapReduce framework is Hadoop 0.20.2. The network speed is 1 G bps. In the experiments, this paper compares the algorithm with four algorithms: MRNaive algorithm (which is a naive MapReduce algorithm), MRLevel algorithm (which parallelizes the computation of cuboids in each level), and MRGBLP and MRPipeSort (which are the MapReduce version of GBLP and PipeSort, respectively).

3.2   Varying the Number of Tuples

Figure 3 shows the elapsed time for cube computation by varying the number of tuples, where we increase the number of tuples from 20 million to 100 million. As shown in the figure, MRNaive algorithm execution time increases significantly as the data size increase. For the other algorithms, the difference of cube computation time is not significant. However, MRPipeLevel algorithm shows the smallest execution time and MRLevel algorithm shows a similar rate with MRPipeLevel algorithm.

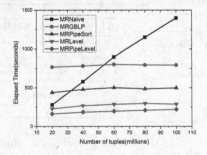

Fig. 3. The elapsed time of different methods with regarding to number of tuples

3.3 Varying the Number of Dimensions

Figure 4a shows the elapsed time obtained by varying the number of dimensions. In this experiment, we set the number of tuples to 50 billion and we increase the number of dimensions from three to nine by one. As shown in Fig. 4a, MRPipeLevel algorithm is fastest in all dimensions and MRNaive algorithm is slowest. In the case of 9 dimensions, MRPipeSort and MRLevel algorithm does not work because they emit too much data.

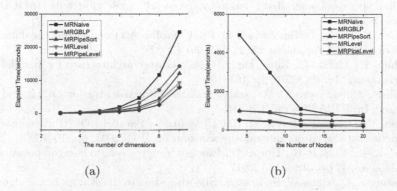

(a)   (b)

Fig. 4. The elapsed time of different methods with regarding to the number of dimensions and nodes

3.4 Varying the Number of Nodes

Figure 4b is a comparison among algorithms as increasing the number of nodes. Figure 4b measured the execution time of each algorithm for 50 billion tuples with 5 dimensions as increasing the number of nodes from 4 to 20. From the result, 5 dimensions as increasing the number of nodes from 4 to 20. From the result, we can observe that the execution time of every algorithms decreases as the number of nodes increases. MRPipeLevel algorithm is fastest in all nodes and MRNaive algorithm reduces the execution time the most significantly. Up to 12 nodes, the execution of every algorithms is reduced in a meaningful degree. But for more than 16 nodes, the computation time is enhanced in smaller degree, because it is getting saturated.

4   Conclusion

In this work, we extend the notion of data cubes using the skyline operation as the aggregated function. Moreover, to deal with big data problems, we propose several scalable methods to store, materialize and query data cubes generated by the skyline functions. For the future works, we will explore several more aggregation methods on the communication problem to improve further on the quality of communication.

Acknowledgement. This work was supported by Natural Science Foundation of China (No. 61170003).

References

1. Borzsony, S., Kossmann, D., Stocker, K.: The skyline operator. In: ICDE 2001, pp. 421–430 (2001)
2. Gray, J., Chaudhuri, S., Bosworth, A., et al.: Data cube: a relational aggregation operator generalizing group-by, cross-tab, and sub-totals. In ICDE'96, pp. 152–159
3. Dehne, F., Eavis, T., Hambrusch, S., Rau-Chaplin, A.: Parallelizing the data cube. Distrib. Parallel Databases **11**(2), 181–201 (2002)
4. Dehne, F., Eavis, T., Rau-Chaplin, A.: A cluster architecture for parallel data warehousing. In: ISCC'01, pp. 161–168
5. Goil, S., Choudhary, A.: A parallel scalable infrastructure for OLAP and data mining. In: IDEAS'99, pp. 178–186
6. Chen, Y., Dehne, F., Eavis, T., Rau-Chaplin, A.: Parallel ROLAP data cube construction on shared-nothing multiprocessors. In: IPDPS'03, pp. 10–18
7. Tan, K.-L., Eng, P.-K., Ooi, B.C.: Efficient progressive skyline computation. In: VLDB, vol. 1, pp. 301–310 (2001)
8. Kossmann, D., Ramsak, F., Rost, S.: Shooting stars in the sky: an online algorithm for skyline queries. In: PVLDB'02, pp. 275–286
9. Papadias, D., Tao, Y., Greg, F., Seeger, B.: Progressive skyline computation in database systems. TODS **1**, 41–82 (2005)
10. Bartolini, I., Ciaccia, P., Patella, M.: Efficient sort-based skyline evaluation. TODS **33**(4) (2008). 31
11. Borzsony, S., Kossmann, D., Stocker, K.: The skyline operator. In: ICDE'01, pp. 421-430
12. Chomicki, J., Godfrey, P., Gryz, J., Liang, D.: Skyline with presorting. In: ICDE, vol. 3, pp. 717–719 (2003)
13. Godfrey, P., Shipley, R., Gryz, J.: Algorithms and analyses for maximal vector computation. VLDBJ **16**(1), 5–28 (2007)
14. Zhang, S., Mamoulis, N., Cheung, D.W.: Scalable skyline computation using object-based space partitioning. In: SIGMOD'09, pp. 483–494
15. Pei, J., Jin, W., Ester, M., Tao, Y.: Catching the best views of skyline: a semantic approach based on decisive subspaces. In: PVLDB, pp. 253–264
16. Pei, J., Yuan, Y., Lin, X., Jin, W., Ester, M., Liu, Q., Wang, W., Tao, Y., Yu, J.X., Zhang, Q.: Towards multidimensional subspace skyline analysis. TODS **31**(4), 1335–1381 (2006)
17. Yuan, Y., Lin, X., Liu, Q., Wang, W., Yu, J.X., Zhang, Q.: Efficient computation of the skyline cube. In: PVLDB'05, pp. 241–252
18. Zhao, Y., Deshpande, P., Naughton, J.F.: An array-based algorithm for simultaneous multidimensional aggregate. In: SIGMOD'97
19. Beyer, K., Ramakrishnan, R.: Bottom-up computation of sparse and iceberg cubes. In: SIGMOD'99
20. Han, J., Pei, J., Dong, G., Wang, K.: Efficient computation of iceberg cubes with complex measures. In: SIGMOD'01, pp. 1–12

A Solution for Mining Big Data Based on Distributed Data Streams and Its Classifying Algorithms

Guojun Mao[(⊠)] and Jiewei Qiao

Information School, Central University of Finance and Economics,
Beijing, China
maximmao@hotmail.com

Abstract. With the advent of the era of big data, a require to discover valuable knowledge from big data is being one of focuses. However, big data is a term which has been described with the features of alarming velocity, super volume and various data structures, and so how to express big data to effectively mine is becoming a key problem. Aiming at requires of big data analyses, this paper will construct the concept of DDS (Distributed Data Stream), and build the mining model and some key algorithms. The experiments show integrating these algorithms under our model can get higher mining accuracies to distributed data streams.

Keywords: Big data · Data mining · Distributed data stream · Micro-cluster

1 Introduction

Big data is a term for data sets that are larger than traditional data. It has been described with the features of alarming velocity, super volume and structural diversities [1]. With development of the Internet and its applications, big data is becoming an inevitable data phenomenon and resulting in a new data age called as Big Data Era [2]. Yet such an era should not blind us to the fact that this is an important shift about the role of data in the world, which is changing our thinking and activities. Actually, big data is being generated by everything around us at all times, and their advantages in breadth and depth can help us to make better decisions with greater operational efficiency, lower cost and smaller risk.

In fact, many problems of mining big data are being focused on how to represent big data. Of course, it is so complex that it is difficult to make a general data structure to represent all kinds of big data. However, it is feasible to search out typical technical features from a type or group of big data applications.

In general, there are two techniques to support to analyze such big data: the first is data stream mining, which can help extracting knowledge patterns from continuous data records; and the second is distributed computing, which can help integrating local patterns into global patterns. Therefore, the concept of the DDS (Distributed Data Stream) which has appeared in recent years will become an ideal data format to represent most kinds of big data.

Y. Tan et al. (Eds.): DMBD 2017, LNCS 10387, pp. 263–271, 2017.
DOI: 10.1007/978-3-319-61845-6_27

Mining distributed data streams is an ongoing task, and many problems are still studying. As far as information exchange between nodes is considered, there are two basic processing strategies: (1) transferring all local data from multiple local nodes into the central node, where the global patterns could be found out; (2) transferring mined local patterns in local nodes to the central node, where integrate the global patterns. The first strategy should suffer from not only a large transferring cost between nodes but also very a heavy processing burden in the central servers. The second strategy can prevent a large communication cost happening, so it is more expecting and challenging. Figure 1 gives a data flow chart to mine global patterns in a distributed stream by the second strategy.

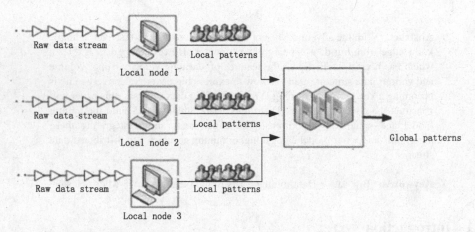

Fig. 1. A possible processing path for mining distributed data streams.

2 Related Work

Some literatures have introduced concepts, strategies and frameworks for mining distributed data streams. The two of them are directly related to this paper. Parthasarathy et al. discussed the problems and processing architectures to mine distributed data streams [3]. Bhaduri et al. pointed out that mining distributed data deals with distributed computing, communication and related human-factors [4].

There have been a few of efforts to design algorithms to mine distributed data streams. Guerrieri et al. proposed an algorithm for clustering distributed data streams called DS-means [5]. Gianluigi et al. presented an adaptive distributed ensemble approach to mine data streams [6]. Gaber et al. gave a classifying model in distributed wireless sensor networks [7]. Street and Kim proposed an ensemble method to mine data streams called SEA [8]. Gibbons et al. stated the availability of sliding window technique for mining data streams [9], and Yang et al. thought that sliding window techniques should be sorted out into two types of data-based and time-based ones [10].

In addition, Masud et al. designed a structure, named *micro-cluster* [11], and used it to do classifying to data streams. Similarly, Xu et al. also presented a learning method based on micro-clusters [12]. Our work in this paper will make use of the concept and methods of micro-cluster mining.

3 Terminology

In this section, we will give some terms to well discuss our methods.

Definition 1 (*Data stream*). Given a time series $T = <t_1 t_2, ..., t_i, ... >$, a data stream on T is represented as $S = <r_1, r_2, ..., r_i, ... >$, where r_i is the data record collected on time point t_i ($i = 1, 2, ...$).

Definition 2 (*Distributed data stream*). Given a time series $T = <t_1, t_2, ..., t_i, ... >$ and node number n, a distributed data stream on T is defined as $DS = \{S^1, S^2, ..., S^n\}$, where S^k ($k = 1, 2, ..., n$) is a single data stream on T collected from the kth node in the investigated distributed system.

Definition 3 (*History window*). Given a time series $T = <t_1, t_2, ..., t_i, ... >$ and a data stream $S = <r_1, r_2, ..., r_i, ... >$ on T. Let a time interval $[i, k)$ ($i < k$) in T, the history window on time point k is defined as the data sequence $<r_i, r_{i+1}, ..., r_{k-1} >$. For the distributed data stream $DS = \{S^1, S^2, ..., S^n\}$, its history window on k, denoted as H_k, is comprised of all history windows of its local data streams on k.

Based on these concepts, we can describe a mining model for distributed data streams. As is Fig. 2 shown, if a mining time point is t, then the knowledge patterns M (t) on t should be incrementally reasoned from both the current collected data chunk D (t) on H_t and the patterns of its last mining results $M(t-1)$.

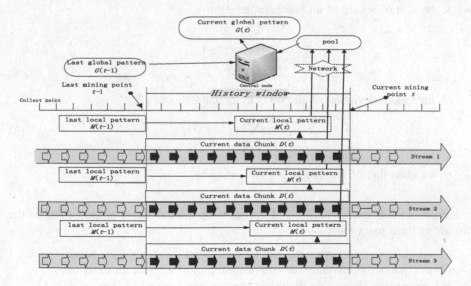

Fig. 2. A process to incrementally mine a distributed data stream.

4 Classifying Distributed Data Streams

In many data stream mining applications, the goal is to predict the class labels for new instances, called classifying data streams. In general, classifying distributed data streams should be divided into three relatively separate phases:

- *Local mining*: classifying a local data stream at the mining point;
- *Pattern transferring*: sending local classifying results to the central node;
- *Global mining*: integrating local classifiers in the central node.

4.1 Local-Miner

We have noted that Aggarwal et al. presented an idea of data summary statistics for mining data streams [13]. According to this idea, some micro-clusters, valuable statistical values for mined clusters, can be abstracted. Obviously, such micro-clusters can become a better choice of local pattern expresses for mining distributed data streams.

Definition 4 (*Micro-cluster structure*). Given a dataset $X = \{x\}$, its micro-cluster is defined as the 5-tuple $< N, C, S, D, F >$:

- N: the number of points in the dataset, i.e. $N = |X|$;
- C: the central point in the dataset, i.e.

$$C = \left(\sum_{x \in X} x \right) / N \tag{1}$$

- S: the sum of square of all points in the dataset, i.e.

$$S = \left(\sum_{x \in X} x^2 \right)^{1/2} \tag{2}$$

- D: the square deviation of all points in the dataset, i.e.

$$D = \left(\sum_{x \in X} (x - C)^2 \right) / N \tag{3}$$

- F: the class flag of the dataset X

When we use micro-clusters as local patterns, a local miner at least need do the following three main jobs:

- *Chunk-collector*: collecting the current data chunk by given time point,;
- *Cluster-generator*: dividing the dada chunk into a series of clusters;
- *MS-abstractor*: abstracting the micro-clusters from mined clusters.

Algorithm 1 describes the main tasks of a local miner.

Algorithm 1. Local-miner.
```
input:
t: mined time point;
D: data chunk collected on t;
k: the number of clusters.
output:
M: the set of micro-clusters mined.
process:
get k clusters C through k-means algorithm;
FOR each c in C
    calculate c.N, c.C, c.S, c.D by Definition 4;
    flag c.F with the most label in instances of c;
    integrate c.N, c.C, c.S, c.D  and c.F into vector m;
    insert m into M;;
ENDFOR
return M.
```

4.2   Global-Miner

Based on requires to mine distributed data streams, a global miner at least should have the following two functions:

- *MS buffer pool*: collecting micro-clusters from all local nodes.
- *MS-maintainer*: making use of both all current micro-clusters from local nodes and the micro-cluster set at the last mining point to incrementally maintain global micro-clusters.

Here, a basic work is to search two micro-clusters that can be united.

Definition 5 (*Operator to unite two micro-clusters*). Given two micro-clusters m_1 and m_2, then a new micro-cluster can be obtained through an operator called *Unite*: if $m_1.F = m_2.F$, then $m_3 = Unite(m_1, m_2.)$; otherwise break. It can be done step by step as follows:

$$m_3.N \leftarrow m_1.N + m_2.N \tag{4}$$

$$m_3.C \leftarrow (m_1.N \times m_1.C + m_2.N \times m_2.C)/m_3.N \tag{5}$$

$$m_3.S \leftarrow \sqrt{(m_1.S^2 + m_2.S^2)} \tag{6}$$

$$m_3.D \leftarrow (m_1.S^2 + m_2.S^2 - 2 \times m_3.C \times (m_1.C \times m_1.N + m_2.C \\ \times m_2.N))/m_3.N + m_3.C^2 \tag{7}$$

$$m_3.F \leftarrow m_1.F \tag{8}$$

Using Definition 5, we can integrate a micro-cluster into the existing set of micro-clusters in the central node. Where, a key problem is how to search the best two micro-clusters to unite. We will use the minimum square deviation standard to search two united micro-clusters. That is, given a set of micro-clusters M, if m_1 and m_2 in M satisfy the following, then them will be first united:

$$\min\{unite(m_1, m_2).D \mid m_1, m_2 \in M\} \tag{9}$$

Algorithm 2 describes the process to update micro-clusters in the central node.

Algorithm 2. Global miner.
```
input:
t: investigating time point;
M*: micro-clusters from all local nodes on t;
M: existing set of clusters before t in the central
node;
L: top threshold for number of clusters.
output:
M: the set of maintained micro-clusters on t in the
central node.
process:
FOR   each m* in M*
  FOR each m in M
    M ← M ∪ {m*};
    Lₘ ←Lₘ +1;
    IF Lₘ <=L THEN break;
    b ← the largest number of machine;
    FOR each m₁ in M
      FOR each m₂ in M
        IF  unite(m₁,m₂).D < b
          THENs₁ ←m₁; s₂ ←m₂; b←Unite(m₁,m₂).D;
    p ←unite(s₁, s₂);
    M ← M ∪ {p};
    M ← M −{s₁}−{s₂};
    Lₘ ←Lₘ −1;
  ENDFOR
ENDFOR
return M.
```

5 Experiments

To show the effectiveness of our approaches, we designed some experiments. All experiments are done in the environment with 3 local nodes and a central node. The raw data used in these experiments are from the dataset *pen-digits* [14]. It is a data set for recognition of hand written digits with 16 numeric attributes and 9 class labels. We use about 10000 instances of *pen-digits* as raw static dataset. The experiments are based on the changing of different control parameters. Table 1 gives the main control parameters and their meanings.

Table 1. Control parameters and meanings in experiments

Symbol	Meaning	Scope
Speed	Data collecting peed in local nodes	$10 \sim 100$(sec.)
Interval	Collecting time interval in local nodes	1/100(sec.)
%Unlab	Unlabeled sample ratio collected	$5 \sim 50(\%)$
$|H|$	Size of history window	$5 \sim 50$(sec.)
k	Number of clusters when k-means executing	$10 \sim 100$

Experiment 1 (*Accuracy evaluation on different time intervals between the chunks*). Fixing clustering number in k-means $k = 30$ and unlabeled Ratio *%unlab* = 25%, Fig. 3 shows the accuracy changes with different history window sizes in 1000 s.

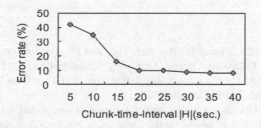

Fig. 3. Fixing $k = 30$ and *%unlab* = 25%, error rate changes with different history window sizes

As is Fig. 3 shown, increasing the time lengths of chunks can improve mining accuracy, but it also is certain to increase costs in computing and transferring. Therefore, we cannot always expect increasing lengths of chunks in an unlimited way. In fact, this is a problem to balance accuracy with costs. However, our method has a good scalability. That is, when the size is increasing to a suitable value, like 20 in Fig. 3, it can hold a better accuracy, and it can no longer enhance accuracies in an obvious scope by increasing again window size to larger.

Experiment 2 *(Adaptability evaluation for unlabeled data)*. Fixing $k = 30$ and $|H| = 20$. Figure 4 shows the experiment results with various ratios of unlabeled instances in training data streams in 1000 s.

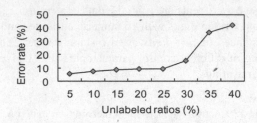

Fig. 4. When fixing $k = 30$ and $|H| = 20$, error rate changes with different unlabelled data ratios.

Since we investigate the problem of classifying both labeled and unlabeled data, the influence of unlabeled data to classifying accuracy should exist in our methods. It is natural to make accuracy decreasing when the radio of unlabeled instances increases, but such a decreasing trend is controllable. As is Fig. 4 shown, when unlabelled data ratios are not larger than 25%, mining accuracies are above 90%.

References

1. James, M., Michael, C., Brad, B.: Big data: The Next Frontier for Innovation, Competition and Productivity. Technology Report, McKinsey Global Institute (2011)
2. Viktor, M., Kenneth, C.: Big data: A Revolution That Will Transform How We Live, Work and Think. John Murray Publishers (2013)
3. Parthasarathy, S., Ghoting, A., Otey, M.E.: A survey of distributed mining of data streams. In: Aggarwal, C.C. (ed.) Data Streams. Advances in Database Systems, vol. 31, pp. 289–307. Springer, Heidelberg (2007)
4. Bhaduri, K., Das, K., Sivakumar, K., Kargupta, H., Wolff, R., Chen, R.: Algorithms for distributed data stream mining. In: Aggarwal, C.C. (ed.) Data Streams. Advances in Database Systems, vol. 31, pp. 309–331. Springer, Heidelberg (2007)
5. Guerrieri, A., Montresor, A.: DS-Means: Distributed data stream clustering. In: Kaklamanis, C., Papatheodorou, T., Spirakis, Paul G. (eds.) Euro-Par 2012. LNCS, vol. 7484, pp. 260–271. Springer, Heidelberg (2012). doi:10.1007/978-3-642-32820-6_27
6. Gianluigi, F., Clara, P., Giandomenico, S.: An adaptive distributed ensemble approach to mine concept-drifting data streams. In: Proceedings of the IEEE Conference on Tools with Artificial Intelligence (ICTAI 2007), pp. 183–187. IEEE Press (2007)
7. Gaber, M., Shiddiqi, A.: Distributed data stream classification for wireless sensor networks. In: Proceedings of the ACM Symposium on Applied Computing (SAC 2010), pp. 1629–1630. ACM Press (2010)

8. Street, W., Kim, Y.: A streaming ensemble algorithm (SEA) for large-scale classification. In: Proceedings of the ACM Conference Knowledge Discovery and Data Mining (KDD 2001), pp. 377–382. ACM Press (2001)
9. Gibbons, B., Tirthapura, S.: Distributed streams algorithms for sliding windows. In: Proceedings of the ACM Symposium on Parallel Algorithms and Architectures (SPAA 2002), pp. 63–72. ACM Press (2002)
10. Yang, Y., Mao, G.: A self-adaptive sliding window technique for mining data streams. In: Proceedings of the International Conference on Information and Multimedia Technology (ICIMT 2010), pp. 56–60. IEEE Press (2010)
11. Masud, M., Gao, J., Khan, L., Han, J., Thuraisingham, B.: A practical approach to classify evolving data streams: training with limited amount of labeled data. In: Proceedings of the IEEE International Conference on Data Mining (ICDM 2008), pp. 929–934. IEEE Press (2008)
12. Xu, W., Qin, Z., Ji, L., Chang, Y.: A feature weighted ensemble classifier on stream data. In: Proceedings of the IEEE Conference on Computational Intelligence and Software Engineering (CISE 2009), pp. 1–5. IEEE Press (2009)
13. Aggarwal, C., Han, J., Wang, J., Yu, P.: A framework for clustering evolving data streams. In: Proceedings of the the the 29th VLDB Conference, Berlin, Germany, pp. 81–92 (2003)
14. Alimoglu, F., Alpaydin, E.: Combining multiple representations and classifiers for pen-based handwritten digit recognition. In: Proceedings of the IEEE Conference on Document Analysis and Recognition, pp. 637–640. IEEE Press (1997)

Data Analysis

Design of a Quality-Aware Data Capture System

R. Vasanth Kumar Mehta[1(✉)] and Shubham Verma[2]

[1] Department of Computer Science and Engineering, SCSVMV University,
Kanchipuram 631561, Tamil Nadu, India
vasanth.mehta@kanchiuniv.ac.in
[2] Evive Software Analytics Pvt. Ltd., Bengaluru 560034, Karnataka, India
shubham.verma@evivehealth.com

Abstract. Data analytics is an ever-growing field which provides insights, predictions and patterns from raw data. The outcome of analytics is greatly affected by the quality of input data on which the analytics is done. This paper explores the design of a quality-aware data capture system, which uses Data Mining Techniques and algorithms, specifically a decision-tree based approach for data validation and verification, with an objective of identifying data quality issues right at a stage when data enters the system by providing appropriate feedback through a carefully designed user-interface.

Keywords: Data quality mining · Data analytics · Data validation

1 Introduction

Analytics plays a major role in all aspects of an organization. It is the driving force behind understanding competition, increasing profits, developing marketing strategies, iterating over the core products and services, and improving overall business processes.

Data Analytics is used in various domains. In this paper, the authors have studied the use of data analytics for providing personalized healthcare engagement solutions through predictive analysis along with behavioral economics. The application necessitates that huge amounts of data have to be captured and processed through complex business processes. During the development of the data analytics pipeline, several data quality issues were identified - such problems vary from missing values, to duplications, inconsistencies, incompleteness, lack of precision and outdated data. Data is said to lack quality if it is either syntactically incorrect or semantically incorrect.

1.1 Literature Survey

Berti-Équille proposed a framework for Quality-Aware Data Mining [1]. In such an approach, the Data Mining Algorithms are modified to ensure that they take into consideration the Data Quality Issues. For example, the authors have proposed changes to the classic Association Rule Mining Task where various Data Quality factors like Freshness, Accuracy, Completeness and Consistency are ensured. These dimensions

© Springer International Publishing AG 2017
Y. Tan et al. (Eds.): DMBD 2017, LNCS 10387, pp. 275–282, 2017.
DOI: 10.1007/978-3-319-61845-6_28

are represented as association rules of the form x $\rightarrow$ y by using a Fusion. In such an approach, the data on which the KDD process is carried out is not modified to increase its quality. However, a Quality-Aware data mining approach is undertaken so that the mining process is conscious of the inherent quality issues in data, and the output patterns are produced along with a quality measure. Dasu and Johnson describe about Exploratory Data Analysis that involves the use of simple statistical techniques for exploring and understanding the data. However, the EDA approach involves underlying assumptions of normality and symmetry, which may not always be true in real world scenarios. Various statistical measures are very sensitive to outliers [2]. Tamraparani Dasu et al. highlighted the issue of data sets being federated from many databases containing thousands of tables and tens of thousands of fields, with little reliable documentation about primary keys or foreign keys [3]. Jiawei Han et al. suggests the methods like Data Cleaning, Data Integration, Data Transformation and Data Reduction for enhancing data quality through Data Preprocessing [4].

2   Objectives

2.1   Preventing Fat-Fingering of Data

The most common problem with applications with a data entry component is lack of proper validation for the data-entry fields. Some of the applications even blindly accept the data entered without any checks for missing values or data validation. Some applications do provide validations but due to a poor user experience, they fail to convey the exact problem to the user, leaving him confused and frustrated.

One of the objectives of the proposed system is empowering the user with appropriate feedback during the data entry process by providing error messages and suggesting means to correct the error. This feedback is given by the validation engine using the decision-tree approach.

2.2   Preventing Outliers

Like data errors, it is important to avoid any outliers in the dataset for the base model. It becomes hard to identify outliers when data comprises of many closely related variables. In this paper, a simple approach is proposed for detecting outliers early-on with help of feature ranking and decision trees.

2.3   Preventing Bad Modeling of Data

The previous two objectives lead us to a more general objective which is to prevent bad modeling of data. Presence of errors and outliers in the dataset affects the performance and outcome of the model. This can be easily avoided if both errors and outliers are prevented from entering the pipeline. Any system cannot guarantee zero errors, but it can be minimized using effective approaches.

3   Experiment Setup

The experiment was performed on existing system architecture in place at Evive Health. The main entities are Data Providers, Databases, ETLs, Business Processes and the Web App.

The main application is written in Java along with Polyglot storage provided by databases such as Cassandra, Redis, MongoDB etc. It is complemented by other small services running separately for dedicated business processes. ETL is where the bulk loading and pre-processing of data takes place. The web application consists of different user interfaces through which users enter the data.

4   Proposed System

The entire process is divided into two phases – The Data Profiling Phase and the Data Input and Verification Phase.

4.1   Preventing Bad Modeling of Data

a. In any operational system, the primary data modeling is done in which the input fields (or features) along with the types (numeric, categorical etc.) are identified.
b. Various syntactic and semantic constraints are imposed on the features, as is usually done in a well-planned and designed data project.
c. Based on the schema designed above, the input values are fed into the system. Here, it is to be ensured that the data entered is almost correct. If getting real-world data and ensuring its quality at this stage is found difficult, some sample manufactured data is sufficient.
d. Such data entered is used as the input to the Feature Selection Algorithm, based on which the importance of Features is obtained along with the Decision-Tree for classification.

4.1.1   Feature Selection Algorithm

The Decision Tree is a hierarchical model for supervised learning whereby the local region is identified in a sequence of recursive splits. Hall proposed a new attribute weighted method [5] based on the degree to which they depended on the values of other attributes. A random sampling is made from the training set and the following operations are performed for dependency computation to estimate the degree to which an attribute depends on others:

a. An unpruned decision tree is constructed from the training data.
b. The minimum depth d at which the attribute is tested in the tree is noted. The deeper an attribute occurs in a tree, the more dependent it is. The depth of the root is assumed to be 1 and the dependency D of an attribute can be measured using:

$$D = 1/\sqrt{d} \tag{1}$$

where d is the minimum depth at which the attribute is tested. Nodes with more dependency will have a lesser D value and nodes with lesser dependency will have a larger D value. If an attribute is not tested, we can assume its dependency as zero.

Make a vector of the dependencies of all the attributes. This dependency vector is now an indicator of the relative dependencies of the attributes.

4.1.2   Data Input and Verification Phase

a. Based on the Feature Rank Vector, the input system is now to be designed such that the most critical Feature is accepted as input first, followed by the next ranked feature, and so on in the decreasing sequence of the Vector.
b. Each time an input is taken, this input is verified to see if it confirms to the Decision Tree.
c. When an exceptional condition is met, it is ascertained from the user if it is an error in entry, or whether it is a genuine case of deviation. In case of the former, the entry is corrected, and the Decision Tree is not violated. However in case of the latter, if the violation or deviation happens and the user confirms the deviation as reflecting the real-world condition, then two conditions arise depending on the level at which the deviation occurs.
d. If the deviation is at the upper levels of the tree, it indicates a significant difference from the general behavior of the data, and hence, it can be flagged as an outlier. In case the deviation is closer to the lower levels, the difference is only in insignificant features, which do not generally have a bearing on the further classification of data. Hence, such deviation has no significant impact, and can either be replaced with the value of the particular branch, or can be left unchanged.

5   Experimental Results and Analysis

5.1   Analytics Before Introducing Validation

The data used for experiment is a subset of the real world data collected and processed at Evive Health. Only few features have been taken under consideration for the performing the experiment. Some features such as personal information have been removed to ensure anonymity (Table 1).

Table 1.   Features present in database

Feature name	Feature type	Description
Upin	Quantitative	Unique ID of person
Age	Quantitative	Age of person
Gender	Categorical	Gender of person
Ethnicity	Categorical	Ethnicity of person
GL_level	Quantitative	Glucose levels of person (mg/dL)
WC_level	Quantitative	Waist circumference of person (inches)
Outcome	Categorical	Feature to predict

In the first stage, analytics is performed on the dataset before introducing any validations. This involved data binning for GL_level and WC_level – each of the values were assigned a corresponding value of – SAFE, ALARM, DANGER. The logic for binning is as follows (Table 2):

Table 2. Levels for WC_level and GL_level

WC_level	Male	Female	Level
	[22, 40]	[22, 35]	SAFE
	[41, 65]	[36, 65]	ALARM
	[66, 1000]	[66, 1000]	DANGER
GL_level	Range		Level
	[25, 400]		SAFE
	[401, 650]		ALARM

Running GLM (Generalized Linear Model) on training dataset resulted in the following coefficients (Table 3):

Table 3. Coefficients from GLM on training data

	Estimate	Std. error	z-value	Pr(>|z|)
(Intercept)	−0.9223	0.3196	−2.885	0.00391
Age	0.0051	0.0054	0.952	0.34128
genderM	0.0610	0.1016	0.600	0.54818
ethnicityET02	−0.8415	0.2578	−3.265	0.00110
ethnicityET04	0.1101	0.6170	0.179	0.85833
GL_levelSAFE	−0.0975	0.1040	−0.938	0.34847
WC_levelSAFE	−0.2366	0.1057	−2.238	0.02520

GLM was performed on the combined training and test dataset. Examining the coefficients, it was found that two new levels were introduced which were not initially present in the training dataset (Table 4):

Table 4. Coefficients from GLM on training and test data

	Estimate	Std. error	z -value	Pr(>|z|)
(Intercept)	−1.0148	0.3031	−3.349	0.00081
Age	0.0068	0.0051	1.339	0.18073
genderM	0.0363	0.0961	0.378	0.70575
ethnicityET02	−0.8404	0.2452	−3.426	0.00061
ethnicityET03	13.1849	229.565	0.057	0.95420
ethnicityET04	−0.1135	0.6127	−0.185	0.85310
GL_levelSAFE	−0.1171	0.0976	−1.199	0.23035
WC_levelDANGER	0.9800	0.4521	2.167	0.03020
WC_levelSAFE	−0.1930	0.0992	−1.945	0.05179

For 'ethnicity' there was a new level - 'ET03' and for 'WC_level' there was a new level - 'DANGER'. This suggests that either all possible levels were not exhaustively captured in the training dataset or the test dataset had some foreign values which could be errors or outliers. As there was no validation present at the data entry level, these new levels were inevitable.

5.2   Introducing Validation

The first step in introducing validation was to calculate the feature rank vector from the training dataset. Following was obtained after the running the feature ranking algorithm on the dataset (Table 5):

Table 5. Feature ranking vector

Feature	Avg. estimated dependency	Feature rank
Ethnicity	0.97	1
Age	0.68	2
WC_level	0.52	3
GL_level	0.52	4
Gender	0.41	5

In the next step, the test dataset was validated row-by-row, and error identified. The issues with the test dataset were the new levels of WC_level (DANGER) and ethnicity (ET03) (Fig. 1).

```
<simpleError in model.frame.default(Terms, newdata, na.action = na.action, xlev =
attr(object,    "xlevels")): factor WC_level has new level DANGER>
```

Fig. 1. Error due to new level of WC_level in test data

Plotting a simple histogram for WC_level made this clear as some values were quite apart from the rest of the data points (Fig. 2):

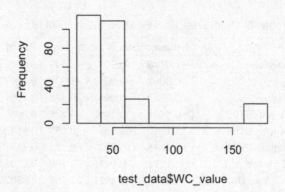

Fig. 2. Histogram showing new levels of WC_LEVEL

Further investigation revealed that some of the users had entered the WC_level values in 'centimeters' instead of 'inches'. The data-entry UI did not have out-of-range validations and thus allowed for entry of erroneous values. These values were fixed by converting them back to inches, adding validation to the UI and verified by another histogram (Fig. 3):

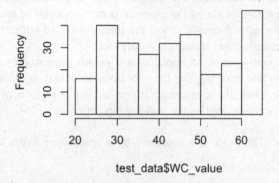

Fig. 3. Histogram for corrected values of WC_LEVEL

GLM was performed again on the combined dataset after fixing the WC_level errors. There was a significant improvement in the coefficient for WC_levelSAFE (Fig. 4):

```
WC_levelSAFE    -0.192980    0.099224   -1.945 0.051789 .
                             (insignificant at 5% confidence)
vs

WC_levelSAFE    -0.213246    0.098731   -2.160 0.030782 *
                (reasonable improvement after fixing DANGER)
```

Fig. 4. Comparison of coefficients, before and after validation

The only remaining issue with test data now was the 'ET03' level for 'ethnicity'. There are two approaches to handle this:

i. If it is a possible valid observation, this needs to be overcome by taking larger and representative sample, which is outside the scope of this paper.
ii. Otherwise replace this value with the most frequent value for 'ethnicity' and re-run the GLM on combined training and test dataset (Fig. 5).

```
WC_levelSAFE   -0.236605   0.105705   -2.238   0.02520 *
                                (from train data alone)
vs


WC_levelSAFE   -0.217599   0.098685   -2.205 0.027456 *
       (closest to original model after fixing DANGER and ET03)
```

Fig. 5. Comparison of coefficients after replacing 'Ethnicity' values with average

6 Conclusion

The importance of data validation cannot be stressed enough when it comes to analytics as it directly impacts the quality of results produced. Most of such validations can be achieved at the data-entry level itself by providing appropriate feedback to the user. The validations can be predicted and suggested by an effective data quality mining approach, instead of rules made by the programmers. This helps in ensuring the quality of the dataset with a refined sample space, minimal errors and outliers, hence reducing the efforts in the cleaning and transformation process.

Also, the user experience of the application is greatly enhanced by assisting the user through the process and ensuring that the user does not make any inadvertent errors at the data-entry stage. Overall, the approach suggested in this paper is of immense benefit to the analytics process as well as the end-user experience.

Acknowledgements. This work was supported by a grant from Evive Software Analytics Pvt. Ltd.

References

1. Berti-Équille, L.: Quality awareness for managing and mining data. Doctoral dissertation, Université de Rennes 1 (2007)
2. Dasu, T., Johnson, T.: Exploratory Data Mining and Data Cleaning. Wiley, New York (2003)
3. Dasu T., Johnson T., Muthukrishnan S., Shkapenyuk V.: Mining database structure; or, how to build a data quality browser. In: Proceedings of the 2002 ACM SIGMOD International Conference on Management of Data, pp. 240–251. ACM (2002)
4. Jiawei, H., Micheline, K.: Data Mining: Concepts and Techniques. Morgan Kaufmann, San Francisco (2006)
5. Hall, M.: A decision tree-based attribute weighting filter for naive Bayes. Knowl.-Based Syst. **20**(2), 120–126 (2007)

Data Architecture for the Internet of Things and Industry 4.0

José Ignacio Rodríguez Molano$^{(\boxtimes)}$,
Leonardo Emiro Contreras Bravo, and Eduyn Ramiro López Santana

Universidad Distrital Francisco José de Caldas, Bogotá, Colombia
{jirodriguez, lecontrerasb, erlopezs}@udistrital.edu.co

Abstract. This paper analyzes Internet of Things (IoT), its use into manufacturing industry, its foundation principles, available elements and technologies for the man-things-software communication already developed in this area. And it proves how important its deployment is. In that sort of systems, information process. Is related to manufacturing status, trends in energy consumption by machinery, movement of materials, customer orders, supply data and all data related to smart devices deployed in the processes. This paper describes a proposal of data architecture of the Internet of things applied to the industry, a metamodel of integration (Internet of Things, Social Networks, Cloud and Industry 4.0) for generation of applications for the Industry 4.0.

Keywords: Industrial internet of things · Data mining · Cloud

1 Introduction

Internet of Things has reached so much development and importance that several reports foresee it as one of the technologies of higher impact until 2025 [1, 2]. Thousands of millions of physical elements and objects will be equipped with different types of sensors and actuators, all of them connected in real-time to internet through heterogeneous networks. This will generate high amounts of data-flows [3] that have to be stored, processed and shown to be interpreted in an efficient and easy manner. Here is where IoT and Cloud computing integration permits that this huge amount of data can be stored in internet, keeping resources, services and data available and ready to use and for end-to-end service providing [4] both in the business and in the consumer field, from anywhere. It provides the virtual integration infrastructure for storage devices, analysis tools, monitoring and platform. Most interactions in internet are human-to-human (H2H). However, more and more ubiquitous and heterogeneous objects can connect to Internet. In the future, it is expected that more objects be connected than people [5, 6]. Objects can interact with each other, send data and perform certain actions according to certain information. The Internet of Things has been catalogued as heterogeneous and ubiquitous objects, all interconnected between them and communicating with each other via the Internet [7].

The Internet of Things counts on sensors or scattered objects to generate information from any accessible site or inside a machine. Thus, it requires the interconnection of these heterogeneous objects via the Internet [8, 9]. This will lead to a future

© Springer International Publishing AG 2017
Y. Tan et al. (Eds.): DMBD 2017, LNCS 10387, pp. 283–293, 2017.
DOI: 10.1007/978-3-319-61845-6_29

in which it is not only used for communication between people, but also between human and machine, and even between different machines (M2M) [10]. Here is found the importance of Smart Objects: physical objects with an embedded system that allows information processing and communication with other devices, and also the processes based on a certain action or event [11]. However, all of these complex systems present a problem when connecting the Smart Objects due to the differences between software and hardware used for each process [12].

Due to the enlargement of the IoT and the Cloud Computing application development, technological problems decrease and an expansion of services is created in an industrial context [13]. This paper outlines the need to integrate the IoT, sensors, actuators, social networks and computing in the cloud. It will enable to build the Industrial Internet of Things (IIoT) up.

2   Architecture of Industrial Internet of Things IIOT

The development of Internet of Things technology for industrial applications, the implementation of sensors and detection intelligence advantages has attracted great attention. With those implementations, it has been achieved an efficient monitoring and control in reducing costs and power consumption in the production of goods and/or services [14, 15] in industries. The maintenance of these machines and systems is controllable and automatable with the help of sensors and wireless devices incorporated in machines and industrial systems [16].

The architectures proposed have not converged to a reference model [17] or a common architecture. In current literature, several models can be found, as can be seen in Fig. 1 [18]. The basic model has three layers (layer application, network and perception). It was designed to address specific types of communication channels and does not cover all of the underlying technologies that transfer data to an IoT platform. Other models based on four layers (layer of sensors, network, services and interfaces) are designed in the context of industry 4.0 [19].

2.1   Proposal for Industrial Internet of Things Architecture

The Fig. 2, in the context of Industry 4.0, a prototype of Industrial Internet of Things platform in 5 layers is presented pursuing the integration of sensors, actuators, networks, cloud computing and technologies of the Internet of Things:

Explains the integration of the data layer with the other layers of the architecture:

- Sensing layer:

This layer is composed by several types of devices. It directly determines the implementation and production of the specific data type. From a functional perspective, it is responsible for various activities such as manufacturing, transport, mobility, logistics, and data collection from sensors or other devices. The detection layer is integrated with the available hardware objects to detect the state of the things.

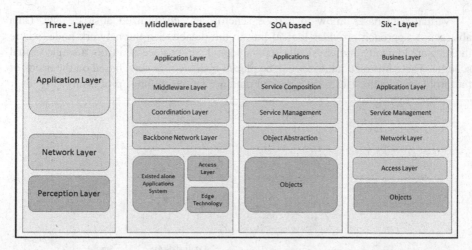

Fig. 1. IoT architectures [18].

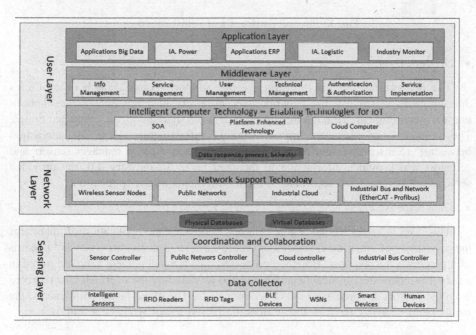

Fig. 2. Proposal of integration architecture

- Databases layer:

Composed of physical databases and virtual databases. The physical databases use a common set of SQL and nonSQL (object). It enables easy integration with external applications, so that should not be dependent. In this architecture, there is little reliance on proprietary database features, such as stored procedures and triggers. IoT data come from sensors and devices, where the data can be collected and processed in real time.

Virtual data bases allow to expose schema and customized abstract data. Data abstraction approach provides logical link to the database in the network node, where the data are logically and physically separated and can be accessed from a single virtual schema. These abstract databases can be placed inside containers, based on the needs of the problem domain. Here, the acquired data are stored without processing (Fig. 3).

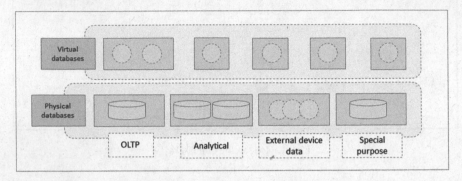

Fig. 3. Physical databases and virtual databases

- Network layer:

The network layer is the infrastructure that supports connections between things through cable or wireless. IoT network layer, connects all things and enables them to know its environment. Through this layer all data are shared with all connected things. The network layer adds data from existing IT infrastructure. Here, the interaction between the Sensor layer and the User layer takes place. Also, it manages sensors and actuators and provides information to the next layer. In industry 4.0, the provided services are personalized according to the requirements of the application.

- Data response layer:

This layer is a data set that can be assigned to devices and applications. It maintains the persistence of other layers. It focuses on providing automatic responses and it learns as it processes the response. Here, the processed data are maintained. Data are stored in a way that makes the physical database updated as stated by the user. All data received from the sensors are processed in the network node. The cloud-based components, the components near IoT devices and sensors are paired logically (Figs. 4 and 5).

- User layer:

In this layer, the API are used for application design. Within industry 4.0, they preferably are ERP applications, which allow the monitoring of raw material, equipment failures, quality control and programming and production. Here, several services

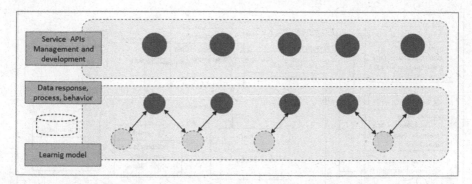

Fig. 4. Data response layer.

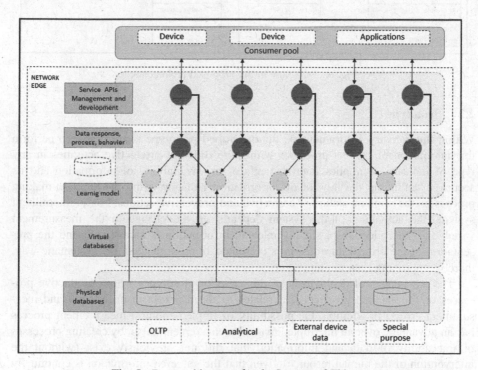

Fig. 5. Data architecture for the Internet of Things.

are provided, such as data compilation, transmission data processing. This layer is based on the Middleware technology, which is important in enabling of IoT services and applications, assisting in the recycling of software and hardware.

The information that is captured by the different sensors is stored in the cloud and in physical databases. This information is processed by Big Data tools and facilitates decision making to the user. Figure 6 shows the flow of information in the system.

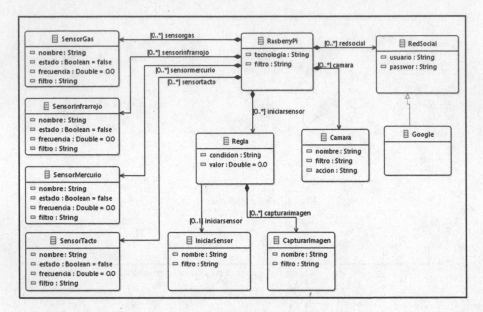

Fig. 6. Data flow in the proposed architecture

2.2 Prototype

Within the industry 4.0 framework, the developed prototype is applicable to perform the automation of manual processes within the different productive activities in any type of industry. It implies improvements of the availability of information and its security, facilitating the flow of the information and improving the decision-making process on the implementation of corrective and preventive actions. For an intermittent system (the activation of the system depends on the change in the measurement variable). The collection of slightly relevant information is almost zero and the presentation of the critical information is presented in an organized and systematic way, thereby improving data mining.

The advantages of deployment of this class of prototypes within productive processes or for tracking products [23] relates to the decrease in uncertainty and measurement errors. In addition, the possibility of implementing a measurement process has an advantage over an interrupted manual measurement. Also, by creating processes of traceability of data and information that did not previously exist (without the intervention of the human resource) given that the gathered information is continually pushed towards the Big Data or cloud computing for its consolidation.

The communication environment, subject of the surveillance system, allowed to monitor several variables. For this purpose, four sensors were used. They allowed to determine the flow of products in a production plant. The presence of gas in the complex, the alignment and stability of the equipment and the existence of contact between a person and the prototype. The information on changes in the status of sensors was processed in a microcomputer Raspberry Pi [24].

A communication object-object between the Raspberry Pi and a mobile device was deployed with the purpose of establishing the synchronization of data directly to the user via smartphone or tablet. It was achieved through the connection established between the Raspberry Pi and the storage server in the cloud [25].

The prototype has the following features:

- Use of technologies of sensors: The prototype uses different sensors for the detection of "strange" elements in the working environment. Then it initiates capture and data communication. The sensor acts like a link between what happens in the process and the user.
- Data collection at any time: the prototype does not restrict the capture of images and is always aware of each movement in the area where the system is deployed.
- Data Transfer: it is possible to transfer data within the communication network that is composed by the user (via a mobile device) and the physical system (Raspberry Pi) through cloud computing.
- User interface: the prototype can communicate with the user through well-known platforms or server storage.
- Functional Independence: the prototype works automatically independently of the user, since the capture of data and their transfer functions are carried out only with the activation of a sensor and a wireless network.
- Communication with the mobile device: the system does not require direct intervention to the hardware or software to operate. The communication is conducted under the principles of communication machine-to-machine, in which the micro-computer Raspberry Pi and the sensors are the prototype and the mobile device is the terminal that enables communication with the user through a server and a communication network. The collected data are transferred from one machine to another in the same way, disregarding which sensor is in use.
- Total availability of data: the system generates a communication network that allows the access to the user in any place and consultation of data at any time, due to the connection of the physical prototype (Raspberry Pi) with the server online.

3   Tests of the Prototype and Results

Tests to the monitoring system were performed with two sensors, making in each test the connection of each individual sensor to the Raspberry Pi. The tests were performed on a production line of ceramics in a period of 24 h of work.

For the functional validation of the system devices were evaluated in parallel. In the first round, the devices already in place in the production line. It generated a failure report, stopping the process. In the second round the prototype recorded the failure and generated the report without stopping the process because it was functioning in parallel.

3.1   Stops Caused by Failures in the Production Flow

The data permitted to verify the actual condition of the stops in the production line. Some registered stops were not true failures, systems recorded them as a failure because a change in the speed of the process. The data generated by the prototype also recorded ordinary stops of and quality control stops. It forced adjustments in the decision making parameters (Fig. 7).

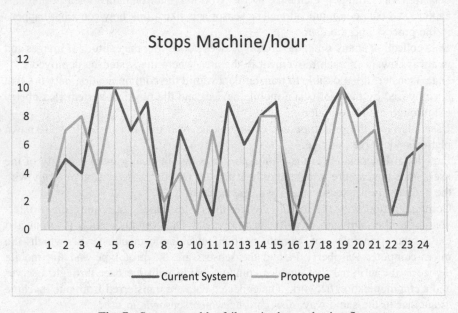

Fig. 7. Stops caused by failures in the production flow

3.2   Stops Caused by Misalignment and Instability of the Computer

The current computer does not count with this type of information, therefore it permits that the process generates failures, affecting the product. Also, it affects the equipment itself, because it subsequently generates a greater fault. The information provided by the prototype allowed to check and adjust the computer without generating any stop in the process.

The test of the monitoring system with the sensor of gas detection was carried out. Into the environment in which they were performed there was no presence of gases LP, methane or smoke. Consequently, signals sent by the sensor to the Raspberry Pi did not activate the collecting system nor synchronization (Fig. 8).

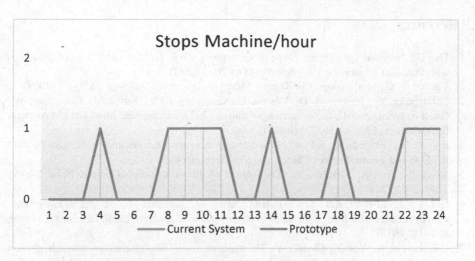

Fig. 8. Stops caused by misalignment and instability of the computer

4   Conclusions and Future Work

The technology itself is not the main obstacle for the implementation of IIoT; the interfaces of interoperability of the systems belonging to different providers are, that hinders its adoption. The achievement of interoperability requires standardization of interfaces for interaction between different components of the system.

IIoT is the combination of computing technologies, communication and micro-electronics. Thus, there are many interfaces between hardware, software and network components. In the context of the industry 4.0, the proposed architecture let establish communication networks between objects and people that permit the flow of information in two ways, with minimal intervention in the network. Communication networks focus on direct and comprehensive transmission of data.

The change in the way of gathering and storing the information, as much as the use of sensors, requires a change in the technology. IIOT can register automatically, precisely and timely several parameters of the production process. The traditional industrial production performs the communication between machines via the M2M technology, but the IIOT can achieve connections between people, machines and physical objects. Nevertheless, in the IIOT environment, the function and the yield of communication devices are different. Some applications need a high performance in real-time, while others do not. Some application tasks are executed periodically while others are activated by events. These characteristics increase the complexity of the actual applications of the IIOT.

Future research should focus on:

- Safety of data and system accuracy
- Standardization of technology and interoperability of systems
- Actual deployments

References

1. The US National Intelligence Council: Disruptive Civil Technologies: Six Technologies with Potential Impacts on US Interests out to 2025 (2008)
2. Vermesan, O., et al.: Internet of Things: Strategic Research Roadmap, pp. 1–50 (2009)
3. Chonggang, W., Mahmoud, D., Mischa, D., Qingyang, H.R., Xufei, M., Honggang, W.: Guest editorial special issue on internet of things (IoT): architecture, protocols and services. IEEE Sens. J. **13**(10), 3505–3510 (2013)
4. Biswas, R., Giaffreda, R.: IoT and cloud convergence: opportunities and challenges. In: 2014 IEEE World Forum Internet Things, pp. 375–376 (2014)
5. Atzori, L., Iera, A., Morabito, G.: The internet of things: a survey. Comput. Netw. **54**(15), 2787–2805 (2010)
6. Hui, T.K.L., Sherratt, R.S.: Towards disappearing user interfaces for ubiquitous computing: human enhancement from sixth sense to super senses. J. Ambient Intell. Humaniz. Comput. **8**, 1–17 (2016)
7. Hao, L., Lei, X., Yan, Z., ChunLi, Y.: The application and implementation research of smart city in China. In: 2012 International Conference on in System Science and Engineering (ICSSE), pp. 288–292 (2012)
8. Atzori, L., Iera, A., Morabito, G., Nitti, M.: The social internet of things (SIoT) – when social networks meet the internet of things: concept, architecture and network characterization. Comput. Netw. **56**(16), 3594–3608 (2012)
9. González García, C., Pelayo G-Bustelo, B.C., Pascual Espada, J., Cueva-Fernandez, G.: Midgar: generation of heterogeneous objects interconnecting applications. A domain specific language proposal for internet of things scenarios. Comput. Netw. **64**, 143–158 (2014)
10. Roman, R., Zhou, J., Lopez, J.: On the features and challenges of security and privacy in distributed internet of things. Comput. Netw. **57**(10), 2266–2279 (2013)
11. Xu, B., Xu, L.D., Cai, H., Xie, C., Hu, J., Bu, F.: Ubiquitous data accessing method in IoT-based information system for emergency medical services. IEEE Trans. Ind. Inf. **3203**, 1 (2014)
12. Gama, K., Touseau, L., Donsez, D.: Combining heterogeneous service technologies for building an Internet of Things middleware. Comput. Commun. **35**(4), 405–417 (2012)
13. Steiner, W., Poledna, S.: Fog computing as enabler for the Industrial Internet of Things. e i Elektrotech. Inform. **133**, 310–314 (2016)
14. Rahmani, B.: Industrial internet of things: design and stabilization of nonlinear automation systems. J. Intell. Robot. Syst. **86**, 311–323 (2016)
15. Georgakopoulos, D., Jayaraman, P.P.: Internet of things: from internet scale sensing to smart services. Computing **98**(10), 1–18 (2016)
16. Chen, Y., Lee, G.M., Shu, L., Crespi, N.: Industrial internet of things-based collaborative sensing intelligence: framework and research challenges. Sensors (Switzerland) **16**(2), 1–19 (2016)
17. Krco, S., Pokric, B., Carrez, F.: Designing IoT architecture(s): a European perspective. In: 2014 IEEE World Forum on Internet of Things, WF-IoT 2014, pp. 79–84 (2014)
18. Fraga-lamas, P., Fernández-caramés, T.M., Suárez-albela, M., Castedo, L.: A review on internet of things for defense and public safety. Sensors **16**, 1–46 (2016)
19. Wan, J., et al.: Software-defined industrial internet of things in the context of industry 4.0. IEEE Sens. J. **16**, 1–16 (2016)
20. González García, C., Meana-Llorián, D., G-Bustelo, B.C.P., Lovelle, J.M.C.: A review about smart objects, sensors, and actuators. Int. J. Interact. Multimed. Artif. Intell. **4**(3), 7–10 (2017)

21. Núñez-Valdez, E.R., García-Díaz, V., Lovelle, J.M.C., Achaerandio, Y.S., González-Crespo, R.: A model-driven approach to generate and deploy videogames on multiple platforms. J. Ambient Intell. Humaniz. Comput. **8**, 1–13 (2016)
22. García-Díaz, V., Tolosa, J.B., G-Bustelo, B.Cristina Pelayo, Palacios-González, E., Sanjuan-Martínez, Ó., Crespo, R.G.: TALISMAN MDE framework: an architecture for intelligent model-driven engineering. In: Omatu, S., Rocha, Miguel P., Bravo, J., Fernández, F., Corchado, E., Bustillo, A., Corchado, Juan M. (eds.) IWANN 2009. LNCS, vol. 5518, pp. 299–306. Springer, Heidelberg (2009). doi:10.1007/978-3-642-02481-8_43
23. Lee, J., Bagheri, B., Kao, H.: A cyber-physical systems architecture for industry 4.0 based manufacturing systems. Manuf. Lett. **3**, 18–23 (2015)
24. Rodríguez Molano, J.I., Medina, V.H., Moncada Sánchez, J.F.: Industrial internet of things: an architecture prototype for monitoring in confined spaces using a raspberry pi. In: Tan, Y., Shi, Y. (eds.) DMBD 2016. LNCS, vol. 9714, pp. 521–528. Springer, Cham (2009). doi:10.1007/978-3-319-40973-3_53
25. Molano, J.I.R., Betancourt, D., Gómez, G.: Internet of things: a prototype architecture using a raspberry pi. In: Uden, L., Heričko, M., Ting, I.-H. (eds.) KMO 2015. LNBIP, vol. 224, pp. 618–631. Springer, Cham (2015). doi:10.1007/978-3-319-21009-4_46

Machine Learning in Data Lake
for Combining Data Silos

Merlinda Wibowo, Sarina Sulaiman[(✉)],
and Siti Mariyam Shamsuddin

Faculty of Computing, UTM Big Data Centre,
Ibnu Sina Institute for Scientific and Industrial Research,
Universiti Teknologi Malaysia, 81310 Skudai, Johor, Malaysia
merlindawibowo@gmail.com, {sarina,mariyam}@utm.my

Abstract. Data silo can grow to be a large-scale data for years, overlapping and has an indefinite quality. It allows an organization to develop their own analytical capabilities. Data lake has the ability to solve this problem efficiently with the data analysis by using statistical and predictive modeling techniques which can be applied to enhance and support an organization's business strategy. This study provides an overview of the process of decision-making, operational efficiency, and creating the solution for an organization. Machine Learning can distribute the architecture of data model and integrate the data silo with other organizations data to optimize the operational business processes within an organization in order to improve data quality and efficiency. Testing is done by utilizing the data from the Malaysia's and Singapore's Government Open Data on the Air Pollutant Index to determine the condition of air pollution levels for the health and safety of the population.

Keywords: Data silos · Data lake · Machine learning · Big data · Prediction · Air pollutant index · Rough set

1 Introduction

Evolving technology has a major role to the operational processes of an organization. The ongoing process of an organization will involve several different stages (design, material acquisition, manufacturing, distribution, sales, usage, service and others) to obtain the meaningful, accurate, and efficient information [1]. Each stage of this process will require the supply of accurate information as a decision-making process, operational efficiency, and the creation of the desired solution. An organization has variety of data, which can support transactional applications, analytical decision support, and master as a universal business object [1].

Increasing amount of organization data is creating challenges in data management to make the hundreds of data entry into a single business view of data [1]. Lack of storage management system in many organizations, it makes data to become overlapping and indefinite quality. A large number of data in different types can also affect the analytics process of data to deal with uncertainty, prediction and dynamics data [2, 3]. Data on a large scale is a collection of various data assets which is complex and cannot be

© Springer International Publishing AG 2017
Y. Tan et al. (Eds.): DMBD 2017, LNCS 10387, pp. 294–306, 2017.
DOI: 10.1007/978-3-319-61845-6_30

managed efficiently by data processing technology state-of-the-art [4]. Many organizations rely on traditional data warehouses and business intelligence solutions as decision makers to access their data and reports. But this solution will ignore most of the external data sources because it is too large or in a format that is not easily manipulated and stored [5]. We need to determine the best architecture, common metadata, data integration and so on for optimizing operational business processes within an organization in order to improve data quality and more efficient [6]. It aims to improve the quality of data, use data as a competitive advantage, manage change, comply with work regulations, and adapted to the standards of work [7].

This research will describe the background of study trough definition of data silos, data lake concept, and application. Next part will give the explanation about the modification of data lake architecture with machine learning techniques for combining the data silos. Experiment and results of this study are described and followed by the conclusion of the study at the end of this paper.

2 Data Silos Concept

Repository of data that available under an organization called Data Silos [8, 9]. Evolving organization will certainly affect the growing amount of data. The data collected by each organization will be different. It depends on technical priorities, cultural, and responsibilities that come naturally. In addition, these data can be in the form of unstructured data in heterogeneous formats such as documents, video, images and others that do not have the scheme and come from a variety of sources [10]. Information from an organization is certainly not only from the body of an organization only but needs the support from outside sources an organization. This data collection can be allowed to develop the ability of an organization in the own analytic capacity and capability. Therefore, data silos is big data because it already has characteristics four "V" [11, 12]: Volume, Variety, Velocity, and Veracity.

Organizations must understand the needed insight to make strategic decisions and a better operation. An organization needs to predict the future to identify trends and correlations that encourage changes that are more useful in a business that changing wheel rapidly [13]. Data processing needs to be done in an integrated and systematic way to generate accurate and up-to-date [1, 14]. This process is a challenge for an organization to analyze data and brings a high value for decision-making as an effective solution that is influential for the future. Future prediction becomes more important than the simple visualization of current or historical perspective.

In fact, there are some fundamental issues that occur in the analytical data access. There are pressures, damage the user experience, the power of decision without the data, and the ability to access the data [15]. In addition, unstructured data will complicate the process of further analysis of data such as data mining process. Data mining can help to deduce the meaning of the data, which has been designed to work in accordance with the orientation of a database schema with structured data [16]. There are several ways to solve the problem of data silos, namely the origin of data, analysis of data management, enterprise, consumption data, data normalization, includes data, data curation, data access, cloud and machine learning [11, 14].

3 Data Lake Concept and Application

Data lake is a new concept that has the ability to secure, convert and process the data, which make the data can be consumed with speed and value required by the user even though that operation is impossible to run [17]. This concept is a simply storage repository that can store data regardless of size, schema, format, and complexity [18–20]. It can catalog, indexing and metadata management that needed for its own purposes and provide information to the invention of data and calculations analysis [19]. Data lake can be used for log management, knowledge management, business intelligence, cloud services, application hosting and relational databases, which can provide the advantage of minimizing costs and risks.

Data lake has included in database semantic or conceptual models using same standards and technologies. This concept can create a hyperlink internet, adding a layer of context data that provides information on the meaning of data and data relationships with other data [19]. It can also combine with SQL, NoSQL databases, online analytical processing and online transaction processing capabilities [20]. Different from warehouse traditional, data lake does not require a rigid schema or data manipulation (structured data, not structured or semi-structured) and size but based on an order of arrival data. Thus, the data lake lacks a formal schema-on-write but uses various tools which apply schema-on-read.

Data lake is center of the big data movement and often associated with Hadoop object-oriented storage [21]. Hadoop ecosystem will implement scalable and parallel processing framework that process data on a large scale in a subtle way and almost impossible to lose any data [19]. All organizations which take advantage of large datasets and diverse, it will be difficult to manage the increased volume, velocity and variety of the latest information. Data lake is an emerging approach to extract and put all the relevant data (e.g., logs and sensor data, social media, document collections, images, video, audio and another data useful for integrated analysis) without losing any data that relevant to the analysis, now or in the future [5, 21].

Data lake will store all relevant data which considered for analysis and digest in raw format corresponding to any data model. Data lake will organize data to promote better, more efficient access and will reuse the data management process or introduce new tools that improve search knowledge and general knowledge of data content. Many recent works have been used this concept in the crucial sectors such as banking, healthcare, retail, digital risk management, and other enterprises [5, 21, 22]. The systems used to restrict the flow of data into the warehouse as well as their use in organization, which focused on providing automation and tools to enable less skilled workers to clean, integrate and link data [21–23].

4 Machine Learning Techniques in Data Lake Architecture

Machine learning will focus on the development of computer programs that can teach themselves to grow and change when given a new data. Machine learning is a type of artificial intelligence that has the ability to learn from the data, without explicit and will follow the instructions that have been programmed [24]. There are some examples of machine learning products that have the ability to perform large data processing such as Apache Spark, Hadoop, Cloudera Manager, HBase, NoSQL Database, MapReduce and others [25]. Machine learning will assist in finding a solution, optimize performance by using sample data or previous experience to gain new insights, reveal new patterns and the result of production are more accurate [11, 26]. Figure 1 depicts the modification of data lake architecture with machine learning techniques from existing architectures [5, 27].

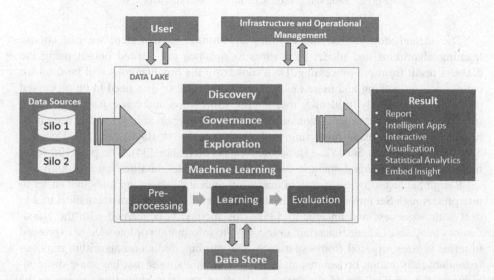

Fig. 1. Machine learning in data lake architecture for combine data silos.

The data source is consists of several data silos with large scale and different parameters. Data lake will process the data source through several phases to ensuring its operational availability, integrity, access control, authentication and authorization, monitoring and audit, business continuity, and find the importance solution [22]. Data discovery will describe the data and determination data. Data governance will capture and contextualize data by cataloging and indexing and manage incremental improvements to the metadata [21]. Exploration is creating new models of this data or modifies the existing model and can be combined with machine learning process. There are several steps in the machine learning process. First step is pre-processing, which convert the data into a form that can be used as input to the study through data cleaning, integration, transformation, construction, and reduction. In Fig. 2 depicts the data integration architecture for combining data silos in machine learning process.

In identification process, data silos will be recognized of data types, data structure, content and semantic. Data silos will align the structure of entities and define the relationships between the data silos. The harmonization process is unification of data types, value, scaling, format, and dimensions to merge the data based on previously transformation rules. The result of the merge data silos will be presented in combined view and can be used for other processes.

Fig. 2. Data integration scheme for combine data silos.

The second step in machine learning is learning. In this step, we can choose learning algorithms and model parameters to produce the desired output using the datasets result from pre-processing. Data silos from the organizations will be used for better decision-making and more efficient operations. These data need to be processed promptly and correctly to identify meaningful information and can extract knowledge from the data itself for organization needs. Various intelligent solutions have been used to analyze the data using predicting method such as Naïve Bayes [28, 29], Decision Tree [30, 31], Rough Set [32, 33], and K-Nearest Neighbor [34]. The powerful technique is needed for this domain in order to have accurate classification is Rough Set.

Rough Set has its own strength in managing data and generating rules that easier to interpret. Rough Set introduced by Zdzislaw Pawlak in 1980's is a mathematical tool to deal with vagueness and uncertainty [35]. This method is concerned with the classificatory analysis of vague, uncertain or incomplete information or knowledge expressed in terms of data acquired from experience. The attribute reduction algorithm removes redundant information or features and selects a feature subset that has same discernibility as the original set of the features. To lead the best prediction accuracy, the selected features can describe the decision as well as the original whole features set. Decision rules extracted by these algorithms are concise and valuable, which can be benefited in data mining by enlightening some hidden knowledge from the data [36]. Rough Set is chosen in this study because the important features can be identified based on reducts computation and thus will eliminate insignificant features in data silos.

The last step in machine learning process is evaluation, which determines the performance of the learning model. The results of whole data processing will always be stored in data storage and anytime can be used by users, which related to infrastructure and operational management of the organization. The outcomes of this process can be various forms such as intelligent applications, interactive visualization, predictive analysis, and statistical reports depending on the needs of an organization [27].

5   Air Pollutant Index Challenges and Opportunity

Air pollution is part of significant health hazard worldwide. According to the existing research, about 1.2% of total global annual deaths related to air pollution [37], which half of this amount comes from developing countries. In 2015, Malaysia and Singapore had bad haze which came from forest fires in Indonesia. The extreme weather conditions caused by El Nino can increase the temperature of the sea in the Southern Ocean [38]. In Indonesia, El Niño has delayed the rainy season, causing drought across the country, impacting the water supply and the harvest of rice and other crops. Under the dry conditions, Indonesia's forests and peat lands became into tinderboxes. The rainy season is being late and being shorter than normal because of El Niño. This Phenomenon happened again in early 2016.

Air pollution has been formed as a result of climate change could damage the epidermis, affects the immune reaction, and by mixing pollen, can increase the likelihood of allergic diseases such as asthma, allergic rhinitis and allergic conjunctivitis [39]. The study of human health effects caused by air pollution in developing countries is essential in providing more detailed information to evaluate the impact on health and the environment. Several previous studies have focused on the human health effects of pollutants single reflecting damage air pollution overall for human health [40]. Air Pollutant Index is a standard value of the level of air pollution which was developed by the Environmental Protection Agency (USEPA) to provide information that is more easily understood. API has a goal to be able to predict how long or heavy fog which has a thickness or may have a particular pattern, it is done with proper modeling software that includes weather, wind speed, and air mass concentration. According to USEPA, standard API value divided into several levels, namely: Good Level (<50), Moderate Level (51−100), Unhealthy Level (101−200), Very Unhealthy Level (201−300), and Hazardous Level (>300).

6 Experiment and Result

Experiments were performed to present the results of combined data silos by using a machine learning technique in data lake. This experiment use Talend Studio, Weka and Rosetta as simulated tools for data silos management system.

6.1 Data Pre-processing

In this research was used two data silos set, there are 24574 data from Malaysia's Government Open Data about Air Pollution Index (API) in Malaysia and 23760 data from Singapore's Government Pollutant Standard Index (PSI) in 2014 until 2015. The data samples in real value format as shown in Tables 1 and 2.

In Tables 1 and 2 show the pollutant index which has attributes date, time, state, region, and API. The data in Tables 1 and 2 are the raw data that still needs data pre-processing. This process includes the preparation, cleaning and selection process to obtain better data quality. In many applications, an outcome which is represented by a Decision Attribute based on standard API value that can be shown in Tables 3 and 4.

6.2 API Data Integration Process

The data will be integrated with several different resources. In this case, the data in Table 3 will be combined with data from Table 4. Transformation result of merging the source will provide a more useful format. The two pieces of combined data

Table 1. Sample data of Malaysia's Air pollutant index dataset.

Example	Date	Time	State	Region	API
1	26/11/2014	1:00AM	Kedah	Langkawi	25
2	26/11/2014	1:00AM	Pulau Pinang	USM	51
3	05/06/2015	1:00AM	Selangor	Shah Alam	38
4	05/06/2015	1:00AM	Johor	Kota Tinggi	47
5	02/08/2015	9:00AM	Sarawak	Sri Aman	31
…	…	…	…	…	…
…	…	…	…	…	…
245752	10/06/2016	1:00PM	Sabah	Sandakan	29

Table 2. Sample data of Singapore's pollutant standard index dataset.

Example	Date	Time	State	API
1	25/11/2014	1:00AM	South	45
2	25/11/2014	1:00AM	East	53
3	25/11/2014	2:00AM	South	43
4	27/12/2014	4:00AM	East	29
5	27/12/2014	4:00AM	South	28
…	…	…	…	…
…	…	…	…	…
23760	10/06/2016	1:00PM	West	37

Table 3. Sample of decision system for Malaysia's air pollutant index dataset.

Example	Date	Time	State	Region	API	Decision
1	26/11/2014	1:00AM	Kedah	Langkawi	25	Good
2	26/11/2014	1:00AM	Pulau Pinang	USM	51	Moderate
3	05/06/2015	1:00AM	Selangor	Shah Alam	38	Good
4	05/06/2015	1:00AM	Johor	Kota Tinggi	47	Good
5	02/08/2015	9:00AM	Sarawak	Sri Aman	31	Good
…	…	…	…	…	…	…
…	…	…	…	…	…	…
245752	10/06/2016	1:00PM	Sabah	Sandakan	29	Good

organization will be reorganized, restructured, and integrated with data from several internal and external resources illustrates in Fig. 3.

Selected data certainly affect the amount of data. The amount of data will be reduced because there is a lot of data which is certainly not irrelevant. There is 10427 ignored data with unknown instances. The result of combined data silo is reported in Table 5.

Table 4. Sample of decision system for singapore's pollutant standard index dataset.

Example	Date	Time	State	API	Decision
1	25/11/2014	1:00AM	South	45	Good
2	25/11/2014	1:00AM	East	53	Moderate
3	25/11/2014	2:00AM	South	43	Good
4	27/12/2014	4:00AM	East	29	Good
5	27/12/2014	4:00AM	South	28	Good
...	...	...	...	...	...
...	...	...	...	...	...
23760	10/06/2016	1:00PM	West	37	Good

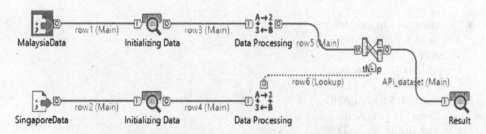

Fig. 3. Data integration scheme for combined API data.

Table 5. Combination data of Malaysia and Singapore API.

Example	Date	Time	State	Region	API	Decision
1	11/26/2014	1:00AM	Kedah	Langkawi	25	Good
2	06/15/2015	10:00AM	Johor	Muar	25	Good
3	04/25/2015	10:00AM	West Singapore	Jurong	42	Good
4	04/25/2015	10:00AM	South Singapore	Bukit Merah	40	Good
5	02/08/2015	9:00AM	Selangor	Banting	51	Moderate
...	...	...	...	...	...	...
...	...	...	...	...	...	...
248658	10/06/2016	12:00PM	Sabah	Sandakan	29	Good

6.3 The Prediction Method for API Dataset Using Rough Set

This research is done using a soft computing prediction method, Rough Set. Rough Set consists of data preparation, discretization, reduct computation, generation rules, and classification. In this research, the rules are generated from combination of API data for Malaysia and Singapore. This data needs to be discretized in certain interval prior to classification using Rough Set. The discretization technique that has been chosen is Boolean Reasoning and the reducer selected is Genetic Algorithm [32, 33]. Next, the dataset is divided into training and test sets with 10-fold cross-validation. This splitting

Table 6. Sample rules for Malaysian and Singapore API dataset with highest support.

No	Rule	LHS Support	RHS Support	LHS Coverage	RHS Coverage	Rule Length
1	API([*, 51)) => Decision(good)	11265	11265	0.17416	1.0	1
2	Time(10:00AM) AND State(Miri) => Decision (good)	125	125	0.01542	1.0	2
3	Date(2014-12-07) AND Time(1:00AM) AND State (Sarawak) => Decision (good)	14	14	0.00032	1.0	3
4	Date(2015-01-06) AND State(Selangor) AND API ([51,100]) => Decision (moderate)	9	9	0.00042	1.0	3
5	Date(2015-03-26) AND Time(1:00PM) AND State(Serawak) AND API ([51,100]) => Decision (moderate)	5	5	0.00023	1.0	4

dataset has been chosen to provide less biased estimation of the accuracy for the large scale dataset [41]. Rough Set produced reducts and rules for the classifier. Table 6 shows the sample of generated rules.

The sample generated rules explanation and its rule statistics as given in Table 6. The rules classify API into meaningful information for safety and health of population.

Rule 4: Date(2015-01-06) AND State(Selangor) AND API([*,51]) => Decision (moderate).

Based on the value given in the rule condition, the rule can be interpreted as:

IF Date is 6 January 2015, AND State in Selangor AND Air Pollutant Index (API) between 51 and 100 then the air condition is moderate.

The description of the rule statistics, the rule support is 11265, represents 11265 objects in the training data set that match with the rule condition. The rule accuracy is 1, represents the number of Right Hand Support (RHS) divided by the number of Left Hand Support (LHS) is 11265/11265 = 1. The conditional coverage is 0.11265; its represents the fraction of the records that satisfied the IF condition of the rule. It is obtained by dividing the support of the rule by the total number of records in the training data set. The decision coverage is 1.0, and it is the fraction if the training records that satisfied the THEN conditions. It is obtained by dividing the support of the rule by the number of records in the training data set that satisfied the THEN condition. The rule length is defined as the number of conditional elements in the IF part. In Rule 1 there is one attribute being used as conditional elements; API.

The generate rules from Malaysia and Singapore dataset can help to determine which areas have appropriate air condition. Moreover, it can help the government to makes predictions for time and date when disasters will happen and the right decision for the safety and health of the population in the future.

6.4 Analytics Result

The analysis results of the combination of Malaysia and Singapore API dataset using Rough Set approach can select significant data from the main database based on the generated rules. The selected dataset is displayed to chart [42]. This is intended to make it easier to present the result of analysis. Based on the Malaysian Air Quality Guidelines (MAAGs) and NEA (National Environment Agency) that have been adapted to the recommendations of the World Health Organization (WHO), Fig. 4 depicts APIs in each state in Malaysia and Singapore with their API conditions with maps chart.

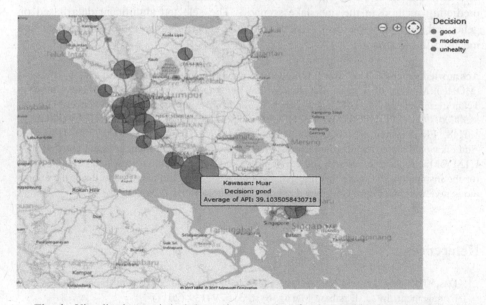

Fig. 4. Visualization statistical data using map chart for the combination of API data

These results can also send warnings to the state which the measured air quality is not appropriate to the population as it can damage the epidermis, affect the immune response, and by mixing pollen, may increase the possibility of allergic diseases such as asthma, allergic rhinitis and allergic conjunctivitis [33]. It is also possible to provide the warning alarm in areas that have the highest API and governments can also be aware of these conditions.

7 Conclusion

The improving of the quality of the data organization is considered will influence on the processing of information from the data. The challenges consist of data selection, description, maintenance, and governance. In order to increase profits at the operational processes of an organization, technical limitations should be minimized. Data lake can help to determine the best architecture, common metadata, data integration and so on for optimizing operational business processes within an organization in order to improve data quality and more efficient. The overall data lake process trough data discovery, governance, explore and machine learning. The use of machine learning techniques using soft computing approach can maximize the data management process in data lake that used the scenario of data integration process with the different data sources, both within a domain and across domains. It aims to improve the quality of data, use data as a competitive advantage, manage change, comply with work regulations, and adapted to the standards of work.

Testing is done by using the data from the Malaysia's and Singapore's Government Open Data about the Air Pollutant Index to determine the condition of air pollution levels for the health and safety of the population. The both of data have been combined be a combination of the new information and have been analyzed using Rough Set as a predicting method in the data lake scenario. The selected significant data based on generated rules will be displayed on the map chart that is more informative and meaningful information.

Acknowledgements. This work is supported by Ministry of Higher Education Malaysia (MOHE), Ministry of Science, Technology and Innovation Malaysia (MOSTI), and Universiti Teknologi Malaysia (UTM). This paper is financially supported by E-Science Fund, R. J130000.7928.4S117, PRGS Grant, R.J130000.7828.4L680, GUP Tier 1 UTM, Q.J130000.2528. 13H48, FRGS Grant, R.J130000.7828.4F634 and IDG Grant, R.J130000.7728.4J170. The Authors would like to express their deepest gratitude to the Research Management Centre (RMC), UTM for the support in research and development and Soft Computing Research Group (SCRG) for the inspiration and make this research success. Authors would also like to thank the anonymous reviewers who have contributed enormously to this work.

References

1. Das, T.K., Mishra, M.R.: A study on challenges and opportunities in master data management. Int. J. Database Manag. Syst. **3**, 129–139 (2011)
2. Bilal, M., Oyedele, L.O., Akinade, O.O., Ajayi, S.O., Alaka, H.A., Owolabi, H.A., Bello, S. A.: Big data architecture for construction waste analytics (CWA): a conceptual framework. J. Build. Eng. **6**, 144–156 (2016)
3. Bobek, S., Nalepa, G.J.: Uncertain context data management in dynamic mobile environments. Future Gener. Comput. Syst. **66**, 110–124 (2016)
4. Philip Chen, C., Zhang, C.Y.: Data-intensive applications, challenges, techniques and technologies: a survey on Big Data. Inf. Sci. **275**, 314–347 (2014)

5. Khanna, A., Bhatt, H., Hardik, M., Sakhuja, R.: Data lake for socio-knowledge platforms. Int. J. Eng. Appl. Sci. Technol. **1**, 110–113 (2016)
6. Milanov, G., Njeguš, A.: Using scrum in master data management development projects. J. Emerg. Trends Comput. Inf. Sci. **3**, 1446–1452 (2012)
7. Indrajani, I.: Master data management model in company: challenges and opportunity. ComTech. **6**, 514–524 (2015)
8. Konverse: Big Data Lake (2015)
9. Margaret, R.: Data Silo Definition (2015). http://searchcloudapplications.techtarget.com/definition/data-silo
10. Fenwick, T., Brunsdon, D., Seville, E.: Reducing the Impact of Organisational Silos on Resilience (2009). http://www.resorgs.org.nz/images/stories/pdfs/silos.pdf
11. Yaqoob, I., Hashem, I.A.T., Gani, A., Mokhtar, S., Ahmed, E., Anuar, N.B., Vasilakos, A. V.: Big data: from beginning to future. Int. J. Inf. Manage. **36**(6), 1231–1247 (2016)
12. Gandomi, A., Haider, M.: Beyond the hype: big data concepts, methods, and analytics. Int. J. Inf. Manage. **35**(2), 137–144 (2015)
13. Eygm, K.W., Peng, T.: big data changing the way businesses. Int. J. Simul. Syst. Sci. Technol. **16**, 28 (2014)
14. Lomotey, R.K., Deters, R.: Terms extraction from unstructured data silos. In: Proceedings of 2013 8th International Conference on System of Systems Engineering: SoSE in Cloud Computing and Emerging Information Technology Applications, SoSE 2013, pp. 19–24 (2013)
15. Krensky, P.: Winter is not coming: eliminating data silos and ending information hoarding. 10 (2015)
16. Lowans, B., Perkins, E.: Market Guide for Data-Centric Audit and Protection (2014)
17. Shahrokni, A., Soderberg, J.: Beyond information silos challenges in integrating industrial model-based data. CEUR Workshop Proc. **1406**, 63–72 (2015)
18. Laskowski, N.: Data lake governance: a big data do or die (2016). http://searchcio.techtarget.com/feature/Data-lake-governance-A-big-data-do-or-die
19. Kottursamy, K., Raja, G., Padmanabhan, J., Srinivasan, V.: An improved database synchronization mechanism for mobile data using software-defined networking control. Comput. Electr. Eng. 1–11 (2016)
20. Hegde, S.D., Ravinarayana, B.: Survey paper on data lake. Int. J. Sci. Res. **5**(7) (2016)
21. Miloslavskaya, N., Tolstoy, A.: Big data, fast data and data lake concepts. Procedia Comput. Sci. **88**, 300–305 (2016)
22. Terrizzano, I., Schwarz, P., Roth, M., Colino, J.E.: Data wrangling: the challenging journey from the wild to the lake. 7th Biennial Conference on Innovative Data Systems Research, CIDR 2015 (2015)
23. Means, A.: The Role of Data Lakes in Healthcare. North America, Perficient (2014)
24. Tan, P.: Introduction to Data Mining. Pearson Education, Boston (2006)
25. Budiharto, W.: Machine learning & computational intelligence. Penerbit Andi (2016)
26. Zhou, L., Pan, S., Wang, J., Vasilakos, A.V.: Machine learning on big data: opportunities and challenges. Neurocomputing (2017)
27. Nugroho, A.S.: Pengantar Support Vector Machine (2007)
28. Sun, S., Luo, C., Chen, J.: A review of natural language processing techniques for opinion mining system. Elsevier B.V. (2016)
29. Tsangaratos, P., Ilia, I.: Comparison of a logistic regression and Naïve Bayes classifier in landslide susceptibility assessments: the influence of models complexity and training dataset size. CATENA **145**, 164–179 (2016)

30. Mayilvaganan, M., Kalpanadevi, D.: Comparison of classification techniques for predicting the performance of students' academic environment. In: International Conference on IEEE Communication and Network Technologies (ICCN 2014), pp. 113–118 (2014)

31. Gray, G., McGuinness, C., Owende, P.: An application of classification models to predict learner progression in tertiary education. In: IEEE International Advance Computing Conference (IACC), pp. 549–554 (2014)

32. El-Alfy, E.-S.M., Alshammari, M.A.: Towards scalable rough set based attribute subset selection for intrusion detection using parallel genetic algorithm in MapReduce. Simul. Model. Pract. Theory., 1–12 (2016)

33. Fang, B.W., Hu, B.Q.: Probabilistic graded rough set and double relative quantitative decision-theoretic rough set. Int. J. Approx. Reason. **74**, 1–12 (2016)

34. Jiang, L., Li, C., Wang, S., Zhang, L.: Deep feature weighting for naive Bayes and its application to text classification. Eng. Appl. Artif. Intell. **52**, 26–39 (2016)

35. Pawlak, Z.: Rough sets. Int. J. Comput. Inf. Sci. **11**, 341–356 (1982)

36. Rissino, S., Lambert-Torres, G.: Rough Set Theory–Fundamental Concepts, Principals, Data Extraction, and Applications. In: Ponce, J., Karahoca, A. (eds.) Data Mining and Knowledge Discovery in Real Life Applications. InTech Publisher (2009)

37. Velez, F., Agrawal, S.: Data lakes: Discovering, Governing and Transforming Raw Data into Strategic Data Assets. Persistent System (2016)

38. Murray, C.J., Vos, T., Lozano, R., Naghavi, M., Flaxman, A.D., Michaud, C., Bridgett, L.: Disability-adjusted life years (DALYs) for 291 diseases and injuries in 21 regions, 1990–2010: a systematic analysis for the global burden of disease study 2010. Lancet **380**(9859), 2197–2223 (2013)

39. Bank, World: Indonesia's Fire and Haze Crisis (2015). http://www.worldbank.org/en/news/feature/2015/12/01/indonesias-fire-and-haze-crisis

40. Kim, H., Park, Y., Park, K., Yoo, B.: Association between pollen risk indexes, air pollutants, and allergic diseases in korea. Osong Public Heal. Res. Perspect. **7**, 172–179 (2016)

41. Kohavi, R.: A study of cross-validation and bootstrap for accuracy estimation and model selection. In: Proceedings of International Joint Conference on AI, pp. 1137–1145 (1995)

42. Verbert, K., Govaerts, S., Duval, E., Santos, J.L., Van Assche, F., Parra, G., et al.: Learning dashboards: an overview and future research opportunities. Personal and Ubiquitous Computing **18**(6), 1499e1514 (2014)

Development of 3D Earth Visualization for Taiwan Ocean Environment Demonstration

Franco Lin[1(✉)], Wen-Yi Chang[2], Whey-Fone Tsai[2],
and Chien-Chou Shih[1]

[1] Taiwan Ocean Research Institute, National Applied Research Laboratories,
No. 196, Henan 2nd Road, Qianjin District, Kaohsiung 801, Taiwan
{0803007, raymondshih}@narlabs.org.tw
[2] National Center for High-Performance Computing, National Applied Research
Laboratories, No. 7, R&D 6th Road, Science Park, Hsinchu 300, Taiwan
{c00wyc00,9303115}@narlabs.org.tw

Abstract. This paper presents a three-dimensional earth representation system for visualization of Taiwan Ocean data, based on the mixed MySQL and HBase databases. The Client-Server architecture is used to provide Web GIS services in this system. In addition, the WebGIS and WebGL are adopted to enhance 3D visualization of the ocean data. In order to achieve better data-access performance, the architecture of HBase combined with MySQL is used to accommodate different data types and sizes in the back end of system. Finally, we implement the information of sea-going onboard survey, including concentrations of chlorophyll a salinity and temperature, shore-based surface current data and PM2.5 data, onto the 3D-GIS display platform for environmental decision making in the future.

Keywords: Three-dimensional · 3D visualization · HBase · MySQL

1 Introduction

Extreme weather caused by global warming has resulted in many natural disasters all over the world. For the hazard mitigation purposes in Taiwan, it is very important to collect and use the governmental monitoring information (including Environmental Protection Bureau, Central Weather Bureau, Taiwan Ocean Research Institute) for the environmental decision support. However, as the growth of time, the single-device database and file system may become insufficient due to the explosion of data volume. Under such a BigData era, the well-known HDFS and HBase in Hadoop [1] ecosystem get more and more popularity. Wu [2] used the VM (Virtual Machine) system to test the environment of Hadoop. Lee [3] developed a visualization platform based on the distributed SQL servers for the environmental monitoring. General speaking, DAS (Disk Array Storage) devices and SQL-like databases still have some superior features, so a mixed database system may be a good alternative for engineers.

© Springer International Publishing AG 2017
Y. Tan et al. (Eds.): DMBD 2017, LNCS 10387, pp. 307–313, 2017.
DOI: 10.1007/978-3-319-61845-6_31

Data visualization is a very important task for helping people to understand data. In the past, limited by the computer's hardware and network bandwidth, only the standalone GIS (Geographic Information System) is used to display the various geographic information. Now, in the Internet era, the Web-based GIS have become the mainstream to provide Web-information services. In general, the front end of the WebGIS includes Leaflet, OpenLayers, Google Map, and the back end is the GeoServer that provides layer information and controller. In addition, the geographic information database, such as PostgreSQL, HBase, etc., are widely used for data storage. Recently, because HTML5 gradually breaks the browser restrictions, users become able to use the syntax of OpenGL to write web pages, called WebGL. Therefore, the Computer Graphics by WebGL start to emerge in the field of WebGIS with a progress from 2D GIS to 3D GIS display.

Recently, there are several national research institutions that have converted their original display platforms into 3D GIS platforms, such as NASA [4]. As shown in Fig. 1(a), NASA has the world's largest satellite imagery information and develops the WebGIS platform to show their satellite imagery. Similarly, as shown in Fig. 1(b), NOAA Visualization Lab [5] displays the 3D atmospheric information on its official website. Specifically, due to that NOAA has a dynamic flow field technology, the dynamic GIS display makes users much easier to understand the current atmospheric information. Besides, Hidenori Watanave's team [6] also uses the 3D earth dynamic technology to show the Japanese 318 earthquake and the earthquake victims of the action record.

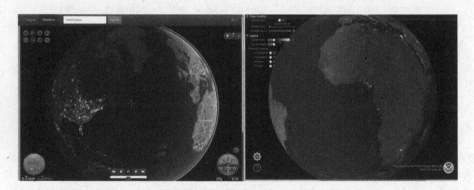

Fig. 1. (sssa) NASA Web; (b) NOAA Web

2   System Representation Technology

2.1   System Architecture

In this paper, as shown in Fig. 2, the Client-Server architecture is used to provide Web GIS services. In addition to the information generated by TORI (Taiwan Ocean Research Institute) [7], the Open Data released by the government is also collected and post-processed to store in Data Warehouse via QA/QC (Quality Assurance/Quality Control). Due to the growth of data, the NoSQL's database is also adopted in this study,

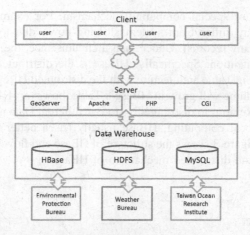

Fig. 2. System architecture

which combined Hadoop HDFS with HBase to store Big Data. For data visualization, the Cesium [8] is adopted in the front end as a display module.

2.2 Data Types

In 2016, the ocean database developed by TORI had grown to 94TB. In this year, it is expected to break 100TB. Below shows the types of data:

1. Cruise: When each research vessel sails, all navigational status is recorded by the instruments, including the time of flight, latitude and longitude, wind speed, wind direction, water temperature, water depth and other information. Some underwater vehicle information are also available.
2. Satellite imagery: The advantage of satellite imagery is that it can obtain a wide range of spatial data in a short time, which avoids human to reach a dangerous place for observation. Telemetry technology is also widely used in navigation, agriculture, meteorology, resources, environment, planetary science and so on. Now, the basic water temperature, chlorophyll data have been included in the database.
3. TOROS: TOROS (Taiwan Ocean Radar Observing System) has set up 12 longrange high-frequency radar and five standard radar around Taiwan, observing the surface velocity of Taiwan's coastal surface.
4. Drifter: Drifters are used to observe the atmosphere, ocean flow and hydrological conditions in the South China Sea and Taiwan coast for a long time. Such information can be provided in a real-time manner.

2.3 Mixed Databases

In the Big Data era, the traditional file systems and relational databases are gradually unable to handle the expansion of data volume, even though through hardware upgrades or the use of special computing mechanism. For example, the traditional database MySQL may suffer from system slowing down due to the big data (the Table just has too many records). Under such a circumstance, the Hadoop ecosystem has drawn a great attention. Specifically, HBase is the distributed database used in Hadoop. HBase is a big Table architecture with the distributed HRegion. Just like other database, the data volume will grow in HBase's Table, however, HBase can divide the Table into several HRegions to distributedly store in different Data nodes. Hence, based on the concept of local computing, HBase usually has a better performance while accessing big data. Figure 3 shows the structure of HBase and flowchart, which reveals the high scalability and distributed mechanism of HBase.

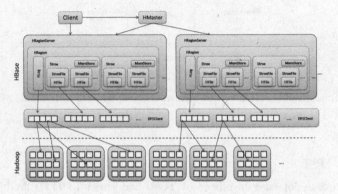

Fig. 3. HBase flowchart

HBase is NoSQL's data structure and based on the concept of Key-Value to access data. Different from the traditional relational database, there is no index in the design. Instead, through the Row, Column Family, Column Qualifier in the Table, records can be queried. In the present study, taking the Cruise data as an example, Table 1 explains the difference between HBase and MySQL databases as follows:

Table 1. Data structure of HBase compared with MySQL for the same Cruise data

	Column family-1		Column family-2	
HBase table				
Row key	Lat	Lng	Velocity	Direction
2016-12-25	25.12	121.524	23.5	45
2016-12-24	25.37	122.547	21.6	50

MySQL Table					
ID	Date	Lat	Lng	Velocity	Direction
1	2016-12-25	25.12	121.524	23.5	45
2	2016-12-24	25.37	122.547	21.6	50

1. Table: Like the Table in SQL, HBase has the same attributes of the data.
2. Row: It is the most basic unit in HBase database, and RowKey is the only value to query the data stored in a column.
3. Column Family: Several Columns can be combined as a Column Family, and a Table can have several Column Families.

3   Performance Experiment

The proposed 3D-GIS display platform was developed on Windows7 OS with Intel Core i5-6400 2.7Ghz CPU and 16 GB RAM. The front-end web pages mainly use Javascript and html, and the back-end programming language uses PHP5. The Database uses MySQL and HBase. The system uses Cesium as a front-end package. Figure 4 shows the visualization of 3D-GIS data for coastal environment monitoring. In Fig. 4(a), the route of ship voyage can be displayed along with the coastal information, In Fig. 4(b), the surface velocity measured by TOROS can real-time display on the map. In Fig. 4(c), the monitoring PM 2.5 values can be exactly located in a Taiwan map. Furthermore, Fig. 4(d) shows the distribution of chlorophyll in the third quarter of 2016.

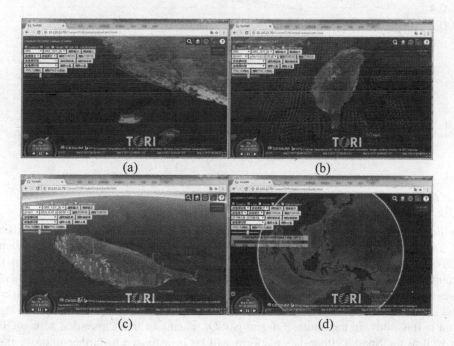

(a)   (b)

(c)   (d)

Fig. 4. 3D visualization of environmental data

In order to compare the computational efficiency in different database systems, Fig. 5 shows the execution time along with the query data number in MySQL and HBase. In the figure, one can see that when the number of query data is less than 700000, the search speed in MySQL is better than that in HBase. The reason for that could attribute to the extra inherent latency of distributed computing in HBase. On the contrary, when the data number is more than 700000, HBase shows a great advantage due to the distributed computing. In this case, the speedup can reach 1.17 over 1200000 records. Therefore, the experimental result suggests that HBase is suitable for big data, on the other hand, MySQL is suitable for small data. In the present 3D-GIS display system, both HBase and MySQL are used as mixed databases to accommodate different data types and data sizes.

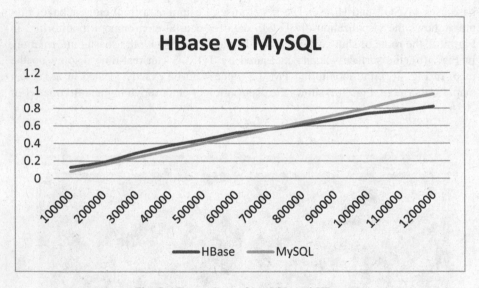

Fig. 5. Comparison of MySQL and HBase

4 Conclusion

In the present study, a 3D-GIS display system was proposed to present the oceanic data of Taiwan for environmental decision-support purposes. In the front end, the system combines WebGIS and WebGL to enhance 3D visualization of the ocean data, which architecture is different from the traditional 2D WebGIS. In the back end, the system uses mixed databases, namely HBase and MySQL, to accommodate different data types and sizes. Through an experimental test, HBase outperforms MySQL for a relative big data in this study. In the future, different NoSQL databases like Cassandra, MongoDB as well as the Spark [9] big-data platform will be investigated to further improve the performance of our 3D-GIS display system.

Acknowledgments. Financial support from the Ministry of Science and Technology, Taiwan, under Grants MOST 105-2634-E-492-001 is highly appreciated. We are also grateful to the National Center for High-Performance Computing for computer time and Taiwan Ocean Research Institute.

References

1. Hadoop. http://hadoop.apache.org/
2. Wu, C.-H., Tsai, W.-F., Lin, F., Chang, W.-Y., Lin, S.-C., Yang, C.-T.: Big data development platform for engineering applications. In: 2016 IEEE Big Data Conference-Big Spatial Data Workshop, Washington D.C., pp. 2699–2702. IEEE (2016)
3. Lee, J.-G., Tsai*, W.-F., Lee, L.-C., Lin, C.-Y., Lin, H.-C., Tsuang, B.-J.: In-place query driven big data platform: applications to post processing of environmental monitoring. J. Concurr. Comput. Pract. Exp. (2017)
4. NASA. http://explorer.worldwind.earth
5. NOAA. http://www.nnvl.noaa.gov/weatherview/index.html
6. Hidenori Watanave's team. http://www.nnvl.noaa.gov/weatherview/index.html
7. TORI (Taiwan Ocean Research Institute). http://www.tori.org.tw/
8. Cesium. https://cesiumjs.org/
9. Spark. http://spark.apache.org/

Exploring Potential Use of Mobile Phone Data Resource to Analyze Inter-regional Travel Patterns in Japan

Canh Xuan Do[(✉)] and Makoto Tsukai

Hiroshima University, 1-4-1 Kagamiyama, Higashi-Hiroshima, Hiroshima
739-8527, Japan
canh.doxuan@gmail.com, mtukai@hiroshima-u.ac.jp

Abstract. In Japan, Inter-Regional Travel Survey gives rich information to
researchers and transportation planners. The current survey data was conducted
in 2010, and the newest survey data collected in 2015 will be available soon.
This national survey is mainly based on the on-site questionnaire survey which
requires an enormous budget and spends so much time to finalize and publish
the data result. Recently, ubiquitous mobile computing and the big data give us
new opportunities for exploring a new type of data resource besides the traditional survey data. This study clarifies the deviation of cell phone data at
aggregated origin-destination level of inter-regional trip flows, compared with
the traditional on-site passenger survey. Also, the mechanisms of inter-regional
trip generation are explained through travel patterns by a classification tree
analysis, one of the big data mining classification algorithms.

Keywords: Mobile phone data · Inter-regional Travel Survey data ·
Classification tree · Origin-destination trips · Travel patterns · Japan

1 Introduction

The fast growth of expanding cities along with the need for long-distance travel among
regions has led a requirement to improve or find a new way to estimate inter-regional
travel demand. To meet these challenges, there are various methods such as improving
traditional four-step models and more recent activity-based models developed to utilize
available computational resources. These models usually use methods of statistical
sampling in local [1, 2] or national travel surveys [3–6] to analyze and infer trip
characteristics between areas of a city or regions of a country.

Inter-regional travel surveys are typically administered by government or regional
planning organizations and are integrated with public data such as national census with
detail demographic characteristics of their residents, made available by city, state, and
federal agencies. These surveys are designed to select representative samples in population carefully, so they are relatively expensive for surveying, and required much
more time in the post-survey processing. As a result, the time between two consecutive
surveys is four years or more in even the most developed cities (e.g., 1995 American
Travel Survey (ATS) with a new version so-called National Household Travel

© Springer International Publishing AG 2017
Y. Tan et al. (Eds.): DMBD 2017, LNCS 10387, pp. 314–325, 2017.
DOI: 10.1007/978-3-319-61845-6_32

Survey [6], Inter-regional Travel Survey in Japan [4], and so forth). The appearance of ubiquitous mobile phone computing has led to a noticeable increase in new big data resources capturing cell phone user activities in nearly real time and provided solutions to the conventional travel demand models. New data sources are collected by new providers such as large telecommunications companies with their own applications and network providers. Compared to the traditional survey, these big data sources provide large and long-time samples at low cost. Along with new opportunities, however, new mobile phone data (hereafter, called MOBI data) comes with new challenges of estimation, integration, and validation of existing models since they often lack or miss relevant contextual, social demographic information due to privacy reasons and have their interior noise and biases. Although these issues are significant obstacles to accommodate them to existing models, their use for regional transportation planning has the potential to decrease the period of survey conducting, increase survey coverage, and reduce survey costs. Therefore, it is essential to evaluate and propose a new method to integrate the new data sources into the traditional modeling.

To make these new data sources useful for inter-regional planning, we should clarify their biases and limitations. Moreover, we need to evaluate the appropriate level of implementing new data sources as an input data for inter-regional travel demand modeling. Many studies have been explored using these new massive, passively collected data such as individual survey tracking and stay extraction [7], origin-destination flows estimation and validation [8–10], traffic speed estimation [11, 12], and activity modeling [13]. Nevertheless, these studies generally present alternatives for only urban planning, while there is a small number of studies utilizing new data resources into modeling inter-regional travel demand [14]. Thus, here in this study, we clarify the characteristics of MOBI data by several comparisons of origin-destination (hereafter, called OD) pair travel flows and try to explore inter-regional travel patterns.

This paper is organized into following sections. The next section demonstrates the summary of data collection. Section 3 shows the methodology use in this study. The comparisons and the results of exploring travel patterns are presented in the following section. Section 5 illustrates the conclusions, limitations and future research.

2 Data Summary

2.1 Net Passenger Transportation Survey in Japan

In Japan, Inter-regional Travel Survey or Net Passenger Transportation Survey (from now on, called NPTS) has been carried out since 1990 in every five years by Ministry of Land, Transportation, Infrastructure and Tourism (MLIT) [4]. This survey collects passenger traffic at certain cross-sections and then aggregates the data into the inter-regional OD tables with its expansion factors. Individual characteristics of travel are captured such as departing point (origin) and arrival point (destination) of each transportation mode on the entire route if travelers transfer. It collects the questionnaire sheet from a respondent, one out of about a million samples, who uses inter-regional trains, express buses, airlines, cars or ships. The purpose of NPTS is to provide basic information for inter-regional transportation infrastructure planning. MLIT provides the

OD tables on MLIT website [15]. After the survey in 2000, trip information at the individual level with the corresponding expansion factor became available. Inter-regional net passenger traffic data used in this study was extracted from 2010's NPTS. The OD table records passenger trips between 207 areas of residential areas (origins) and destinations. The database of regional resources and demographics was compiled from 2011 Japan National Census.

NPTS provides two types of OD tables with a daily OD and an annual OD trips table. For comparison with MOBI data, in this study, we use the daily OD trips table. NPTS data has total 2,323,497 observations with 9,153,592 inter-regional daily trips.

2.2   Mobile Phone Data

The new MOBI data source is provided by NTT DOCOMO, the largest cell phone service provider in Japan. This company had more than 70 million mobile phone subscriptions as of 2016 [16] while the population of Japan was over 127 million as of October 1, 2015, based on the national census in Japan [17]. This company developed a new kind of small area statistics named Mobile Spatial Statistic (MSS), which is used to make estimations of the population of a small area by using the operations records from a mobile terminal network. The MSS's coverage of age is from 15 to 79, which is the generation of active mobile phone users in Japan. It also accounts for around 80% of the total population (2010 Population Census). Another reason why MSS selects this age range is that cell phone penetration rates are dominant in these ages, resulting in enough sample size for MSS to provide accurate estimates.

This study used the MOBI data collected in two days, one in holidays and one in weekdays in October 2015. There are 255,232 observations to be collected, which is equivalent to near 473 million trips over two days. The total number of trips covers intra-regional trips and inter-regional trips in 207 zones.

3   Methodology

The following two subsections review the algorithm for transforming MOBI data into OD matrices, describing some indices in MOBI trip characteristics used to compare with NPTS data regarding the number of inter-regional trips. The last subsection describes a classification method to find inter-regional travel pattern of trip generation.

3.1   OD Inter-regional Travel Flows Matrices

Current methods, which have been employed to estimate the trips between two places, fall into two categories: (1) four-step travel demand estimation methods; and (2) activities based approaches. The MOBI data records the successive activity trajectory through the mobile phone base stations. It provides an opportunity to transform the raw data with billions of points into an OD matrix of flows. Though a mobile phone user should have traveled between two different points at different times, we do not know the precise departure time of their trip. Thus, to extract meaning locations, termed as *stays*, we assume that origin points are collected at mid-night time when mobile phone users would stay in their homes or hotels, and destination points are collected at noon of the day after they traveled. In this way, a full OD matrix is made by summing the trip volume computed for all users between all pairs of ODs.

3.2   Indices in MOBI Trip Characteristics

For comparisons between two kinds of data sources, we put the NPTS data as a standard reference. Therefore, from the standard reference, we clarify the deviation of MOBI data at aggregated OD level in the number of trips in the three following cases: (1) traveling between OD pairs; (2) generating from origin areas; and (3) distributing to destination zones. In this study, we use three indices, including *Pearson's correlation coefficient*, *Spearman's rank correlation coefficient*, and *deviation index* to measure these differences.

First, we employ the *Pearson's correlation coefficient*, *r*, which provides an indication of how closely a set of MOBI data values and a set of NPTS data values agree in relative terms. It is noted that a perfect correlation ($r = 1$) does not imply perfect reliability of MOBI data values – all MOBI data values may be biased in a consistent direction.

Second, *Spearman's rank correlation coefficient*, ρ, provides an indication of similarity between the ranks of two sets of values. It ranges from -1 to $+1$. The use of ranks means that, if the order of MOBI data is correct, the index will be high.

Finally, we introduce *deviation index* as an index for quantitatively evaluating the size of different values of two datasets. This index is used to indicate how much they deviate from the ideal state (i.e., when both are equal) (see Fig. 1).

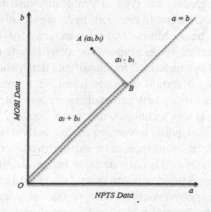

Fig. 1. Intuitive illustration of deviation index

Assume that for each pair of comparative values, *i*, the value in NPTS dataset is a_i, the value in MOBI dataset is b_i, and the average of the two values is $\mu_i = (a_i + b_i)/2$.

The deviation index, δ_i, for each pair of comparative values is defined as follows

$$\delta_i = \frac{b_i - \mu_i}{\mu_i} \tag{1}$$

Substituting μ_i into Eq. 1 gives us the new equation for deviation index as follow

$$\delta_i = \frac{b_i - a_i}{a_i + b_i} \tag{2}$$

Figure 1 describes an intuitive understanding of the concept, with deviation index, δ_i, represented by the ratio between the distance from the point $A(a_i, b_i)$ to its orthogonal projection point B on the line $a_i = b_i$ to the distance from the origin point O to the point B. As seen from the Eq. 2, the deviation index is normalized from -1 to 1. It can be inferred that deviation index is nearer to zero indicating the smaller difference between two values. Values are near -1 or 1, which means larger differences. Positive deviation indexes suggest that the MOBI data values are greater than the NPTS data values and vice versa. In case of $a_i = b_i = 0$, we also set the deviation index $\delta_i = 0$.

3.3 Decision Tree Analysis – A Classification Method

In the field of data mining, there are many classification methods such as decision tree, rule-based classifiers, and so forth. In this study, decision tree method is selected among other classification methods because of its outstanding advantages over other methods. First, the decision tree is a nonparametric approach which does not require any prior assumptions regarding the type of probability distributions satisfied by the class and other attributes. Second, decision trees are relatively easy to interpret, especially smaller-sized trees. Moreover, the accuracies of the decision trees are comparable to other classification techniques for many simple datasets.

Decision tree method is a technique for classifying data patterns into tree structures and consists of predicting a certain outcome based on a given input. In regression models, the statistical relationship between explanatory and outcome is assumed before starting the analysis, but in this technique, it is not necessary. Ture, Tokatli and Kurt [18] also stated that the outstanding advantage of this method is to discover and give a better explanation of the relationships between the predictors that would make it possible to predict the outcome. Also, the decision tree model often inputs all predictor variables, and then the representation of the outcome and predictor relationship is made with more concise and perspicuous results. A tree structure consisting of three sub-elements: a root node, internal nodes, and terminal nodes - the main output of a decision tree. The development of a decision tree includes three stages: a tree growing, a tree pruning and an optimal tree selection.

Decision tree consists of three groups such as Classification and Regression Tree (e.g., its name is often as CRT, CART, or C&RT), Chi-Square Automatic Interaction Detector (CHAID), and Quick-Unbiased-Efficient Statistical Tree (QUEST) [18]. Of these groups, the CHAID analysis has not been widely applied in travel and tourism research field [19–23]. Van Middelkoop, Borgers and Timmermans [21] used an Exhaustive CHAID to identify heuristic principles for transport mode choices. The results proved that the methodology could be applied to understand tourist behavior. Bargeman, Chang-Hyeon, Timmermans and Van der Waerden [19] investigated the relationships between holidays choice behaviors and socioeconomic variables to classify respondents into different clusters by using a combination of CHAID and log-linear analyses. Chen [23] concluded that CHAID would be a useful tool to put forward the subdivision methodology in travel and tourism research. Chen [20] made a CHAID analysis to classify each actionable group by demographic and trips characteristics.

CHAID analysis has a variety of advantages. First, it can produce non-binary trees, which is different with CRT and QUEST. Another advantage of CHAID is, at each node, the procedure of splitting and merging pairs of categories of the predictor variables considering their differences from the outcome and using Chi-square test to measure these differences. A modification of CHAID method, called Exhaustive CHAID, was originally proposed by Biggs, De Ville and Suen [24]. Compared to the earlier CHAID, the optimal split procedure of Exhaustive CHAID was improved by continuously testing all possible category subsets. Fundamentally, it is a decision tree based on the Chi-square test, which is built by repeatedly splitting a parent node into two or more child nodes.

4 Empirical Study

4.1 Trip Weight

In order to focus on long-distance trips, the intra-zonal trips within each prefecture and three metropolitan areas (e.g., Tokyo, Osaka, and Nagoya metropolitans) are excluded to eliminate. In this paper, the number of study zones is 194 out of 207 zones since isolated islands with no choice in mode are also excluded. Thus, both data utilized in this study has a total of 36,346 OD pairs among 194 areas. Moreover, both datasets include trips in weekday and holiday, therefore, to calculate the number of daily trips, we introduce a daily trip weight to keep the balance between holiday trips and weekday trips in the dataset. The trip weight, m, is calculated as follow

$$m = \frac{N_h}{N_w} \tag{3}$$

where N_h is the number of holidays in Japan (i.e., national holidays and non-working days) and N_w is the number of weekdays in the autumn (from September to November 2010 for NPTS data, and from September to November 2015 for MOBI data).

Since the coverage age of MOBI data ranges from 15 to 79, all observations in NPTS data which are out of this age range are removed. Thus, NPTS data has 2,202,466 observations with near 7.8 million inter-regional trips, and MOBI data has 91,742 observations with over 8.2 million inter-regional trips.

4.2   Comparison of MOBI Data with NPTS Data

As mentioned in Subsect. 3.2, to clarify overall characteristics in MOBI data for aggregated OD trips, we first computed *Pearson's correlation index* and *Spearman's rank correlation index* as in Table 1. Compared to two other cases, the r value of trips among OD pair zones is the lowest, which means there is different between two data. This finding would be clarified in the following part of this subsection. The values of ρ are near one indicating the small difference of the number of trips in MOBI data compared to that of NPTS data.

Table 1. Summary of Pearson's correlation and Spearman's rank correlation coefficient

Aggregated trips	r	ρ
(1) OD pair zones	0.602	0.918
(2) From origin zones	0.955	0.942
(3) To destination zones	0.961	0.947

To be more detail comparison, the deviation indexes are calculated to exam the differences in other aspects. In order to understand the difference of OD pair trips, the heat map is made in Fig. 2a. In this figure, the direction of left-to-right or bottom-to-top represented for the north-to-south direction of Japan. The blue points mean that the number of OD pair trips in MOBI data is smaller than that of NPTS data (i.e., deviation index is near or equal minus one). Also, the opposite is true for the red points (i.e., deviation index is near or equal one). These two types of points mean that there is big different between two datasets while white points indicate that there is no different or the difference is very small.

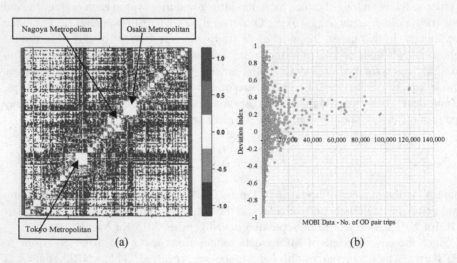

(a)   (b)

Fig. 2. Heat map (a) and density (b) of deviation indexes for the differences of OD pairs trips. (Color figure online)

As shown in Fig. 2a, most OD pairs are dominated by white and blue pixel points. Note that there are several large white areas in the diagonal because OD trips in three metropolitans are excluded in both NPTS data and MOBI data. Blue pixels concentrate around three metropolitans or in OD pairs where trips are generated from or distributed to the three metropolitans. Also, the red, green, and purple points are scattered along the diagonal. It can be inferred that MOBI data seems to over- or under-estimate at for trips around the densely inhabited areas and short-distance trips.

Figure 2b illustrates the density of deviation indexes and shows the relationship between the number of OD pair trips in MOBI data and deviation indexes. The graph indicates that the density is negatively skewed, which provides another evidence of over-estimation as seen in Fig. 2a. This figure, also, shows that scattering in deviation indexes decrease to near 0.2 as the number of OD pair trips increases. In other words, there is an average of 20% differences between two datasets. This reflects that MOBI data can capture more inter-regional trips in urbanized or the densely inhabited areas.

For evaluating the differences regarding trip generation and trip distribution, Fig. 3a and b illustrates the density of deviation indexes by origin and destination zones, respectively. This graph shows that there is a small difference between two datasets in term of trip distribution at origins and destinations. As seen these figures, most of the zones have smaller deviation indexes ranging from −0.2 to 0.2, indicating the small difference between two datasets in term of aggregated trips at origin or destination zones. However, some large cities such as Tokyo, Osaka, and Sapporo have large deviation indexes, indicating the less reliability of MOBI data in those areas.

4.3   Travel Patterns by a Decision Tree Analysis

To explore travel patterns, the Exhaustive CHAID was employed with 36,346 OD pair flows observations. The number of observations is consistent with Van Middelkoop, Borgers and Timmermans [21] because a CHAID-based algorithm requires a dataset with approximately 150 to 200 observations per one predictor variable. To put it simply, with 68 predictors, we need only 13,600 observations for CHAID analysis. As in Figs. 4 and 5, the objective variable is represented into two categories (i.e., "YES" = there is OD pair travel flow, and "NO" = there is no OD pair travel flow).

At a glance, travel pattern of trip generation is different. This may be because the number of zero OD travel flows in MOBI data is much higher. This is consistent with the previous result in Subsect. 4.2 (i.e., many blue points are seen in the OD pairs with deviation index that is near or equal negative one).

In general, most of the child nodes have low trip generation rate in MOBI data, while this is opposite in NPTS. More specifically, there is equivalent or similar at the strongest predictors between two trees in term of travel distance. The strongest predictor in the tree of NPTS data is the rail travel time variable, while rail travel cost variable is chosen in the tree of MOBI data. These two variables are closely related to travel distance. Therefore, based on these two variables, we defined three groups by the levels of travel distance (see detail in Figs. 4 and 5). However, in the second level of both trees, the difference appears. In the tree of NPTS data, the division is mostly depended on regions' demographic characteristics such as the number of workers of the

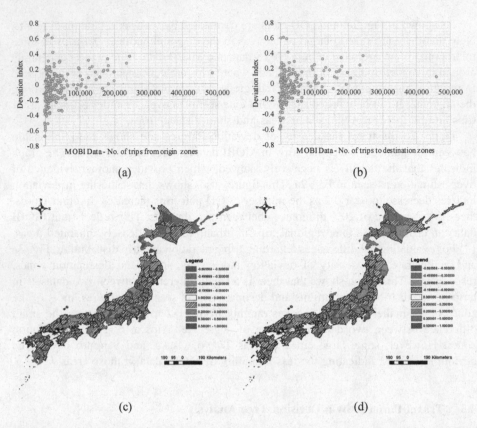

Fig. 3. Density and spatial distribution of deviation indexes by (a; c) origin zones and (b; d) destination zones.

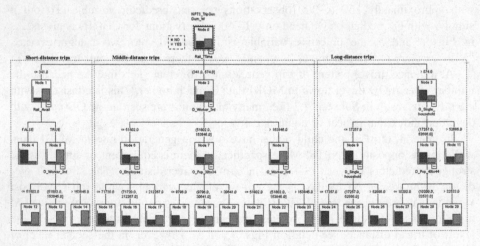

Fig. 4. Travel patterns of trip generation in NPTS data.

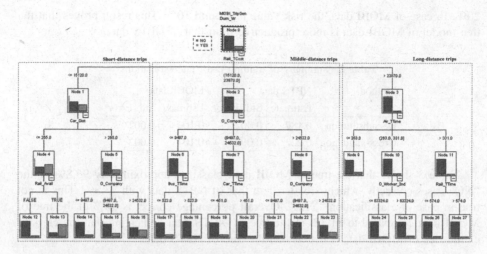

Fig. 5. Travel patterns of trip generation in MOBI data.

tertiary industry sector in destination zone, while that of the tree of MOBI data are variables of level of travel services (e.g., car distance, air travel time). This phenomenon is also seen in the lowest level of two trees. Regarding NPTS data, in the group of short-distance trips, trip generation is higher when there are railway connections between OD pairs and the number of tertiary sector workers in destinations is greater. Moreover, in the group of middle-distance trips, the more number of tertiary sector workers in destinations as well as in origins, and the more number of employees and people aged from 30 to 34 is, the more OD pair trips are made. In the group of long-distance trips, the number of single households in origins and destinations and the number of people aged from 40 to 44 are significant effects on trip generation rate. In the tree of MOBI data, trip generation rate is so low in both the group of middle-distance trips and long-distance trips. In the group of short-distance trips, the highest trip generation rate is seen in Node 13 where travel distance by care is lower than 265 km, and railway connection is available between two areas.

In conclusion, compared to trip generation rate at Node 0, the higher trip generation rate is seen in the three groups of OD travel distance in NPTS data (e.g., Node 12 to 14, Node 17, Node 20 to 23, Node 11, and Node 29), while it is only seen in groups of short-distance and middle-distance trips (e.g. Node 13,15,16, and Node 23). Furthermore, in the tree of MOBI data, trip generation rate drops approximately zero in the nodes for long-distance trips. It also happens in Node 12 in the short-distance trips group as well as in Node 18 and 20 in the middle-distance trips groups.

In order to evaluate the performance of two decision trees (or two prediction models), re-substitution and cross-validation test are used. In cross-validation test, our data is divided into ten portions stratified on dependent variable levels. The training error rates are estimated and shown in Table 2. The training error estimate of around 0.2 in case of NPTS data indicates that the category predicted by the NPTS tree model is wrong for 20% of the case or the risk of misclassifying trip generation is around

20%. In case of MOBI data, the risk value is around 10%. This result proves that the tree model of MOBI data is more productive than that of NPTS data.

Table 2. Summary of training error rate estimations

Method	NPTS data		MOBI data	
	Estimate	Std. Error	Estimate	Std. Error
Re-substitution	0.187	0.002	0.100	0.002
Cross-validation	0.223	0.002	0.107	0.002

Table 3 shows that the tree of MOBI data classifies approximately 96.8% of the "NO" cases correctly, which is better than that of NPTS data with 73.3%. This is not true in case of classification "YES". Overall percentage correct is 90% in the tree of MOBI data compared to 81.3% in the tree of NPTS data, indicating that the classification tree of MOBI data is better than the tree of NPTS data.

Table 3. Summary of decision tree coincidence matrix

Method	Tree result for NPTS data			Tree result for MOBI data		
	NO	YES	Correct (%)	NO	YES	Correct (%)
Actual NO	11,560	4,201	73.3%	30,523	1,024	96.8%
Actual YES	2,855	19,020	86.9%	2,731	3,358	55.1%
Overall (%)	38.3%	61.7%	81.3%	88.4%	11.6%	90.0%

5   Conclusions

In this paper, we clarify the characteristics of MOBI data by comparing with 2010 NPTS data as a reference. MOBI data is much similar with NPTS data in case of the number of aggregated trips from origin zones and to destination zones than in case of the number of OD pairs trips and travel pattern of trip generation.

However, the significant different between two data may be caused by the difference in the year of two surveys collected; MOBI data is observed in 2015 while NPTS data was collected in 2010. Thus, to clarify the difference more accurately, we should make another comparison with the 2015 NPTS data to come. Also, in the future, we will find new ways of accurate comparisons by controlling some side effects such as differences in socio-demographic characteristics of respondents and then finding another appropriate method to estimate origin-destination location more precisely.

References

1. Smith, M.E.: Design of small-sample home-interview travel surveys. Transp. Res. Record **701**, 29–35 (1979)
2. Daganzo, C.F.: Optimal sampling strategies for statistical models with discrete dependent variables. Transport. Sci. **14**, 324–345 (1980)

3. Stopher, P.R., Greaves, S.P.: Household travel surveys: where are we going? Transport. Res. A Pol. **41**, 367–381 (2007)
4. Ministry of Land, Infrastructure, Transport and Tourism (MLIT). www.mlit.go.jp/common/001005633.pdf
5. Bureau of Transportation Statistics - U.S. Department of Transportation. http://www.transtats.bts.gov/DatabaseInfo.asp?DB_ID=505&Link=0
6. Federal Highway Administration - U.S. Department of Transportation. https://www.nationalhouseholdtravelsurvey.com/
7. Asakura, Y., Hato, E.: Tracking survey for individual travel behaviour using mobile communication instruments. Transport. Res. C Emer. **12**, 273–291 (2004)
8. Caceres, N., Wideberg, J., Benitez, F.: Deriving origin destination data from a mobile phone network. IET Intell. Transp. Sy. **1**, 15–26 (2007)
9. Jing, W., Dianhai, W., Xianmin, S., Di, S.: Dynamic OD expansion method based on mobile phone location. In: Fourth International Conference on Intelligent Computation Technology and Automation, pp. 788–791. IEEE (2011)
10. Iqbal, M.S., Choudhury, C.F., Wang, P., González, M.C.: Development of origin–destination matrices using mobile phone call data. Transport. Res. C Emer. **40**, 63–74 (2014)
11. Bar-Gera, H.: Evaluation of a cellular phone-based system for measurements of traffic speeds and travel times: a case study from Israel. Transport. Res. C Emer. **15**, 380–391 (2007)
12. Zhan, X., Hasan, S., Ukkusuri, S.V., Kamga, C.: Urban link travel time estimation using large-scale taxi data with partial information. Transport. Res. C Emer. **33**, 37–49 (2013)
13. Reades, J., Calabrese, F., Ratti, C.: Eigenplaces: analysing cities using the space-time structure of the mobile phone network. Environ. Plann. B Plann. Des. **36**, 824–836 (2009)
14. Bindra, S.: Using cellphone OD data for regional travel model validation. In: 15th TRB Planning Applications Conference (2015)
15. Ministry of Land, Infrastructure, Transport and Tourism (MLIT). http://www.mlit.go.jp/sogoseisaku/soukou/sogoseisaku_soukou_fr_000018.html
16. NTT Docomo, Inc. https://www.nttdocomo.co.jp/english/corporate/ir/binary/pdf/library/annual/fy2015/p05_e.pdf
17. Japan Statistics Bureau. http://www.stat.go.jp/english/index.htm
18. Ture, M., Tokatli, F., Kurt, I.: Using Kaplan-Meier analysis together with decision tree methods (C&RT, CHAID, QUEST, C4.5 and ID3) in determining recurrence-free survival of breast cancer patients. Expert Syst. Appl. **36**, 2017–2026 (2009)
19. Bargeman, B., Chang-Hyeon, J., Timmermans, H., Van der Waerden, P.: Correlates of tourist vacation behavior: a combination of CHAID and loglinear logit analysis. Tourism Anal. **4**, 83–93 (1999)
20. Chen, J.S.: Market segmentation by tourists' sentiments. Ann. Tourism Res. **30**, 178–193 (2003)
21. Van Middelkoop, M., Borgers, A., Timmermans, H.: Inducing heuristic principles of tourist choice of travel mode: a rule-based approach. J. Travel. Res. **42**, 75–83 (2003)
22. Welte, J.W., Barnes, G.M., Wieczorek, W.F., Tidwell, M.C.: Gambling participation and pathology in the United States - a sociodemographic analysis using classification trees. Addict. Behav. **29**, 983–989 (2004)
23. Chen, J.S.: Developing a travel segmentation methodology: a criterion-based approach. J. Hosp. Tour. Res. **27**, 310–327 (2003)
24. Biggs, D., De Ville, B., Suen, E.: A method of choosing multiway partitions for classification and decision trees. J. Appl. Stat. **18**, 49–62 (1991)

REBUILD: Graph Embedding Based Method for User Social Role Identity on Mobile Communication Network

Jinbo Zhang[1,2], Yipeng Chen[1,2], Shenda Hong[1,2], and Hongyan Li[1,2(✉)]

[1] Key Laboratory of Machine Perception, Peking University,
Ministry of Education, Beijing, China
lihy@cis.pku.edu.cn
[2] School of Electronics Engineering and Computer Science,
Peking University, Beijing, China

Abstract. Inferring users' social role on a mobile communication network is of significance for various applications, such as financial fraud detection, viral marketing, and target promotions. Different with the social network, which has lots of user generated contents (UGC) including texts, pictures, and videos, considering the privacy issues, mobile communication network only contains the communication pattern data, such as message frequency and phone call frequency as well as duration. Moreover, the profile data of mobile users is always noisy, ambiguous, and sparse, which makes the task more challenging. In this paper, we use the graph embedding methods as a feature extractor and then combine it with the hand-crafted structure-related features in a feed-forward neural network. Different with previous embedding methods, we consider the label info while sampling the context. To handle the noisy and sparsity challenge, we further project the generated embedding onto a much smaller subspace. Through our experiments, we can increase the precision by up to *10%* even with a huge portion of noisy and sparse labeled data.

1 Introduction

User profiling, which inferss users' essential attributes, such as age, gender, and interests, has been a holy grail in enabling efficient information services. Among which, inferring users' social role (a student, a teacher or a civil servant) is especially vital to a large range of applications. For example, in China, full-time students are banned from applying for credit cards according to financial policy. However, for personal reasons, some students disguise themselves as employees to fake the policy. As the phone number is typically the essential and crucial part in the apply procedure, the detection of student applicant using mobile communication data will alleviate the *Credit Reference Center* (the authorized credit report maintainer of China) and financial institutes the suffering of such financial fraud.

© Springer International Publishing AG 2017
Y. Tan et al. (Eds.): DMBD 2017, LNCS 10387, pp. 326–333, 2017.
DOI: 10.1007/978-3-319-61845-6_33

As of 2015, the mobile phone subscriptions have reached 15.1 billion[1]. Profiling such significant amounts of users are deserving and challenging. Most of previous user profiling approaches target on social network and highly rely on user-generated content (e.g. tweets or photos). For example, [3] classify user locations from their tweets and Pennacchiotti [5] infer user's labels like political orientation from profile features, linguistic content features, and social network features. These methods hardly tailor with the mobile network, where it is hard for individuals to crawl user generated data. Sampled data are often obtained directly from operators, including call histories, the number of short messages (SMS), and anonymized profile data.

Moreover, Zeng [10] used label propagation model to infer user affiliation based on social activities on the Internet. Dong [2] used a probabilistic model which integrated users' social properties and network features to infer users' age and from the mobile social network. Li [4] proposed a unified discriminative influence model which integrated network and tweets to infer user's location. These methods rely on the distribution of particular attributes such as age, gender, and location. The distribution of social roles of mobile users is different from the distributions of these attributes. Thus, existing methods that use probabilistic model cannot be directly applied to infer mobile users' social roles.

What's more, the data analysis is challenged by few issues. First, the data attributes are insufficient, which yields the richness of the data is low to leverage many advanced methods. Second, the attribute labels are noisy and incomplete. For instance, some users provide wrong profile data or the data are changed over time. Third, although the data can form a structured graph, but its inherent sparsity and lack of attributes make the existing studies proposed for social networks inapplicable.

Our method utilizes the emerging deep learning methods on the massive network data. Specifically, we firstly exploit the embedding method to represent each user using low dimensional embedding vectors. In contrast with existing embedding methods, we consider the label information of nodes when sampling the context while generating the embedding vector and then project the embedding into a much smaller subspace, where it can resist with sparse and noisy data. We then feed the extracted embedding together with the previous existing feature vectors into a feed forward neural network and achieve superior results compared with 'state of the art' methods.

2   Network Embedding Based Methods for User Social Role Identity

Let $G = (V, E)$, where V represents the users in the network and $E \subseteq (V \times V)$ is the set of edges between two users, each represents the communication closeness of the corresponding two users. And $V = V^L \cup V^N$, where $V^L(V^N)$ is the set of labeled (unlabeled) users respectively. Each edge $e \in E$ is associated

---

[1] http://www.itu.int/en/ITU-D/Statistics.

with a weight $\omega_{uv} > 0$. Moreover, for each user in the network, we compute the degree centrality, average neighbor degree, local clustering coefficient, and embeddedness as the structure feature, which show different aspects of network properties. We also use label propagation and homophily theory to extract connection feature. Briefly, we assume that $\forall v_i \in V$, an attribute vector X_i, where $|X_i| = m$.

What's more, $G_L = (V, E, \Omega, X, Y)$ is a partially labeled network, with attribute $X \in \mathbb{R}^{|V| \times m}$, m is the dimension of the attribute vector, and $Y \in \mathbb{R}^{|V| \times |\ell|}$, where ℓ is the set of labels. For each $v_i \in V$, we have the attribute vector of v, denoted as X_i, and the embedding vector of v, denoted as E_i. Our goal is to learn the hypothesis $f : (X, E) \rightarrowtail \ell$.

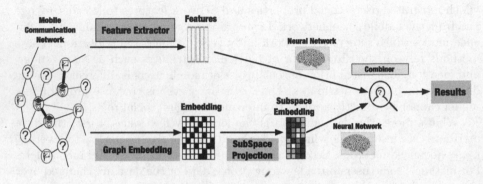

Fig. 1. The framework of *REBUILD*

As showed in Fig. 1, our methods include the *Feature Extractor*, *Graph Embedding*, *Subspace Projection Part* (dealing with noisy and sparse problem), the *Neural Network* for the two parts and *Combiner Part* (a softmax layer).

In this work, we use *LINE* [8] as the primary procedure of generating the representation of vertices. *LINE* is based on the idea that vertices sharing many connections to other vertices are similar to each other. For each directed edge (i, j), the probability of 'context' v_j given the vertex v_i is defined as:

$$p(v_j | v_i) = \frac{\exp(\vec{u}_j^{'T} \cdot \vec{u}_i)}{\sum_{v_k \in V} \exp(\vec{u}_k^{'T} \cdot \vec{u}_i)}. \tag{1}$$

where V is all the 'context' (the whole node set) of v_i, $\vec{u}_i$ is the latent representation of v_i as a vertex and ($\vec{u}_j^{'}$ is the latent representation of v_j as a 'context') and the dimension is d. By learning $\{\vec{u}_i\}_{i=1..|V|}$ and $\{\vec{u}_j^{'}\}_{i=1..|V|}$ that minimize the sum of all the edges using Eq. 1, each vertex of the network will be represented as a d-dimensional vector $\vec{u}_i$.

However, it's very time consuming when calculating Eq. 1 and some methods use *Negative Sampling* and transform Eq. 1 into: $\sum_{(i,j,\gamma) \in C} \log \sigma(\gamma \cdot \vec{u}_j^{'T} \cdot \vec{u}_i)$, where $\sigma(x) = 1/(1 + \exp(-x))$ is the sigmoid function and $\mathbb{C}$ is the sampled edges set.

The framework of our method is based on the feed-forward neural networks. Specifically, the $n - th$ layer of the neural network is defined as: $z^{(n+1)} = W^{(n)}a^{(n)} + b^{(n)}, a^{(n+1)} = f(z^{(n+1)})$ where $z^{(n+1)}$ is the input of the layer n and $a^{(n+1)}$ is the output of layer n, and f is a family of cunctions, here, we used the rectified linear unit function, namely $f(z^{(n+1)}) = \max(0, z^{(n+1)})$.

As illustrated in Fig. 1, we apply k layers on the attribute vector and l layer on the embedding e. The attribute vector and embedding vector are the initial input of the neural network. Moreover, we will use one more layer to get the final result:

$$p(y|x, e) - \frac{\exp[z^{(l+1)}(x)^T, z^{(k+1)}(e)^T]\omega_y}{\sum_{y'} \exp[z^{(k+1)}(x)^T, z^{(l+1)}(e)^T]\omega_{y'}} \tag{2}$$

where, $[\cdot, \cdot]$ denote concatenation of row vectors and ω is the model parameter.

Combined with Eq. 2, the overall objective function will be:

$$-\frac{1}{L}\sum_{i=1}^{L}\log p(y_i|x_i, \overrightarrow{u}_i) - \lambda \sum_{(i,j,\gamma)\in\mathbb{C}} \log \sigma(\gamma \cdot \overrightarrow{u}_j^{'T} \cdot \overrightarrow{u}_i) \tag{3}$$

where the first term is for the loss of label prediction and the latter one is for the context prediction. And the context is denoted as (i, j, γ), where γ represent whether it's positive ($\gamma = 1$) or negative $\gamma = -1$ sample.

We adopt the *stochastic gradient algorithm* (ASGD) [7] to optimize Eq. 3. As illustrated in Algorithm 1, the two terms of Eq. 3 will be updated interactively. In each iteration, *Line 2* to *Line 4* will sample a mini-batch of labeled instances and then update the parameters with specified learning ratio. From *Line 5* to *Line 7* the second part of Eq. 3 will be updated by sampling a batch of 'context' (the way that we used to sample will be discussed in Sect. 3). This loop will repeat until the stop criteria satisfied (reach the pre-defined iteration times or converge).

3 Optimization

3.1 Embedding Generation Considering Label Info

As previous stated, existing methods seldom consider label info. To inject the label info into the sampling procedure, we modify the context generation part of *LINE* and sample two types of context: the first type of context is based on the graph structure, which encodes the structure information, and the second type of context is based on the labels, which we use to inject label information into the embedding. The intuition of our sampling method is that for an edge (v_i, v_j), if the label of the edge is equal, then the edge is a positive sample, otherwise a negative one. Then, we can minimize the objective function on the newly mixture context. In this way, the sampled labeled pair server as the guide for the generation of embedding vectors.

Algorithm 1. Training Algorithm

Input: Partial Labeled Graph G_L, number of samples T_1, T_2, λ,ratio τ
Output: Model Parameters
1 **repeat**
   /*first Batch to Update the first term of *Eq. 3*   */
2   Sample T_1 labeled instances i
3      $\mathcal{L}_1 = -\frac{1}{T_1}\sum_{i=1}^{T_1}\log p(y_i|x_i, e_i)$
4   Calculate and Update $\frac{\partial \mathcal{L}_1}{\partial \omega}$ with pre-defined ratio τ
   /*Second Batch to Update the Second term of *Eq. 3*   */
5   Sample T_2 'context' (i, j, γ) using method in *Sect. 3.1*
6      $\mathcal{L}_2 = -\lambda \sum_{(i,j,\gamma)\in\mathbb{C}}\log \sigma(\gamma \cdot \overrightarrow{u}_j'^T \cdot \overrightarrow{u}_i)$
7   Calculate and Update $\frac{\partial \mathcal{L}_2}{\partial e}$ with pre-defined ratio τ
8 **until** *Stopping*;

3.2 Adapting Embedding with Subspace Projections

As only a small volume of noisy labeled data is available in the communication network setting, the previous method is prone to over-fitting. To handle with this, we project the previous embedding onto a smaller subspace. The intuition of making such projection is since we only have limited labeled data, instead of learning the underlying representation in a larger dimensional space, we express the representation in the subspace, where the embedding can better fit the complexity of our task. Moreover, we can update the subspace embedding immediately with the labeled data, thus, the labeled instances can be better captured in the subspace. For those unlabeled ones, as the assumption that the labeled instances are representative of the whole data (including the unlabeled ones), a reasonable assumption, which has been validated in many previous works [1], which means that in the subspace, there is some other points (generated by labeled instance) that are sufficiently close enough to those unlabeled points.

Let $E \in R^{|V| \cdot |d|}$ denote the original embedding matrix obtained, e.g. with the structured skip-gram model described previously. We define the adapted embedding in the subspace by using a projection factorization $E \cdot P$, where $P \in R^{|d| \cdot |s|}$, with $s \ll e$.

4 Experiment

In this section, we validate our methods in three parts: the effectiveness of our methods in considering label info while sampling context, the feasibility of projecting the embedding into subspace and the performance of our framework in inferring users' social role.

Our data is extracted from real-world mobile communication records in China which span three months. The data consists 15,000,000 distinct users, who generate 140,000,000 distinct calling records and totally 77,000,000 text messaging records. Each record contains IDs of calling and called part, frequency of communication behaviors, and duration of communication behaviors. The users are divided into three kinds of users based on the data provided by the provider: specifically, including 77,000 students users, 77,000 non-student users and the social roles of others are unknown.

To reach optimal performance for each baseline algorithm, we set negative samples $\Lambda = 5$, context window size $K = 10$ for *DeepWalk*. The neural network has three layers and is implemented using *Theano*. Furthermore, we tuned the dropout ratio which reduces error in the optimization process and the neurons were activated by the sigmoid function.

4.1 Effectiveness of Subspace Projection

We firstly randomly select part of labeled users and construct a subset of data with different ratios of labeled users. Then we project the embedding onto different subspaces vary from 0 (no projection) to 15 ($s = 15$) and repeat the results five times and show the average results in Fig. 2a. From the results, we can conclude that, as the increase percentage in labeled users, the *F-score* will increase gradually. And specifically, when the subspace size is 10 and 15, it performs better than others. However, when the subspace increases to 15, the performance will diminish. The reason of performance loss lies in as the space increase, the projected instance will hardly become a cluster and thus, make the total performance downgrade.

To testify the capability of subspace projection on noisy data, we randomly transform part of labeled users into incorrectly labeled users. From Fig. 2b, we can observe that as the fraction of incorrectly labeled users increases, the accuracy will overall decrease. And similar to the previous one, the subspace size of 5and 10 perform better than others. Even when the noisy users increase to 40%, this method can still handle with a relatively high accuracy (over 80%). In conclusion, this method can handle the situation where the data are sparse and incorrectly labeled users exist. The subspace size with ten is a better choice and thus, we will use $s = 10$ hereafter in the later experiments.

4.2 Performance of *REBUILD*

To demonstrate the effectiveness of our method, we compare our method against several *state-of-art* methods:

KNN: As users likely contact closely with those have the same social role, so many methods adopt the cluster methods. We can infer users' social roles based on the information of their neighbors.

Community Detection: Community detection is often used to infer users' information. We use the method of [6] to infer users' social.

Label Diffusion: Label Diffusion is often used to infer users' information. We use the method of [11] to infer users' social roles.

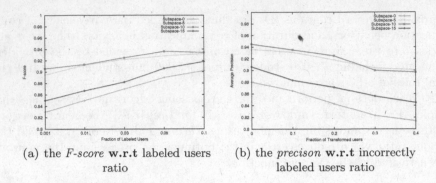

(a) the *F-score* **w.r.t** labeled users ratio

(b) the *precison* **w.r.t** incorrectly labeled users ratio

Fig. 2. The performance of our subspace projection

MCB: Chen et al. [9] propose a method by using user extracted features and then use logistic regression to do this work. Moreover, the features we used in our work are the same with [9].

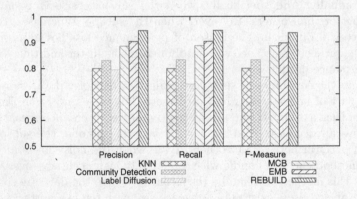

Fig. 3. Performance of different methods on social role identity

As showed in Fig. 3, our methods outperform the state-of-art methods by nearly 10%. Specifically, as the existence of noisy and sparse data, the precision of *state-of-art* methods is below *90%*, and our methods can achieve over *95%*, which we believe is an acceptable result considering this problem.

5   Conclusion

In this work, we analyze the characteristic of user social role identity under the mobile communication setting and propose corresponding methods using the embedding methods. Our method mainly has two advantages: on the one hand, we consider the label info while sampling, on the other hand, we project the embedding into subspace to further deal with the noisy and sparse problem. Through extensive experiments, we validate the proposed methods.

We exploit the possibility of combing embedding methods with existing work. In the future, we plan to solve the efficiency challenge in embedding. As compared with existing social role identity methods, our methods are in some way slower. This is, though, for the reason of the existing drawbacks of embedding methods, we wanna to propose some faster methods to solve this problem even in an on-line manner.

Acknowledgement. This work was supported by Natural Science Foundation of China (No. 61170003).

References

1. Astudillo, R.F., Amir, S., Lin, W., Silva, M., Trancoso, I.: Learning word representations from scarce and noisy data with embedding sub-spaces. In: ACL 2015 (2015)
2. Dong, Y., Yang, Y., Tang, J., Yang, Y., Chawla, N.V.: Inferring user demographics and social strategies in mobile social networks. In: KDD 2014, pp. 15–24 (2014)
3. Hecht, B., Hong, L., Suh, B., Chi, E.H.: Tweets from Justin Bieber's heart: the dynamics of the location field in user profiles. In: CHI 2011, pp. 237–246 (2011)
4. Li, R., Wang, S., Deng, H., Wang, R., Chang, K.C.C.: Towards social user profiling: unified and discriminative influence model for inferring home locations. In: Proceedings of the 18th ACM SIGKDD International Conference on Knowledge Discovery and Data Mining, pp. 1023–1031. ACM (2012)
5. Pennacchiotti, M., Popescu, A.M.: Democrats, republicans and starbucks afficionados: user classification in twitter. In: KDD 2011, pp. 430–438 (2011)
6. Raghavan, U.N., Albert, R., Kumara, S.: Near linear time algorithm to detect community structures in large-scale networks. Phys. Rev. E **76**(3), 036106 (2007)
7. Recht, B., Re, C., Wright, S., Niu, F.: Hogwild: a lock-free approach to parallelizing stochastic gradient descent. In: NIPS 2011 (2011)
8. Tang, J., Qu, M., Wang, M., Zhang, M., Yan, J., Mei, Q.: Line: large-scale information network embedding. In: WWW 2015, pp. 1067–1077 (2015)
9. Chen, Y., Li, H., Zhang, J., Miao, G.: Inferring social roles of mobile users based on communication behaviors. In: Cui, B., Zhang, N., Xu, J., Lian, X., Liu, D. (eds.) WAIM 2016. LNCS, vol. 9658, pp. 402–414. Springer, Cham (2016). doi:10.1007/978-3-319-39937-9_31
10. Zeng, G., Luo, P., Chen, E., Wang, M.: From social user activities to people affiliation. In: ICDM 2013, pp. 1277–1282 (2013)
11. Zhou, D., Bousquet, O., Lal, T.N., Weston, J., Schölkopf, B.: Learning with local and global consistency. In: NIPS 2004 (2004)

Study a Join Query Strategy Over Data Stream Based on Sliding Windows

Yang Sun, Lin Teng[✉], Shoulin Yin, Jie Liu, and Hang Li

Software College, Shenyang Normal University, No. 253, HuangHe Bei Street,
HuangGu District, Shenyang 110034, China
17247613@qq.com, 1532554069@qq.com, 352720214@qq.com,
nan127@sohu.com, 1451541@qq.com

Abstract. Data stream sliding window is a common query method, but traditional approaches can get inaccurate data results. Due to the blocked character of join and the infinite character of data stream, there must be some constraints on join operations. In order to solve these constraints, this paper proposes a strategy based on load shedding techniques for sliding window aggregation queries over data stream for basic query processing and optimization. The new strategy supports multiple streams and multiple queries. Through experimental tests, the results show that the new strategy based on load shedding techniques can greatly improve query processing efficiency. Finally, we make a comparison to other join query strategies over data stream to verify the new method. Join query strategy over data stream based on sliding windows is an effective method, which can effectively reduce query processing time.

Keywords: Multiple queries · Data stream · Sliding windows · Join query strategy

1 Introduction

With the development of network, recently a new class of data-intensive applications has become widely recognized. The data in the applications is modeled best not as persistent relations but rather as transient data streams [1–3]. Their continuous arrival in multiple, rapid, time-varying, possibly unpredictable and unbounded streams appear to yield some fundamentally new research problems. Taking some block operations for example, it can't be used over data streams directly. Data stream multi-sliding (DSMS) should be developed to solve the key problems. Wang [4] proposed similarity query based on Longest Common Sub-Sequence (LCSS) over data stream window query processing algorithm based on possible solution domain optimization strategy. Dallachiesa [5] extended the semantics of sliding window to define the novel concept of uncertain sliding windows and provide both exact and approximate algorithms for managing windows under existential uncertainty.

In normal case, there are standard SPJ queries over data streams. Corresponding to select and project, join operations are more complex. Because of the blocked character of join and the infinite character of data stream, there must be some constraints on join operations. As a result, sliding windows [6, 7] are introduced. This paper proposed a

© Springer International Publishing AG 2017
Y. Tan et al. (Eds.): DMBD 2017, LNCS 10387, pp. 334–342, 2017.
DOI: 10.1007/978-3-319-61845-6_34

strategy for basic query processing and optimization which supported multiple streams and multiple queries. Finally, we made experiments and some comparisons to other join query methods to verify the effectiveness of our method. The results showed that the using of indexing techniques can greatly improve query processing efficiency.

2 Materials and Methods

The Join operations based on sliding window are widely studied, some researchers have given various join methods, and in this section some related works will be briefly introduced. Join processing is a complex query processing operations, many join methods are given in the literature. Here mainly introduced are symmetrical hash join [8], the extended ripple hash join [9] and XJoin [10].

2.1 Indexed Join Based on Sliding Windows

(1) Partition of sliding windows

Sliding windows generally have two parameters, window size and number of hops. Partitioning method is used on the sliding window, according to the number of hops the whole sliding window is divided into several regions, namely: the sliding window W_i is divided into a number of BW_i accordance with the number of hops h_i, each BW_i is called the basic window. Noted, each basic window is actually a number of hops a sliding window, the size is completely equal to the number of hops. Next, benefits of this way will be introduced.

After sliding windows are divided, entire data on the window can be indexed. Here, there are two cases with join query processing of the sliding window: the equi-join and Non-equi-join(also called the scope join). These two cases will be discussed, but mainly about scope join. As mentioned previously, the current study are almost based on equi-join. Therefore, this paper focus on the scope join.

(2) Set index

The method is described in detail before, the division of sliding window method, and each divided basic window is set indexed. After the index is created on the sliding window, the join operation is performed based on index, which can improve the efficiency. The indexing method and join operation using the index will be introduced on the following.

First, a simple explanation is given in Fig. 1, it describes the partition of the sliding window and indexing methods using this strategy.

In Fig. 1 the process is this: here sliding window size is set to 6 s, while the size of hops is 2 s. Then the window can be divided into 6/2 = 3 BW regions. In 7 and 8, are tuples in a new number of hops when sliding windows arrive their updating time (the 7th and 8th second), when the hops arrive, tuples of number of hops in 1 and 2 expire (because the window size is 6), an expired hops need to be removed. Every time a

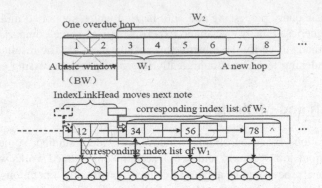

Fig. 1. Establishing indices over partitioned sliding windows.

window's updating time arrives, query processing system adds tuple of a number of hops to the window, these new tuples form a new basic window, and the corresponding index is created for the new basic window. Then verify and delete the expired tuples of the oldest one of the basic windows, as while as the corresponding index of this expired basic window.

(3) Index-join

After the steps above, tuples in every basic window of data stream sliding window are indexed. Thus, the join query using the index is a batch method.

Each time a window's updating time arrives, tuples within the new scope of hops will be generated, and also will cause the expiration of tuples in the oldest hops. In the above diagram, take R stream as an example, if tuples within the new scope of hops generate, they are first used to search for indexes in all the basic windows of S stream, and the join results are matched in output. Then these tuples are added to current sliding window of R stream. These tuples constitute a new basic window, generate the index of this basic window. And to verify and delete all the data tuples in expired and oldest one of basic windows, while their corresponding indexes have expired too, simply delete the entire index tree.

Thus, considering the case which join between the two streams R and S, take R stream generating tuples in one hop as an example (the same situation for the stream S, it is symmetric).

After this process, all the tuples between new hops of R stream and current window of S stream are joined and construct the results. With this method, using incremental and batched way, processes tuples in each new hops. Under normal circumstances, the results are immediately returned to the query users without interval. Results can be immediately removed after returned to the user, thus release the memory resources, and such results are the final query results; but if there are other queries beyond the join results, then the intermediate results need to be cached. Whether to output immediately or make temporary cache depends on actual conditions mark.

2.2   Index Sharing and Aggregation Query

(1) Multi-linked index query

In stream applications, the system usually has many continuous queries, and if two (or more) queries have the same hops, then indexes between queries can be shared. Only one index is maintained for different queries instead of each query for every query, the sharing can improve efficiency. .

Under the same query hop, window sizes are still needed to be checked: window with same size and window with different size. First, consider the same hop and the same window size, as two queries in the following.

Q1: SELECT *
FROM Stock1[SLIDINGRANGE(10,2)], Stock2[SLIDINGRANGE(12,4)]
WHERE A.price>B.price
Q2: SELECT Stock1.sid, Stock2.sid
FROM Stock1[SLIDINGRANGE(10,2)], Stock2[SLIDINGRANGE(12,4)]
WHERE A.price>B.price

These two queries are about two stocks, within which SLIDINGRANGE denotes that constraints of window is based on sliding window join, the two parameters are the window size and hop (update frequency) size. These two queries have the same join conditions and window constraints, indexes are created respectively for stream Stock1 and Stock2, according to the constraints of the window, each window can be divided into $10/2 = 5$ BW regions, for each region a red-black tree index is created. Query Q1 and Q2 can share the index while executing join query processing. Each time tuples in one stream is used to scan the corresponding index in another stream, the join result is returned to two queries at the same time, and then they continue to execute the rest of query separately (the join result can be returned to Q1 directly, while projection operation should be made to the result for Q2). And then consider the same hop and the different window size, as three queries in the following:

Q1: SELECT *
FROM Stock1[SLIDINGRANGE(10,2)], Stock2[SLIDINGRANGE(12,4)]
WHERE A.price>B.price
Q2: SELECT *
FROM Stock1[SLIDINGRANGE(8,2)], Stock2[SLIDINGRANGE(12,4)]
WHERE A.price>B.price
Q3: SELECT *
FROM Stock1[SLIDINGRANGE(6,2)], Stock2[SLIDINGRANGE(12,4)]
WHERE A.price>B.price

These queries' condition are the same, but the window size of stream Stock1 is different (Stock2 can also be different, the situation is completely similar to Stock1), the largest window size is set to 10, the indexes are maintained for those size 10 windows,

Range of tuples are contained in the process of query, result of different range are returned to the appropriate query.

(2) **Aggregation query over data stream**

Aggregation over data streams are similar to traditional database systems, include MAX, MIN, SUM, COUNT and AVG.

Aggregation operations in traditional database generally require all data to be seen before outputting results, even if the memory limit all the data to be seen in one time, multi-installment algorithm is capable of reading data from the secondary storage. But in data stream environment, memory capacity is limited, secondary storage is not an option because real-time requirements. For the unbounded data, if results output must wait until the end of the stream, then aggregation operation will never get data output. In essence, the aggregation and join is the same type of query processing, changing traditional blocked property is unavoidable, so some special processing should be done in aggregation operations over data stream.

Similarly, to solve aggregation query blocked property is exactly the same to the method that in join operation, is still sliding window constraints in data stream. Under normal circumstances, the aggregation queries over data stream are also referring to the sliding window aggregation query over data stream.

This processing is a aggregation operation with window assisting, since the aggregation operation can not handle all the data on stream (the same as join operations), then the data scope must be limited. Operation window used by aggregation is basically the same with window used by join operation.

Aggregation queries over data streams can also be divided into two categories according to the nature of operation: MAX and MIN aggregation queries are extremum operation; COUNT, SUM, and AVG are accumulate operation. Aggregation query and join query has similarities, aggregation query processing methods are similar to join query ones. Similarly, the sliding window partition method described above can also be used here. Generally indexes meet aggregation queries will also be created for tuples of data stream, which can improve the aggregation query processing efficiency. Join and aggregation are both blocked queries over data stream, this section focuses on method of blocked join query processing over stream. The core idea is to use indexed join over stream. Performance evaluation in the following will be able to explain benefits of using this indexed join strategy.

For index sharing and aggregation blocking query processing over stream, they are not focus of this paper, so this section only gives a brief introduction and description.

3 Results and Discussion

In this section, we conduct experiments by using JAVA language under 2.2 GHz CPU, 2 GB RAM, WINDOWS XP system. Running time is within 45 s. For blocking join query processing, nested loop join method and the indexed based on sliding window join method are implemented respectively, compare the efficiency of the two, get execution time of both in the same case to make performance comparison. Specific process goes like this: system generate data tuples, simulate joining between two synchronous streams, these two streams have same sliding window parameters: window size is 6 M, updating frequency is 2 s. Every time updating happens, two streams produce tuples of a hop size respectively, until tuples of 6 hops are generated. In the following experiments, except special caption, the stream rate is 800 tuples/s, range of joining tuples is set to [0,128), joining condition for two streams are Stream1. Attr>Stream2.attr. Execution time of nested loop join method (NLJ) and indexing method (INDEX) are measured respectively and compared as Fig. 2.

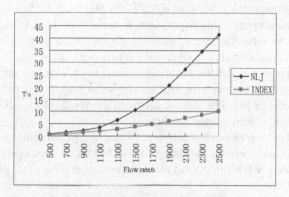

Fig. 2. Different stream speed.

Selectance of join operator is 0.07. It can be seen from Fig. 2, in the case of low selectance, indexing method benefits more with stream speed increasing. Especially, when flow rate is 1300 t/s, the time of NLJ and INDEX are approximately 3 s and 4 s. Then time of NLJ increases dramatically, however, that of INDEX goes up stability. Impact induced by various ratio of sliding window size and number of hops is given in the end, as shown in Fig. 3.

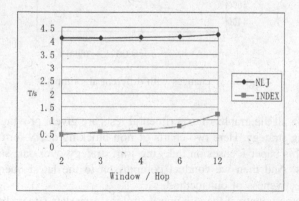

Fig. 3. Different window size/hop.

In this experiment, parameter are set like this: sliding window size is 12 s, data of only one sliding window is generated, change the ratio of sliding window size and number of hops, from 2 to 12 (hops change from 6 to 1), The stream rate is 2000 tuples/s, selectance is 0.078. In this experiment, method of changing range of joining tuples is used to control the selectance of join. Sliding window parameters are changed also to test various ratio of sliding window size and number of hops.

In this section, we also evaluate performance of the proposed join query strategy over data stream based on sliding windows (abbreviated as JQSW) through comparison experiments. We compare JQSW with SQLSW-PS [4], NUSW [5] and DSMFP-Miner [11] on various parameters. This experiment is written as JAVA language and conducted 10 times. We use the size of element sliding window W, element similarity threshold ε, LCSS similarity threshold Δ and the number of elements in query sequence $|Q|$ to make comparison under the same conditions. In order to get the real results, we also use the sea surface temperature data set shown at http://www.pmel.noaa.gov/tao/data_deliv as reference [4]. The date is set from 1.JAN.2000 to 1.JUL.2016. In this part, we only consider temperature attribute.

First, we study the relationship between W and the performance of the four algorithms as Fig. 4. X-axis denotes the change of W, Y-axis denotes operation time. From Fig. 4, we know that W has direct ratio relations with operation time. Second, we study the relationship between $|Q|$ and the performance of the four algorithms as Fig. 5. X-axis denotes the change of $|Q|$, Y-axis denotes operation time. The curve has the similar trend to Fig. 4. It is obviously that performance of the four algorithms is different, however, the new algorithm JQSW is high effective.

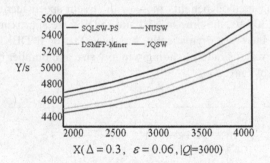

Fig. 4. W changes with different algorithms.

In this section all the graphs of experimental tests are given, proving the advantage of using indexing strategy. Here two charts of non-blocking query for comparison are given, because this paper focuses on indexing join strategy over data stream based on sliding windows. And then we conduct comparison to the latest query strategies to illustrate the effectiveness of our method.

During the experiments, if the join query strategy over data stream based on sliding windows is tested for window size, only for running time, it will greatly increase the performance and quality of the method. But the window size comparison is obviously.

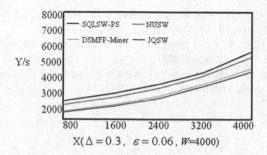

Fig. 5. |Q| changes with different algorithms.

In addition, its convergence time is very short comparing to other query methods with different parameters. Therefore, it is a better choice for data stream query.

4 Conclusion

This paper describes the query processing techniques of indexing join, mainly focus on join query processing based on sliding windows, and it also introduces partition of sliding windows and indexing technique used to increase efficiency of blocking query processing. Experimental results show that the using of indexing techniques can greatly improve query processing efficiency.

References

1. Fisher, D., Chandramouli, B., DeLine, R.: Tempe: an interactive data science environment for exploration of temporal and streaming data. MSR Technical report MSR-TR-2014-148 (2014)
2. Rachmawati, E., Supriana, I.: Khodra, M.L.: Review of local descriptor in RGB-D object recognition. Telkomnika **12**, 1132–1141 (2014)
3. Sclocco, A., Leeuwen, J.V., Bal, H.E.,: A real-time radio transient pipeline for ARTS. In: IEEE Global Conference on Signal and Information Processing. IEEE (2015)
4. Wang, S., Wen, Y., Zhao, H.: Study on the similarity query based on LCSS over data stream window. In: IEEE International Conference on E-Business Engineering, pp. 68–73 (2015)
5. Dallachiesa, M., Jacques-Silva, G.: Gedik B,: Sliding windows over uncertain data streams. Knowl. Inf. Syst. **45**, 159–190 (2015)
6. Sund, T., Møystad, A.: Sliding window adaptive histogram equalization of intraoral radiographs: effect on image quality. Dentomaxillofacial Radiol. **35**, 133–138 (2006)
7. Golab, L., Dehaan, D., Demaine. E.D., et al.: Identifying frequent items in sliding windows over on-line packet streams. In: ACM SIGCOMM Conference on Internet Measurement, pp. 173–178. ACM (2003)
8. Mirzadeh, N., Koçberber, Y.O., Falsafi, B.: Sort vs. Hash join revisited for near-memory execution. In: Proceedings of the 5th Workshop on Architectures and Systems for Big Data (ASBD 2015) (2015)

9. Han, X., Li, J., Gao, H.: Efficiently processing (p, ε)-approximate join aggregation on massive data. Inf. Sci. **278**, 773–792 (2014)

10. Gomes, J., Choi, H.A.: Adaptive optimization of join trees for multi-join queries over sensor streams. Inf. Fusion **9**, 412–424 (2008)

11. Zhou, F.F., Yang, J.R.: Mining maximal frequent patterns over data stream based on time decaying. Appl. Mech. Mater. **605**, 3835–3838 (2014)

Data Mining

L2 Learners' Proficiency Evaluation Using Statistics Based on Relationship Among CEFR Rating Scales

Hajime Tsubaki[✉]

GITI/Language and Speech Science Research Laboratories,
Waseda University, Tokyo, Japan
hjm.tsubaki@gmail.com

Abstract. In this paper, aiming at an objective evaluation of second language (L2) learners' proficiencies, it was tried to predict the learners' language proficiency using 94 statistics. The statistics were extracted automatically and manually from English conversation data collected from groups of Japanese English learners at educational institutions and were classified into 5 subcategories. To estimate the learners' English proficiencies represented as Central European Framework of Reference (CEFR) Global Scale scores, canonical correlation analysis was performed on the statistics and the 5 subcategories, and their correlations to CEFR Global Scale scores were analyzed. As the result of the analysis, 24 statistics were selected for predicting the learners' English proficiencies. The estimation experiment was carried out using a neural network trained by data set of 135 learners and the 24 statistics matrixes in cross-validation. An overall correlation of 0.894 was shown between the predicted proficiency scores and the L2 learners' actual CEFR Global Scale scores. These results confirmed the usefulness of the 24 statistical measures out of the beginning set of 94 measures in the objective evaluation of L2 language proficiency.

Keywords: Second language learners' proficiency evaluation · CEFR · Canonical correlation analysis · Neural network

1 Introduction

There are a lot of methods which are proposed for evaluating second language (L2) learners' proficiency. In such educational fields, data mining is usually available in many research cases. Data mining in these areas plays important roles to find, extract and analyze information and patterns on the L2 learners' learning from speech and text data [1–4]. Furthermore, these learners' proficiency can be predicted by a combination of data mining methods and L2 learners' data [5–8]. These results expect to make it possible to provide effective improvements in studying and teaching approach for the L2 learners and language teachers.

Conventionally, many researches analyze correlations between evaluation values assessed by human raters and statistics extracted from rating items and data for the L2 learners' proficiency. Then, using multivariate regression methods, statistical estimation models are built based on the rating items as independent variables and the L2

© Springer International Publishing AG 2017
Y. Tan et al. (Eds.): DMBD 2017, LNCS 10387, pp. 345–352, 2017.
DOI: 10.1007/978-3-319-61845-6_35

learners' proficiency as a dependent variable. However, in case that there are too many independent variables, a phenomenon called multicollinearity can be caused by a highly-correlational relationship in two or more independent variables to a dependent variable, and predicted values can become unstable.

In this study, from the viewpoint of correspondence relations among rating items and categories as independent variables, the independent variables were narrowed in number and the L2 learners' language proficiencies were predicted. 94 statistics classified into 5 subcategories for rating the L2 learners' language proficiencies and English proficiencies represented as Central European Framework of Reference (CEFR) Global Scale scores were used. A canonical correlation analysis was performed on the statistics and the 5 subcategories, and correspondence relations among the each 5 subcategories and their correlations to the CEFR Global Scale scores were analyzed in order to select useful statistics for estimation of the L2 learners' CEFR Global Scale scores. As the result of the analysis, 24 statistics were chosen and their usefulness as statistical measures was researched by predicting experiment using neural network and multiple regression models.

The study proceeds as follows. In the following sections, analysis data, methods for statistics extraction for second language learners' proficiency estimation parameters and predicting experiment for their proficiency scores are indicated in the next Sect. 2. In Sect. 3, the analyzed and experimental results are showed. Finally, in Sect. 4, the findings and discussed further works are summed up.

2   Research Data, Analysis Method and Experimental Design

2.1   Research Data

The data set for this study is English conversation data on Japanese English learners' groups in educational institution. The data set were collected and constructed as follows; the participants are 135 students from seven schools from among three kinds of educational institutions, namely, two junior high schools, two senior high schools, and three universities in Japan. They are divided into a total of 45 groups having three students as shown in the Table 1. Each group is comprised of 3 students who are randomly selected from junior high, senior high and university students. Three students are interacted orally as a group for 5 min on given topic such as Family, Friends, Hobbies, English, and Culture. Their English conversations are recorded in video format, transcribed and extracted as 94 statistics as Number of Tokens, Types and part-of-speech, Speaking duration time, Number of syllables and so on for each student.

Proficiency rating for the students is conducted by 10 raters who are all Japanese teachers of English holding a minimum of a master's degree in the field of English education or applied linguistics, and teaching English at either high schools or universities. The rating is conducted for 6 categories which are Global and 5 Subcategories, such as "Range", "Accuracy", "Fluency", "Interaction" and "Coherence" in the CEFR Rating Scales. The rating criteria is 7 levels of the CEFR - Below A1, A1, A2, B1, B2, C1, C2 - on both the Global Oral Assessment Scale and Oral Assessment Criteria Grid. Below A1 corresponds to lower ability or "Not good" and sequentially

Table 1. Number of participants in each educational institution

Institutions	Public schools		Private schools		Total participants	Total groups
	Participants	Groups	Participants	Groups		
Junior high schools	27	(9)			27	(15)
			18	(6)	18	
Senior high schools	24	(8)			24	(15)
			21	(7)	21	
Universities	15	(5)			15	(15)
			30	(10)	30	
Total	66	(22)	69	(23)	135	(45)

C2 applies to higher ability or "Excellent" in general proficiency rating. These categorical scores of Below A1 to C2 obtained from the raters' ratings are changed into numerical values by Rasch model, one of normalizing methods which is generally used in educational fields [9]. The normalized CEFR Global Scale score distribution is shown in the Fig. 1.

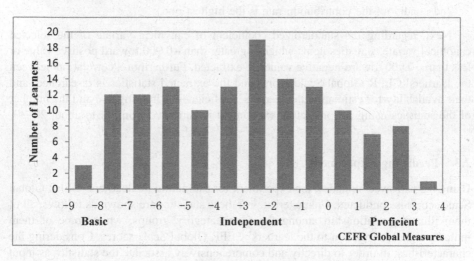

Fig. 1. Normalized CEFR Global Scale score distribution

2.2 Analysis Method

Canonical correlation analysis is conducted to 10 combinations consisting from each 2 category groups of the 5 subcategories in order to analyze category characteristics and relationships between the 5 subcategories as shown in the Table 2. Canonical correlation analysis is one of multivariate analysis methods for synthesize other variables by weighting in order to maximize correlation between 2 groups consisting of one or more variables. Using canonical correlation vectors from the analysis, relationship between variables of the 2 groups can be identified and measured.

Table 2. Combinations of the 5 subcategories

Subcategory	Range	Accuracy	Fluency	Interaction
Range				
Accuracy	R-A			
Fluency	R-F	A-F		
Interaction	R-I	A-I	F-I	
Coherence	R-C	A-C	F-C	I-C

In this study, the analysis is carried out for relations between variables of the 5 subcategories using 3 canonical correlation analysis outputs such as canonical variate, standardized coefficient of canonical variate and canonical correlation coefficient. In accordance with these outputs, corresponding relationships among statistics of the 5 subcategories are characterized and extracted statistically. Based on the canonical variate, variates satisfying the following 2 conditions are selected.

1st condition; the canonical correlation coefficient is statistically significant.
2nd condition; the contribution rate is the highest one.

Next, regarding the standardized coefficient of canonical variate of the selected canonical variate, statistics items which is greater than +0.900 toward positive value or less than −0.900 toward negative value are extracted. Furthermore, correlation between the learners' CEFR Global Scale scores and the extracted statistics is researched, and their availability for estimating the learners' proficiency is investigated and the number of the statistics using for predicting experiment is narrowed from 94 to about 20–30.

2.3   Predicting Experiment

Using the narrowed statistics, the experiment for estimating the learners' CEFR Global Scale scores is conducted. Characteristics of these statistics are shown as follows; all of them illustrate relationship among the 5 subcategory groups, while some of them indicate higher correlation to the learners' CEFR Global Scale scores. Considering the characteristics, in order to directly and comprehensively associate the statistics as input with the learners' CEFR Global Scale scores as output, a neural network is employed. The estimated scores are compared with the learner's actual CEFR Global Scale scores for verifying usefulness as statistical measures of L2 learners' proficiencies. In addition, an estimating experiment using a multiple regression model is conducted for comparing difference between non-linear and linear prediction.

3 Analysis and Experimental Results

3.1 Canonical Correlation Analysis Result

According to Table 3, 24 statistics are narrowed from the 94 statistics and there are 2 kinds of statistics as the result of the canonical correlation analysis. One shows correlation to the learners' CEFR Global Scale scores, the other does not. As a general tendency seen at upper rank being greater than 0.600 in the correlation, there are 6 statistics.

Table 3. Narrowed statistics based on the canonical correlation analysis

Statistics Name	Subcategory Combination	Subcategory	Correlation to CEFR GLOBAL Scale scores	Standardized coefficient
Number of words	A-C	Coherence	0.8516	-0.9698
Tokens/Clause	R-A	Range	0.7199	-1.0563
Tokens	R-F	Range	0.7091	0.9856
Total number of words	A-F	Fluency	0.7091	1.6004
Number of syllables including dysfluency	A-F	Fluency	0.7017	-1.3200
Number of words/topic	A-C	Coherence	0.6586	1.7803
Speaking time including pause time (sec)	A-F	Fluency	0.5787	1.6224
	F-I	Fluency	0.5787	-1.2766
	F-C	Fluency	0.5787	-1.5328
Clause	R-A	Range	0.4717	-1.3926
Tokens/C-unit	R-A	Range	0.4550	2.6163
Formulaic/C-unit	R-A	Range	0.4283	-2.1706
More than 100/token	R-I	Range	0.4146	-0.9713
Formulaic clause/token	R-I	Range	0.2393	2.0359
Article	A-C	Accuracy	0.2375	-0.9954
Self-correction per 100 words	A-F	Accuracy	0.1312	1.2818
Total number of topic moves	R-C	Coherence	0.1211	-0.9965
Noun	A-C	Accuracy	0.0919	-1.4078
Continuing question per turn	I-C	Coherence	0.0891	-0.9416
Ask for info/opinion	I-C	Interaction	0.0115	1.0618
Individual turn-taking	A-C	Coherence	-0.0017	-0.9265
TOTAL	A-I	Accuracy	-0.0082	1.3068
1-100/token	R-A	Range	-0.0263	-2.0437
Number of turn/topic	A-C	Coherence	-0.0531	-1.0453
Verb	A-C	Accuracy	-0.0991	-1.0728
Number of turn-taking	A-C	Coherence	-0.1137	1.0490

As for the standardized coefficient of canonical variate shown in the Table 3, Tokens of Range, Total number of words of Fluency and Number of words/topic of Coherence indicate toward positive value. Meanwhile, Number of words of Coherence, Tokens/Clause of Range and Number of syllables including dysfluency of Fluency show toward negative value. This result can say that utterance quantity works positively, though unnecessary utterance repetition does negatively in human rating.

Regarding the canonical variate, there are 2 kinds of following patterns as shown in the Fig. 2.

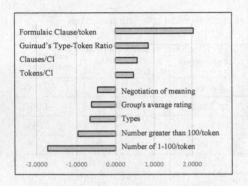   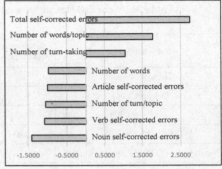

(a) The first pattern: Range - Interaction   (b) The second pattern: Accuracy- Cohehrnce

Fig. 2. Standardized coefficient vectors of canonical variate

- The first pattern

This pattern satisfies the following conditions.

1st condition;   the contribution rate of the canonical variate exceeds 50% in the total amount.

2nd condition;   the canonical correlation coefficient of the canonical variate is greater than 0.800.

For example, the result of the analysis between Range and Interaction corresponds to this pattern. Figure 2 (a) shows its standardized coefficient vectors of canonical variate. Formulaic Clause/token and Guiraud's Root Type-Token Ratio of Range indicate toward highly positive value. Meanwhile, Number of 1–100/token and Number being greater than 100/token of Range show toward negative value. This result can say that varieties of vocabulary is more important than utterance quantity from the point of view of Range in Interaction.

- The second pattern

This pattern fulfills the following conditions.

1st condition;   the contribution rate of the canonical variate is less than 50% in the total amount.

2nd condition;   the canonical correlation coefficient of the canonical variate is greater than 0.550.

For instance, the result of the analysis between Accuracy and Coherence is fit to this pattern. Figure 2(b) indicates its standardized coefficient vectors of canonical variate. Self-corrected errors ratio of total part-of-speech of Accuracy and Number of words/topic and turn-taking of Coherence indicate toward highly positive value. On the other hand, Self-corrected errors ratio of noun, verb and article of Accuracy and Number of words and turn/topic of Coherence show toward negative value. This result can say that native-like English usage and utterance in accordance with given topics are more important than mere utterance quantity and that errors of basic vocabulary give negative effect to human raters.

Based on these analysis results, the statistics not showing correlation to the learners' Global Scale scores have important roles in the 5 subcategories of learners' English conversation. Thereby the statistics were used for parameters of the learners' proficiency estimation.

3.2 Predicting Experiment Result

The 24 statistics which shows correlation to the learners' Global Scale scores and/or relationship among the 5 subcategories were used as input information on the neural network. By dividing the research data into three sets, each 2 data sets were used for training data and the other was done for test data in cross-validation. The learners' Global Scale scores were estimated by using the neural network based on the above-mentioned input information. The correlation with the actual values of the estimated scores showed the value of 0.894. Additionally, another estimating experiment using the research data and a multiple regression model was conducted. The correlation with the actual values of the predicted scores indicated the value of 0.870 as shown in Fig. 3. In non-linear and linear prediction of the learners' Global Scale scores, these results confirmed the usefulness of the 24 statistics as L2 language proficiency measures.

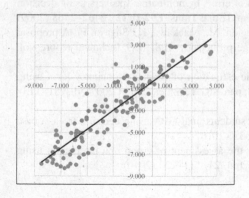
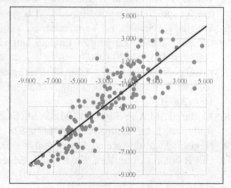

(a) The neural network estimation   (b) The multiple regression estimation

Vertical axis: actual CEFR Global Scale score
Horizontal axis: estimated scores

Fig. 3. Correlation between actual and estimated CEFR Global Scale scores

4 Summary and Future Works

To estimate the L2 learner's proficiency, the applicability of statistics extracted from English conversation data on Japanese English learners' groups in educational institutions by canonical correlation analysis was examined. By using neural network and multiple regression models, the learners' English proficiency scores were estimated by the 24 statistics based on the 5 subcategories, such as "Range", "Accuracy", "Fluency", "Interaction" and "Coherence" in the CEFR Rating Scales. The estimated scores of the learners showed higher correlation being greater than 0.850 to their actual scores. These results indicate the usefulness of the statistics for the estimation of L2 learners' proficiency that have been obtained from canonical correlation analysis. As it is confirmed that the 24 statistics narrowed from the total 94 statistics can be used for L2 learners' proficiency evaluation, the statistics are expected for applying to not only numeric rating but also categorical evaluation used in decision tree method.

References

1. Hirabayashi, K., Nakagawa, S.: Automatic evaluation of English pronunciation by Japanese speakers using various acoustic features and pattern recognition techniques. In: Proceedings of INTERSPEECH, pp. 598–601 (2010)
2. Wang, H., Waple, C.J.: Kawahara, T.: Computer assisted language learning system based on dynamic question generation and error prediction for automatic speech recognition. J. Speech Commun. **51**(10), 995–1005 (2009)
3. Nakamura, S., Matsuda, S., Kato, H., Tsuzaki, M., Sagisaka, Y.: Objective evaluation of English learners' timing control based on a measure reflecting perceptual characteristics. In: Proceedings of IEEE ICASSP, pp. 4837–4840 (2009)
4. Nakamura, S., Kato, H., Sagisaka, Y.: Effects of Mora-timing in English rhythm control by Japanese learners. In: Proceedings of INTERSPEECH 2009, pp. 1539–1542 (2009)
5. Wang, H., Kawahara, T.: Effective prediction of errors by non-native speakers using decision tree for speech recognition-based CALL system. IEICE Trans. **E92-D**(12), 2462–2468 (2009)
6. Yasuda, K., Sumita, E., Yamamoto, S., Yanagida, M., Maekawa, K., Sugaya, F.: A proposal for automatically gauging of English language proficiency. IPSJ SIG Technical report, vol. 2003-NL-155, 65–70 (2003)
7. Kosuke, S., Masashi, S., Yuji, M.: Automatic estimation of English proficiency level using corpora. Information Processing Society of Japan SIG Technical report 2007-NL-181, 113–119 (2007)
8. Al-Barrak, M.A., Al-Razgan, M.: Predicting students final GPA using decision trees: a case study. Int. J. Inf. Educ. Technol. **6**(7), 528–533 (2016)
9. Negishi, J.: Multi-faceted Rasch analysis for the assessment of group oral interaction using CEFR criteria. Ann. Rev. Engl. Lang. Educ. Jpn. **21**, 111–120 (2010)

Research Hotspots and Trends in Data Mining: From 1993 to 2016

Zili Li[(⊠)] and Li Zeng[(⊠)]

Center for National Security and Strategic Studies (CNSSS),
National University of Defense Technology, Changsha 410073, China
zilili@163.com, crack521@163.com

Abstract. Data mining, which is also referred to as knowledge discovery in databases, means a process of nontrivial extraction of implicit, previously unknown and potentially useful information from data in databases. This paper was to explore a Bibliometric approach to quantitatively assessing current research hotspots and trends on Data Mining, using the related literature in the Science Citation Index (SCI) database from 1993 to 2016. It shows that the research of Data Mining in 2016 was in the mature period with a maturity of 85.55%, the total of 11071 articles covered 131 countries(regions) and Top 3 countries(regions) were USA(2311, 21.37%), China(1474, 13.63%) and Taiwan (904, 8.36%). In addition, Top 10 keywords are found to have citation bursts: big data, social network, particle swarm optimization, data warehouse, gene expression, self-organizing map, intrusion detection, recommender system, bioinformatics, svm. This study provided scholars in the data mining research, as well as research hotspots and future research directions.

Keywords: Data mining · Bibliometric analysis · Research hotspots

1 Introduction

Data mining, which is also referred to as knowledge discovery in databases, means a process of nontrivial extraction of implicit, previously unknown and potentially useful information (such as knowledge rules, constraints, regularities) from data in data-bases [1]. In fact, Data Mining is a relatively new field of research, which has gained huge popularity in these days.

In recent years, the literature associated with data mining has grown rapidly [2, 3]. There are many comprehensive revision on the data mining, such as: manufacturing [4], particle swarm optimization [5], customer relationship management [6], bee Data Mining [7], educational data mining [8], time series data mining [9], but few studies have used bibliometrics and a visualization approach to conduct deep mining and reveal the data mining field. Liao [10] reviews data mining techniques and their applications and development, through a survey of literature and the classification of articles, from 2000 to 2011. Dai [11] applied the optimized bibliometric method to evaluate global scientific production of data mining papers of the Science Citation Index (SCI). Wang [12] presents the evolution of the intellectual structure in tourism destination literature as determined by means of bibliometric and social network

© Springer International Publishing AG 2017
Y. Tan et al. (Eds.): DMBD 2017, LNCS 10387, pp. 353–365, 2017.
DOI: 10.1007/978-3-319-61845-6_36

analysis of 17 552 citations of 414 articles published in Social Sciences Citation Index and Sciences Citation Index journals from 1955 to 2011. Gu [13] explore the foundational knowledge and research hotspots of big data research in the field of healthcare informatics by conducting a literature content analysis and structure analysis.

In this paper, the Bibliometric analysis tools Tviz and Citespace [14] was performed by investigating annual scientific outputs, distribution of countries, institutions, journals, research performances by individuals and subject categories to offer another perspective on the development of Data mining research. Moreover, innovative methods such as Keyword co-citation analysis, semantic clustering and Keyword Frequent Burst Detection were applied to provide insights into the global research hotpots and trends from various perspectives which may serve as a potential guide for future research.

2   Data and Methods

We collected the bibliographic records from the Web of Science (WOS) of Thomson Reuters on January 31, 2017 and determined the time frame of this analysis to 1993 and 2016. The ultimate query string about data mining looked like this: TOPIC: ("Data Mining") Refined by: RESEARCH AREAS: (COMPUTER SCIENCE) Indexes = SCI-EXPANDED Timespan = 1993–2016. The query resulted in 11071 bibliographic records. The whole bibliographic records were then downloaded for subsequent analysis. Then we used a Bibliometric approach to quantitatively assessing current research hotspots and trends on Data Mining Results.

3   Result and Discussion

3.1   Characteristics of Article Outputs

Figure 1 shows the number of articles and the prediction of the maturity of the research about Data Mining between 1993 and 2016. Black curve stands for the annual number of publications about Data Mining. From the curve, we found that a substantial interest in data mining research did not emerge until 1996, although a few articles related to data mining were published previously. And the highest annual number occurred in the years of 2005, 2006 and 2016. The red curve stands for the cumulative number of publication in the field of Data Mining. According to the theory of technology maturity, the cumulative number of the publication will be presented as an S-curve in general [15]. By using this theory, the cumulative number of data mining articles was accurately approximated by a logistic model $(y = 12362/(1 + e^{530.352 - 0.2639x}), r^2 = 0.9972)$, where x and y denote the year and article number respectively, and r^2 is the Goodness of Fit. According to this, we can divide the development of Data Mining into four stages: infant period (before 1997), growth period (1997–2007), mature period (2008–2020) and Decline period (after 2020). According to the above stage division, the research of Data Mining in 2016 was in the mature period with a maturity of 85.55%.

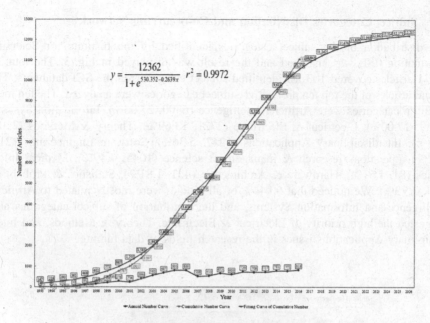

$$y = \frac{12362}{1+e^{530.352-0.2639x}} \quad r^2 = 0.9972$$

Fig. 1. Variation of article numbers

3.2   Characteristics of Document Type

The distribution of the document type was displayed in Fig. 2. Overall, the papers about Data Mining involving a total of six document types, and the details are as follows: (1,641,53.45%), article (8237,74%), proceedings paper (2332,21%), editorial material (225,2%), review (230,2%), others (47,1%). The distribution of document type suggested the high priority of article issues and proceedings paper in data mining research.

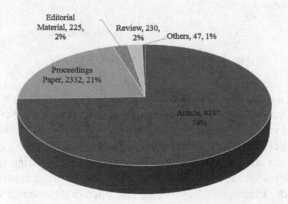

Fig. 2. Distribution of document type

3.3   Subject Categories Distribution and Co-occurring Network

The distribution of the subject categories identified by the Institute for Scientific Information (ISI) was analyzed and the result was displayed in Fig. 3. The total of 11071 articles covered 103 ISI identified subject categories in the SCI databases. The annual articles of the top ten productive subject categories were analyzed. The ten most common categories were Artificial Intelligence (6300, 27.06%), Information Systems (3959, 17.00%), Electrical & Electronic (2721, 13.69%), Theory & Methods (2202, 8.68%), Interdisciplinary Applications (1247, 5.36%), Software Engineering (1216, 5.22%), operations research & management science (1045, 4.49%), Medical Informatics (81, 1.51%), Hardware & Architecture (421, 1.81%), Statistics & Probability (404, 1.73%). We noticed that 60.44% of all articles were mostly related to Artificial Intelligence and Information Systems, and the distribution of subject categories also suggested the high priority of Electrical & Electronic, Theory & Methods, and Interdisciplinary Applications issues in the research fields of data mining.

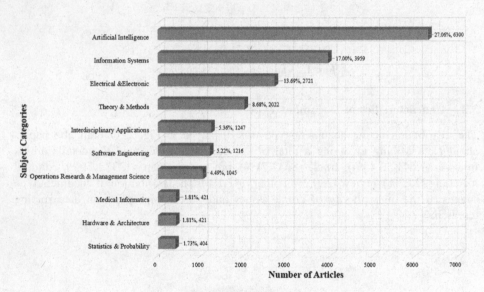

Fig. 3. Distribution of subject categories

We visualized a subject categories co-occurring network applying a threshold to the network between centrality in the network of subject categories. Network centrality measures the relative importance of nodes within networks and could be used as an indicator of a subject category's position within the network [16]. We can find that the Artificial Intelligence and engineering took part in more co-occurring relationship, and Information Systems took the central position in the co-occurring network, followed by Theory & Methods, Electrical &Electronics and operations research & management science (Fig. 4).

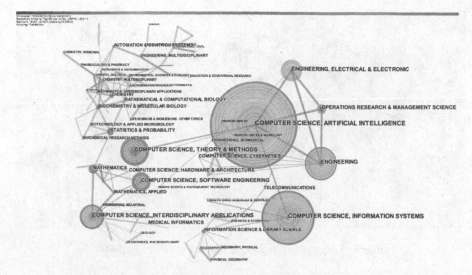

Fig. 4. Subject categories co-occurring Network (the thickness of each link represents the intensity of co-occurring, the size of each node represents the number of total articles, and the purple color denotes the high betweenness centrality node)

3.4   Geographic Distribution Map of Countries and International Collaboration

Data on geographic information were generated from author affiliations. Figure 5 shows the geographic distribution of countries (regions) in the field of "Data Mining". Overall, The total of 11071 articles covered 131 countries(regions) and ten most common countries(regions) were USA(2311, 21.37%), China(1474, 13.63%), Taiwan (904, 8.36%), Spain(487, 4.5%), South Korea(426, 3.94%), Australia (413, 3.82%), Germany(397, 3.67%), Canada(391, 3.62%), England(373, 3.45%), Italy(372, 3.44%), India(301, 2.78%), Japan(295, 2.73%), (283, 2.62%), France (169, 1.56%).

Figure 6 depicts a network consisting of 115 nodes and 116 links on behalf of the collaborating countries between 2000 and 2016. As can be seen, the major contribution of the total output mainly came from two countries, namely, USA and China. Clearly, USA is the largest contributor publishing 2311 papers. In other words, USA has a dominant status in the field of data mining, which produced about which produced about one fifth of world's total during this period. An interesting observation is that there are certain countries which have relatively low frequency but have high value of centrality among all other countries. SWEDEN leads other countries, which are shown as node rings in purple in Fig. 6. This is followed by papers originating from NORTH IRELAND, MACEDONIA, PORTUGAL, ITALY and so on. In other words, they are pivotal nodes in the network with the highest betweenness centrality. In addition, nine countries are found to have citation bursts: USA (39.2247), JAPAN (22.7672), GERMANY (12.5448), GREECE(9.2354), NORTH IRELAND(7.9009), SOUTH KOREA (7.1758), FINLAND (6.7726), TURKEY (6.0673), AUSTRIA

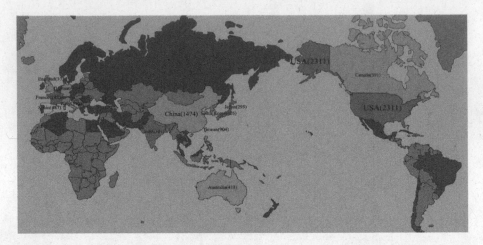

Fig. 5. Geographic distribution map of countries (regions)

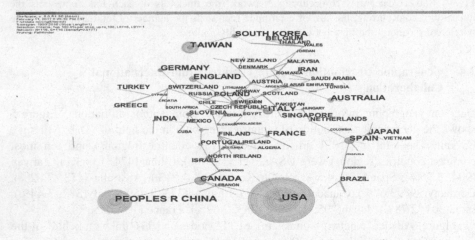

Fig. 6. Cooperation network of Countries on data mining research papers

(5.7766) and CUBA (5.4665), suggesting that they have abrupt increases of citations, and the details are listed in the Table 1.

3.5   Institutions Distribution and Co-occurring Network

Overall, a total of 4914 research institutes in the world were engaged in Data Mining during 1993 and 2016. Figure 7 lists the top ten of them: Chinese Acad Sci (109 articles), Natl Taiwan Univ (86 articles), Univ Hong Kong (83 articles), IBM Corp (74 articles), Univ Illinois (73 articles), Natl Cent Univ (72 articles), Natl Chiao Tung Univ (72 articles), Natl Chiao Tung Univ (72 articles), Harbin Inst Technol (68 articles) and Univ Texas (66 articles).

Table 1. Top ten countries with strongest citation bursts.

Countries	Year	Strength	Begin	End	1993 - 2016
FINLAND	1993	6.7726	1993	2001	
USA	1993	39.2247	1995	2001	
GERMANY	1993	12.5448	1997	2002	
NORTH IRELAND	1993	7.9009	1997	2004	
JAPAN	1993	22.7672	1998	2003	
AUSTRIA	1993	5.7766	2000	2003	
SOUTH KOREA	1993	7.1758	2005	2006	
GREECE	1993	9.2354	2007	2009	
TURKEY	1993	6.0673	2009	2013	

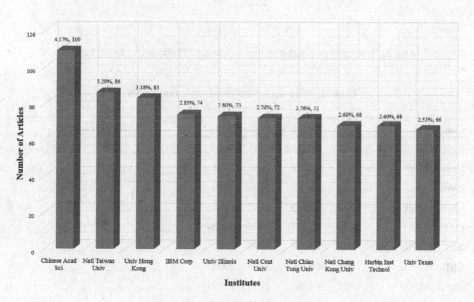

Fig. 7. Distribution of institutes

Figure 8 shows the visualization of the distribution of institutions. In order to show the core institutions of this field, we filter out the institutions with small number of publications and get an institute co-occurring network with 723 nodes and 606 links. Obviously, Chinese Acad Sci in China takes the first place with a frequency of 109 articles. The second place is Natl Taiwan Univ with a frequency of 86 articles. Apart from that, there are still other institutions participating in data mining research, such as

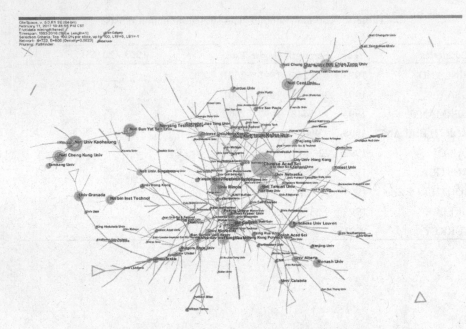

Fig. 8. Institutes co-occurring network with 723 Nodes and 606 Links

Table 2. Top ten institutes with citation bursts.

Countries	Year	Strength	Begin	End	1993 - 2016
IBM Corp	1993	19.0654	1997	2006	
Univ Ulster	1993	7.6014	1997	2004	
Kyoto Univ	1993	6.5962	1998	2000	
Univ Helsinki	1993	6.1642	1997	2005	
CALTECH	1993	5.5867	1996	2000	
Natl Univ Singapore	1993	5.5038	1998	2000	
CSIRO	1993	4.5159	1998	2002	
Bar Ilan Univ	1993	4.486	1997	1999	
George Mason Univ	1993	3.9344	1998	2001	
Univ Calif Irvine	1993	3.2768	1997	2002	

Univ Hong Kong, IBM Corp and so on. We also notice that China' institutes such as Chinese Acad, Natl Taiwan Univ, Univ Hong Kong and Harbin Inst Technol were on the top of the list. In addition, ten institutes are found to have citation bursts: IBM Corp (19.0654), Univ Ulster (7.6014), Kyoto Univ (6.5962), Univ Helsinki (6.1642), CALTECH (5.5867), Natl Univ Singapore (5.5038), CSIRO (4.5159), Bar Ilan Univ

(4.486), George Mason Univ (3.9344) and Univ Calif Irvine (3.2768), suggesting that they have abrupt increases of citations, and we listed the details in the Table 2.

3.6   Research Hotspots and Emerging Trends of Data Mining

Figure 9 shows the document co-citation network derived from the whole datasets. In this network, there are 378 unique nodes and 1360 links. These nodes represent cited references from the collected articles, and the links in the network represent co-citation relationships. Each link colors correspond directly to each time slice. For example, blue links describe articles that were co-cited in 1993, and the most recent co-citation relationships are visualized as orange or red links. Larger node size simply that the article is an important one within the knowledge domain. Second, red rings around a node represent a citation burst. Third, purple rings indicate nodes that have a relatively high betweenness centrality in the network. Table 3 presents the top ten high burst articles which can represented the research hotspots of data mining.

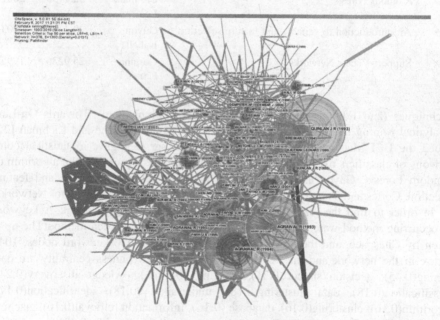

Fig. 9. Document co-citation network

Table 3 lists the top 10 high cited articles which represent the research hotspots in the field of Data Mining. Witten [17] discussed Data mining: Practical machine learning tools and techniques. Agrawal and Srikant [18] proposed a Fast Algorithms for Mining Association Rules. Fayyad [19] discussed the advances in knowledge discovery & data mining. Han [20] writes a book named "Data Mining: Concepts and

Table 3. Top ten high burst articles.

NO.	Title	First author	Burst strength	Year
1	Data mining: Practical machine learning tools and techniques	Witten, I.H	66.0847	1995
2	Fast Algorithms for Mining Association Rules	Agrawal, Rakesh	59.5165	2004
3	Advances in Knowledge Discovery & Data Mining	FAYYAD U M	53.4643	1996
4	Data Mining: Concepts and Techniques (2nd Edition)	HAN J	44.3352	2006
5	Towards On-Line Analytical Mining in Large Databases, SIGMOD Record.	Han, J	43.0357	1998
6	UCI Machine Learning Repository	Bache, K.	40.8589	2013
7	Statistical Comparisons of Classifiers over Multiple Data Sets.	Demsar, Janez	36.3458	2006
8	Random Forests	Breiman, Leo.	32.9436	2001
9	An introduction to variable and feature selection	Guyon, Isabelle	28.4353	2003
10	Support-Vector Networks	Corinna Cortes	25.9236	1995

Techniques (2nd Edition)". Han [21] discussed the problem of "Towards On-Line Analytical Mining in Large Databases, SIGMOD Record". Bache and Lichman [22] issued the UCI Machine Learning Repository. Demsar [23] gave a statistical comparisons of classifiers over multiple data sets. Breiman [24] proposed the algorithm of Random Forests. Guyon and Elisseeff [25] an introduction to variable and feature selection. Cortes and Vapnik [26] proposed the algorithm of Support-Vector Networks.

In order to find the research hotspots about Data Mining in detail, A keyword co-occurring method were used, and Fig. 8 show the result of such method(The up is given by Citespace, and the down is by Tviz). There are 630 keyword nodes, 1044 vertex in the network and the keywords with high betweenness centrality are data mining(0.45), decision support(0.31), learning (0.27), knowledge discovery(0.24), classification(0.18), data clustering(0.18), uncertainty (0.18), identification(0.17), algorithm(0.16), clustering(0.16), database (0.16), information retrieval(0.16), machine learning(0.15), information (0.15), neural network (0.14), model (0.13), association rule(0.13), optimization(0.13), bia(0.13), knowledge discovery (0.12), agent (0.12), feature selection(0.11), design(0.11), feature extraction (0.11) and so on (Fig. 10).

In addition, twenty keywords are found to have citation bursts: big data (28.7044), social network(16.1266), particle swarm optimization (13.1973), data warehouse (11.6965), gene expression(11.2986), self-organizing map (11.0229), intrusion detection (9.9739), recommender system(8.9949), bioinformatics (8.8154), svm (8.5505), fuzzy logic(7.6128), rough set theory(7.2184), outlier detection (6.967), ontology (6.7049), information visualization(6.544), web usage mining (5.9487), visual data

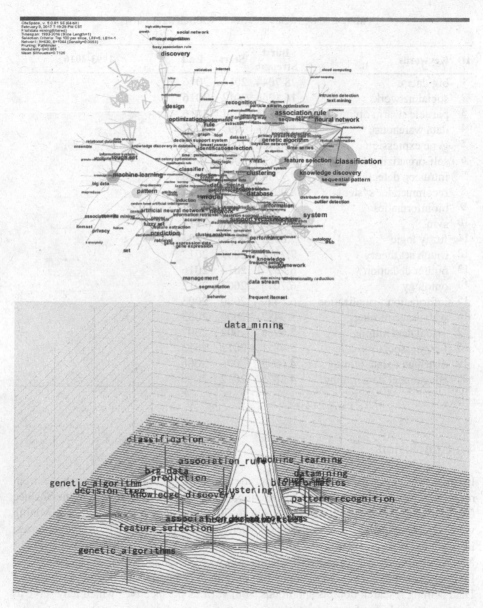

Fig. 10. Keyword co-occurring network

mining (5.6966), fuzzy association rule(5.4851), distributed data mining(4.6597) and educational data mining(4.3875), suggesting that they have abrupt increases of citations, and we listed the details in the Table 4.

Table 4. Top twenty keywords with citation bursts.

ID	Keywords	Burst Strength	Start	End	1993-2016
1	big data	28.7044	2014	2016	
2	social network	16.1266	2012	2016	
3	particle swarm optimization	13.1973	2011	2016	
4	data warehouse	11.6965	1997	2007	
5	gene expression	11.2986	2003	2007	
6	self-organizing map	11.0229	2003	2007	
7	intrusion detection	9.9739	2004	2008	
8	recommender system	8.9949	2013	2016	
9	bioinformatics	8.8154	2003	2007	
10	svm	8.5505	2011	2012	
11	fuzzy logic	·7.6128	2001	2004	
12	rough set theory	7.2184	2009	2012	
13	outlier detection	6.967	2014	2016	
14	ontology	6.7049	2007	2012	
15	information visualization	6.544	2000	2003	
16	web usage mining	5.9487	2000	2003	
17	visual data mining	5.6966	2002	2007	
18	fuzzy association rule	5.4851	2012	2013	
19	distributed data mining	4.6597	2006	2008	
20	educational data mining	4.3875	2012	2016	

4   Conclusion

In this paper, we used a Bibliometric method to quantitatively assessing current research hotspots and trends on Data Mining, using the related literature in the Science Citation Index (SCI) database from 1993 to 2016. Articles referring to Data Mining were concentrated on the analysis of scientific outputs, distribution of countries, institutions, periodicals, subject categories and research performances by individuals. Moreover, the innovative methods such as Keyword co-citation analysis and Keyword Frequent Burst Detection were applied to provide a dynamic view of the evolution of Data Mining research hotspots and trends from various perspectives which may serve as a potential guide for future research.

References

1. Fayyad, U.M.: Knowledge discovery in databases: an overview. In: ILP (1991)
2. Bonabeau, E., Dorigo, M., Theraulaz, G.: Data Mining - From Natural to Artificial Systems. Oxford University Press, New York (1999)

3. Martens, D., Baesens, B., Fawcett, T.: Editorial survey: data mining for data mining. Mach. Learn. **82**(1), 1–42 (2011)
4. Harding, J.A., Shahbaz, M., Srinivas, Kusiak, A.: Data mining in manufacturing: a review. J. Manufact. Sci. Eng. Trans. ASME **128**(4), 969–976 (2006)
5. Poli, R., Kennedy, J., Blackwell, T.: Particle swarm optimization. Data Min. **1**(1), 33–57 (2007)
6. Ngai, E.W.T., Xiu, L., Chau, D.C.K.: Application of data mining techniques in customer relationship management: a literature review and classification. Expert Syst. Appl. **36**(2), 2592–2602 (2009)
7. Karaboga, D., Akay, B.: A survey: algorithms simulating bee data mining. Artif. Intell. Rev. **31**(1–4), 61–85 (2009)
8. Romero, C., Ventura, S.: Educational data mining: a review of the state of the art. IEEE Trans. Syst. Man Cybern. Part C-Appl. Rev. **40**(6), 601–618 (2010)
9. Fu, T.C.: A review on time series data mining. Eng. Appl. Artif. Intell. **24**(1), 164–181 (2011)
10. Liao, S.H., Chu, P.H., Hsiao, P.Y.: Data mining techniques and applications - a decade review from 2000 to 2011. Expert Syst. Appl. **39**(12), 11303–11311 (2012)
11. Dai, L., Ding, L.X., Lei, Y.W., Tian, Y.G.: A Study of data mining trend through the optimized bibliometric methodology based on SCI database from 1993 to 2011. Appl. Math. Inf. Sci. **6**(3), 705–712 (2012)
12. Wang, C.H., Chen, S.C.: Bibliometric and social network analysis for data mining: the intellectual structure of tourism. J. Test. Eval. **42**(1), 229–241 (2014)
13. Gu, D.X., Li, J.J., Li, X.G., Liang, C.Y.: Visualizing the knowledge structure and evolution of big data research in healthcare informatics. Int. J. Med. Inform. **98**, 22–32 (2017)
14. Chen, C.: CiteSpace II: detecting and visualizing emerging trends and transient patterns in scientific literature. J. Am. Soc. Inf. Sci. Technol. **57**(3), 359–377 (2006)
15. Rogosa, D., Brandt, D., Zimowski, M.: A growth curve approach to the measurement of change. Psychol. Bull. **92**(3), 726 (1982)
16. Freeman, L.C.: Centrality in social networks conceptual clarification. Soc. Netw. **1**(3), 215–239 (1979)
17. Witten, I.H., Frank, E.: Data mining: Practical Machine Learning Tools and Techniques, 2nd edn. Morgan Kaufmann, San Francisco (2005)
18. Agrawal, R., Srikant, R.: Fast Algorithms for Mining Association Rules (1994)
19. Fayyad, U.M., et al.: Advances in knowledge discovery & data mining. Technometrics **40**(1), xviii (1996)
20. Han, J., et al.: Data Mining: Concepts and Technique, 2nd edn. Morgan Kaufmann, Amsterdam (2006)
21. Han, J.: Towards on-line analytical mining in large databases, SIGMOD Record. Sigmod Rec. **27**(27), 97–107 (1998)
22. Bache, K., Lichman, M.: UCI Machine Learning Repository (2013)
23. Demsar, J.: Statistical comparisons of classifiers over multiple data sets. J. Mach. Learn. Res. **7**, 1–30 (2006)
24. Breiman, L.: Random forests. Mach. Learn. **45**(1), 5–32 (2001)
25. Guyon, I., Elisseeff, A.: An introduction to variable and feature selection. J. Mach. Learn. Res. **3**(6), 1157–1182 (2003)
26. Cortes, C., Vapnik, V.: Support-vector networks. Mach. Learn. **20**(3), 273–297 (1995)

Proposal for a New Reduct Method for Decision Tables and an Improved STRIM

Jiwei Fei[1,2], Tetsuro Saeki[1,2(⊠)], and Yuichi Kato[1,2]

[1] Yamaguchi University, 2-16-1 Tokiwadai, Ube, Yamaguchi 755-8611, Japan
tsaeki@yamaguchi-u.ac.jp
[2] Shimane University, 1060 Nishikawatsu-cho, Matsue, Shimane 690-8504, Japan
ykato@cis.shimane-u.ac.jp

Abstract. Rough Sets theory is widely used as a method for estimating and/or inducing the knowledge structure of if-then rules from a decision table after a reduct of the table. The concept of a reduct is that of constructing a decision table by necessary and sufficient condition attributes to induce the rules. This paper retests the reduct by the conventional methods by the use of simulation datasets after summarizing the reduct briefly and points out several problems of their methods. Then, a new reduct method based on a statistical viewpoint is proposed and confirmed to be valid by applying it to the simulation datasets. The new reduct method is incorporated into STRIM (Statistical Test Rule Induction Method), and plays an effective role for the rule induction. The STRIM including the reduct method is also applied for a UCI dataset and shows to be very useful and effective for estimating if-then rules hidden behind the decision table of interest.

1 Introduction

Rough Sets theory was introduced by Pawlak [1] and used for inducing if-then rules from a dataset called the decision table. The induced if-then rules simply and clearly express the structure of rating and/or knowledge hiding behind the decision table. Such rule induction methods are needed for disease diagnosis systems, discrimination problems, decision problems and other aspects. The first step for the rule induction is to find the condition attributes which do not have any relationships with the decision attribute, to remove them and finally to reduce the table. Those processes to obtain the reduced table are useful for efficiently inducing rules and called a reduct. The conventional Rough Sets theory to induce if-then rules is based on the indiscernibility of the samples of the table. The reduct by the conventional method also uses the same concept and various types of indiscernibility, methods to find their indiscernibility and algorithms for the reducts are proposed to date [2–7].

This paper retests the conventional reduct methods through the use of a simulation dataset and points out their problems after summarizing the conventional rough sets and reduct methods. Then a new reduct method is proposed to overcome their problems from a statistical point of view. Specifically, the

© Springer International Publishing AG 2017
Y. Tan et al. (Eds.): DMBD 2017, LNCS 10387, pp. 366–378, 2017.
DOI: 10.1007/978-3-319-61845-6_37

Table 1. An example of a decision table.

U	$(C(1)C(2)C(3)C(4)C(5)C(6))$	D
1	563242	3
2	256124	6
3	116226	1
4	416646	6
...	......	...
$N-1$	151252	2
N	513135	4

new method recognizes each sample data in the decision table as the outcomes of random variables of the tuple of the condition attributes and the decision attribute, since the dataset is obtained from their population of interest. Accordingly, the reduct problem can be replaced by the problem of finding the condition attributes which are statistically independent of the decision attribute and/or its values. The statistical independence can be easily tested, for example, by a Chi-square test using the dataset. The validity of the new reduct method is confirmed by applying it to the simulation dataset. The experiment also gives an idea of improving STRIM (Statistical Test Rule Induction Method [8–11]) to include the reduct function and to induce if-then rules more efficiently. The usefulness of the reduct method and the improved STIRM are also confirmed by applying them to a UCI dataset [12] prepared for machine learning.

2   Conventional Rough Sets and Reduct Method

Rough Sets theory is used for inducing if-then rules from a decision table S. S is conventionally denoted $S = (U, A = C \cup \{D\}, V, \rho)$. Here, $U = \{u(i)|i = 1, ..., |U| = N\}$ is a sample set, A is an attribute set, $C = \{C(j)|j = 1, ..., |C|\}$ is a condition attribute set, $C(j)$ is a member of C and a condition attribute and D is a decision attribute. V is a set of attribute values denoted by $V = \cup_{a \in A} V_a$ and is characterized by an information function $\rho \colon U \times A \rightarrow V$. Table 1 shows an example where $|C| = 6$, $|V_{a=C(j)}| = M_{C(j)} = 6$, $|V_{a=D}| = M_D = 6$, $\rho(x = u(1), a = C(1)) = 5$, $\rho(x = u(2), a = C(2)) = 5$, and so on.

Rough Sets theory focuses on the following equivalence relation and equivalence set of indiscernibility:

$$I_C = \{(u(i), u(j)) \in U^2 | \rho(u(i), a) = \rho(u(j), a), \forall a \in C\}.$$

I_C derives the quotient set $U/I_C = \{[u_i]_C|i = 1, 2, ...\}$. Here, $[u_i]_C = \{u(j) \in U|(u(j), u_i) \in I_C, u_i \in U\}$. $[u_i]_C$ is an equivalence set with the representative element u_i and is called an element set of C in Rough Sets theory [2]. Let be

Line No.	Algorithm to compute a single global covering
1	(input: the set A of all attributes, partition $\{d\}^*$ on U; output: a single global covering R);
2	**Begin**
3	compute partition A^*;
4	$P := A$;
5	$R := \emptyset$;
6	if $A^* \leq \{d\}^*$
7	**Then**
8	**Begin**
9	**for** each attribute a in A **do**
10	**Begin**
11	$Q := P - \{a\}$;
12	compute partition Q^*;
13	if $Q^* \leq \{d\}^*$ then $P := Q$
14	**end** $\{\text{for}\}$
15	$R := P$
16	**end** $\{\text{then}\}$
17	**end** $\{\text{algorithm}\}$

Fig. 1. An example of LEM1 algorithm.

$\forall X \subseteq U$ then X can be approximated like $C_*(X) \subseteq X \subseteq C^*(X)$ by use of the element set. Here,

$$C_*(X) = \{u_i \in U | [u_i]_C \subseteq X\}, \tag{1}$$
$$C^*(X) = \{u_i \in U | [u_i]_C \cap X \neq \emptyset\}, \tag{2}$$

$C_*(X)$ and $C^*(X)$ are called the lower and upper approximations of X by C respectively. The pair of $(C_*(X), C^*(X))$ is usually called a rough set of X by C. Specifically, let $X = D_d = \{u(i) | (\rho(u(i), D) = d\}$ called concept $D = d$ then $C_*(X)$ is surely a set satisfying $D = d$ since $C_*(X) \subseteq X$ and it derives if-then rules of $D = d$ with necessity.

The conventional Rough Sets theory seeks a minimal subset of C denoted with $B(\subseteq C)$ satisfying the following two conditions:

(i) $B_*(D_d) = C_*(D_d)$, $d = 1, 2, ..., M_D$.
(ii) $a(\in B)$ satisfying $(B - \{a\})_*(D_d) = C_*(D_d)$ $(d = 1, 2, ..., M_D)$ does not exist.

$B(\subseteq C)$ is called a relative reduct of $\{D_d | d = 1, ..., M_D\}$ preserving the lower approximation and is useful for finding if-then rules since redundant condition attributes have been already removed from C.

LEM1 algorithm [2] and the discernibility matrix method (DMM) [3] are well known as representative ways to perform reducts. Figure 1 shows an example of LEM1, and A and $\{d\}^*$ at Line 1 of the figure respectively correspond to C and $\{D_d | d = 1, ..., M_D\}$ in this paper. LEM1 from Line 6 to 16 in the figure in principle checks and executes (i) and (ii) for all the combinations of the condition attributes.

DMM [3] at first forms a symmetric $N \times N$ matrix having the following (i, j) element δ_{ij}:

$\delta_{ij} = \{a \in C | \rho(u(i), a) \neq \rho(u(j), a)\}; \exists d \in D, \rho(u(i), d) \neq \rho(u(j), d)$ and $\{u(i), u(j)\} \cap Pos(D) \neq \emptyset, = *$; otherwise.

Here, $Pos(D) = \cup_{d=1}^{M_D} C_*(D_d)$ and $*$ denotes "don't care". Then, a relative reduct preserving the lower approximation can be obtained by the following expression:

$$F^{reduct} = \bigwedge_{i,j:i<j} \bigvee \delta_{ij}. \tag{3}$$

3 Retests of the Conventional Reduct Method

Here we retest the ability of the reducts obtained through LEM1 and DMM by use of a simulation dataset. Figure 2 [8–11] shows a way of how to generate simulation datasets. Specifically let (a) generate the condition attribute values of $u(i)$, that is, $u^C(i) = (v_{C(1)}(i), v_{C(2)}(i), ..., v_{C(|C|)}(i))$ by the use of random numbers with a uniform distribution and (b) determine the decision attribute value of $u(i)$ without NoiseC and NoiseD for a plain experiment, that is $u^D(i)$ by use of if-then rules specified in advance and the hypotheses shown in Table 2 and repeat the (a) and (b) processes by N times. Table 1 shows an example dataset generated by the use of those procedures with the following if-then rule $R(d)$ specified in advance:

$$R(d): \text{ if } Rd \text{ then } D = d \quad (d = 1, ..., M_D = 6), \tag{4}$$

where $Rd = (C(1) = d) \wedge (C(2) = d) \vee (C(3) = d) \wedge (C(4) = d)$.

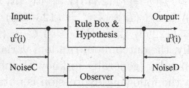

Fig. 2. A data generation model for a decision table contaminated with noise.

Table 2. Hypotheses with regard to the decision attribute value.

Hypothesis 1	$u^{C(i)}$ coincides with $R(k)$ and $u^{D(i)}$ is uniquely determined as $D = d(k)$ (uniquely determined data)
Hypothesis 2	$u^{C(i)}$ does not coincide with any $R(d)$ and $u^{D(i)}$ can only be determined randomly (indifferent data)
Hypothesis 3	$u^{C(i)}$ coincides with several $R(d)$ $(d = d1, d2, ...)$ and their outputs of $u^{C(i)}$ conflict with each other. Accordingly, the output of $u^{C(i)}$ must be randomly determined from the conflicted outputs (conflicted data)

The results of retesting both methods using the $N = 10,000$ dataset showed $F_{LEM1}^{reduct} = F_{DMM}^{reduct} = C(1) \wedge C(2) \wedge C(3) \wedge C(4) \wedge C(5) \wedge C(6)$ while the results were expected to be $F^{reduct} = C(1) \wedge C(2) \wedge C(3) \wedge C(4)$ from the rules (4) specified in advance. The retest experiment was repeated three times by changing the generated dataset and obtained the same results.

These results are clearly derived from the indiscernibility and/or discernibility caused by the element set which could not distinguish the differences between samples by the if-then rules (see Hypothesis 1 in Table 2) or those obtained by chance (see Hypothesis 2 and 3 in Table 2).

4   Proposal of Statistical Reduct Method

As mentioned in Sect. 3, the conventional reduct methods are unable to reproduce the forms of reducts specified in advance from the decision table due to a lack of abilities adaptive to the indifferent and conflicted samples in datasets despite the fact that real-world datasets will have such samples. This paper studies this problem with reducts from the view of STRIM (Statistical Test Rule Induction Method) [8–11]. STRIM regards the decision table as a sample set obtained from the population of interest based on the input-output system as shown in Fig. 2. According to a statistical model, $u(i) = (u^C(i), u^D(i) = (v_{C(1)}(i), v_{C(2)}(i), ..., v_{C(|C|)}(i), u^D(i))$ is an outcome of the random variables of $A = (C, D) = (C(1), C(2), ..., C(|C|), D)$ (hereafter, the names of the attributes are used as the random variables). Then, the following probability model will be specified: For any j, $P(C(j) = v_{C(j)}(k)) = p(j, k)$, $\sum_{k=1}^{M_{C(j)}} p(j, k) = 1$. For any $j1 \neq j2$, $C(j1)$ and $C(j2)$ are independent of each other for simplicity. According to the rules specified in (4), if $C = (1, 1, 2, 3, 4, 5)$ (hereafter (112345) briefly), for example, then $P(D = 1 | C = (112345)) = 1.0$ by use of Hypothesis 1 in Table 2. If $C = (123456)$ then $P(D = 1 | C = (123456)) = 1/M_D = 1/6$ by use of Hypothesis 2. If $C = (112256)$ then $P(D = 1 | C = (112256)) = 1/2$ by use of Hypothesis 3. Generally, the outcome of random variable D is determined by the outcome of C, if-then rules (generally unknown) and the hypothesis shown in Table 2. Consequently, the following expression is obtained:

$$P(D = l, C = u^C(i)) = P(D = l | C = u^C(i))P(C = u^C(i)). \qquad (5)$$

Here, $P(D = l | C = u^C(i)$ is the conditional probability of $D = l$ by $C = u^C(i)$ and very dependent on the if-then rules to be induced.

In the special case, if $C(j)$ does not exist in the condition part of the if-then rules of $D = l$, then the event $D = l$ is independent of $C(j)$, that is $P(D = l, C(j)) = P(D = l | C(j))P(C(j)) = P(D = l)P(C(j))$. This independence between $D = l$ and $C(j)$ can be used for a reduct of the decision table for the concept $D = l$. The problem of whether they are independent or not can be easily dealt with using a statistical test of hypotheses by the use of $\{u(i) = (v_{C(1)}(i), v_{C(2)}(i), ..., v_{C(|C|)}(i), u^D(i)) | i = 1, ..., N\}$. Specifically, specifications and testing of the following null hypothesis $H0(j, l)$ and its alternative hypothesis $H1(j, l)$ $(j = 1, ..., |C|, l = 1, ... M_D)$ were implemented:

Table 3. Example of contingency table by a statistical reduct ($N = 3000$, $df = 5$).

$D = 1$ $|U(D = 1)| = 503$	C					
	1	2	3	4	5	6
1	**158**	**150**	**135**	**118**	68	89
2	69	76	88	74	79	76
3	63	63	57	81	94	96
4	69	77	74	71	93	71
5	76	67	72	78	84	84
6	68	70	77	81	85	87
χ^2	74.00	78.92	45.84	28.22	4.00	4.83
p-values	1.51E−14	1.41E−15	9.34E−9	3.30E−05	0.550	0.437

$H0(j, l)$: $C(j)$ and $D = l$ are independent of each other.
$H1(j, l)$: $C(j)$ and $D = l$ are not independent of each other.

This paper adopts a Chi-square test since it is a standard method for testing the independence of two categorical variables by use of the contingency table $M_{C(j)} \times 1$. The test statistic χ^2 of $C(j)$ vs. $D = l$ is

$$\chi^2 = \sum_{k=1}^{M_{C(j)}} \frac{(f_{kl} - e_{kl})^2}{e_{kl}}, \tag{6}$$

where, $f_{kl} = |U(C(j) = k) \cap U(D = l)|$, $U(C(j) = k) = \{u(i)|\rho(u(i), C(j)) = k\}$, $U(D = l) = \{u(i)|u^D(i) = l\}$, $e_{kl} = n\hat{p}(j, k)\hat{p}(D, l)$, $n = \sum_{k=1}^{M_{C(j)}} \sum_{l=1}^{M_D} f_{kl}$, $\hat{p}(j, k) = \frac{f_{k.}}{n}$, $\hat{p}(D, l) = \frac{f_{.l}}{n}$, $f_{k.} = \sum_{l=1}^{M_D} f_{kl}$, $f_{.l} = \sum_{k=1}^{M_{C(j)}} f_{kl}$. χ^2 obeys a Chi-square distribution with degrees of freedom $df = (M_{C(j)} - 1)$ under $H0(j, l)$ and testing condition [13]: $n\hat{p}(j, k)\hat{p}(D, l) \geq 5$. This paper proposes a reduct method to adopt only the $C(j)$s of $H0(j, l)$ that were rejected and to construct a decision table for $D = l$ composed by them, since the test of the hypotheses cannot control type II errors, but only type I errors by a significance level. This paper names the proposed method the statistical reduct method (SRM) to distinguish it from the conventional methods.

A simulation experiment was conducted to confirm the validity of the proposed method using the decision table of the samples of $N = 10,000$ used in Sect. 2, and the following procedures:

Step 1: Randomly select samples by $N_B = 3000$ from the decision table ($N = 10,000$), and form a new decision table.
Step 2: Apply SRM to the new table, and calculate χ^2 every $C(j)$ by $D = l$.

Table 3 shows an example of the contingency table for the case of $D = 1$ vs. $C(j)$ ($j = 1, ..., 6$) and the results of a Chi-square test of them with $df = (M_{C(j)} - 1)$, and suggests the following knowledge:

(1) The p-values of $C(5)$ and $C(6)$ are quite high compared with the other condition attributes and indicate that $C(5)$ and $C(6)$ are independent of $D = 1$, that is, they are redundant and should be removed from the viewpoint of reduct.

(2) The frequencies $f_{kl=1}$ of $C(1) = 1$, $C(2) = 1$, $C(3) = 1$ and $C(4) = 1$ are relatively high compared with those of the rest of the same $C(j)$ $(j = 1, ..., 4)$. Accordingly, the combinations of $C(j) = 1$ $(j = 1, ..., 4)$ will most likely construct the rules of $D = 1$, which coincides with the rules specified in (4).

The above knowledge of (1) and (2) was also confirmed for the case of $D = l$ $(l = 2, ..., 6)$ and coincided with the specifications of Rules (4), and thus through them the validity and usefulness of SRM have been confirmed.

5   Proposal of Improved STRIM

STRIM has been proposed as a method to induce if-then rules from decision tables by use of two stages [8–11]. The first stage is that of searching rule candidates by the following procedures:

Step 1: Specify a proper condition part of trying if-then rules:

$$CP(k) = \bigwedge_j (C(j_k) = v_k). \tag{7}$$

Step 2: Test the condition part on the null hypothesis ($H0$) that $CP(k)$ is not a rule candidate and its alternative hypothesis ($H1$) specifying a proper significance level. Specifically, use a test statistic $z = \frac{(n_d + 0.5 - np_d)}{\sqrt{np_d(1-p_d)}}$ which obeys the normal distribution $N(0, 1^2)$ on $H0$, if $np_d \geq 5$, $n(1 - p_d) \geq 5$ (testing condition [14]). Here, $p_d = \frac{1}{|M_D|}$, $n_d = \max(n_1, n_2, ..., n_{M_D})$, $n = \sum_{m=1}^{M_D} n_m$, $n_m = |U(CP(k)) \cap U(D = m)|$, $U(CP(k)) = \{u(i)|u^C(i)$ satisfies $CP(k)\}$, $U(D = m) = \{u(i)| u^{D=m}(i)\}$.

Step 3: If $H0$ is rejected then add the trying rule to the rule candidates.

Step 4: Repeat from Step 1 to Step 3 changing the trying rule systematically until the patterns of it are exhausted.

The basic notion of STRIM is that the rule makes a bias in the distribution of decision attribute values $(n_1, n_2, ..., n_{M_D})$. It should be noted that $P(D|CP(k)) = P($ if $CP(k)$ then $D)$ corresponding to (5), and can be estimated by $(n_1, n_2, ..., n_{M_D})/n$ using the sample set.

The second stage is that of arranging the rule candidates having an inclusion relationship by representing them with $CP(k)$ of the maximum bias.

However, the conventional STRIM [8–11] did not have such a reduct function studied in Sect. 4 so that it had to search $CP(k)$s in (7) including even $C(j)$ to be reduced and induced many kinds of rule candidates to burden the second stage. The knowledge from (1) and (2) studied in Sect. 4 can drastically squeeze the

Line No.	Algorithm to induce if-then rules by STRIM with a reduct function
1	int main(void) {
2	int rdct_max[|CV|] = {0, ... ,0}; //initialize maximum value of C(j)
3	int rdct[|CV|] = {0, ..., 0}; //initialize reduct results by D = l
4	int rule[|C|] = {0, ..., 0}; //initialize trying rules
5	int tail = -1; //initial vale set
6	input data; // set decision table
7	for (di = 1; di<= |D|; di++) { // induce rule candidates every D = l
8	attribute_reduct(rdct_max)
9	set rdct[ck] ; // if (rdct_max[ck]==0) {rdct[ck] = 0;} else {rdct[ck] = 1;}
10	rule_check(rcdct, redct_max, tail, rule); // the first stage process
11	}// end of di
12	arrange rule candidates // the second stage
13	}// end of main
14	int attribute_reduct(int rdct_max[]) {
15	make contingency table for D = l vs. C(j)
16	Test H0(j,l);
17	if H0(j,l) is rejexted then set rdct_max[j,l] = jmax else rdct_max[j,l] = 0; // jmax:the attribute vale of the maximum frequency
18	}// end of attribute_reduct
19	int rule_check(int rdct[], int rdct_max[], int tail,int rule[]) { // the first stage process
20	for (ci = tail+1; cj<|C|; ci++) {
21	for (cj = 1; cj <= rdct[ci]; cj++) {
22	rule[ci] = rdct_max[cj]; // a trying rule sets for test
23	count frequency of the trying rule; // count n1, n2, ...
24	if (frequency>= N0) { //sufficient frequency ?
25	if (|z|>3.0) { //sufficient evidence ?
26	add the trying rule as a rule candidate
27	}// end of if |z|
28	rule_check(ci,rule)
29	}// end of if frequency
30	}// end of for cj
31	rule[ci] = 0; // trying rules reset
32	}// end of for ci
33	}// end of rule_check

Fig. 3. An algorithm for STRIM including a reduct function.

search space without idle $C(j)$s and/or its values. Figure 3 shows an algorithm for the improved STRIM described in a C language style including the reduct function "attribute_reduct()" (Line 14–18) studied in Sect. 4. The first stage (Line 7–11) is executed every $D = l$ in a function "rule_check()" (Line 19–33) after operating the reduct (Line 8). The algorithm develops the patterns of trying rules implemented by the dimension "rule[]" (Line 4), for example, for $D = 1$ as $(100000) \rightarrow (110000) \rightarrow (111000) \rightarrow (111100) \rightarrow (110100) \rightarrow (101100) \rightarrow (10010) \rightarrow (010000) ... \rightarrow (001100) \rightarrow (000100)$ by the operation of "rule[ci] = rdct_max[cj]" (Line 22) and the recursive call of "rule_check(ci,rule)" (Line 28) since "rdct_max[] = $[1, 1, 1, 1, 0, 0]$" and "rdct[] = $[1, 1, 1, 1, 0, 0]$" at Line 9 have been obtained (see Table 3). Accordingly, the number of trying rule patterns for $R(d)$ specified in (4) is $(2^4 - 1) \times 6 = 90$. If the function of reduct is not implemented, the number is $(6^6 - 1) \times 6 = 279,930$, which burdens the second stage with a heavy load. From here, the effectiveness of the improved STRIM can be seen.

Table 4. An example of rule candidates.

Trying $CP(k)$	$C(1)C(2)C(3)C(4)C(5)C(6)$	$f = (n_1, n_2, n_3, n_4, n_5, n_6)$	z
1	004400	$(2, 1, 1, 101, 2, 1)$	21.45
2	002200	$(0, 91, 0, 2, 0, 0)$	21.43
...	...	...	...
5	005500	$(1, 0, 1, 2, 78, 2)$	19.07
6	440000	$(2, 0, 0, 76, 2, 1)$	18.70
...	...	...	...
12	001100	$(63, 2, 0, 0, 2, 2)$	16.7
13	000400	$(71, 62, 68, 168, 74, 73)$	9.63
14	004000	$(74, 67, 79, 170, 72, 73)$	9.32
15	303000	$(5, 4, 39, 5, 7, 6)$	9.26
16	400000	$(69, 74, 61, 154, 65, 61)$	8.90
17	404000	$(13, 6, 13, 45, 4, 6)$	8.86
...	...	...	...
20	400400	$(14, 8, 7, 48, 11, 11)$	8.57
...	...	...	...

Table 4 shows examples of the rule candidates arranged in descending order from z value obtained from the distribution of decision attribute values $f = (n_1, n_2, ..., n_{M_D})$ by applying the improved STRIM to the dataset shown in Table 3. We can see the following from the table:

(1) The trying rule $CP(k = 1) = 004400$ makes bias of the distribution of D like $f = (2, 1, 1, 101, 2, 1)$ at $D = 4$ intensively. Accordingly, the candidate is a rule for $D = 4$. The intensity of the bias can be measured by the z value as mentioned in Step 2.

(2) There are inclusion relationships between rule candidates, for example, $U(CP(1) = 004400) \subset U(CP(14) = 004000)$, $U(CP(17) = 404000) \subset U(CP(14) = 004000)$, while the z value of $CP(1) >$ that of $CP(14)$ and the z value of $CP(14) >$ that of $CP(17)$.

The second stage at Line 12 in Fig. 3 arranges their candidates and represents them with only $CP(1)$, which happens to coincide with the specified rule in (4). Table 5 shows the last results induced through the first and second stages with D, f, p-value, accuracy and coverage besides $CP(k)$. Here, accuracy and coverage are defined as follows:

$$\text{accuracy} = \frac{|U(CP(k)) \cap U(D = d)|}{|U(CP(k))|}, \quad \text{coverage} = \frac{|U(CP(k)) \cap U(D = d)|}{|U(D = d)|},$$

and they are often used for showing the indexes of the validity of the induced rules in Rough Sets theory. The improved STRIM induced all of twelve rules

Table 5. Estimated rules for the decision table in Table 3 by improved STRIM.

Trying $CP(k)$	$C(1)C(2)C(3)$ $C(4)C(5)C(6)$	D	$f = (n_1, n_2, n_3,$ $n_4, n_5, n_6)$	p-value(z)	Accuracy	Coverage
1	004400	4	$(2, 1, 1, 101, 2, 1)$	2.09E−102(21.45)	0.94	0.20
2	002200	2	$(0, 91, 0, 2, 0, 0)$	3.37E−102(21.43)	0.98	0.19
3	110000	1	$(91, 3, 2, 1, 10)$	6.57E−92(20.3)	0.93	0.18
4	330000	3	$(1, 1, 89, 3, 0, 0)$	6.33E−91(20.19)	0.95	0.17
5	005500	5	$(1, 0, 1, 2, 78, 2)$	2.15E−81(19.07)	0.93	0.16
6	440000	4	$(2, 0, 0, 76, 2, 1)$	2.92E−78(18.70)	0.94	0.15
7	003300	3	$(1, 4, 77, 0, 0, 1)$	6.70E−77(18.52)	0.93	0.15
8	550000	5	$(1, 1, 2, 0, 75, 3)$	9.15E−77(18.51)	0.91	0.15
9	660000	6	$(0, 3, 1, 3, 0, 76)$	5.32E−76(18.41)	0.91	0.15
10	006600	6	$(3, 3, 3, 3, 0, 73)$	8.69E−67(17.22)	0.86	0.15
11	220000	2	$(0, 60, 1, 2, 0, 2)$	2.46E−63(16.76)	0.92	0.12
12	001100	1	$(63, 2, 0, 0, 2, 2)$	3.63E−63(16.74)	0.93	0.13
13	303000	3	$(5, 4, 39, 5, 7, 6)$	9.56E−21(9.27)	0.59	0.076

Table 6. An arrangement of Car Evaluation dataset of UCI.

Unified attribute value	$C(1)$: buying	$C(2)$: maint	$C(3)$: doors	$C(4)$: person	$C(5)$: lug boot	$C(6)$: safety	D: class (freq.)
1	vhigh	vhigh	2	2	small	low	unacc (1210)
2	high	high	3	4	med	med	acc (383)
3	med	med	4	more	big	high	good (69)
4	low	low	5more	–	–	–	vgood (65)

specified in advance from the decision table with $N = 3,000$, and also one extra rule. However, there are clear differences between them in the indexes of accuracy and coverage.

6 An Example of Application for an Open Dataset

This paper applied SRM for the "Car Evaluation" dataset included in the literature [12]. Table 6 shows the summaries and specifications of the dataset: $|C| = 6$, $|C(1)| = 4,..., |\{D\}| = 4$, $N = |U| = |C(1)| \times, ..., \times |C(6)| = 1,728$ which consists of every combination of condition attributes' values and there were no conflicted or identical samples. The frequencies of D extremely incline toward $D = 1$ as shown in Table 6.

Table 7 shows the results obtained by SRM and suggests the following:

(1) Given $1.0E - 5$ as the critical p-value, $C(3)$ is commonly redundant at $D = l$ ($l = 1, ..., 4$).
(2) With regard to the if-then rule of $D = 1$, $C(5)$ is redundant besides $C(3)$. In the same way, so are $C(2)$ and $C(5)$ at $D = 2$, as well as $C(5)$ at $D = 3$.

Table 7. Results by SRM for Car Evaluation dataset.

		$C(1)$	$C(2)$	$C(3)$	$C(4)$	$C(5)$	$C(6)$
$D=1$	χ^2	22.94	19.24	2.58	111.00	8.81	118.81
	p-value	4.16E−05	2.44E−04	4.62E−01	7.88E−25	1.22E−02	1.59E−26
$D=2$	χ^2	11.77	10.77	3.19	192.56	6.52	194.25
	p-value	8.21E−03	1.30E−02	3.64E−01	1.53E−42	3.85E−02	6.59E−43
$D=3$	χ^2	84.33	84.33	0.39	34.70	0.26	36.26
	p-value	3.61E−18	3.61E−18	9.42E−01	2.92E−08	8.78E−01	1.34E−08
$D=4$	χ^2	70.20	28.60	4.23	33.08	37.69	130.00
	p-value	3.87E−15	2.72E−06	2.38E−01	6.57E−08	6.53E−09	5.90E−29

Table 8. Examples of contingency table and χ^2 test by SRM ((a): $D = 1$ vs. $C(j)$, (b): $D = 4$ vs. $C(j)$ ($j = 1, ..., 6$)).

(a) $D = 1$

$V_{C(i)}$	$C(1)$	$C(2)$	$C(3)$	$C(4)$	$C(5)$	$C(6)$
1	***360***	***360***	326	***576***	450	***576***
2	324	314	300	312	392	357
3	268	268	292	322	368	277
4	258	268	292	–	–	–
χ^2	22.94	19.24	2.58	111.00	8.81	118.81
p-value	4.16E−05	2.44E−04	4.62E−01	7.88E−25	1.22E−02	1.59E−26

(b) $D = 4$

$V_{C(i)}$	$C(1)$	$C(2)$	$C(3)$	$C(4)$	$C(5)$	$C(6)$
1	–	–	10	–	–	–
2	–	13	15	30	25	–
3	26	***26***	20	35	***40***	***65***
4	***39***	***26***	20	–	–	–
χ^2	70.20	28.60	4.23	33.08	37.69	130.00
p-value	3.87E−15	2.72E−06	2.38E−01	6.57E−08	6.53E−09	5.90E−29

Table 8 shows examples of the contingency tables of $D = 1$ (a) and $D = 4$ (b), and their χ^2 by SRM, and suggests the following knowledge:

(1) With regard to the if-then rules of $D = 1$, (a) the frequencies of $C(1) = 1$, $C(2) = 1$, $C(4) = 1$ and $C(6) = 1$ are distinctively high. Accordingly, the if-then rules of $D = 1$ are supposed to be constructed by the combinations of them as shown in the knowledge (2) studied in Sect. 4.

(2) In the same way, the if-then rules of $D = 4$ are constructed by the combinations of $C(1) = 4$, $C(2) = 4$, $C(5) = 3$ and $C(6) = 3$.

Table 9. Estimated rules of $D = 1$ (a) and $D = 4$ (b) for Table 6

(a) $D = 1$

Trying $CP(k)$	$C(1)C(2)C(3)$ $C(4)C(5)C(6)$	D	$f = (n_1, n_2,$ $n_3, n_4)$	p-value(z)	Accuracy	Coverage
1	000100	1	$(576, 0, 0, 0)$	3.51E−56(15.75)	1.00	0.476
2	000001	1	$(576, 0, 0, 0)$	3.58E−56(15.75)	1.00	0.476
3	110000	1	$(108, 0, 0, 0)$	2.52E−12(6.90)	1.00	0.089

(b) $D = 4$

Trying $CP(k)$	$C(1)C(2)C(3)$ $C(4)C(5)C(6)$	D	$f = (n_1, n_2,$ $n_3, n_4)$	p-value(z)	Accuracy	Coverage
1	400003	4	$(52, 33, 20, 39)$	1.08E−50(14.93)	0.271	0.600
2	000033	4	$(88, 64, 0, 40)$	7.93E−37(12.62)	0.208	0.615
3	000303	4	$(49, 96, 12, 35)$	3.84E−37(10.73)	0.182	0.538

Corresponding to Tables 8, 9 shows estimated rules of $D = 1$ (a) and $D = 6$ (b). With regard to rules of $D = 1$, the improved STRIM clearly induces their rules. To express those rules by use of the original notation in Table 6, if person = "2" ∨ safety = "low" ∨ buying = "vhigh" ∧ maint = "vhigh" then class = "unacc" is obtained with accuracy = 1.0 and coverage = $1,008/1,210 \approx 0.833$. In the same way, three examples of the trying rule of $D = 4$ satisfying the testing condition $n\hat{p}(j, k) \geq 5$ (for $D = 4$, $n \geq \frac{5}{0.04} = 125$) are shown in Table 9 (b) although their $n_d = \max(n_1, n_2, ..., n_{M_D})$ is not satisfied at $D = 4$ (in the table $D = 4$ is forcibly entered). The first rule that if buying = "low" ∧ safety = "high" then class = "vgood" is thought to be proper since the p-value is the best and the indexes of accuracy and coverage are moderate among the trying rules for $D = 4$. Both estimated rules for $D = 1$ and 4 coincide with our common sense.

7   Conclusions

The Rough Sets theory has been used for inducing if-then rules from the decision table. The first step in inducing the rules is to find reducts of the condition attributes. This paper retested the conventional reduct methods LEM1 [2] and DMM [3] by a simulation experiment after summarizing the conventional Rough Sets theory and pointed out their problems. Then, this paper proposed a new statistical reduct method (SRM) to overcome the problems of the conventional method from the view of STRIM [8–11]. STRIM including SRM was developed and its validity and usefulness were confirmed by a simulation experiment and application to an open dataset of UCI for machine learning. The improved STRIM should be recognized to be particularly useful for not only reducts of condition attributes but also inducing if-then rules.

References

1. Pawlak, Z.: Rough sets. Int. J. Inf. Comput. Sci. **11**(5), 341–356 (1982)
2. Grzymala-Busse, J.W.: LERS — a system for learning from examples based on rough sets. In: Słowiński, R. (ed.) Intelligent Decision Support — Handbook of Applications and Advances of the Rough Sets Theory. Theory and Decision Library, vol. 11, pp. 3–18. Kluwer Academic Publishers, Amsterdam (1992)
3. Skowron, A., Rauser, C.M.: The discernibility matrix and functions in information systems. In: Słowiński, R. (ed.) Intelligent Decision Support — Handbook of Applications and Advances of the Rough Sets Theory. Theory and Decision Library, vol. 11, pp. 331–362. Kluwer Academic Publishers, Amsterdam (1992)
4. Pawlak, Z.: Rough set fundamentals; KFIS Autumn Coference Tutorial, pp. 1–32 (1996)
5. Ślęzak, D.: Various approaches to reasoning with frequency based decision reducts: a survey. In: Polkowski, L., Tsumoto, S., Lin, T.Y. (eds.) Rough Set Method and Applications, vol. 56, pp. 235–285. Physical-Verlag, Heidelberg (2000)
6. Bao, Y.G., Du, X.Y., Deng, M.G., Ishii, N.: An efficient method for computing all reducts. Trans. Jpn. Soc. Artif. Intell. **19**(3), 166–173 (2004)
7. Jia, X., Shang, L., Zhou, Z., Yao, Y.: Generalized attribute reduct in rough set theory. Knowl.-Based Syst. **91**, 204–218 (2016). Elsevier
8. Matsubayashi, T., Kato, Y., Saeki, T.: A new rule induction method from a decision table using a statistical test. In: Li, T., Nguyen, H.S., Wang, G., Grzymala-Busse, J., Janicki, R., Hassanien, A.E., Yu, H. (eds.) RSKT 2012. LNCS (LNAI), vol. 7414, pp. 81–90. Springer, Heidelberg (2012). doi:10.1007/978-3-642-31900-6_11
9. Kato, Y., Saeki, T., Mizuno, S.: Studies on the necessary data size for rule induction by STRIM. In: Lingras, P., Wolski, M., Cornelis, C., Mitra, S., Wasilewski, P. (eds.) RSKT 2013. LNCS (LNAI), vol. 8171, pp. 213–220. Springer, Heidelberg (2013). doi:10.1007/978-3-642-41299-8_20
10. Kato, Y., Saeki, T., Mizuno, S.: Considerations on rule induction procedures by STRIM and their relationship to VPRS. In: Kryszkiewicz, M., Cornelis, C., Ciucci, D., Medina-Moreno, J., Motoda, H., Raś, Z.W. (eds.) RSEISP 2014. LNCS (LNAI), vol. 8537, pp. 198–208. Springer, Cham (2014). doi:10.1007/978-3-319-08729-0_19
11. Kato, Y., Saeki, T., Mizuno, S.: Proposal of a statistical test rule induction method by use of the decision table. Appl. Soft Compt. **28**, 160–166 (2015). Elsevier
12. Asunction, A., Newman, D.J.: UCI machine learning repository. University of California, School of Information and Computer Science, Irvine (2007). http://www.ics.edu/~mlearn/MlRepository.html
13. Walpole, R.E., Myers, R.H., Myers, S.L., Ye, K.: Probability and Statistics for Engineers and Scientists, 8th edn., pp. 374–377. Pearson Prentice Hall, New Jersey (2007)
14. Walpole, R.E., Myers, R.H., Myers, S.L., Ye, K.: Probability and Statistics for Engineers and Scientists, 8th edn., pp. 361–364. Pearson Prentice Hall, New Jersey (2007)

Post-marketing Drug Safety Evaluation Using Data Mining Based on FAERS

Rui Duan[1], Xinyuan Zhang[2], Jingcheng Du[2], Jing Huang[1],
Cui Tao[2(✉)], and Yong Chen[1(✉)]

[1] University of Pennsylvania, Philadelphia, PA, USA
ychen123@mail.med.upenn.edu
[2] University of Texas Health Science Center at Houston, Houston, TX, USA
Cui.Tao@uth.tmc.edu

Abstract. Healthcare is going through a big data revolution. The amount of data generated by healthcare is expected to increase significantly in the coming years. Therefore, efficient and effective data processing methods are required to transform data into information. In addition, applying statistical analysis can transform the information into useful knowledge. We developed a data mining method that can uncover new knowledge in this enormous field for clinical decision making while generating scientific methods and hypotheses. The proposed pipeline can be generally applied to a variety of data mining tasks in medical informatics. For this study, we applied the proposed pipeline for post-marketing surveillance on drug safety using FAERS, the data warehouse created by FDA. We used 14 kinds of neurology drugs to illustrate our methods. Our result indicated that this approach can successfully reveal insight for further drug safety evaluation.

Keywords: Data mining · Post-marketing surveillance · Zero-truncated negative binomial regression model

1 Introduction

Healthcare is going through a big data revolution [1]: healthcare data can now be collected in scenarios far beyond traditional clinical trials. For example, the US Food and Drug Administration (FDA) has initiated a drug safety surveillance system, known as FDA Adverse Event Reporting System (FAERS), where health providers and patients can report adverse events (AEs) and medication errors post marketing. The FAERS database contains 6,156,081 reports from the past 13 years and offers a great opportunity to conduct research on post-marketing surveillance of drug safety at the population level. However, with opportunities come challenges. For example, the safety data could be gathered from different resources, such as clinical trials or spontaneous reporting systems; processed under different data standards; contained in different formats including free text or images; and reported by people with different

---

R. Duan and X. Zhang—Contribute equally to the paper.

© Springer International Publishing AG 2017
Y. Tan et al. (Eds.): DMBD 2017, LNCS 10387, pp. 379–389, 2017.
DOI: 10.1007/978-3-319-61845-6_38

levels of domain knowledge including physicians, caregivers, and patients. With different levels of granularities, standards, and even formats of the data, it is important to develop methods to normalize and clean the data for further analyses. It is critical to advance our knowledge on the benefits and risks of healthcare interventions by processing and mining drug safety information using large healthcare data (e.g., FAERS) accurately and efficiently for proper statistical analysis [2]. In this paper, we present a data mining and statistical analysis pipeline to study the drug safety using FAERS data. We devised data mining techniques and statistical models to account for the features of FAERS, and provided visualization tools to present the safety results in an intuitive and comparative fashion so that results can be useful in the medical decision-making process. We use 14 first-line drugs for treating acute mania as a use case to demonstrate our pipeline.

FAERS documented the voluntary inputs on adverse events post drug administration. The term AE refers to uncomfortable, or dangerous effects a drug may have [3]. The reports submitted to FAERS are evaluated by the Center for Drug Evaluation and Research (CDER) and the Center for Biologics Evaluation and Research (CBER) to monitor the safety of FDA-approved drugs. FDA uses FAERS to construct post-marketing surveillance programs. If a safety concern is identified in FAERS, further action will be taken.

The growing number of treatment options has generated an increasing need for scientifically rigorous comparisons of multiple treatments to inform healthcare decision-making. In a systematic review for efficacy and safety of antimanic drugs in adults, Cipriani et al. reviewed 67 randomized controlled trials (16,073 participants) published from 1980 to 2010. This study compared the 14 first-line pharmacological drugs at therapeutic doses [4]. Through a network meta-analysis, they quantified the relative effectiveness (measured as the mean change on mania rating scales) of these 14 drugs. In addition, they also compared the safety of these drugs by ranking dropout rates within 3 weeks after initiating treatment. From a pharmaco-epidemiological perspective, the dropout rates within 3 weeks can be an inadequate surrogate for drug safety for the following reasons: first, 3 weeks may be not long enough for the long-term AEs to appear; second and perhaps more importantly, many of the AEs are rare and have different degrees of severity (e.g., suicide due to depression) [5]. This information is difficult to uncover using clinical trials with limited sample sizes.

In this paper, we explore the feasibility and usefulness of the post-marketing safety surveillance database by mining and analyzing the FAERS data. We find that our results on the relative safety of these 14 antimanic drugs based on FAERS data are qualitatively in agreement with previously reported results based on systematic review of randomized controlled trials. However, the analysis based on the FAERS data provides considerably more insights on different types of adverse effects with respect to different system organ classes, death, as well as total number of AEs each patient experienced. These findings can contribute important evidence on relative safety, and can be potentially combined with the pre-marketing findings to better aid medical decision-making.

The proposed analytical pipeline is generally applicable to large healthcare data, although we describe the methods using antimanic drugs in FAERS data for concreteness. More precisely, we describe a methodological framework to extract safety data, pre-process them, and conduct statistical analysis on comparative effectiveness. We hypothesize that the proposed procedure can effectively produce useful insights from large-scale datasets. The proposed pipeline is ready to implement, and results are easy to interpret. It can be used to select drugs that are important for future investigation, and is also applicable to other types of data warehouses.

2 Methodology

2.1 Data Extraction Method

To make the data clean and standardized for further analysis, we followed Banda's work [6] to normalize FAERS data by removing duplicate records and mapping the drug name to RxNorm [7]. For this study, we focused on the FAERS reports from 01/01/2004 to 12/31/2015. There are 6,156,081 unique records in total. Even though we have matched the drug names to the same standard, there is still a mixture of brand names and generic names for the drug name list in FAERS data. We further linked the brand names to the corresponding generic names based on the Drugs.com database [8]. For example, the generic name Lithium has Eskalith, Lithobid, Lithonate, Lithotabs, and Eskalith-CR as its brand names. After data extraction and normalization, the details of the records for the 14 antimnic drugs are as follows Table 1:

Table 1. The number of records extracted from FAERS for 14 antimanic drugs.

Drug name	Number of records
Aripiprazole(ari)	43,126
Asenapine(ase)	5,872
Carbamazepine(cbz)	30,557
Gabapentin(gbt)	102,774
Haloperidol(hal)	18,182
Lamotrigine(lam)	53,957
Lithium(lit)	19,638
Olanzapine(olz)	43033
Quetiapine(qtp)	77,682
Risperidone(ris)	52,758
Topiramate(top)	32,623
Valproate(val)	18,679
Ziprasidone(zip)	13,757
Placebo(pbo)	8,869

2.2 Data-Processing Methods

Since we wanted to use adverse reaction as a variable for analysis, we selected "drug name list" and "reaction-PT list" from the FAERS database. "PT" refers to "preferred term in Medical Dictionary for Regulatory Activities (MedDRA), which serves as a medical terminology used during the pharmaceutical regulatory process [9]. MedDRA also provides a hierarchical structure of the terms where PT is the second lowest level descriptor for AEs. In addition, we included "age" and "sex" information for future analysis. Since there are too many terms on the PT level, we further linked the PT terms to the System Organ Class (SOC) level, which is the highest level of descriptors in MedDRA. There are 26 SOCs in total. These SOCs (details are in Table 2) describe medical symptoms on the organ system level, which gives an overview of how these drugs targeting mental illness can influence physical health [10]. In the final step, we labeled different levels of severities to each report. The levels of severities include death, life-threatening, hospitalization, disability, congenital anomaly, required intervention to prevent permanent impairment/damage, and other serious medical event. After this data processing, we created 14 matrixes for the 14 drugs containing AEs, age, sex, and severity level for each report.

Table 2. Definitions of system organ classes on internationally agreed order [11]

System organ class	Order
Infections and infestations	1
Neoplasms benign, malignant and unspecified (incl cysts and polyps)	2
Blood and lymphatic system disorders	3
Immune system disorders	4
Endocrine disorders	5
Metabolism and nutrition disorders	6
Psychiatric disorders	7
Nervous system disorders	8
Eye disorders	9
Ear and labyrinth disorders	10
Cardiac disorders	11
Vascular disorders	12
Respiratory, thoracic and mediastinal disorders	13
Gastrointestinal disorders	14
Hepatobiliary disorders	15
Skin and subcutaneous tissue disorders	16
Musculoskeletal and connective tissue disorders	17
Renal and urinary disorders	18
Pregnancy, puerperium and perinatal conditions	19
Reproductive system and breast disorders	20
Congenital, familial and genetic disorders	21
General disorders and administration site conditions	22
Investigations	23
Injury, poisoning and procedural complications	24
Surgical and medical procedures	25
Social circumstances	26

2.3 Statistical Method

We proposed a statistical analysis pipeline for comparative effectiveness research that accounts for the unique features of the AE outcomes obtained from the FAERS database. More precisely, the dimension of FAERS datasets was large in terms of both number of AE reports and types of AEs. Specifically, the number of columns (corresponding to the types of AEs) varied from 1,176 to 8,246 for each antimanic drug. The number of patients reported AEs for each drug varied from 5,872 to 77,682. Another feature of the FAERS data was that the data matrices were very sparse since each patient generally only had several AEs reported. We included both exploratory analyses and confirmatory data analyses. The exploratory data analyses included direct comparisons of proportions of classes of AEs among 14 drugs using grouped bar plots, and evaluation of model assumptions such as over-dispersion. The confirmatory data analysis included fitting a modified generalized linear regression model on the total number of AEs reported by each patient.

Exploratory Data Analysis:
To account for the high dimensionality of types of AEs reported for each drug, we collapsed them into 26 categories based on system organ classification [10]. The definitions of the 26 system organ classes were listed in Table 2. The proportion of patients who had at least one report within each system organ class is calculated. These proportions were compared across drugs for each system organ classes. Figure 1 presented grouped bar plots to visualize the differences among drugs. In addition to different types of system organ classes, we also compared the all-cause death proportion for each drug using bar plots, and it was calculated by the number of patients who died dividing the total number of the people who have reported at least one AE. This all-cause death proportion might not be a direct reflection of the drug safety, but it might provide complementary information that is of interest.

Another strategy to account for the high dimensionality of types of AEs was to simply calculate the total number of AEs per subject. We can compare the pattern of the total number of AEs per subject across different drugs. To explore the distribution of the total number of AEs per subject, we calculated the empirical variance and compared it with sample mean for each drug. Such exploration can be used to investigate the distribution assumption of the Poisson model based on the fact that Poisson model assumes the mean and variance of the observed data to be equivalent.

Confirmatory data analysis:
For each patient, we summarized the total number of AEs reported. Since a patient needed to have at least one report to be included in the dataset, the total number of AE for each patient was a positive integer. When comparing the empirical mean and variance of the total number of AE, the variance was substantially larger than the mean. This indicates the over-dispersion of the data, which is likely due to co-occurrences of AEs. To model the total number of AEs, we used a zero truncated negative binomial regression to account for the facts that the data are positive and over-dispersed [12, 13]. The model assumes the total number of AEs for every patient follows a zero-truncated-binomial distribution which has the probability mass function:

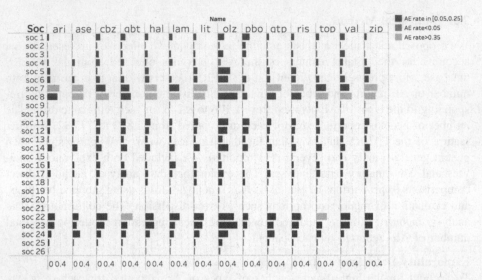

Fig. 1. Proportions of patients having at least one AE in each SOC class for 14 antimanic drugs.

$$\Pr(Y_i = k) = \frac{f(k; \mu_i, \alpha_i)}{1 - f(0; \mu_i, \alpha_i)} \tag{1}$$

where Y_i denotes the total number of AEs for the i-th patient and $f(y; \mu_i, \alpha_i)$ is the probability mass function of the negative binomial distribution with mean μ_i and over-dispersion parameter α_i.

To quantify the rate ratio of AE per subject across different drug user groups, we proposed to use the following generalized linear regression model:

$$\log(\mu_i) = \beta_0 + \beta_1 x_{1,i} + \beta_2 x_{2,i} + , \ldots, \beta_{13} x_{13,i}, \tag{2}$$

where placebo is used as the reference group and $x_{1,i}, x_{2,i}, \ldots, x_{13,i}$ are the dummy variables created for the other 13 antimanic drugs. The variable $x_{j,i}$ takes value 1 if the i-th patient takes the drug j. The coefficient β_j is the log rate ratio of AE comparing drug j to the placebo. The Wald test can be used to test each $\beta_j = 0$, which is to test whether the AE rate of drug j is the same as the placebo. To test any of the two drugs having the same AE rate, the likelihood ratio test can be constructed while the parameters under the null are estimated with the constrain $\beta_j = \beta_{j'}$. The test statistic follows a chi-square distribution with one degree of freedom when the null hypothesis is true. The global likelihood ratio test in this case is to test all the drugs having the same AE rate, which follows a chi-square distribution with 13 degrees of freedom. This regression was fitted using R package "VGAM" and the log rate ratios were estimated with 95% confidence intervals. The drugs were then ranked based on the rate ratios.

3 Results

The analytic results suggested that the post-marketing results of AEs have some agreements with the pre-marketing results of drug acceptability, while also showing differences in some aspects. The premarketing acceptability is measured by the dropout rate in clinical trials, and the causes for dropout may include multiple factors. Therefore, the dropout rate is related to drug safety but not a direct measure of drug safety. Figure 1 displayed the proportions of patients who had at least one report of AE in this class for each of the 14 drugs for each one of the system organ classes. It showed that for most of these antimanic drugs, the most frequent AEs are related to psychiatric disorders (SOC 7) and nervous system disorders (SOC 8), which is consistent with previous findings [14]. Also, Haloperidol, which was showed to be the drug with highest efficacy with middle-ranked acceptability (7th of the 14 drugs) based on the pre-marketing clinical results [4]. Based on the post-marketing data (i.e., FAERS data) and our analysis, the relative rank of the AEs rate for Haloperidol was also lying in the middle of the 14 drugs. More precisely, for all the SOC classes, the relative rank varied from the 3rd in SOC 18 and SOC 23 to the 11th for SOC 25, among 14 drugs. We also found sizable differences between the results from FAERS and the findings from clinical studies. For instance, relative rankings in safety from pre-marketing clinical trials showed that Gabapentin had the highest dropout rate; in our analysis, Gabapentin had relatively low AE rates for SOC 7 and SOC 8 among all the 14 drugs. In addition, Olanzapine was reported as the drug with lowest dropout rate based on premarketing clinical trials, but it was found to have the highest AE rate in SOC 5, 6, and 23 among all drugs.

The patients having mental disorders are subject to certain risk of suicide related death [15]. On the contrary, some severe adverse events of the drugs might also cause death. The proportion of all-cause death calculated from the FAERS dataset for each drug can be a useful measure for drug safety, and is presented in Fig. 2. More precisely, the proportion of all-cause death for each drug was calculated by the number of patients who died divided by the total number of the people who have reported at least one AE. It was showed in Fig. 2 that Haloperidol had the highest death rate (17.26%) while Asenapine had the lowest death rate (4.81%). Under the assumption of no ascertainment bias, which means that the populations under different drug treatments have similar characteristics, the death rate can be considered as a combination of both drug safety and efficacy. However, if the drugs were assigned to the patients selectively based on their conditions (e.g., the patients with relatively severe disorder were more likely to be assigned to Haloperidol), this death rate comparison may not directly reflect drug efficacy or safety.

The results for the zero-truncated negative binomial regression model were shown in Fig. 3. Each black dot in the plot was the estimated ratio of number of AEs per patient for each of the 13 drugs compared to Placebo, which served as the control group. The corresponding line segments showed the 95% confidence intervals of the rate ratio estimates. The p-value for the global test was less than 0.001, suggesting significant difference for at least two of the drugs.

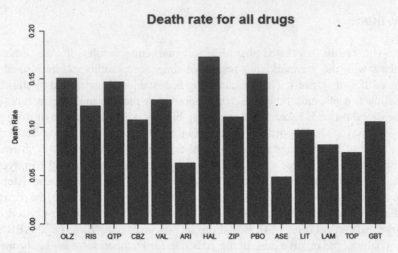

Fig. 2. All-cause death proportions for the 14 antimanic drugs.

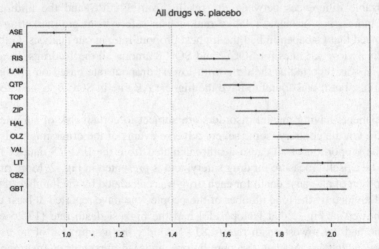

Fig. 3. Estimated rate ratios with 95% confidence intervals for the 13 antimanic drugs compared to placebo.

Among all the 13 drugs, only Asenapine had a ratio smaller than 1, which suggested that Asenapine has smaller expected number of AEs compared to the Placebo. However, since the p-value for testing the difference was 0.152, we cannot claim that Asenapine is significantly better than the Placebo in terms of number of AEs. The rest of the drugs all had rate ratios significantly greater than 1, meaning that these drugs on average would cause more AEs per person compared to Placebo.

In Fig. 3, we also ranked the 14 drugs based on the point estimates of the rate ratio (Placebo has a ratio equals to 1), and the rank was listed on the vertical axes of Fig. 3 from the top to the bottom. Our result indicated that Asenapine was the drug associate with smallest number of AEs per patient, placebo was in the second place, and Gabapentin was the drug associated with the largest number of AEs per person. The pre-marketing results also supported that Gabapentin had the highest dropout rate in clinical trials, and it was possible that too many AEs might be a reason for the high dropout rate [4].

4 Discussion and Conclusion

While the revolution of healthcare data greatly increases the amount of available information, extracting the data and conducting proper analysis become new challenges. With many events reported after the administration of FDA-approved drugs to the FAERS database, the safety of drugs can be evaluated using post-marketing information. In this study, we extracted data from the FAERS database to investigate post-marketing drug safety of 14 antimanic drugs. We prepared datasets according to information needed for further analysis by cleaning and normalizing the original database. To account for the unique features of the FAERS data, we proposed a statistical analysis pipeline for drug safety comparison. To explore the major types of AEs of each drug, we classified the reported events according to the system organ classification. The proportion of patients reporting at least one event for each drug and each SOC class was calculated. These proportions help us better understand the possible AEs of a certain drug, and also allow comparison across all drugs.

A truncated negative binomial regression was fitted to the total number of AEs, and the ratios of the expected event counts compared to the placebo were reported with 95% confidence interval. The 14 drugs were then ranked based on the estimated ratios and compared to the results from pre-marketing clinical trials.

Our study showed that for most of these antimanic drugs, the most frequent AEs are related to psychiatric disorders (SOC 7) and nervous system disorders (SOC 8). The drug that has the highest or lowest proportion for a certain SOC class varies. Olanzapine, which is reported to have the highest acceptability based on pre-marketing clinical trials, is found to have the highest proportion of AEs in SOC 5, 6 and 23. Haloperidol, which is reported to have the highest efficacy, ranks in the middle of the 14 drugs. The model fitting results show that Asenapine is the drug associated with the least number of AEs per person and Gabapentin associated with the most number of AEs per person. The pre-marketing clinical trials usually report the dropout rate as an indicator for the acceptability of the drugs. Sometimes people tend to evaluate drug safety based on the acceptability. However, because the reasons for dropouts are not clearly specified, dropout rates do not completely correlate to drug safety. The findings in our study therefore provide complementary information for drug safety evaluation by analyzing the post-marketing AE data. The important contribution of our study is that we create a data-mining and statistical analysis pipeline to compare multiple drugs in terms of their reported AEs through the FAERS database. Similar works for comparing multiple drugs are based solely on the counts of events for each drug or only one specific AE such as bleeding, while our work is designed for all the AEs combined and specifically for the FAERS data [16, 17].

Due to nature of the self-reported FAERS data, the statistical analysis in our study may have limitations. Unlike data from clinical trials, the FAERS data are reported when patients have AEs after the treatment. Therefore, the causal relationship between the drugs and the reported events is not certain. In addition, the population of patients taking each drug might be different. For example, based on the pre-marketing results, doctors might prescribe the drug according to the health conditions of patients, which may cause ascertainment bias in our results. Also, there is no certainty linking between one AE and one product. For drugs that treat serious mental disorder, death cannot be simply considered as an extreme severe AE since patients are under certain risk of suicide related death. Further, whether an event will be reported or not depends on many factors, making it more difficult to evaluate the drug safety.

In the future, through the methods of data mining and statistical analysis, the FAERS data can be used to gain more insight into drug safety and AEs. For example, if the time interval between taking the drug and each report can be extracted, we may use the patterns of AEs to predict the risk of having certain extremely severe AEs such as death or hospitalization, which will help make decision in clinical practice and monitor severe events. Furthermore, our workflow can be applied to a variety of data mining tasks in medical informatics.

In conclusion, the results of analyzing the post-marketing AEs for the 14 antimatic drugs are largely consistent with the pre-marketing conclusions and can help further understand the pre-marketing results. More importantly, analyzing post-marketing data provides a more comprehensive evaluation of drug safety. Rigorous frameworks that can effectively combine the pre-marketing data (e.g., clinical trials data) with the post-marketing data (e.g., FAERS data) can potential generate new research hypotheses, help better understanding of the association of drug use and AEs, and provide better evidence in aiding medical decision-making.

Acknowledgements. This research was partially supported by the National Library of Medicine of the National Institutes of Health under Award Number R01LM011829, the National Institute of Allergy and Infectious Diseases of the National Institutes of Health under Award Number R01AI130460, and the UTHealth Innovation for Cancer Prevention Research Training Program Pre-doctoral Fellowship (Cancer Prevention and Research Institute of Texas grant # RP160015).

References

1. Nambiar, R., Bhardwaj, R., Sethi, A., et al.: A look at challenges and opportunities of big data analytics in healthcare. In: 2013 IEEE International Conference on Big Data, pp. 17–22 (2013)
2. Hazell, L., Shakir, S.A.W.: Under-reporting of adverse drug reactions. Drug Saf. **29**(5), 385–396 (2006)
3. Food and Drug Administration (FDA), Questions and Answers on FDA's Adverse Event Reporting System (FAERS). http://www.fda.gov/Drugs/GuidanceComplianceRegulatory Information/Surveillance/AdverseDrugEffects/

4. Cipriani, A., Barbui, C., Salanti, G., et al.: Comparative effectiveness and acceptability of antimanic drugs in acute mania: a multiple-treatments meta-analysis. Lancet **378**, 1306–1315 (2011)
5. Brent, D.A., Emslie, G.J., Clarke, G.N., et al.: Predictors of spontaneous and systematically assessed suicidal adverse events in the treatment of SSRI-resistant depression in adolescents (TORDIA) study. Am. J. Psychiatry **166**(4), 418–426 (2009)
6. Banda, J.M., Evans, L., Vanguri, R.S., Tatonetti, N.P., Ryan, P.B., Shah, N.H.: A curated and standardized adverse drug event resource to accelerate drug safety research. Sci Data **3**, 160026 (2016)
7. Xu, J., Wu, Y., Zhang, Y., Wang, J., Lee, H.J., Xu, H.: CD-REST: a system for extracting chemical-induced disease relation in literature, Database (Oxford) (2016)
8. Drugs.com. https://www.drugs.com/
9. Medical Dictionary for Regulatory Activities. http://www.meddra.org/how-to-use/support-documentation
10. Du, J., Cai, Y., Chen, Y., et al.: Trivalent influenza vaccine adverse symptoms analysis based on MedDRA terminology using VAERS data in 2011. J. Biomed. Semant. **7**(1), 13 (2016)
11. Medical Dictionary for Regulatory Activities. http://www.who.int/medical_devices/innovation/MedDRAintroguide_version14_0_March2011.pdf
12. Cameron, A.Colin, Trivedi, P.K.: Regression Analysis of Count Data. Cambridge University Press, Cambridge (1998)
13. Hilbe, J.M.: Negative Binomial Regression. Cambridge University Press, Cambridge (2007)
14. Tomé, A.M., Filipe, A.: Quinolones. Drug Saf. **34**(6), 465–488 (2011)
15. Kelleher, I., Ramsay, H., DeVylder, J.: Psychotic experiences and suicide attempt risk in common mental disorders and borderline personality disorder. Acta Psychiatr. Scand. **135**(3), 212–218 (2017)
16. Hoffman, K.B., Dimbil, M., Erdman, C.B., Tatonetti, N.P., Overstreet, B.M.: The Weber effect and the United States Food and Drug Administration's Adverse Event Reporting System (FAERS): analysis of sixty-two drugs Approved from 2006 to 2010. Drug Saf. **37**(4), 283–294 (2014)
17. Southworth, M.R., Reichman, M.E., Unger, E.F.: Dabigatran and postmarketing reports of bleeding. New England J. Med. **368**(14), 1272–1274 (2013)

Crowd Density Estimation from Few Radio-Frequency Tracking Devices: I. A Modelling Framework

Yenming J. Chen[1(✉)] and Albert Jing-Fuh Yang[2]

[1] Professor of Logistics Management,
National Kaohsiung 1st University of Science and Technology,
Kaohsiung, Taiwan
yjjchen@nkfust.edu.tw
[2] Marketing and Distribution Management,
National Kaohsiung 1st University of Science and Technology,
Kaohsiung, Taiwan
jfyang@nkfust.edu.tw

Abstract. This study proposes a modeling framework based on few radio frequency (RF) tracking devices, such as smartphones or beacon. The proposed framework aims to estimate crowd density continuously where vision analysis is unreliable and the relation between pedestrian speed and density can be at least specified. The crowd density estimated through the modelling framework can not only be used for evacuation commanding at emergency times, but also can be used for commercial usage at a normal time and building/facility layout improvement during design time. In the proposed framework, the application level maps input data spaces into feature spaces. The model level applies multiple data models to increase the accuracy of the estimated states. Moreover, the abstract level fuses the heterogeneous parameters estimated from the model level. The models we included in the framework are cellular automata models, ferromagnetic models, social force models, and complexity models. The model parameters are estimated by Markov Chain Monte Carlo (MCMC) and particle swarm optimization (PSO) methods. The fusion algorithm factory instantiates a data assimilation approach and a continuous receiver operating characteristic (ROC) estimator.

Keywords: Pedestrian density distribution · RF tracking device · Parsimony of sensor · Markov Chain Monte Carlo · Data assimilation · Particle swarm optimization

1 Introduction

Crowd or pedestrian density represents important information for assessing crowd situations in certain critical environments. Computerized crowd density information can be estimated through surveillance cameras. However, visual

© Springer International Publishing AG 2017
Y. Tan et al. (Eds.): DMBD 2017, LNCS 10387, pp. 390–398, 2017.
DOI: 10.1007/978-3-319-61845-6_39

devices frequently suffer from problems on scalability and reliability. Fortunately, mobile devices, such as smartphones or beacons, have become increasingly viable as alternative estimation agents to complement the drawbacks of visual devices [1]. The major problem for such media is the need for people's voluntary willingness to be tracked. Another problem is that the number of racketeers is far less than the total number of people in a crowd [11]. These problems are challenging in many aspects. This study is the first attempt to propose a modelling framework that continuously estimates the crowd density distribution on the basis of only few RF tracking devices.

Our principle of sensor parsimony is useful for many pervasive applications. Visual approaches suffer from various limitations in spite of the maturity of image processing technology in crowd distribution analysis. Video devices are normally used on a small scale because digital cameras and dedicated supporting infrastructure are expensive and requiring massive deployment to cover a surveillance area. Reliability could also be a problem for visual-based devices in certain critical situation. Surveillance cameras may not function properly in emergency situations, such as escaping from fire, evacuating from earth quake, human safety monitoring, traffic control, and smart guiding in public. For example, the light condition may be insufficient to capture a clear image. The massive data transmission between cameras and servers may also overload the limited communication capacity at emergency time. The recorded image is also unavailable for calculating crowd density for non-emergency usage, because consumers' personal information is protected by law. Voluntary tracking could be one of the few available choices under privacy law restriction. An alternative approach to acquiring crowd information is through the ever-popular radio-frequency (RF) devices. Smartphones have almost become a must-have for all individuals. The RF devices could be reliable and scalable tracking device if the position information of every pedestrian in target space could be obtained [3].

The major challenge in applying RF tracking devices is on how to pursue individuals to reveal their private information. Wireless users only share their private information on a voluntary basis. Only a small fraction of pedestrians willingly reveal their location information. Estimate that only under a hundred devices registering their locations in a particular space among thousands of people is achievable. We tackle the challenging problem by superimposing multiple dynamical models. If people in a space do not move, most non-tracked individual's locations are difficult to know from few tracked devices and, in such situation, may need to resort to other auxiliary methods, such as acoustic Doppler waves, if necessary. However, the more they move the more information we have and the density estimation becomes more accurate [5–7]. The two upper pictures in Fig. 1, show scant possibilities in obtaining density information through few persons with RF devices in a highly sparse or crowded situation. However, in the two lower pictures in Fig. 1, the persons wearing RF tracking devices are represented by circles and the relative movement can help to estimate the crowd density.

The model we first included in the framework are cellular automata (CA) models, ferromagnetic models, social force models, and complexity models. We

Fig. 1. Relations among speed, density, and tracking devices. Individual tracking devices are shown in circles and the number of devices is far less than the total number of pedestrians.

apply Markov Chain Monte Carlo (MCMC) and particle swarm optimization (PSO) for state parameter estimation. The parameters estimated from multiple models are fused by data assimilation (DA) and a continuous ROC estimator, which robustly relaxes the maximum a posteriori alue in the optimization algorithm. This study proposes a modelling framework that can connect real-world situations to mathematical models. The scenario simulation provided by the framework can also increase the specificity of modelling process.

2   Modelling Framework

To address the complicated modelling problem for the real-world pedestrian behavior, we provide a new and efficient means and fill the gap between real-world and theoretical problems.

2.1   General Concept

We propose a three-level framework that performs computations on an abstract-model level over a unified space, collects estimations on a model level over an algorithmic space, and converts real-world signals to an application levels over a feature space.

Applications at the top level manifest a property of variety in appearance, which make the underlying algorithms non-reusable across different kinds of real-world situations. The framework exploits a conversion that maps versatile

.application domains to a standard feature space using suitable kernel functions. The process of feature extraction pertains to a mapping $\Phi : \mathbb{R}^N \mapsto \mathcal{F}$ from the space of input data $\mathbb{R}^N$ to the space of feature $\mathcal{F}$. For example, a typical sound signal contains ten thousands of samples in a second, and directly performing an algorithm on this high-dimension is difficult. We use Fourier kernels to decompose this signal from a time domain into a frequency domain. We then assign a set of significant frequency components to a vector in a low-dimension feature space. In the application of pedestrian flow, some algorithms directly take pedestrians' sampled positions and velocities as input while some algorithms need to transform the detailed information into an aggregate quantity, such as average speed or oscillation frequency. Functions in this level mainly provide common input data in case the algorithms entail the same information.

2.2   Framework in Model Level

We attempt to extend the estimation by superimposing or fusing as many known relations as possible to tackle the ill-posed problem in crowd density estimation from limited observations [4]. Figure 2 shows our attempt to incorporate multiple relations in this model level. Each model embodies the relationship of internal states to observations. To isolate changing applications, we do not specify the states and observations. We may later assign them to density distribution $\rho(x, y, t)$ and velocity field $\mathbf{v}(x, y, t)$ for coordinates (x, y) at time t.

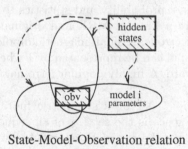

State-Model-Observation relation

Fig. 2. Models superpositioning in our fused assimilation framework

Given a model with known states and observations, the model parameters still possess a large degree of freedom. Each set of model parameters forms a subspace. Figure 2 shows that smaller intersection of such subspaces result in higher likelihood that a unique parameters set could be determined. We apply MCMC and PSO to determine the unknown parameters. The states can be estimated accurately if another set of observations can be acquired through alternate means (Fig. 3). For example, pedestrian velocity can be obtained through RF devices, surveillance cameras, or Doppler microwaves. The probability of obtaining a unique solution thus increases.

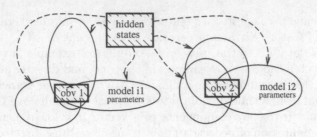

Fig. 3. Fusion through distinct observations

Various crowd models can be instantiated in this level. Several main assumptions distinguish the models [9,12]. The major difference among crowd models is at the modeling scale. In the microscopic scale, approaches involve CA, agent-based method (ABM), and Markov random field (MRF). In the macroscopic scale, fundamental diagram, social force, fluid dynamics, and game theory exist. Some models assume homogeneous behaviors among pedestrians, whereas others allow heterogeneous behaviors. Models can also build on discrete or continuous space and time. In terms of psychological response, normal or emergency situations account for another important modelling attribute. Models of pedestrian dynamics can build on normal and emergency situations, and their moving patterns can flow through various topological spaces.

Each pedestrian in MRF models is considered as a cell on a grid. These models mainly use means of probability and statistics to study crowd behavior. Ferromagnetic models are a kind of MRF and originated from the quantum mechanical spinning of electrons. A small magnetic dipole moment is associated with the spin. Thus, the spin can be represented by 1 when pointing upward and -1 when pointing downward. A highly popular ferromagnetic model is the Ising model.

Let $\Lambda = \{j \in \mathbb{Z} : |j| \leq N\}$ be a symmetric finite hypercube on integer set $\mathbb{Z}$. The configuration space is the set Ω_Λ of all sequence $\omega = \{\omega_j\}_{j \in \Lambda}$, i.e., $\Omega_\Lambda = \{-1,1\}^\Lambda$. We denote the set of Borel σ-field by $\mathfrak{B}(\Omega_\Lambda)$. Let ρ be the measure $\frac{1}{2}\delta_{-1} + \frac{1}{2}\delta_1$ and $\pi_\Lambda \mu_\rho$ be the product measure on $\mathfrak{B}(\Omega_\Lambda)$ with identical one-dimensional marginal ρ. The ρ satisfies $\pi_\Lambda \mu_\rho\{F\} = \mu_\rho\{\pi_\Lambda^{-1} F\}$ for all $F \in \mathfrak{B}(\Omega_\Lambda)$. For each $\omega \in \Omega_\Lambda$, $\pi_\Lambda \mu_\rho\{\omega\} = 2^{-|\Lambda|}$, where $|\Lambda| = 2N + 1$ for this $\mathbb{Z}$ case.

The coordinate mappings on Ω_Λ, defined by $Y_j(\omega) = \omega_j$, are called the spin random variables at the site j. The *Hamiltonian* or *interaction energy* of a spin configuration $\omega \in \Omega_\Lambda$ is defined as $H_{\Lambda,h}(\omega) = -\frac{1}{2}\sum_{i,j \in \Lambda} \Phi(\{i,j\})\omega_i \omega_j - h\sum_{j \in \Lambda} \omega_j$. We call Φ a *ferromagnetic interaction potential* and assume that Φ is a non-negative function on $\mathbb{Z}$ which is symmetric and translation invariant. The corresponding *potential* $V(\cdot) = \sum_\Lambda \Phi(\cdot)$. The parameter h is a real number that gives the strength of an external magnetic field that acts at each site in Λ.

Let $\beta = 1/T > 0$ be the inverse absolute temperature. The ferromagnetic model is defined by the probability measure $\mu_{\Lambda,\beta,h}$ on $\mathfrak{B}(\Omega_\Lambda)$ as follows:

$$\mu_{\Lambda,\beta,h}\{\omega\} = \exp\left[-\beta H_{\Lambda,h}(\omega)\right]\pi_\Lambda\mu_\rho\{\omega\} \cdot \frac{1}{Z(\Lambda,\beta,h)}, \tag{1}$$

where the partition function is defined as $Z(\Lambda,\beta,h) = \int_{\Omega_\Lambda} \exp[-\beta H_{\Lambda,h}(\omega)]\pi_\Lambda\mu_\rho$ $d\omega = \sum_{\omega\in\Omega_\Lambda} \exp[-\beta H_{\Lambda,h}(\omega)]\frac{1}{2^{|\Lambda|}}$. The measure $\mu_{\Lambda,\beta,h}$ is called a *finite-volume Gibbs state* on Λ.

A probability measure μ is defined on Ω_Λ with strictly positive values for finite cylinder sets. The conditional probabilities of the form $\mu[\omega(x) = 1|\,\omega(\cdot)$ on $\Lambda\backslash x]$ depend only on the values of ω at the neighbors of x and are invariant under graph isomorphism. The set of all MRF's is denoted by $\mathcal{G}$. An MRF is an infinite Gibbs state with a homogeneous nearest neighbor pair potential Φ and *vice versa* [8]. Let $0 < p < 1$, and $0 < q < 1$. We consider the matrix $M = \begin{bmatrix} p & 1-p \\ 1-q & q \end{bmatrix}$ as the transition matrix of the Markov chain with two states. Let $\pi = \{\pi(-1), \pi(1)\}$ be the unique stationary distribution (i.e., $\pi M = \pi$).

In order to predict the crowd density inside a topology Λ, i.e., a corridor, we need to determine two parameters from the Ising ferromagnetic model: temperature T and external intensity h. Assume that the time dependent temperature distribution over the corridor is given in the sense of average. The crowd density can be modeled as a stochastic process, where each random variable depends on its vicinity. The probability measure of crowd density is obtained at a certain point and time if a finite number of equilibrium states exist. We can then evaluate the density distribution and provide an index for the cost of minimum cost algorithm Based on the temperature information.

We need the conditional probability of each spot for the calculation of the nearest neighbor pair potential. Obtaining the pedestrian count with infinitely high resolution is impossible in the real world application. We can only estimate the nearest neighbor pair potential from the boundary condition, i.e., RF tracking devices or smartphone apps installed on few volunteers.

The macroscopic crowd analysis can roughly start from a relationship between density and speed. We refer to density as an overall average scalar and speed as a non-directional quantity in terms of the entire flock of crowd to achieve a loose and simple analysis. We will also discuss models that facilitate vector velocities and density distribution for the x-y positions and time course.

Fundamental diagram basically characterizes a speed-density relation [2]. Pedestrian speed decreases as the density increases When crowd density falls within a certain range. The relation is achieved because neighboring pedestrians may influence each other on the moving velocity [7]. Free motion should be maintained without interference when the density is sufficiently low at a normal situation. The relation is as follows:

$$v(\rho) = \rho_0\left\{1 - \exp\left[-\gamma\left(\frac{1}{\rho} - \frac{1}{\rho_{max}}\right)\right]\right\}, \tag{2}$$

where $v_0 = 1.34\,\text{m/s}$ is about the free walking speed at flow density, ρ_{max} is the maximal crowd density that completely blocks human movement, and γ is the scaling constant. The relation is also suitable to fit into empirical data because of the simplicity of relation.

Crowd density is an influencing factor of pedestrian speed and can be uniquely derived from speed only under certain special conditions. In general case, crowd flocking patterns, obstacles, walking conditions, as well as demographics and cultural aspects significantly affect the fundamental diagram.

The social force model is a macro-level model in which pedestrians follow a rule of social force to perform the interaction. The model mimics the force field in the physical world. The model also assumes that the acceleration of pedestrian depends on the superimposition of personal (pers), social (soc), and physical (phys) force fields. That is, $\mathbf{F}^{(pers)}$, $\mathbf{F}^{(soc)}$, and $\mathbf{F}^{(phys)}$ [10,13]. For a pedestrian of mass m_j with velocity $\mathbf{v}_j$, the basic equation of motion is given as

$$\frac{d\mathbf{v}_j}{dt} = \mathbf{f}_j^{(pers)} + \mathbf{f}_j^{(soc)} + \mathbf{f}_j^{(phys)}, \tag{3}$$

where $\mathbf{f}_j^{(\cdot)} = \frac{1}{m_j}\mathbf{F}_j^{(\cdot)}$ is the specific force. The force pertaining to social interaction is defined as $\mathbf{f}_j^{(soc)} = \sum_{l \neq j} \mathbf{f}_{jl}^{(\cdot)}$, which represents the specific forces due to other pedestrians. There is another force $\mathbf{f}_j^{(pers)}$ keeping pedestrian j moving on its own preferred velocity $\mathbf{v}_j(0)$ and is defined as $\mathbf{f}_j^{(pers)} = \frac{\mathbf{v}_j(0) - \mathbf{v}_j}{\tau_j}$ for an acceleration time τ_j.

The social force analogizes the territorial effect on the private sphere. In psychology, people feel uncomfortable when strangers enter their private spheres. Thus, a repulsive force emerges to separate people. An exponential form is assumed for easy quantification of the forces. The force between person j and l is given as

$$\mathbf{f}_{jl}^{(soc)} = A_j \exp\left[\frac{R_{jl} - \Delta r_{jl}}{\xi_j}\right] \mathbf{n}_{jl}, \tag{4}$$

where pedestrians possess disks of radius R_{jl} and distance r_{jl}. The normal vector $\mathbf{n}_{jl}$ is a vector that points from j to l, representing the direction of $\mathbf{f}_{jl}^{(soc)}$. The quantity A_j is a scaling factor, and ξ_j is the range of the interactions.

3   Conclusions

On the basis of few RF tracking devices, this study proposes a modelling framework to continuously estimate crowd density distributions, provided that vision analysis is unavailable and the speed-density relation specified by the fundamental diagram is at least maintained. The designated events to drive the flock movement include tour walking (shopping or museum), spontaneous walking (subway or bus station), and ordered/unordered evacuation (drill/fire). The crowd flow forming patterns included in the modelling framework were jamming on the bottleneck, congestion, or stairs, shock wave, oscillation, lane formation, intersections, competition, and various irrational phenomena (panic, herding, and stampede).

The applications of this framework are suitable for the situations where the principle of sensor parsimony is prominent. The estimated crowd density obtained by the modelling framework can be used for commercial usage at normal times, for evacuation commanding at emergency times, and for building/facility layout improvement during design times.

This study proposes a modelling framework that can connect real-world situations to mathematical models. The scenario simulation provided by the framework can also make the modelling process precise. On our next paper, detailed parameter calibration algorithms will be presented. Pedestrian behaviors will be investigated empirically on the next stage.

References

1. Blanke, U., Troster, G., Franke, T., Lukowicz, P.: Capturing crowd dynamics at large scale events using participatory GPS-localization. In: 2014 IEEE Ninth International Conference on Intelligent Sensors, Sensor Networks and Information Processing (ISSNIP), pp. 1–7. IEEE (2014)
2. Helbing, D., Johansson, A., Al-Abideen, H.Z.: Dynamics of crowd disasters: an empirical study. Phys. Rev. E **75**(4), 046109 (2007)
3. Hui, S.K., Fader, P.S., Bradlow, E.T.: Path data in marketing: an integrative framework and prospectus for model building. Market. Sci. **28**(2), 320–335 (2009)
4. Kjærgaard, M.B., Wirz, M., Roggen, D., Tröster, G.: Detecting pedestrian flocks by fusion of multi-modal sensors in mobile phones. In: Proceedings of the 2012 ACM Conference on Ubiquitous Computing, pp. 240–249. ACM (2012)
5. Liu, S., Yang, L., Fang, T., Li, J.: Evacuation from a classroom considering the occupant density around exits. Phys. A Stat. Mech. Appl. **388**(9), 1921–1928 (2009)
6. Löhner, R.: On the modeling of pedestrian motion. Appl. Math. Model. **34**(2), 366–382 (2010)
7. Lv, W., Song, W., Ma, J., Fang, Z.: A two-dimensional optimal velocity model for unidirectional pedestrian flow based on pedestrian's visual hindrance field. IEEE Trans. Intell. Transp. Syst. **14**(4), 1753–1763 (2013)
8. Preston, C.: Gibbs States on Countable Sets. Cambridge University Press, London (1974)
9. Schadschneider, A., Klingsch, W., Klüpfel, H., Kretz, T., Rogsch, C., Seyfried, A.: Evacuation dynamics: empirical results, modeling and applications. In: Meyers, R.A. (ed.) Encyclopedia of Complexity and Systems Science, pp. 3142–3176. Springer, New York (2009)
10. Seyfried, A., Steffen, B., Lippert, T.: Basics of modelling the pedestrian flow. Phys. A Stat. Mech. Appl. **368**(1), 232–238 (2006)
11. Wirz, M., Roggen, D., Troster, G.: User acceptance study of a mobile system for assistance during emergency situations at large-scale events. In: 2010 3rd International Conference on Human-Centric Computing (HumanCom), pp. 1–6. IEEE (2010)

12. Zheng, X., Zhong, T., Liu, M.: Modeling crowd evacuation of a building based on seven methodological approaches. Build. Environ. **44**(3), 437–445 (2009)
13. Zhong, J., Hu, N., Cai, W., Lees, M., Luo, L.: Density-based evolutionary framework for crowd model calibration. J. Comput. Sci. **6**, 11–22 (2015)

Text Mining

Mining Textual Reviews with Hierarchical Latent Tree Analysis

Leonard K.M. Poon[1]([⊠]), Chun Fai Leung[2], and Nevin L. Zhang[2]

[1] Department of Mathematics and Information Technology,
The Education University of Hong Kong, Hong Kong SAR, China
kmpoon@eduhk.hk
[2] Department of Computer Science and Engineering, The Hong Kong University
of Science and Technology, Hong Kong SAR, China
cfleungac@connect.ust.hk, lzhang@cse.ust.hk

Abstract. Collecting feedback from customers is an important task of any business if they hope to retain customers and improve their quality of service. Nowadays, customers can enter reviews on many websites. The vast number of textual reviews make it difficult for customers or businesses to read directly. To analyze text data, topic modeling methods are usually used. In this paper, we propose to analyze textual reviews using a recently developed topic modeling method called hierarchical latent tree analysis, which has been shown to produce topic hierarchy better than some state-of-the-art topic modeling methods. We test the method using textual reviews written about restaurants on the Yelp website. We show that the topic hierarchy reveals useful insights about the reviews. We further show how to find interesting topics specific to locations.

Keywords: Review text mining · Hierarchical latent tree analysis · Topic modeling · Yelp Dataset Challenge · Latent tree models

1 Introduction

Collecting feedback from customers is an important task of any business if it hopes to retain customers and improve their quality of service. Nowadays, customers of online services and customers of goods and services can enter reviews on many websites, e.g. Amazon, Yelp, and TripAdvisor. Such reviews provide abundant feedback data for machine learning techniques to learn about the user and product characteristics and to improve business opportunities.

Two forms of data are contained in the feedback. The first form of data, ratings, are easily accessible and has widely been used in recommender systems research. The second form of data, textual reviews, are unstructured and harder to analyze. They can however provide important information unavailable from ratings. For example, the reviews can explain users' preference. By reading the reviews, customers can know what are the important aspects for consideration and focus on the aspects important to them. Businesses can also decide how to

© Springer International Publishing AG 2017
Y. Tan et al. (Eds.): DMBD 2017, LNCS 10387, pp. 401–408, 2017.
DOI: 10.1007/978-3-319-61845-6_40

improve their products by learning users' perception on different aspects of their products or some similar ones. We focus on textual reviews in this paper.

The vast number of textual reviews make it difficult for customers or businesses to read directly. To analyze text data, topic modeling methods are usually used. These methods aim to detect topics in a collection of documents and classify the documents into those detected topics. *Latent Dirichlet allocation* (LDA) has been a popular method for topic modeling [1,3]. Its basic version produces a single level of topics. LDA becomes slow when the number of topic is set to be high. Therefore, LDA is not very suitable for finding fine-grained topics. Some recent methods have extended LDA to produce topic hierarchies, but they have seldom been used to analyze textual reviews.

Recently, *hierarchical latent tree analysis* (HLTA) has been proposed for hierarchical topic detection [4,8]. It uses tree-structured probabilistic models called *latent tree models* (LTMs) for topic modeling. The resulting LTM has a deep model structure and contains multiple levels of latent variables, with each representing a topic. The model yields a topic hierarchy that contains general topics at higher levels and specific topics at lower levels. Chen et al. [4] show that HLTA can produce topic hierarchy of higher quality than two other state-of-the-art LDA-based methods, namely nested Chinese restaurant process [2] and nested hierarchical Dirichlet processes (nHDP) [9]. HLTA has also been shown to have a running time comparable to nHDP on a large data set. Although HLTA has been shown effective for topic detection for academic papers and news articles [4,8], it has not been used to analyze textual reviews.

In this paper, we propose to use HLTA to analyze textual reviews. We test the method using textual reviews written about restaurants on the Yelp website. We show that the topic hierarchy reveals useful insights about the reviews. We further show how to find interesting topics specific to locations.

2   Topic Modeling

Consider a collection $\mathcal{D} = \{d_1, \ldots, d_N\}$ of N documents. In topic modeling, the documents are often represented as *bags of words*. Specifically, suppose M words are included in the vocabulary $\mathcal{V} = \{w_1, \ldots, w_M\}$. Each document d can be represented as a vector $d = (c_1, \ldots, c_M)$, where c_i represents the count of word w_i occurring in the document d. The aim of topic modeling is to detect a number K of topics $z_1, \ldots, z_K$ among the documents $\mathcal{D}$, where K can be given or learned from data. The topic model defines a distribution over words for each topic. A topic is often characterized by representative words based on the distribution.

Latent Dirichlet Allocation (LDA) is a popular method for topic modeling [1, 3]. LDA assumes each document d to belong to each topic z_k of the K topics according to a distribution $P(Z = z_k|d)$ over a topic variable Z. In other words, $\sum_{k=1}^{K} P(z_k|d) = 1$. This kind of model is known as *mixed membership model*. For each topic z_k, LDA defines a conditional distribution $P(w_i|z_k)$ over words w_i. A topic z_k can then be characterized by the most probable words according to $P(w_i|z_k)$.

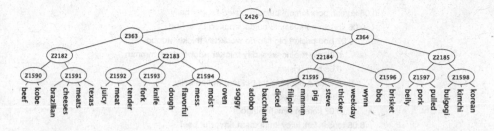

Fig. 1. Subtree of the latent tree model obtained from textual reviews about restaurants on Yelp. The oval nodes represent topic variables whereas the text nodes without borders represent word variables. All variables have two states.

3   Hierarchical Latent Tree Analysis

A *latent tree model* (LTM) [5, 13] is a tree-structured probabilistic graphical model [7]. Figure 1 shows an example of LTM. When an LTM is used for topic modeling, the leaf nodes represent the observed word variables W, whereas the internal nodes represent the unobserved topic variables Z. All variables are binary. Each word variable $W_i \in W$ indicates the presence or absence of the word $w_i \in \mathcal{V}$ in a document. Each topic variable $Z_i \in Z$ indicates whether a document belongs to the i-th topic.

For technical convenience, an LTM is often rooted at one of its latent nodes and is regarded as a Bayesian network [10]. Then all the edges are directed away from the root. The numerical information of the model includes a marginal distribution for the root and one conditional distribution for each edge. For example, edge $Z1590 \to$ beef in Fig. 1 is associated with probability $P(\text{beef}|Z1590)$. The conditional distribution associated with each edge characterizes the probabilistic dependence between the two nodes that the edge connects. The product of all those distributions defines a joint distribution over all the latent variables Z and observed variables W. Denote the parent of a variable X as $pa(X)$ and let $pa(X)$ be an empty set when X is the root. Then the LTM defines a joint distribution over all observed and latent variables as follows:

$$P(W, Z) = \prod_{X \in W \cup Z} P(X|pa(X)). \tag{1}$$

Given a document d, the values of the binary word variables W are observed. We use $d = (w_1, \ldots, w_M)$ to denote also those observed values. Whether a document d belongs to a topic $Z \in Z$ can be determined by the probability $P(Z|d)$. The LTM gives a *multi-membership model* since a document can belong to multiple topics. Unlike in LDA, the topic probabilities $P(Z|d)$ in LTM do not necessarily sum to one.

For topic modeling, an LTM has to be learned from the document data $\mathcal{D}$. This requires learning the number of topic variables, the connection between the variables, and the probabilities in the model. We use the method PEM-HLTA [4] to build LTMs for topic modeling. The method builds LTMs level by

0.06 korean pork kimchi bulgogi pulled tender belly
0.06 korean pork kimchi bulgogi pulled belly bbq
0.07 bbq brisket pig filipino weekday thicker adobo
0.08 pig filipino weekday thicker adobo steve mmmm
0.03 bbq brisket
0.06 korean pork kimchi bulgogi pulled belly
0.07 pork pulled belly
0.07 korean kimchi bulgogi
0.05 tender fork juicy knife meat flavorful beef
0.09 beef meats cheeses kobe brazilian texas
0.04 beef kobe
0.03 meats cheeses brazilian texas
0.05 tender fork juicy meat knife flavorful moist
0.05 fork knife
0.05 tender juicy meat
0.07 flavorful moist oven soggy dough mess

Fig. 2. Topic hierarchy extracted from the subtree shown in Fig. 1. Each topic is characterized by at most seven descendent words most relevant to the topic. The lowest level contains topics about different kinds of food, including Philippine Adobo, BBQ brisket, pork belly, Korean food (kimchi and bulgogi), Kobe beef, etc. The number before a topic indicates the probability of the topic in the data.

level and is thus also known as *hierarchical latent tree analysis* (HLTA). Intuitively, the co-occurrence of words are captured by the level-1 latent variables, whose co-occurring patterns are captured by latent variables in higher levels. In the extracted topic hierarchy, the topics on top are more general and the topics at the bottom are more specific.

To extract the topic hierarchy from an LTM, we follow the tree structure in the model and use each internal node to represent a topic. A topic is characterized by the words most relevant to them. Specifically, we compute the mutual information (MI) [6] between a topic variable and each of its descendent word variable. Then we pick at most seven descendent words with the highest MI to characterize the topic. As an example, Fig. 2 shows the topic hierarchy extracted from the subtree shown in Fig. 1.

4   Mining Textual Reviews on Yelp

Yelp[1] is a website that allows users to write reviews about local businesses. Some of the review data are made available through the Yelp Dataset Challenge. In our experiment, we used the 2015 version of Yelp data. The data set contains 1.57 million reviews for businesses located in 10 cities in the US, Canada, UK, and Germany. The data set contains different categories of businesses. We included the largest category of businesses by keeping only those categories that carry any of these words: "food", "pubs", "bar", "cafe", or "restaurant". After filtering, the data set contains 1.15 million reviews on food-related businesses.

---

[1] http://www.yelp.com/.

Some of the reviews are not written in English. Hence, we used a Java library[2] to detect non-English reviews and translated them into English using the Google Cloud Translation API[3]. Since one of our aims is to detect location-specific topics, we removed the 10 city names from the review. After that, we select as features the 4000 words with highest TF-IDF, which is given by:

$$\text{tf-idf}(w) = \frac{1}{\ln \text{df}(w)} \sum_{d \in \mathcal{D}} \text{tf}(w, d). \tag{2}$$

where the term frequency $\text{tf}(w, d)$ is defined as the number of occurrences of word w in document d, and the document frequency $\text{df}(w)$ is defined as the number of documents that contain word w.

After the above preprocessing steps, we converted the textual reviews into a binary document-term matrix, with each entry indicating the presence or absence of a word in a document. We passed the matrix as input to the PEM-HLTA method[4] to build an LTM. We then extracted a topic hierarchy from the resulting LTM.

Other than the textual reviews, the Yelp data also contains the city of businesses for the reviews. The city can be considered as an additional variable in the data set. The dependency between the city variable C and a topic Z can be quantified by the normalized mutual information [11], which is given by:

$$NMI(C; Z) = \frac{MI(C; Z)}{\sqrt{H(C)H(Z)}}. \tag{3}$$

where $MI(C; Z)$ is the mutual information between C and Z and $H(\cdot)$ is the entropy of a variable [6]. These quantities can be computed from $P(c, z)$, which in turn is estimated by the empirical distribution $P(c, z) - \frac{1}{N} \sum_{k=1}^{N} I_{c_k}(c) P(z|d_k)$, where d_k are the textual reviews in the data, c_k is the city label for the k-th review, and $I_{c_k}(c)$ is an indicator function having value of 1 when $c = c_k$ and 0 otherwise.

The topics with highest $NMI(C; Z)$ are the ones most relevant to location. Sometimes we may be interested in a particular city j. We can replace the city variable C with a binary variable C_j such that $C_j = 1$ if $C = j$ and $C_j = 0$ otherwise. Then the topics most specific to city j can be given by those with highest $NMI(C_j; Z)$.

5   Results

The topic hierarchy produced by PEM-HLTA has 5 levels and contains a total of 1202 topics. The top-level topics are shown in Fig. 3. They include topics related to bars, ambience, service, and several kinds of food such as appetizer/dinner/dessert, breakfast, Thai food, Korean food, Chinese food, etc. Under the bars topic, there are also sub-topics related to drinks, music, sport games, and video games. The topic hierarchy obtained is obviously meaningful.

---

[2] https://github.com/optimaize/language-detector.
[3] https://cloud.google.com/translate/.
[4] https://github.com/kmpoon/hlta.

┌ 0.14 bar drinks drink friends bartender fun bartenders
├ 0.14 bar drinks drink friends bartender fun bartenders
│ ├ 0.15 bar drinks drink friends bartender fun bartenders
│ └ 0.20 hour sliders happy specials daily mini slider
├ 0.04 club band crowd upstairs hip downstairs bands
│ ├ 0.04 crowd upstairs downstairs stairs older dj dress
│ └ 0.03 club band hip bands hop men women
├ 0.04 sports game watch tv flat play games
│ ├ 0.04 tv flat play games screen screens football
│ └ 0.03 sports game watch tvs video poker fans
├ 0.08 quickly room restaurant water away pleasant appropriate
├ 0.15 tables ordered wife appreciated came table meal
├ 0.20 left appetizer dinner dessert care entree friend
├ 0.20 dish dishes four know group different without
├ 0.20 nice place open old outside spice spicy
├ 0.05 butter bread holy peanut pudding fennel basket
├ 0.05 thai tea pad curry sticky iced teas
├ 0.18 time first apart next every kind top
├ 0.08 pepper sauce sweet portion salt light potato
├ 0.04 cheese ricotta pretzel pretzels mac fondue calzone
├ 0.08 korean pork kimchi medium rare bulgogi pulled
├ 0.09 questions asked advantage told said answer order
├ 0.08 ended getting gone even menu little reach
└ 0.07 standard granted thanks mentioned selling fit eyes

Fig. 3. Top-level topics obtained from textual reviews about restaurants on Yelp. The first top-level topic is related to bars. It is expanded to show topics related to different aspects of bars at lower levels in the hierarchy.

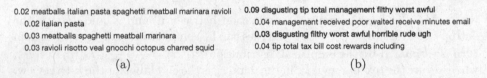

0.02 meatballs italian pasta spaghetti meatball marinara ravioli
  0.02 italian pasta
  0.03 meatballs spaghetti meatball marinara
  0.03 ravioli risotto veal gnocchi octopus charred squid

0.09 disgusting tip total management filthy worst awful
  0.04 management received poor waited receive minutes email
  0.03 disgusting filthy worst awful horrible rude ugh
  0.04 tip total tax bill cost rewards including

(a)   (b)

Fig. 4. The part of topic hierarchy (a) that represents different kinds of Italian food; and (b) that is related to service and ambience.

The sub-topics of bars also reveal what aspects customers are concerned with when visiting bars.

The LTM resulted contains 1202 internal nodes representing topics and 4000 leaf nodes representing words. Due to space constraint, we show only part of the model in Fig. 1. The topic hierarchy extracted from this subtree is shown in Fig. 2. The lowest level contains topics related to different kinds of food.

The topic hierarchy obtained contains more general topics at higher levels and more specific topics at lower levels in general. This can be illustrated by two parts of the topic hierarchy. In Fig. 4(a), the topics at the lower level represent different kinds of Italian food: pasta; spaghetti; ravioli, risotto, and gnocchi. The topic at the higher level represents general Italian food. In Fig. 4(b), the hierarchy groups some topics related to service and ambience together, including one about (poor) management, one about rude service and filthy environment, and one about tip and rewards. The three topics are grouped together possibly because customers consider quality of service and environment when giving tips.

Table 1 lists the level-1 topics with the highest NMI with the city variable. They represent topics most relevant to location. The last column shows the city of which the businesses have the highest number of reviews belonging to the topic. We see the topics are specific to certain locations. For example, Bellagio in the second topic is a resort in Las Vegas, and the Fountains of Bellagio is an attraction there. Foie gras in the fourth topic means liver of a goose and is a popular French cuisine. Hence, it is reasonable to have that topic more often in the French speaking city Montreal. The third topic contains the word Wisconsin and thus is relevant to its capital Madison. The third topic also contains the word poutine, which is a Canadian dish originating in Quebec. It is also relevant to another city Montreal. However, PEM-HLTA cannot distinguish between the two topics. The other two topics buffet and crab legs occur most in reviews about businesses in Las Vegas. This is possibly because Las Vegas is a popular tourist destinations and it provides a large variety of food.

We can also find topics most specific to a particular city as explained in Sect. 4. Table 2 lists the top 3 level-1 topics specific to Las Vegas, Montreal, and Edinburgh, respectively. Most of them look reasonable.

Table 1. Top five level-one topics most relevant to location.

Rank	Topic words	Related city
1	buffet, buffets	Las Vegas
2	bellagio, fountains	Las Vegas
3	curds, poutine, wisconsin, entirely, fashion, form, state	Madison
4	gras, foie	Montreal
5	crab, legs	Las Vegas

Table 2. Top three level-1 topics most specific to a city. Some less characterizing topic words are omitted to save space.

(a) Las Vegas	(b) Montreal	(c) Edinburgh
1 buffet, buffets	1 gras, foie	1 pub, irish, pint
2 bellagio, fountains	2 maple, syrup, waffles	2 lovely, shared, ...
3 crab, legs	3 curds, poutine, ...	2 post, midnight, ...

6 Related Work

Suresh and Locascio [12] has shown a similar study on Yelp dataset about topic detection and finding hidden cultures. They use LDA as a topic model and cluster businesses on a map by topic distribution. Unlike their work, we do not use clustering to find out the representative topic of a district. Instead, we calculated the association to find out the representative topics of a city among the whole dataset. HLTA has been applied for analyzing text data from academic papers and news articles [4,8]. However, it has not been used on the Yelp data. Besides, the topics detected by HLTA have not been associated with other non-word attributes for further analysis.

7   Conclusion

In this paper, we propose to use HLTA for mining textual reviews. We test the method using reviews written about restaurants on Yelp. The resulting topic hierarchy shows some interesting topics and a meaningful hierarchy. We further identify some reasonable topics most relevant to location and most specific to certain cities.

Acknowledgment. The work was supported in part by the Education University of Hong Kong under grant RG90/2014-2015R and in part by Hong Kong Research Grants Council under grants 16202515 and 16212516.

References

1. Blei, D.M.: Probabilistic topic models. Commun. ACM **55**(4), 77–84 (2012)
2. Blei, D.M., Griffiths, T.L., Jordan, M.I.: The nested Chinese restaurant process and Bayesian nonparametric inference of topic hierarchies. J. ACM **57**(2), 7:1–7:30 (2010)
3. Blei, D.M., Ng, A.Y., Jordan, M.I.: Latent Dirichlet allocation. J. Mach. Learn. Res. **3**, 993–1022 (2003)
4. Chen, P., Zhang, N.L., Poon, L.K.M., Chen, Z.: Progressive EM for latent tree models and hierarchical topic detection. In: Proceedings of the Thirtieth AAAI Conference on Artificial Intelligence (2016)
5. Chen, T., Zhang, N.L., Liu, T., Poon, K.M., Wang, Y.: Model-based multidimensional clustering of categorical data. Artif. Intell. **176**, 2246–2269 (2012)
6. Cover, T.M., Thomas, J.A.: Elements of Information Theory. Wiley, New York (2006)
7. Koller, D., Friedman, N.: Probabilistic Graphical Models: Principles and Techniques. The MIT Press, Cambridge (2009)
8. Liu, T., Zhang, N.L., Chen, P.: Hierarchical latent tree analysis for topic detection. In: Calders, T., Esposito, F., Hüllermeier, E., Meo, R. (eds.) ECML PKDD 2014. LNCS, vol. 8725, pp. 256–272. Springer, Heidelberg (2014). doi:10.1007/978-3-662-44851-9_17
9. Paisley, J., Wang, C., Blei, D.M., Jordan, M.I.: Nested hierarchical Dirichlet processes. IEEE Trans. Pattern Anal. Mach. Intell. **37**(2), 256–270 (2015)
10. Pearl, J.: Probabilistic Reasoning in Intelligent Systems: Networks of Plausible Inference. Morgan Kaufmann Publishers, San Mateo (1988)
11. Strehl, A., Ghosh, J.: Cluster ensembles – a knowledge reuse framework for combining multiple partitions. J. Mach. Learn. Res. **3**, 583–617 (2002)
12. Suresh, H., Locascio, N.: Autodetection and classification of hidden cultural city districts from Yelp reviews. arXiv:1501.02527 [cs.CL] (2015)
13. Zhang, N.L.: Hierarchical latent class models for cluster analysis. J. Mach. Learn. Res. **5**, 697–723 (2004)

Extraction of Temporal Events from Clinical Text Using Semi-supervised Conditional Random Fields

Gandhimathi Moharasan[1]([envelope]) and Tu-Bao Ho[1,2]

[1] Japan Advanced Institute of Science and Technology,
1-1 Asahidai, Nomi, Ishikawa, Japan
[2] John Von Neumann Institute, VNU-HCM, Ho Chi Minh City, Vietnam
{s1460012,bao}@jaist.ac.jp

Abstract. The huge amount of clinical text in Electronic Medical Records (EMRs) has opened a stage for text processing and information extraction for healthcare and medical research. Extracting temporal information in clinical text is much more difficult than the newswire text due to implicit expression of temporal information, domain-specific nature and lack of writing quality, among others. Despite of these constraints, some of the existing works established rule-based, machine learning and hybrid methods to extract temporal information with the help of annotated corpora. However obtaining the annotated corpora is costly, time consuming and requires much manual effort and thus their small size inevitably affects the processing quality. Motivated by this fact, in this work we propose a novel two-stage semi-supervised framework to exploit the abundant unannotated clinical text to automatically detect temporal events and then subsequently improve the stability and the accuracy of temporal event extraction. We trained and evaluated our semi-supervised model with the selected features of testing dataset, resulting F-measure of 89.76% for event extraction.

Keywords: Temporal event extraction · Electronic medical records · Clinical text · Natural language processing · Machine learning

1 Introduction

Automatic recognition and extraction of essential temporal information from text in natural language have been an active area of research in computational linguistics for past several decades. Moreover the recent prevalence of clinical text in Electronic Medical Records (EMRs) receives a great attention from researchers due to its significance and rich patient clinical information. The huge collection of EMRs provides a vast but still a underutilized rich source of patient medical information in the real world. However exploiting clinical text from EMR is very challenging due to its heterogeneity nature, lack of structure and writing quality [1–3].

© Springer International Publishing AG 2017
Y. Tan et al. (Eds.): DMBD 2017, LNCS 10387, pp. 409–421, 2017.
DOI: 10.1007/978-3-319-61845-6_41

Table 1 shows the example of clinical text and types of temporal information in EMR. In this clinical text, most events happened during the patient stay at the hospital, which is associated with clinical time stamps (date, duration and time). Medical doctor's and nurse's write their diagnosis and treatments details in the form of unstructured clinical text with implicit temporal information. Therefore, clinical text in EMRs reflects the patient's health status from the series of clinical observations, disease symptoms and treatment progressions time to time. Consequently temporal information extraction in clinical text indicates the crucial dimension of time-related information, which is significantly different from general newswire text [3–5]. However, recognizing and extracting these informations are inevitable in order to represent the complete temporal information.

Temporal information in clinical text has several particular characteristics, which pose some challenges for temporal information extraction (TIE). Beside the lack of domain-specific resources, we have to face three main challenges for TIE as the following [3]:

- The relative time expressions,
- Extraction of event durations,
- Extraction of implicit temporal relations (determining the order between events and expressions).

Temporal expressions are important to represent the rest of medical events in clinical timeline. These expressions are often mentioned in clinical text with the usage of relative time. Also the clinical events and relations often mentioned without referring/specifying of the time and order, respectively. For instance, we consider the following sentences:

Table 1. Example of clinical text and types of temporal information in EMR

Types of temporal information	Examples
Temporal expressions	The patient was transferred to the hospital for surgical intervention on 2016-06-11. The *next day* she administrated a drug *one hour before* the surgery.
Temporal events	The patient consulted the toxicologist and he felt that most likely had benzo overdose.
Temporal relations	The patient has a history of frequent severe cold, fever and cough.

The above example shows that the temporal expressions, events and relations have often written with relative time, implicit durations and order. Also to understand the medical concepts from clinical text, it requires the considerable medical domain knowledge.

Though the human can be able to understand the temporal information such as events and relations between the events, whereas it remains the non-trivial task in machine learning and clinical natural language processing (CNLP).

Despite of these limitations, the existing works [5–7] established various methods to extract temporal information with the help of annotated corpora.

Annotating clinical text is an important step for various information extraction tasks. However, annotating clinical text is a time-consuming task often requires the domain knowledge, cost and considerable manpower to annotate the corpus manually [8,9]. Hence obtaining annotated corpora is expensive and Scarce [10]. This time-consuming annotation task often requires the domain knowledge, time, cost and considerable manpower to annotate the corpus manually. Hence obtaining annotated data is expensive and Scarce [10].

Although a lot of medical records are publicly available for research such as MIMIC II and MIMIC III clinical databases, only very few annotated records with temporal information such as I2B2 dataset, SemEval 2015 and SemEval 2016 clinical datasets are available. However, learning a stable and reliable model usually requires plenty of annotated data. Motivated by this fact, in this work we propose a novel two-stage semi-supervised framework to exploit the abundant unannotated clinical text to automatically detect the temporal events. Thus, our contribution in this paper are:

- We proposed a two-stage semi-supervised framework to automatically detect the temporal events from clinical text and gradually increase the number of records in annotated corpora and accuracy of temporal event extraction by exploiting rich, abundant unannotated text.
- We exploited various feature sets such as lexical, syntactic and metamap features for each word and phrase level to extract temporal events
- To avoid the data bias and improve the stability and accuracy of the model, we utilized a clustering algorithm to select a rich and sparse training dataset from abundant unannotated text.

2   Related Work

The research field of temporal information extraction from text was commenced on non-clinical settings such as general newswire domain with the creation of TimeML corpora [11,12]. This corpora have three types of temporal information: temporal expressions, events and temporal relations. In [13], the authors made a review of temporal information extraction tasks and techniques in general domain that follow three approaches: rule-based, machine learning-based, and hybrid approach.

In the recent years the interest of temporal event extraction from clinical text has received much attention from researchers due to wide implementation of EMRs in the hospitals. Many researchers worked on various topics of temporal representation and reasoning with clincal data [14–16].

To expedite the progression of temporal information detection from clinical text, I2B2[1] organized a shared task of temporal relations challenge in the year 2012 [7,8]. Followed by I2B2, various research groups such as SemEval

---

[1] https://www.i2b2.org/NLP/TemporalRelations/.

2015 [17] and SemEval 2016 [18] contributed an annotated corpora on temporal relations and entities to develop the various temporal information extraction methods. In Japan, NII Testbeds and Community for Information access Research (NTCIR-MedNLP) provided the medical text for various medical data exploitation research works [19]

The most techniques for temporal information extraction on clinical text are mainly based on available techniques used in the general newswire text domain. The rule-based methods deeply analyzed the nature of the data and developed the hand-coded rule to extract the event [20], which demands domain knowledge besides much of manual effort and time. In the machine learning-based approach, the feature set is determined through NLP data structures, part-of-speech, n-gram, or external resources [6,20] before applying probabilistic models such as Hidden Markov Models(HMM), Conditional Random Fields(CRF) [5]. The best performing system Xu *et al.* from I2B2 shared task Sun *et al.*, using the hybird approach with hand-coded rules, CRFs and SVMs by exploiting the various feature sets [21].

On the other hand, Jindal *et al.* developed the pipeline approach to extract the events from clinical text. In this approach, they recognized the attributes of the events in first stage, then they implemented Integer Quadratic Program (IQP) for the event extraction. For extracting the temporal expressions, they have adopted the publicly available time expression system HeidelTime [22]. However, this approach have the limitation with accuracy. The evetn accuracy of this method is decreased in comparison to existing approaches [5].

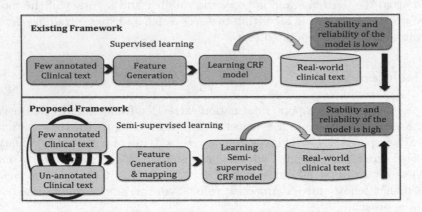

Fig. 1. Problem space and overview

From the above literature review, we discovered that all the existing temporal event extraction work were established with the available small number of annotated temporal corpus. However, a lot of rich unannotated clinical text are available compared to annotated clinical data. Moreover, the temporal annotation task is time-consuming, often requires the domain knowledge and considerable manpower to annotate the corpus manually. Also, unannotated testing corpus is not likely to be same as with their training or testing dataset. Besides that, from the literature review, we discovered unannotated data have significantly contributed to enhance the model when we combine with annotated data [23]. Therefore, we proposed and developed a semi-supervised framework to automatically to detect the temporal events from clinical text by exploiting unannotated text with gradually increasing of the number of annotated text in the corpus, which increases the automatic annotation accuracy as shows in Fig. 1.

3 Semi-supervised Framework for Temporal Event Extraction from Clinical Text

The core of our proposed method relies on semi-supervised framework, which uses the semi-supervised conditional random fields for event recognition as the event extraction considered one of the sequential labeling problem. The semi-supervised models significantly contributing to obtain the stable and reliable model for text processing and classification [23]. Also the undirected probabilistic models such as Hidden Markov Models (HMM), Morkov Random Fields (MRF) have been used for sequential labeling and information extraction in general domain for the past several years. specifically Conditional Random Fields (CRFs) have been exploited to assigns the label sequence to a given set of observed text data sequences [24].

3.1 Proposed Framework

After considering the advantages of semi-supervised learning and the literature review from general and clinical text [3,5], we found that, CRFs outperforms in information extraction compared to other machine learning and rule-based methods. Therefore, we are focusing on developing semi-supervised CRF method to extract events from both annotated and unannotated clinical text [25].

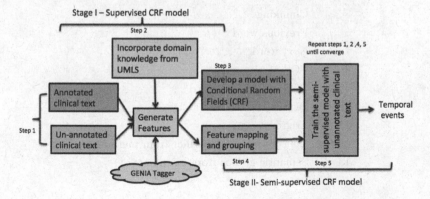

Fig. 2. Proposed framework for temporal event extraction

Figure 2 shows our proposed framework for temporal event extraction from the clinical text using semi-supervised conditional random fields. It consist of two stages. In the first stage (steps 1–3), we generated various feature sets such as lexical, syntactic, semantic features and trained the CRF model[2] [26]. In the second stage (step 1, 2, 4 and 5), we evaluated the effectiveness of feature, to select the best feature set for semi-supervised CRF. To our best knowledge, the proposed approach is the first semi-supervised framework for events extraction that exploits the annotated and unannotated clinical text. Also we introduced the clustering algorithm to select the high quality and sparse data to achieve the fairness of the training data selection from the unannotated clinical text.

3.2 Temporal Event Recognition

Temporal event recognition is observing a particular object/ action or set of objects through clinical timeline. Events in clinical text refers to medical concepts such as symptoms occurred, disease/disorder names diagnosed, medications prescribed by doctors and treatment procedures followed by patients. Therefore, the event extraction task in clinical text demands the domain knowledge as mentioned earlier. In this work, we exploited external knowledge source Unified Medical Language System (UMLS) to leverage the domain knowledge. In the first stage (step 3), we developed and evaluated a supervised conditional Random Fields (CRFs) model to recognize the temporal events with various features sets. Table 2 shows the features used in our proposed framework.

Table 2. Features used for temporal event extraction

S. No	Features used
1	Word
2	Base form of the word
3	Lemma of word
4	POS tag of word
5	Chunking
6	Previous word
7	Next word
8	POS of next word
9	POS of Previous word
10	Bigram
11	Trigram
12	Concept unique identifier from metamap
13	Semantic groups from metamap

---

[2] https://sourceforge.net/projects/crfpp/.

In the second stage of the framework, we utilized the rich quantity of unannotated data to develop the semi-supervised CRF by using the feature set from first stage, and increase the number of records in annotated corpora spirally. Subsequently improve the event extraction accuracy. In this study, we focused on temporal event extraction by the proposed semi-supervised framework. It consists of two steps: *unannotated data selection* for selecting the high quality data from plentiful unannotated data, and *feature generation and mapping* for extending the labels to unannotated training data. After the feature generation and mapping, we trained the semi-supervised CRF (step 5) with Mallet tool[3].

3.3   Selection of Unannotated Training Data

The step 1 of proposed semi-supervised framework requires a training data from an unannotated dataset. As we mentioned earlier, we have an abundant amount of unannotated data. However, we cannot assume that all the patient records in unannotated data are equally distributed among all the disease groups (according to international classification of diseases from WHO) and contains the significant details. In this situation, to maintain the fairness of data selection in the empirical experiment, we leveraged the advantage of clustering algorithm for unlabeled training data selection instead of selecting the records randomly. This approach helps us to achieve the rich and sparse training data from abundant unannotated dataset.

To accomplish the unannotated data selection, we exploited the simple K-means clustering algorithm [27] with TF-IDF. After applying the algorithm we obtained "k" clusters with similar group of patient records. Finally we manually selected "n" training records from each of "k" number of clusters, which contain sparse and rich temporal information. Subsequently, the influence of unannotated data towards the trained model would be high and the model achieves high accuracy and stability.

3.4   Feature Generation and Mapping

In step 2 and 4 of semi-supervised framework, we have created list of feature groups to train and evaluate the performance of the model.

Feature Generation: To generate the features, various external tools have been exploited. These features sets are selected based on the significance of each features. The first group of features contain the lexical features, such as base form, surrounding words, POS tags, and chunking. The lemmatization feature is the only one feature in second group. In the third group, features set is generated from UMLS metathesaurus[4], semantic group and Concept Unique Identifier (CUI) of words.

---

[3] http://mallet.cs.umass.edu/semi-sup-fst.php.
[4] https://metamap.nlm.nih.gov/.

- *Lexical and syntactic features* We have used GENIA Tagger[5] to generate the lexical features such as lemmatization, Part-of-Speech(POS) tags, chunking the phrases. To define the bigram, trigram, near-by words and their POS tags we have exploited Python programming language with NLTK package[6]. The n-gram language model plays the key role as a feature in the label detection.
- *Metamap features:* As we mentioned earlier, the clinical text contains the rich quantity of medical terms and it requires the external medical knowledge system to detect the medical concepts. Thus, we have used UMLS system through metamap tool to get of crucial features such as semantic group and Concept Unique Identifier (CUI) of words from clinical texts.

Feature Mapping: After generating the features from the training dataset, to develop an effective training procedure, we first need to extend the labels for unannotated dataset. Unlike the general text, the clinical text provides the advantage to classify the words into single semantic group from UMLS. Therefore we grouped the features and words based on the semantic groups from UMLS. We selected 35 predominantly contributing groups from 133 semantic groups in UMLS.

4   Experimental Evaluation

4.1   Data Preparation

We obtained the preprocessed and annotated training and testing data from I2B2 temporal relation challenge organizers. This dataset prepared based on the I2B2 guidelines (adopted from THYME guidelines) for clinical text. We used this annotated dataset for our experiments and evaluation.

- **Labeled training data:** 180 annotated records from I2B2 temporal relations shared task, which contains nearly 10,000 sentences.
- **Unlabeled training data:** 900 unannotated records from I2B2 medication and relations shared task, which contains more than 85,000 sentences.
- **Testing data:** 120 annotated records from I2B2 with ground-truth labels

Our experiments consist of 180 annotated records of training set and remaining 120 annotated records of testing set. The training set contains 16,468 and 2368 temporal events and expressions, respectively. The testing set contains 13,594 and 1820 temporal events and expressions, respectively.

We used the 2010-I2B2 relation challenge and 2009-Medication extraction data as an unannotated dataset along with annotated data. After applying the simple K-means clustering algorithm on the unannotated dataset, we obtained 30 clusters and selected 10 records per each clusters. Finally we obtained 300 sparse unannotated training records with rich temporal information. We used 200 records for experiment by slowly increasing the size of unlabeled data, meanwhile we checked the stability of the model with the tested accuracy.

---

[5] http://www.nactem.ac.uk/GENIA/tagger/.
[6] http://www.nltk.org/api/nltk.tag.html.

4.2   Baselines

In the first stage of the proposed framework, we developed the supervised CRF model with lexical features such as base form, part-of-speech tags, chunking and BIO-Events (beginning of event, inside of event, outside of event) as a target label for temporal event extraction. Later to improve the accuracy, we incorporated the domain knowledge by exploiting UMLS through metathesaurus tool. Metamap CUI and semantic group features from metamap helps to improve the performance for event extraction task as the result shown in Table 2. In the second stage of semi-supervised approach, we trained our semi-CRF model with above mentioned features on I2B2 training set and unannotated dataset. Finally, we evaluated our model with I2B2 testing dataset for temporal event extraction.

In the existing works in literature, mostly used only few available annotated training and testing dataset for their empirical experiments. Unlike prevailing experiments, we successfully exploited the large size of rich unannotated dataset along with annotated training and testing dataset for our experiment.

4.3   Results

Our proposed approach of semi-supervised framework for temporal event extraction results on I2B2 testing set is presented in Table 3 (Baseline, CRF with various features and Semi-supervised CRF). We evaluated our model with three measures: Precision, Recall and F-measure.

$$Precision = \frac{|System.Output \cap Annotated.Corpus|}{|System.Output|} \tag{1}$$

$$Recall = \frac{|System.Output \cap Annotated.Corpus|}{|Annotated.Corpus|} \tag{2}$$

$$F - measure = 2 \times \frac{Precision \times Recall}{Precision + Recall} \tag{3}$$

The first stage of supervised CRF model achieved similar results compared to other existing works. For event extraction, our proposed semi-supervised model slightly lagged behind the top performing system by 0.01 F-measure. Please refer

Table 3. Evaluation results of temporal event extraction

Events	Precision	Recall	F-measure
Baseline	69.7	78.1	73.66
CRF + Lexical + syntactic	85.27	72.53	78.39
CRF + Lexical + syntactic + UMLS	85.52	77.56	81.34
Semi-CRF + Random selection	86.41	82.25	84.21
Semi-CRF + Simple K-means	91.56	88.14	89.76
Top performing system from I2B2 [5]	93.74	86.79	90.13

Table 4. Existing results of temporal event extraction

Events extraction - Research group	F-measure	Method
Beihang University; Microsoft Asia	90.13	CRF
Vanderbilt University	90.0	CRF + SVM
The University of Texas, DallasdeSouza	88.0	CRF
University of Manchester	87.0	CRF + Dictionary based
University of Arizona, Tucson	88.0	CRF + SVM + NegEx
Mayo Clinic	85.0	CRF
Prateek Jindal et al., UIUC	87.0	Integer Quadratic Program

Sun *et al.* for the detailed comparison of result analysis [7]. Table 4 provides the comparative analysis of top performing system from I2B2 2012 temporal relations shared task.

The semi-supervised CRF model shows a significant improvement on the results with the influence of unannotated dataset. Apart from the accuracy and performance from testing dataset, we evaluated the stability and reliability of the model by applying on small real-world dataset. Our proposed approach performed well and obtained the high accuracy compared to supervised methods.

5   Discussion

In this work, we proposed and evaluated a novel two-stage semi-supervised framework for temporal event extraction. In the stage I, we developed and evaluated the supervised model for temporal events extraction. To evaluate the accuracy stage I model, we analyzed the obtained results extensively (which kind of medical concepts, difficult to detect through our system).

- Lack of annotated dataset (unseen events not always coinciding with annotated text, unknown abbreviations (SBE, URI, TPN, CK, etc.,)) is a leading reason for less performance in event detection. We focused to develop stage II of semi-supervised framework with utilization of unlabeled dataset to address this problem.
- Our system identified some of the events partially as some symptoms and disease name very long. For example, "connected up with social services", "infarct of the cerebellar hemispheres bilaterally, the right occipital lobe, the right thalamus and bilateral pons". Our method could identify this type events partially, which subsequently affected recall, subsequently overall performance.

After analyzing the stage I results, to improve the recall and overall performance, we focused to implement the semi-supervised framework. In the stage II, we utilized the simple K-means clustering algorithm for data selection and introduced feature mapping to extend the labels for unannotated training data.

We have done our experiment with increasing number of training records and iterations. The results of our experiment clearly shows that the semi-supervised CRFs increases the accuracy of the temporal event prediction. At the same time the precision increased slightly, where as recall increased significantly with the contribution of unannotated clinical text.

Figure 3 demonstrates the performance of the semi-supervised CRF for temporal event extraction. Figure 3(a) and (b) shows the accuracy of event extraction model with 300 and 900 unannotated training records respectively. From the above proposed approach, experimental setup and results, we have concluded that the increasing number of unannotated medical records and iterations, reliably increasing the accuracy of the model and event prediction.

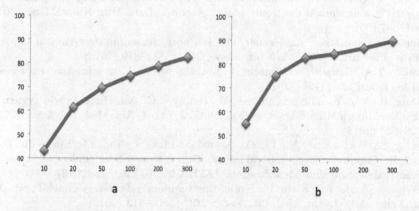

Fig. 3. Performance of temporal event extraction from proposed semi-supervised CRF model with 300 and 900 unannotated training records.

6   Conclusion

In this paper, we have presented semi-supervised approach for the temporal event extraction in clinical text. A novel framework was developed for the semi-supervised approach to extract temporal events from the rich unannotated text by extending the supervised model. The proposed framework has two stages. In stage I, we developed CRFs model on I2B2 annotated data with various features set. We found that Besides that we utilized the simple K-means clustering algorithm to select the rich and sparse data from abundant unannotated data. We reported the achieved results for event extraction from stage I and stage II. Compared to the baseline system, performance of the system with features set is considerably increased due to incorporation of concept unique identifier and semantic group features from UMLS. In stage II, we extended the unannotated data by feature mapping and trained the semi-supervised CRF. In this work we proved that the influence of unannotated clinical text helps to improve the accuracy of temporal event extraction significantly.

Acknowledgments. This work is partially funded by Vietnam National University at Ho Chi Minh City under the grant number B2015-42-02, and Japan Advanced Institute of Science and Technology under the Data Science Project. Also we thank mayo clinic and Informatics for Integrating Biology and the Bedside (I2B2) organizers for providing access to annotated I2B2 temporal relations corpus.

References

1. Roden, D., Xu, H., Denny, J., Wilke, R.: Electronic medical records as a tool in clinical pharmacology: opportunities and challenges. Clin. Pharmacol. Ther. **91**(6), 322–329 (2012)
2. Norn, G., Hopstadius, J., Bate, A., Star, K., Edwards, I.: Temporal pattern discovery in longitudinal electronic patient records. Data Min. Knowl. Disc. **20**(3), 361–387 (2010)
3. Sun, W., Rumshisky, A., Uzuner, O.: Temporal reasoning over clinical text: the state of the art. J. Am. Med. Inf. Assoc. **20**(5), 814–819 (2013)
4. Miller, T.A., Bethard, S., Dligach, D., Lin, C.: Savova. Extracting time expressions from clinical text, G.K. (2015)
5. Tang, B., Wu, Y., Jiang, M., Chen, Y., Denny, J.C., Xu, H.: A hybrid system for temporal information extraction from clinical text. J. Am. Med. Inf. Assoc. **20**(5), 828–835 (2013)
6. Sohn, S., Wagholikar, K., Li, D., Jonnalagaddaa, S., Tao, C., Elayavilli, R.K., Liu, H.: Comprehensive temporal information detection from clinical text: medical events, time, and tlink identification. JAMIA **20**(5), 836–842 (2013)
7. Weiyi, S., Anna, R., Ozlem, U.: Evaluating temporal relations in clinical text: 2012 i2b2 challenge. J. Am. Med. Inf. Assoc. **20**(5), 806–813 (2013)
8. Galescu, L., Nate, B.: A corpus of clinical narratives annotated with temporal information. In: ACM, pp. 715–720 (2012)
9. Styler IV, W.F., Bethard, S., Finan, S., Palmer, M., Pradhan, S., de Groen, P.C., Erickson, B., Miller, T., Lin, C., Savova, G., Pustejovsky, J.: Temporal annotation in the clinical domain. Trans. Assoc. Comput. Linguist. **2**, 143–154 (2013)
10. Sun, W., Rumshisky, A., Uzuner, O.: Annotating temporal information in clinical narratives. J. Biomed. Inf. **46**, s5–s12 (2013)
11. Pustejovsky, J., Hanks, P., Sauri, R., See, A., Gaizauskas, R., Setzer, A., Radev, D., Sundheim, B., Day, D., Ferro, L., et al.: The timebank corpus. In: Corpus linguistics, vol. 2003, p. 40 (2003)
12. Setzer, A., Gaizauskas, R.J.: Annotating events and temporal information in newswire texts. In: Proceedings of LREC-2000, vol. 2000, pp. 1287–1294 (2000)
13. Verhagen, M., Gaizauskas, R., Schilder, F., Hepple, M., Moszkowicz, J., Pustejovsky, J.: The tempeval challenge: identifying temporal relations in text. Lang. Resour. Eval. **43**(2), 161–179 (2009)
14. Zhou, L., Friedman, C., Parsons, S., Hripcsak, G.: System architecture for temporal information extraction, representation and reasoning in clinical narrative reports. Am. Med. Inf. Assoc. **2005**, 869 (2005)
15. Lin, Y.-K., Chen, H., Brown, R.A.: Medtime: a temporal information extraction system for clinical narratives. J. Biomed. Inf. **46**, s20–s28 (2013)
16. Zhu, X., Cherry, C., Kiritchenko, S., Martin, J., De Bruijn, B.: Detecting concept relations in clinical text: insights from a state-of-the-art model. J. Biomed. Inf. **46**(2), 275–285 (2013)

17. Bethard, S., Derczynski, L., Savova, G., Pustejovsky, J., Verhagen, M.: Semeval-2015 task 6: cinical tempeval. In: Proceedings of the 9th International Workshop on Semantic Evaluation (SemEval 2015), pp. 806–814 (2015)
18. Bethard, S., Savova, G., Chen, W.T., Derczynski, L., Pustejovsky, J., Verhagen, M.: Semeval-2016 task 12: clinical tempeval. In: Proceedings of the 10th International Workshop on Semantic Evaluation (SemEval 2016), San Diego, California, June, pp. 962–972. Association for Computational Linguistics (2016)
19. Higashiyama, S., Seki, K., Uehara, K.: Clinical entity recognition using cost-sensitive structured perceptron for ntcir-10 mednlp. Proc. NTCIR **10**, 706–709 (2013)
20. Kovačević, A., Dehghan, A., Filannino, M., Keane, J.A., Nenadic, G.: Combining rules and machine learning for extraction of temporal expressions and events from clinical narratives. J. Am. Med. Inf. Assoc. **20**(5), 859–866 (2013)
21. Xu, Y., Wang, Y., Liu, T., Tsujii, J., Eric, I., Chang, C.: An end-to-end system to identify temporal relation in discharge summaries: 2012 i2b2 challenge. J. Am. Med. Inf. Assoc. **20**(5), 849–858 (2013)
22. Jindal, P., Roth, D.: Extraction of events and temporal expressions from clinical narratives: 2012 i2b2 NLP challenge on temporal relations in clinical data. J. Biomed. Inf. **46**, S13–S19 (2013)
23. Zhu, X.: Semi-supervised learning literature survey. World **10**, 10 (2005)
24. Lafferty, J., McCallum, A., Pereira, F.: Conditional random fields: probabilistic models for segmenting and labeling sequence data. In: Proceedings of the Eighteenth International Conference on Machine Learning, ICML, vol. 1, pp. 282–289 (2001)
25. Jiao, F., Wang, S., Lee, C.H., Greiner, R., Schuurmans, D.: Semi-supervised conditional random fields for improved sequence segmentation and labeling. In: Proceedings of the 21st International Conference on Computational Linguistics and the 44th annual meeting of the Association for Computational Linguistics, pp. 209–216. Association for Computational Linguistics (2006)
26. Moharasan, G., Ho, T.B.: A semi-supervised approach for temporal information extraction from clinical text. In: 2016 IEEE RIVF International Conference on Computing and Communication Technologies, Research, Innovation, and Vision for the Future (RIVF), pp. 7–12. IEEE (2016)
27. Kanungo, T., Mount, D.M., Netanyahu, N.S., Piatko, C.D., Silverman, R., Wu, A.Y.: An efficient k-means clustering algorithm: analysis and implementation. IEEE Trans. Pattern Anal. Mach. Intell. **24**(7), 881–892 (2002)

Characteristics of Language Usage in Inquires Asked to an Online Help Desk

Haruka Adachi and Mikito Toda[✉]

Nara Women's University, Nara 630-8506, Japan
{adachi, toda}@ki-rin.phys.nara-wu.ac.jp

Abstract. We study how to characterize usage of words in inquiries from customers to an online help desk based on their statistical properties such as the number of being used and their correlation in inquires. We also investigate possibility that such statistical analysis enables us to foresee difficulties in dealing with inquiries.

Keywords: Text mining · Online help desk · Entropy · Information

1 Introduction

Data analyses for various social network systems (SNS) are common in recent days. For example, analysis of emotional exchanges are done [1, 2, 3] to detect how people interact with each other using specific terms describing events and emotional expressions. There are also studies on whether there is a discrepancy in the content of dialog between two users by detecting the breakdown of the dialogue between two users [4, 5]. There, three types of dialogue, "utterances that do not collapse" and "utterances that feel uncomfortable but cannot be tolerated" and "utterances that are clearly funny" are prepared. Then, models are constructed using machine learning such as support vector machines and deep learning and they are compared to see how effective those models are.

Another important data is provided by the online help desk. Recently, companies have sections called the online help desk that deals with various inquiries and complaints from customers such as "I cannot understand how to use the product", "I cannot understand how to deal with error messages" and so on. Then, the data of the exchanges between customers and the company accumulates, which includes inquiries from customers, replies of the company, responses from customers, and the number of times customers and the company correspond. Data analysts in the company analyze these accumulated data to improve customer satisfaction.

The purpose of our study is to characterize the usage of words in such online help desk to improve customer service. In particular, we focus on the number of exchanges between customers and the company as an indicator for difficulty of handling the inquiry in the online help desk. When the number of exchanges is smaller, this would mean that the inquiry is relatively easy to deal with. Then, customer satisfaction would be higher because customers spend less time to obtain desired answers. On the other hand, if the number of exchanges is larger, the inquiry would be difficult to handle.

Y. Tan et al. (Eds.): DMBD 2017, LNCS 10387, pp. 422–429, 2017.
DOI: 10.1007/978-3-319-61845-6_42

Then, customers could feel unsatisfied since it takes a longer time to obtain desired answers. If we can foresee difficulty of handling the inquiry based on the words used in it, the company could put more resources to handle it better. Based on such a motivation, we study whether usage of words in inquiries bears certain information concerning difficulty of dealing with them. The data we analyze is provided by KSK Analytics, Inc. [6].

2   Data

Figure 1 is a schematic diagram showing how customers and operators interact in the online help desk. The data of the online help desk consists of inquiries from customers, replies of the company, and responses of customers to the company. These exchanges between customers and the company are accumulated from April 8, 2013 to August 26, 2015, and the total number of inquiries is 11012. For the inquiries and the exchanges following them, the company assigns the identification numbers. The ticket ID specifies the set consisting of an inquiry and the exchanges following it. For each element of the set, the row ID is given in chronological order. Thus, the pair of the ticket and row IDs uniquely specifies an inquiry or an exchange between a customer and the company. We list the data of the online help desk in Table 1.

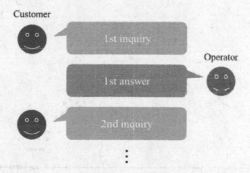

Fig. 1. A schematic diagram showing how customers and operators interact with the online help desk. Customers send inquiries to operators and operators reply to them.

We apply morphological analysis to the texts using an engine MeCab [7]. Here, morphological analysis means dividing sentences into words based on the grammar and dictionaries. The total number of words used is 3441, and we assign an integer n to each word in descending order of the number of being used. When multiple words have the same number of being used, we order them in Japanese lexical order. The number of being used is denoted by $W(n)$ for the n-th word.

Table 1. the data of the online help desk

Column name	Details
Ticket ID	The set of an inquiry and the exchanges following it
Row ID	Chronological order of the exchanges
Text	Sentences in an exchage

3 Analysis

3.1 Basic Statistics

In the following analysis, we focus our attention to the number of exchanges following an inquiry. The reason is that the number is supposed to characterize difficulty of handling the inquiry. That is, we think that the larger the number of the exchanges is. the more difficult and time-consuming to deal with the inquiry is. We denote the number of inquiries by $N(k)$ with the same number k of the exchanges following them. In Fig. 2, we plot $N(k)$ as a function of k. For those inquiries with larger values of k, we see that $N(k)$ exhibits the power law behavior.

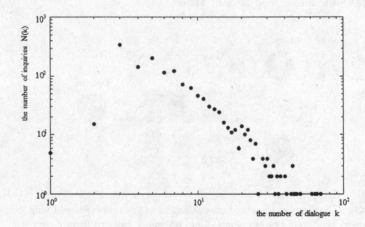

Fig. 2. The number of inquiries $N(k)$ as a function of the number k of the exchanges following them.

In Fig. 3, we depict the number of words $V(w)$ as a function of the number of being used w. There, we can also see the power law behavior. In Fig. 3, however, those words with larger values of $V(w)$ are not necessarily informative concerning difficulty of handling inquiries in which the words are used. The reason is that some of the words are used because of the grammatical structure of the sentences, not because of the meaning of the words. Others frequently appear in the sentences because they are general terms in the field in which the company operates. Thus, the quantity $V(w)$ does not seem to be a good indicator of k, which we think is an indicator for difficulty of handling the inquiries. Therefore, in our study, we try to find other statistical quantities which correlate with k better.

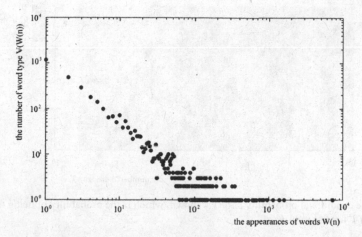

Fig. 3. The number of words $V(w)$ as a function of the number of being used w.

3.2   Conditional Probability of the Words

In our study, we use the Boltzmann entropy and the Kullback–Leibler divergence (KL divergence) [8] to construct an indicator for k, the number of exchanges following an inquiry. In Fig. 4, we display the conditional probability $p(n|k)$, which is the probability of finding the n-th word in those inquiries of k. In Fig. 5, we also show the conditional probability $p(k|n)$, which is the probability of the inquiries of k for the n-th word.

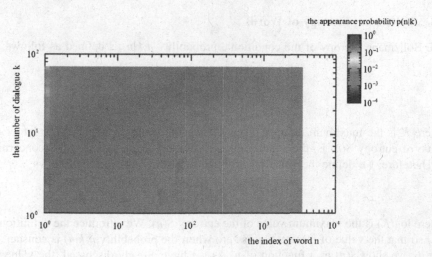

Fig. 4. The conditional probability $p(n|k)$ is shown, which is the probability of finding the n-th word in those inquiries of k.

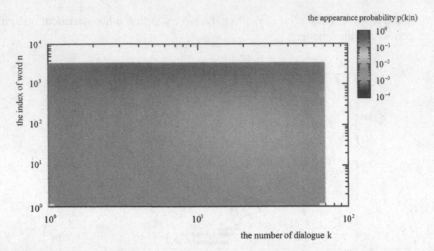

Fig. 5. The conditional probability *p(k|n)* is shown, which is the probability of being used in the inquiries of *k* for the *n*-th word.

In Figs. 4 and 5, we notice that those words with smaller values of *n* tend to be used more uniformly concerning *k* than those with larger values of *n*. This result confirms our finding in Fig. 3 that those words with larger values of *w* are not very informative concerning *k*. However, statistical properties of those words with larger values of *n*, i.e., smaller values of *w* may be an artifact due to smaller numbers of samples. Thus, there can be a tradeoff between the number of being used and possibility of characterizing *k*. In order to handle this problem, various indicators have been proposed such as TF-IDF. Here, we investigate the tradeoff by considering the distributions of the Boltzmann entropy and the KL divergence of the conditional distributions *p(k|n)*.

3.3 Boltzmann Entropy of Words

The Boltzmann entropy of the conditional probability *p(k|n)* is defined as follows

$$S(n) = -\sum_{k=1}^{K} p(k|n) \log p(k|n) \tag{1}$$

where K is the maximum number of exchanges following an inquiry. The larger the value of entropy $S(n)$ is, the smaller the information of the *n*-th word is concerning *k*. Therefore, we define the value of information of the *n*-th word as follows

$$I(n) = \log(K) - S(n) \tag{2}$$

where $\log(K)$ is the maximum value of the entropy $S(n)$. We introduce the definition of $I(n)$ so that the value of information is zero when the probability *p(k|n)* is constant. In Fig. 6, we show *I(n)* as a function of *n*. As we have already discussed, the values of *I(n)* tend to be smaller for smaller values of *n*. On the other hand, the distribution of

$I(n)$ attains its largest value for words of larger values of n. However, it is an artifact since the number of being used for them is just 1.

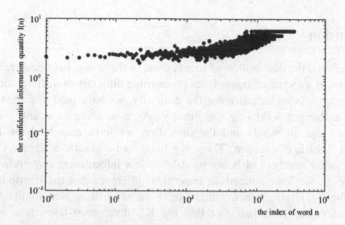

Fig. 6. The distribution of the value of $I(n)$ as a function of n.

3.4   Kullback–Leibler Divergence (KL Divergence)

The KL divergence between the conditional distribution $p(k|n)$ and $p(k)$ is as follows

$$D(n) = \sum_{k=1}^{K} p(k|n) \log \frac{p(k|n)}{p(k)} \tag{3}$$

where $p(k)$ is the probability that inquiries take k exchanges. Figure 7 shows the distribution of $D(n)$ as a function of n. The behavior of $D(n)$ is similar to that of $I(n)$, i.e., the larger n is, the smaller its value is. However, we note an important difference between Figs. 6 and 7. While $I(n)$ does not show much variation as n becomes smaller,

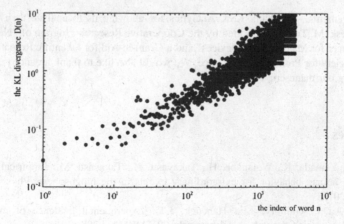

Fig. 7. The distribution of the value of $D(n)$ as a function of n.

$D(n)$ keeps decreasing. Moreover, large gaps of almost two decimal digits exist in Fig. 7 between $n \sim 10^0$ and $n \sim 10^3$. This implies that $D(n)$ is more effective than $I(n)$ as characteristics concerning k, the number of times the inquiry takes to be fully resolved.

4   Conclusion

We have analyzed the distribution of words used in the online help desk to see if their distribution bears a certain characteristics concerning difficulty of dealing with inquiries for the company. As an indicator for the difficulty, we have paid our attention to the number of exchanges following the inquiry. As a measure to quantify correlation between the usage of words and the difficulty, we have estimated the Boltzmann entropy and the KL divergence. Then, we have found similar tendencies that those words with larger numbers of being used have less information concerning the difficulty. However, we have noticed an important difference that the distribution of the values of the KL divergence exhibits larger variation than that of the Boltzmann entropy. This comes from that fact that the KL divergence takes into account the probability of the number of exchanges following inquiries. Thus, the KL divergence is a better indicator concerning the difficulty of handling the inquires. Now, we are planning to find a criterion concerning to what extent our analysis is statistically meaningful. We are also thinking of using the Tsallis statistics since the Tsallis statistics can be more suitable for the analysis of those data which exhibit the power law behavior [9].

We would also like to characterize how difficult and technical those words are which appear in the inquiries which take longer exchanges. One possibility for such characteristics of words is their lengths. We will perform analyses concerning correlation between difficulties of words used in inquires and difficulty of handling inquires. We will also analyze the words used in replies of the company, and see if there exist differences between those replies which used less technical words and those which do more. Such comparison would help the company to improve their customer satisfaction. These studies will be published in a near future.

Acknowledgements. We thank KSK Analytics for allowing us to analyze their data of the online helpdesk. M. Toda is supported by the Cooperative Research Program of "Network Joint Research Center for Materials and Devices" and a Grant-in-Aid for Scientific Research (C) from the Japan Society for Promotion of Science. We would also like to thank an anonymous referee for improving our manuscript.

References

1. Sano, Y., Yamada, K., Watanabe, H., Takayasu, H., Takayasu, M.: Empirical analysis of collective human behavior for extraordinary events in the blogosphere. Phys. Rev. E **87**, 012805 (2013)
2. Kramer, A.D.I., Guillory, J.E., Hancock, J.T.: Experimental evidence of massive-scale emotional contagion through social networks. PNAS **111**, 8788 (2014)

3. Adachi, H., Toda, M.: A network structure of emotional interactions in an electronic bulletin board. In: Takayasu, H., Ito, N., Noda, I., Takayasu, M. (eds.) Proceedings of the International Conference on Social Modeling and Simulation, Plus Econophysics Colloquium 2014. SPC, pp. 311–322. Springer, Cham (2015). doi:10.1007/978-3-319-20591-5_28
4. Higashinaka1, R., Funakoshi, K., Kobayashi, Y., Inaba, M.: The dialogue breakdown detection challenge: task description, datasets, and evaluation metrics. In: 10th Edition of the Language Resources and Evaluation Conference, pp. 3146–3150 (2016)
5. Tokunaga, Y., Inui, K., Matsumoto, Y.: Identifying continuation and response relations between utterances in computer-mediated chat dialogues. J. Nat. Lang. Process. 12(1), 79–105 (2005)
6. KSK Analytics Inc. http://www.ksk-anl.com/
7. MeCab 0.996. http://taku910.github.io/mecab/
8. Cover, T.M., Thomas, J.A.: Elements of Information Theory, 2nd edn. Wiley, Hoboken (2006)
9. Tsallis, C.: Introduction to Nonextensive Statistical Mechanics. Springer, New York (2009)

Deep Learning

A Novel Diagnosis Method for SZ by Deep Neural Networks

Chen Qiao[✉], Yan Shi, Bin Li, and Tai An

School of Mathematics and Statistics,
Xi'an Jiaotong University, Xi'an 710049, China
qiaochen@xjtu.edu.cn

Abstract. Single nucleotide polymorphism (SNP) data are typical high-dimensional and low-sample size (HDLSS) data, and they are extremely complex. In this paper, by using a deep neural network with a loci filter method, multi-level abstract features of SNPs data are obtained. Based on the abstract features, we get the diagnosis results for schizophrenia. It shows that the performance of the deep network is better than those of other methods, i.e., linear SVM with soft margin, SVM with multi-layer perceptron kernel, SVM with RBF kernel, sparse representation based classifier and k-nearest neighbor method. These results indicate that the use of deep networks offers a novel approach to deal with HDLSS problem, especially for the medical data analysis.

Keywords: Deep neural networks · Schizophrenia · Single nucleotide polymorphism · Diagnosis

1 Introduction

HDLSS data have proliferated swiftly in the last few years, especially in medical fields, e.g., genomic data, medical image data and proteomic data, etc. HDLSS data have important impact on learning algorithms. For instance, it is known that when the ratio of the number of training samples to the number of feature measurements is small, classical methods often fail [1]. Generally speaking, the performance of commonly used learning methods degrade since a number of irrelevant and redundant features are involved in them. This phenomenon is called as curse of dimensionality. With the increasing of unnecessary features, the search space becomes larger and more complex, and the generalization ability turns more difficult. In order to overcome such major obstacle, dimensionality reduction techniques are usually employed. That is, the set of features required for describing the data is reduced, and the performance of the learning models is commonly improved. Feature extraction is arguably the most famous dimensionality reduction technique.

Schizophrenia (SZ) is a kind of mental illness, and it is considered to be caused by the interplay of a number of genetic factors and environmental effects. Since schizophrenia can impair normal life and incur huge societal and economic

© Springer International Publishing AG 2017
Y. Tan et al. (Eds.): DMBD 2017, LNCS 10387, pp. 433–441, 2017.
DOI: 10.1007/978-3-319-61845-6_43

costs, early and accurate classification can significantly improve the treatment and reduce associated costs. Recently, many investigation have been put forward to study the corresponding genes associated with SZ [2].

A single nucleotide polymorphism is a gene sequence variation occurring commonly within a population (see, Fig. 1). The dbSNP lists 112,736,879 SNPs in humans in 2014 [8]. For example, multiple genes have been identified as potential causal genetic markers for SZ, including CACNA1C, CACNB2, ZNF323, G72/G30, DISC1, GRIK3, EFNA5, AKAP5 and CACNG2 [3,4]. While, how to use the SNPs data to find the risk biomarkers of diseases is still full of challenging. Most of the existing studies were based on using univariate modeling approaches on specific markers and their effects tend to be quite small. [6] applies support vector machines to SNPs data for the classification of SZ. In [5], a logistic regression procedure is applied to show that specific SNPs are associated with a range of psychiatric disorders, including SZ. By applying sparse models, the SNPs data are used to detect risk genes and classify of SZ [7]. Based on the multi-layer restricted Boltzmann machines (RBMs), classification accuracy was evaluated for the SNPs selected by the significant pathways in [9].

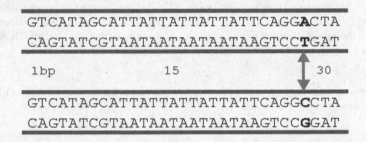

Fig. 1. Single nucleotide polymorphism: a gene sequence variation

A deep belief network (DBN) is a generative graphical model, consisting of one layer of visible units and multiple layers of hidden units, with connections between different layers but not between units within the same layer [10]. A DBN can be viewed as a composition of simple, unsupervised networks such as restricted Boltzmann machines(RBMs). In a DBN, for each upper layer, it takes the output of the previously trained layer as its input, i.e., using the trained hidden units of the previous layer as pre-training initialization. DBN can learn to probabilistically reconstruct its input data, and at the layers it perform feature selection on the inputs. After this learning step, a DBN can be further finetuned in a supervised way by back-propagation, which will then be used for classification. It has been shown in certain domains that a deep network is more effective than a shallow network for feature extraction, in that deep network can also obtain layer-wise abstract features [11].

For performance evaluation, several commonly used classifiers are compared with the proposed method, e.g., linear SVM with soft margin, SVM with multilayer perceptron kernel, k-nearest neighbors method, sparse representation based classifier. The training data we used here are from the Mind Clinical Imaging Consortium (MCIC). They are 184 subjects (including 80 SZ cases and 104 controls) and each subject has 147825 SNPs. In order to make more deep understanding about such HDLSS data, we use the real data as the training data set, and for the testing data, each one is obtained from a randomly selected training data with variation probability of 20% at each locus on the training data. For the testing task, we repeat 50 batches, and in every batch, the size of the testing data is 40.

In order to apply the deep learning approach to SNPs data analysis, we perform several preprocessing steps. For the original SNP array data, each SNP is coded by 0, 1 or 2. In an RBM, for the input data with discrete states, the softmax visible units are generally used. In order to suppress noise in the sampling process, the real-valued probabilities should be used for the reconstruction. If we still use the softmax visible units, it is quite hard to formulate an accurate rule to use three real-valued probabilities instead of 0, 1 or 2 without any noise in this procedure. In order to get good performance, we transform the SNP data into a binary sequence. Therefore, the dimension of each subject is $147825 * 3$. Secondly, we incorporate a loci filter method to select raw features for the training data. For a binary distribution, it has two sharp peaks with values 0 and 1. So we know that if one locus is significant, i.e., it is a potential biomarker, then the distance between the two means of case and control is greater than a threshold. At the same time, the standard deviation of case and control data should be limited. We apply these two principles to the training data to select raw features, and then train a DBN constructed via 3-layer RBMs on unlabeled data in a greedy layer-wise way and use the fine-tuning method on the training data to further adjust the parameters of the network. The 3-level abstract features of the simulation data are thus achieved and stored in the network. In addition, a linear classifier is added to the output layer of the deep network to classify the testing data. The results show that the performance of the deep network followed by a linear classifier performs better than the other classifiers. That is because, the deep network can extract more meaningful abstract features, which work better than the raw data.

In all, in this present, we consider SNPs data with a dimension of nearly One hundred and fifty thousand. For the training data, by combining with a loci filter method, a deep belief network with stacked restricted Boltzmann machines is applied to perform feature extraction for SNPs data. Further, a linear classifier is added on top of the deep network to carry out fine-tuning step. The classification performance of the trained network on the testing data can show it is quite efficient, when comparison with other classifiers. We can see that by using a deep network combining with a loci filter method, layer-wise deep feature representations of SNPs data can be extracted, which provide distinguishable features for better diagnosis of disease. The SNP data may contain complementary information about multiple psychiatric disorders (e.g., schizophrenia, bipolar, unipolar disorders) and the model by DBN with multi-task learning procedure can be generalized for accurate classification of these multiple diseases. In addition, it is promised that the proposed model can be generalized for multi types of genomic and imaging data.

2   Structure of DBN

In the present paper, a DBN constructed by 3-layer RBMs is utilized to perform feature learning task.

The deep belief network (DBN) is a generative graphical model, consisting of one layer of visible units and multiple layers of hidden units, with connections between different layers but not between units within the same layer.

As a kind of unsupervised feature learning method, DBN has been widely used for unsupervised feature abstraction, dimensionality reduction, collaborative filtering, feature learning and topic modeling. It has shown a strong promise in automatically representing the feature space using unlabeled data to increase the accuracy of subsequent classification tasks. While, till to now, DBN has seldom been applied to genomic analysis. This may be due to the extremely high dimensionality of genomic data, the lack of sufficient samples, and the lack of known characteristics (e.g., the locality in genomic data). These factors limit the use of deep learning techniques such as the convolution or pooling, which have been used in image data analysis.

A DBN can be viewed as a composition of simple, unsupervised networks such as RBMs. An RBM is a stochastic artificial neural network, which can find a compressed feature representation of the current input data, and the last layer of the model provides abstract features of the raw data. In an RBM, there are two layers: one is the visible input layer, and the other is a hidden layer. There are connections between the layers but no connection between units within each layer. Since the input of the first layer is the SNPs data, i.e., the information coded as 0, 1 or 2, and usually, for the discrete state data, softmax and multinomial units are used as the visible layer. In order to obtain more accurate feature selection, we convert the SNPs sequences into binary sequences. Thus, we can use the real-valued probabilities of the visible layer in the reconstruction phase directly (this process is formulated in (2)). In [9], we have shown that this transformation improves the performance of RBMs compared to using softmax units. The binary RBM model, i.e., the Bernoulli-Bernoulli RBM (BBRBM), uses binary units for both the visible and the hidden layer. The energy function of the BBRBM is defined as

$$E_\theta(v, h) = -v^T W h - v^T \alpha - h^T \beta. \tag{1}$$

where the visible units $v = (v^{(1)}, v^{(2)}, v^{(3)})$, and $v^{(i)} = (v_1^{(i)}, v_2^{(i)}, \cdots, v_{N_v}^{(i)})^T$ $(i = 1, 2, 3)$ is an N_v-dim vector with each $v_j^{(i)} \in \{0, 1\}$. $h = (h_1, h_2, \cdots, h_{N_h})^T$ is the hidden units with each $h_j \in \{0, 1\}$. $\alpha = (a_1, a_2, \cdots, a_{3*N_v})^T$ is the bias of the visible units, $\beta = (b_1, b_2, \cdots, b_{N_h})^T$ is the bias of the hidden units, and $W = \{w_{ij}\}_{3*N_v \times N_h}$ with each w_{ij} being the connection weight between v_i and h_j. Here Θ is the set of parameters of the BBRBM, especially referring to the connection weight matrix W between the visible units v and hidden units h, the bias α of v and the bias β of h. For every sample, denote the SNP sequence as S, i.e., S is an N_v-dim vector and each $S_j \in \{0, 1, 2\}$. For each $v_j^{(i)}$ ($i = 1, 2, 3$, $j = 1, 2, \cdots, N_v$), it is defined as follows

$$v_j^{(i)} = \begin{cases} 1, S_j = i - 1 \\ 0, S_j \neq i - 1 \end{cases}. \tag{2}$$

For a BBRBM, the distribution of the observed data v_θ, $P_\theta(v)$, is defined as $P_\theta(v) = \frac{1}{Z_\theta} \sum_h e^{-E_\theta(v,h)}$. Let $\mathcal{L}(\theta) = \log P_\theta(v)$, and find one θ^*, such that

$$\theta^* = \arg \max_\Theta \mathcal{L}(\theta). \tag{3}$$

is just the task of training the BBRBM, and this can be achieved by performing stochastic steepest ascent method. In order to avoid calculating the expectation of $v_i h_j$ on the model which is appeared in the partial derivative of $\mathcal{L}(\theta)$, a faster learning approach called contrastive divergence (CD) is used. And the change in the parameters are then given by

$$\Delta w_{ij} = \epsilon \cdot (\langle v_i h_j \rangle_{data} - \langle v_i h_j \rangle_{recon}). \tag{4}$$

$$\Delta a_i = \epsilon \cdot (\langle v_i \rangle_{data} - \langle v_i \rangle_{recon}). \tag{5}$$

$$\Delta b_j = \epsilon \cdot (\langle h_j \rangle_{data} - \langle h_j \rangle_{recon}). \tag{6}$$

In which, ϵ is the learning rate, and $\langle \cdot \rangle$ is the operator of expectation with the corresponding distribution denoted by the subscript.

Furthermore, in order to limit the quickly increasing bounds of the weights, the optimization function is modified to be

$$\max_\Theta \mathcal{L}(\Theta) - \lambda \|W\|_2. \tag{7}$$

where $\| \cdot \|_2$ is the L_2 norm and λ is the parameter to control the bound of W. The purpose of adding the regularization $\|W\|_2$ is to prevent overfitting. By using gradient descent, we can update the parameters in (7).

3   Preprocessing and Classification Results

3.1   Preprocessing the Simulation Data

We use simulation data in this paper. They are from the real data of the Mind Clinical Imaging Consortium (MCIC). The SNPs data were collected from 184 subjects including 80 schizophrenia patients and 104 healthy controls, and each sample size is 147825. Since the dimension of each sample is so high, with quite small sample size, so in order to get better understanding about such data, we do some experiments on simulation data. For every batch, the training data are the real data, and for the testing data, each one is obtained from a randomly selected training data with variation probability of 20% at each locus. For every batch, the size of the testing data is 40, and we repeat this procedure 50 times.

Before classification, we make some preliminary processing on the data. Firstly, by (2), we obtain the binary data where each sample has dimension of 147825 * 3. Next, we notice that for the classification of binary distribution, it has two sharp peaks at value 0 and 1. That is because such kind of data can be considered as an extreme case of the Gaussian mixture distribution with small standard variations around 0 and 1 respectively. Thus, for each locus k of the training data, we define $m_1(k)$ and $\mu_1(k)$ as the mean and the standard deviation of the health subjects, and $m_2(k)$ and $\mu_2(k)$ as the case subjects. Let

$$d_m(k) = |m_1(k) - m_2(k)|. \tag{8}$$

to be the distance between $m_1(k)$ and $m_2(k)$. For the binary distribution, it has two sharp peaks at two values 0 and 1, which can be considered as an extreme case of the Gaussian mixture distribution with small standard variations around 0 and 1 respectively. So $d_m(k)$ can be considered as an index to investigate the distance between centers of the two data sets, i.e., the degree of closeness for the two data sets. If the k-locus is a significant biomarker, $d_m(k)$ should be significant. Further, for standard deviations $\mu_1(k)$ and $\mu_2(k)$, since either of them is an index to show the degree of concentrate for the corresponding kind of data set, so in order to assure that the data should not to be so dispersed, there exists one threshold, such that both of $\mu_1(k)$ and $\mu_2(k)$ are less than this threshold. If these two conditions are satisfied, the potential risk biomarker, i.e., locus k can be identified. For simplicity, for all loci k, the lower bounds of $d_m(k)$ are set to be the same, and so do the thresholds for the two standard deviations. Denote them by D_m and C_μ respectively. In this way, we remove the loci which are not so significant under a given threshold D_m and C_μ, and this method is called as the loci filter method.

3.2   Classification Results for Simulation Data

After the loci filter method, the potential biomarkers from the training data are selected and considered as the raw features for the input of a 3-layer deep belief network. A linear classifier is added on top of the DBN. For the training data, by minimizing the difference between the label and the output from the classifier, we obtain the parameters of the whole network.

The numbers of hidden layer we used here are 1000, 500, 200 respectively; the learning rate is chosen as 0.1; and the penalty parameter λ is 0.0002. For the training data, by minimizing the difference between the label and the output from the classifier, we obtain the parameters of the linear classifier for distinguishing SZ. Figure 2 shows the changes of average accuracy rates as well as the average dimensions with different D_m for 50 batches of testing data. The AR for the classification is the proportion of all the testing samples that are correctly predicted. From (a) and (b) in Fig. 2 we can see, when D_m is small, we can keep enough loci of the samples for further analysis, e.g., we can search for more correlative loci with SZ; but in that case, the AR is low since the noise in the data (i.e., the useless loci for discrimination) overwhelms the information of the risk loci. For the instances of high AR, they correspond to large D_ms, so we can't keep enough dimension of the data and maybe some potential loci correlated with SZ have been missed.

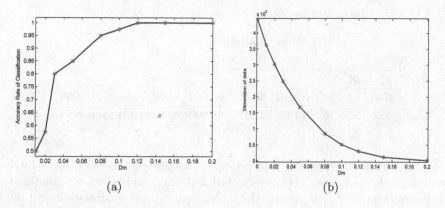

Fig. 2. (a) The accuracy rate of classification for different D_m; (b) The retained dimension of SNPs for different D_m.

In order to keep a balance between deducing the noise caused by the useless loci (which will overwhelm the information of the risk loci), and holding enough dimension of raw data to avoid missing potential biomarkers, we get that $D_m \in [0.025, 0.03]$ is the best choice. In that interval, the average accuracy rate is nearly $70-80\%$ and the dimension of the selected loci is around $2.2-2.9 \times 10^5$. Tables 1 and 2 show the classification performance when $D_m = 0.03$ and $D_m = 0.025$ with different threshold C_μ. Here, in order to compare our results, we also test other classifiers. For $D_m = 0.03$ or $D_m = 0.025$, we further add the threshold T_s to the standard deviations, and the classification performance are shown in Tables 1 and 2. Here, in order to compare our results, we also test other classifiers. CLASSIFIER I, II, III, IV, V, VI refers to linear SVM with soft margin, SVM with multi-layer perceptron kernel, SVM with RBF kernel, k-nearest neighbors method, sparse representation based classifier, DBN with linear classifier followed by fine-tuning method respectively. AR refers to accuracy rate, and DL refers to data length. For each condition and each classifier, the size of the testing data is 40, and the procedure randomly repeats 50 times. On noting that the AR of the raw data is only about 50%, thus, Tables 1 and 2 show the efficiency of the loci filter method no matter by using either the 5 kinds of classifiers. Further, from these two tables, it can be concluded that deep network with linear classifier performs the best.

Table 1. Classification results based on $D_m = 0.03$

		AR of CLASSIFIERS					
		I	II	III	IV	V	VI
T_s	DL						
0.45	$1.1 * 10^5$	0.8500	0.6615	0.7750	0.5655	0.7475	0.8708
0.50	$1.9 * 10^5$	0.8035	0.7350	0.5000	0.6975	0.7890	0.8217

Table 2. Classification results based on $D_m = 0.025$

		AR of CLASSIFIERS					
		I	II	III	IV	V	VI
T_s	DL						
0.45	$1.1 * 10^5$	0.8101	0.6775	0.5500	0.5608	0.6655	0.8257
0.50	$2.1 * 10^5$	0.7375	0.6975	0.5500	0.6933	0.7215	0.7583

Based on the analysis and understanding for simulation data, some further research on the real data (184 subjects with 147825 sample size), e.g., finding the risk biomarkers for SZ, achieving the high classification accuracy for real data, both are under our current investigation.

4   Conclusion

As a kind of unsupervised feature learning method, DBN has been widely used for unsupervised feature abstraction, dimensionality reduction, collaborative filtering, feature learning and topic modeling. It has shown a strong promise in automatically representing the feature space using unlabeled data to increase the accuracy of subsequent classification tasks. Recently, there are some works of applying DBN and deep network methods to medical data analysis, such as identifying networks in neuroimaging analysis and for cancer classification. However, such feature learning methods have seldom been applied to genomic analysis. This may be due to the extremely high dimensionality of genomic data, the lack of sufficient samples, and the lack of known characteristics (e.g., the locality in genomic data). This present is exactly an attempt to apply deep network approach on genomic data. We utilize a loci filter method combing with a deep belief network to extract discriminative features from SNP data and use them for classifying SZ. The results show that the abstract features from the fine-tuned deep network can improve classification accuracy even with a linear classifier. It is promised that DBN can be used for further feature learning, e.g., for searching significant biomarkers and pathways for real data of SZ.

Acknowledgment. This research was supported by NSFC Nos. 11471006 and 11101327, National Science and Technology Program of China (No. 2015DFA81780), the Fundamental Research Funds for the Central Universities (No. xjj2017126) and was partly Supported by HPC Platform, Xi'an Jiaotong University.

References

1. Ahn, J., Lee, M.H., Yoon, Y.J.: Clustering high dimension, low sample size data using the maximal data piling distance. Stat. Sin. **22**, 443–464 (2012)
2. Hjelm, R.D., Calhoun, V.D., Salakhutdinov, R., et al.: Restricted Boltzmann machines for neuroimaging: an application in identifying intrinsic networks. NeuroImage **96**, 245–260 (2014)

3. Ripke, S., O'Dushlaine, C., Chambert, K., et al.: Genome-wide association analysis identifies 13 new risk loci for schizophrenia. Nat. Genet. **45**, 1150–1159 (2013)

4. Luo, X.J., Mattheisen, M., Li, M., et al.: Systematic integration of brain eQTL and GWAS identifies ZNF323 as a novel schizophrenia risk gene and suggests recent positive selection based on compensatory advantage on pulmonary function. Schizophr. Bull. **41**(6), 1294–1308 (2015)

5. Lee, S.H., Ripke, S., Neale, B.M., et al.: Identification of risk loci with shared effects on five major psychiatric disorders: a genome-wide analysis. Lancet **381**(9875), 1371–1379 (2013)

6. Yang, H., Liu, J., Sui, J., et al.: A hybrid machine learning method for fusing fMRI and genetic data: combining both improves classification of schizophrenia. Front. Human Neurosci. **4**, 1–9 (2010)

7. Lin, D., Cao, H., Calhoun, V.D., et al.: Sparse models for correlative and integrative analysis of imaging and genetic data. J. Neurosci. Meth. **237**, 69–78 (2014)

8. National Center for Biotechnology Information, NCBI dbSNP build 142 for human. United States National Library of Medicine (2014)

9. Qiao, C., Lin, D.D., Cao, S.L., Wang, Y.P.: The effective diagnosis of schizophrenia by using multi-layer RBMs deep networks. In: 2015 IEEE International Conference on Bioinformatics and Biomedicine (BIBM), pp. 603–606 (2015)

10. Hinton, G.E., Salakhutdinov, R.R.: Reducing the dimensionality of data with neural networks. Science **313**(5786), 504–507 (2006)

11. LeCun, Y., Bengio, Y., Hinton, G.: Deep learning. Nature **521**(7553), 436–444 (2015)

A Novel Data-Driven Fault Diagnosis Method Based on Deep Learning

Yuyan Zhang, Liang Gao[(✉)], Xinyu Li, and Peigen Li

State Key Laboratory of Digital Manufacturing Equipment and Technology,
Huazhong University of Science and Technology, Wuhan, China
419661350@qq.com, gaoliang@mail.hust.edu.cn,
{lixinyu,pgli}@hust.edu.cn

Abstract. Mechanical fault diagnosis is an essential means to reduce maintenance cost and ensure safety in production. Aiming to improve diagnosis accuracy, this paper proposes a novel data-driven diagnosis method based on deep learning. Nonstationary signals are preprocessed. A feature learning method based on deep learning model is designed to mine features automatically. The mined features are identified by a supervised classification method – support vector machine (SVM). Thanks to mining features automatically, the proposed method can overcome the weakness that manual feature extraction depends on much expertise and prior knowledge in traditional data-driven diagnosis method. The effectiveness of the proposed method is validated on two datasets. Experimental results demonstrate that the proposed method is superior to the traditional data-driven diagnosis methods.

Keywords: Fault diagnosis · Feature mining · Deep learning · SVM

1 Introduction

As is known, mechanical equipment is an especially safe-critical system [1]. With the modern mechanical equipment becoming more complex, the demand for safety is also rising. In order to ensure reliability and improve safety, it is paramount to detect and locate the fault timely, and then implement corresponding maintenance operation.

Fault diagnosis is still a challenging problem both in academia and industries [2]. After the development of four decades, fruitful researches have been published. Existing diagnosis methods can be grouped into 3 categories: model-based, data-driven and hybrid method. Model-based method is based on the model of industrial processes or practical system, which can be obtained by analyzing physical principles [1]. In data-driven methods, features are extracted from the collected signals and identified by some classification technique. Hybrid method is the combination of the first two methods. Among these methods, data-driven methods are the most promising technique and have become the research focus in recent years [3].

Traditional data-driven methods are based on the framework [4]: ① Signal collection ② Manual features extraction ③ Fault type identification. In this framework, feature extraction is the crux. Extraction methods can be categorized into time-domain methods,

© Springer International Publishing AG 2017
Y. Tan et al. (Eds.): DMBD 2017, LNCS 10387, pp. 442–452, 2017.
DOI: 10.1007/978-3-319-61845-6_44

frequency-domain methods and time-frequency methods. In time-domain methods, statistical features are extracted, such as root mean square, skewness. In frequency-domain methods, frequency spectrums are obtained by Fourier transform. However, due to the non-stationary, non-linear and non-Gaussian characteristics, time-domain and frequency-domain methods are insufficient to obtain the most representative features. Because time-frequency methods are able to obtain instant frequency, they have been considered as an effective way to analyze signals' properties. Common time-frequency methods are wavelet packet decomposition (WPD) or empirical mode decomposition (EMD). WPD [5] decomposes raw signals into different sub-bands. Energy or energy entropy are used as the features. EMD [6] can adaptively decompose raw signals into a set of intrinsic mode functions (IMF). Each IMF involves local characteristics of raw signals under different time scale. By analyzing each IMF, fault features can be effectively obtained. There are already lots of studies based on time-frequency method. For instance, Hong employed WPD to filter high-frequency noise and EMD to obtain entropy sequences of the de-noised signals [2]. Soualhi et al. adopted EMD to decompose raw signals. Then the Hilbert marginal spectrum is taken as the feature information [7].

Although the traditional data-driven methods have gained competitive diagnosis accuracy, there exist two common drawbacks. ① Features are extracted manually. Lots of efforts are paid into analyzing signal property and designing fault features. Such process depends on plenty of expertise and domain knowledge. ② For a different diagnosis problem, designed features may be not suitable.

To overcome the above-mentioned drawbacks, this paper proposes a novel data-driven diagnosis method based on deep learning. Firstly, signals are preprocessed by normalizing and whitening. Then, instead of extracting features manually, the proposed method can mine features automatically. In this way, the proposed method releases us from the work of analyzing signal properties and designing features. Finally, features are identified by a supervised classification method – support vector machine (SVM). A set of experiments are carried out to validate the proposed method.

The rest of this paper is organized as follows. Section 2 describes the related work briefly. Section 3 details the proposed method. Section 4 gives experimental results and analysis. General conclusions are presented in Sect. 5.

2 Stacked Sparse Auto-Encoder and SVM

2.1 Stacked Sparse Auto-Encoder

As a deep learning model, stacked sparse auto-encoder (SSAE) [8] holds the ability to mine features automatically. From mathematical perspective, SSAE attempts to learn a nonlinear function. Thus, the input data can be mapped to another coordinate space, in which data corresponding to different patterns are linearly separable. SSAE can be obtained by training and stacking several layers of auto-encoder (AE) in a layer-wise manner.

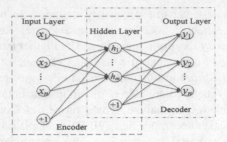

Fig. 1. The structure of auto-encoder

Figure 1 shows the structure of AE, which is a symmetrical 3-layer neural network. Considering an AE with n input units and m hidden units, the value of hidden layer and output layer can be obtained by Eqs. (1) and (2) respectively

$$h = \sigma(Wx + b) \tag{1}$$

$$y = \sigma(W'x + b') \tag{2}$$

where $W \in R^{m \times n}$ and $W' \in R^{n \times m}$ are weights; $b \in R^{m \times 1}$ and $b' \in R^{n \times 1}$ are biases; $\sigma(\cdot)$ denotes nonlinear transformation. x, h and y are respectively the input, hidden and output value. Training AE is to minimize the cost function J

$$
\begin{aligned}
J &= J_1 + \frac{\lambda}{2}J_2 + \beta J_3 \\
&= \frac{1}{2K} \sum_{i=1}^{K} (\|y^i - x^i\|^2) + \frac{\lambda}{2}\left(\sum_{i=1}^{m} \|W_i\|^2 + \sum_{i=1}^{n} \|W_i'\|^2\right) + \beta \sum_{i=1}^{m} \left(\rho \log \frac{\rho}{\hat{\rho}_i} + (1-\rho)\log \frac{1-\rho}{1-\hat{\rho}_i}\right)
\end{aligned} \tag{3}
$$

where J_1 is the error between input and output; K is the number of training samples. J_2 is the weight decay term used for preventing overfitting; λ is the coefficient of J_2. J_3 is the sparse penalty term; ρ is a user-specified parameter; β is the weight of J_3; $\hat{\rho}_i$ is the average activation of i-th hidden unit, which is defined as formula (4).

$$\hat{\rho}_i = \frac{1}{K} \sum_{j=1}^{K} h_i(x^j) \tag{4}$$

SSAE is a multi-layer network model created as a stack of AEs. Figure 2 shows the construction of SSAE with 3 hidden layers. The hidden layer of AE1 is treated as the input layer of AE2. AE2 mines features from the hidden value of AE1. AE3 proceeds in the same way. Finally, the input layers of AE1 is taken as input layer of SSAE. 3 hidden layers are stacked one by one following input layer.

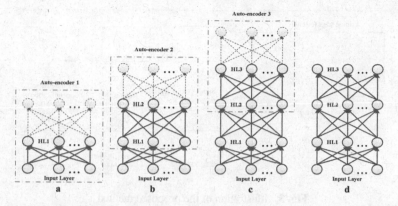

Fig. 2. The construction of SSAE with 3 hidden layers

2.2 Support Vector Machine

Support vector machine (SVM) is a supervised learning algorithm, which has been widely used for solving classification and regression problems. We describe SVM for binary classification problem. It can be extended to multi-class classification by one-vs-one (OVO) or one-vs-all (OVA). Given a training dataset$\{x^i, y^i\}$, $i = 1,...,K$, where x^i is a training sample, y^i is the corresponding label, the objective function of SVM is:

$$\min \frac{1}{2}\|w\|^2 + C\sum_{i=1}^{K} \xi_i \text{ s.t. } y^i(wx^i + b) + \xi_i \geq 1, \xi_i \geq 0, i = 1, 2, \ldots, K \tag{5}$$

where w is the coefficient of hyperplane; b is the bias term; ξ_i is the slack variable; C is misclassification penalty coefficient. By minimizing the above objective function, a new sample can be identified according to the separating hyperplane. For linearly inseparable problem, SVM can be extended by kernel embedding [9].

3 The Proposed Method

This section details the proposed data-driven fault diagnosis method. The illustration is given in Fig. 3. Firstly, signals are pre-processed by normalizing and whitening. Secondly, a SSAE with multiple hidden layers is designed to mine features automatically. Finally, fault patterns are identified by SVM.

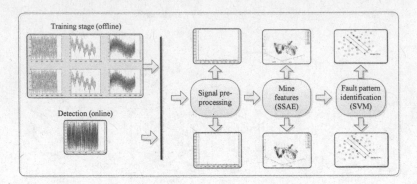

Fig. 3. Illustration of the proposed method

3.1 Pre-processing

Supposing that the training dataset is $\{x^i\}_{i=1}^K$, where $x^i \in \mathrm{R}^{n \times 1}$ is an observation, we normalize them to [0,1]. For clarity, $\{x^i\}_{i=1}^K$ is rewritten as a matrix $X \in R^{n \times K}$.

$$X_{\mathrm{norm}} = \frac{X - \min(X)}{\max(X) - \min(X)} \tag{6}$$

Next, the normalized data is processed by whitening. The aim of whitening is to make the observation less correlated with each other and help SSAE mine the effective features [4]. Whitened data can be obtained as follows

$$X_{\mathrm{white}} = V * D^{-1/2} * V^{\mathrm{T}} * X_{\mathrm{norm}} \tag{7}$$

where V is the matrix of eigenvectors and D is diagonal matrix of eigenvalues of the covariance matrix of X_{norm}. The covariance of any two observations of X_{norm} can be obtained by

$$\mathrm{cov}(x^i, x^j) = \mathrm{E}\big[(x^i - \mathrm{E}(x^i))(x^j - \mathrm{E}(x^j))\big] \tag{8}$$

3.2 Features Mining and Fault Pattern Identification

After preprocessing, the next step is to mine features automatically by SSAE. The main steps of features mining are as follows. (1) For the whitened dataset X_{white}, AE1 of SSAE receives it as the value of input layer. Through minimizing the cost function (Eq. (3)) using L-BFGS algorithm [10], the value of hidden layer is treated as the mined features by AE1. (2) Feed the hidden value of AE1 into AE2. AE2 mines further features in the same way as AE1. (3) After deeply mining features by several AEs, representative features can be obtained.

From mathematical perspective, the feature mining is a process of nonlinear transformations. In this work, sigmoid function is used as the nonlinear transformation function in AE. For simplification, the multiple transformations can be written as

$$X_{feat} = f(X_{white}) \qquad (9)$$

where $X_{feat} \in R^{m \times K}$ denotes the mined features; $f(\cdot)$ denotes the multiple nonlinear transformations.

Extensive studies by Bengio [11] show that the more hidden layers the better features will be got. To balance computational time and performance, SSAE with 5 hidden layers is adopted in this paper. The nodes number of each hidden layer are respectively 100, 80, 60, 40, 20. The dimension of input data of SSAE is the dimension of a whitened observation, and the dimension of mined feature is 20. The mined features are identified by SVM. Considering that they are linearly separable, linear SVM is adopted. For multiple fault patterns problem, One-vs-all is employed.

4 Experimental Study

In this section, experimental results are presented to validate the proposed method. At first, dataset provided by NSF I/UCR Center for Intelligent Maintenance Systems (IMS) is used [12]. Moreover, dataset provided by Case Western Reserve University is employed to further validate the effectiveness of the proposed method [13].

4.1 Dataset 1

Data Description and Parameter setting
Figure 4 shows the test rig and sensors placement. Three run-to-failure tests are carried out. In test 1, inner race defect occurred in bearing 3 and roller element defect in bearing 4. In test 2 and 3, outer race failure occurred in bearing 1 and 3 respectively.

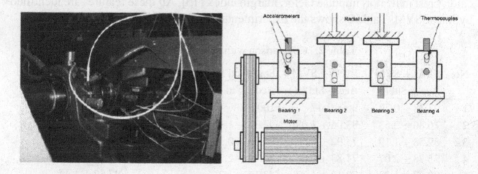

Fig. 4. Test rig and sensor placement

There are 6 fault types: ① degraded inner race(DIR) ② inner race fault(IR) ③ degraded roller element(DR) ④ roller element fault(R) ⑤ degraded outer race(DOR) ⑥ outer race fault (OR). For each fault type, random 20 1-second snapshots are used. For the normal condition (N), 10 snapshots of bearing 4 in test 1 and another 10 snapshots of bearing 1 in test 2 are selected. In each snapshot file, the first 20000 points are divided into 10 phases, each of which has 2000 points. Hence, there are 200 samples under each health condition. With these data, 7 experiments are conducted. 40% samples are selected as training samples and the remaining are testing samples. The parameters setting of SSAE are: $\lambda = 5e-5$; $\rho = 0.05$; $\beta = 5e-4$.

The detailed description is given in Table 1. It should be noted that these data are transformed to frequency-domain before preprocessing. This is because the bearing is rotating component and frequency-domain signals are easy to mine effective features.

Table 1. The detailed description of data used in 7 experiments

No.	Health condition	Classification label	Sample length	No. of training/ testing samples
1	N/IR	1/3	2000	160/240
2	N/DIR/IR	1/2/3	2000	240/360
3	N/R	1/5	2000	160/240
4	N/DR/R	1/4/5	2000	240/360
5	N/OR	1/7	2000	160/240
6	N/DOR/OR	1/6/7	2000	240/360
7	N/DIR/IR/DR/R/DOR/OR	1/2/3/4/5/6/7	2000	560/840

Comparison with Traditional Methods

To show the superiority of the proposed method, comparison experiments with traditional feature extraction methods are carried out. Empirical mode decomposition and wavelet packet decomposition are used to obtain the energy and energy entropy [14]. In addition, 18 statistics are extracted, including standard deviation, root mean square, variance, average, skewness, kurtosis, medium, sum, maximum, minimum, peak-to-peak, maximum gradient, root mean square amplitude, average absolute, shape indicator, crest indicator, impulse factor, margin index [15]. All these features are identified by linear SVM. Table 2 shows the experimental results.

Table 2. Comparison with traditional methods

No.	EMD + SVM Acc ± Std (%)	WPD + SVM Acc ± Std (%)	Statistical features + SVM Acc ± Std (%)	The proposed method Acc ± Std (%)
1	97.38 ± 0.79	96 ± 1.17	100 ± 0	**100 ± 0**
2	76.86 ± 2.53	83.06 ± 1.08	97.33 ± 1.15	**99.72 ± 0.4**
3	89.58 ± 1.19	73.92 ± 2.57	99.96 ± 0.13	**100 ± 0**
4	75.78 ± 2.07	71.83 ± 6.86	80.17 ± 2.24	**99.94 ± 0.18**
5	93.67 ± 1.33	72.12 ± 1.77	83.13 ± 1.36	**97.50 ± 1.44**
6	93.61 ± 1.23	81.17 ± 3.78	83.25 ± 1.94	**98.17 ± 0.95**
7	69.17 ± 0.10	96.19 ± 0	78.85 ± 0.11	**99.26 ± 0.24**

Acc and std denote accuracy and standard deviation respectively. To eliminate the influence of randomness, 10 trials are performed for each experiment. Accuracy is the average of 10 trials. It can be seen that the accuracies got by the proposed method are higher than those with traditional feature extraction methods. From the accuracies obtained by traditional feature extraction methods, it can be found that one feature extraction method may not be suitable for all the datasets.

Comparison with State-of-the-art Methods

In this section, we compare the proposed method with some state-of-the-art methods on IMS dataset. Accuracy comparisons are shown in Table 3. It can be seen that only the accuracy obtained by [16] is higher than ours. However, it doesn't take the degradation into consideration and there are only 3 health conditions. From the results and analysis, it can be concluded that the proposed method is superior to the existing methods.

Table 3. Comparison with some existing methods

Literature	Health conditions	Accuracy (%)
[16]	N, IR, OR	100
[17]	N, IR, OR	93
[18]	N, IR, OR, R	97.5
[19]	N, IR, OR	97
[14]	N, DIR, DOR, DR, IR, OR, R	93
Ours	N, DIR, IR, DR, R, DOR, OR	99.26

4.2   Dataset 2

Data Description

The rolling bearing vibration signals provided by Case Western Reserve University are used to further validate the effectiveness of the proposed method. There are totally 10 health conditions including 9 fault patterns and one normal condition. Drive-end vibration signals under each health condition are collected under four rotating speeds. 300 training and 200 testing samples are randomly selected from them and each has 400 data points. The detailed description is given in Table 4.

Results and Analysis

The testing accuracies of 20 trials are shown in Fig. 5. The minimum is 98.73% and the maximum is 99.6%. The average is 99.25% and standard deviation is 0.29%. From these figures, it can be concluded that the proposed method is effective for mechanical fault diagnosis and the performance is stable in each trial. Scatter plots of the first three principal components for the learned feature are shown in Fig. 6. It can be seen that the learned features characterizing the same health condition are clustered well and each cluster is separated. Although some learned features overlap from a fixed viewpoint, they are linearly separable. This is the reason that the proposed method can obtain higher accuracy than traditional data-driven methods.

Table 4. The detailed data description

Health condition	Severity level (mils)	Sample length	No. of training/ testing samples	Label
Normal	–	400	300/200	1
Inner race fault	7	400	300/200	2
Inner race fault	14	400	300/200	3
Inner race fault	21	400	300/200	4
Outer race fault	7	400	300/200	5
Outer race fault	14	400	300/200	6
Outer race fault	21	400	300/200	7
Ball fault	7	400	300/200	8
Ball fault	14	400	300/200	9
Ball fault	21	400	300/200	10

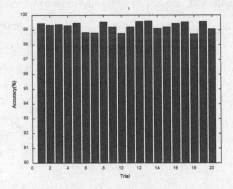

Fig. 5. Testing accuracy of 20 trials

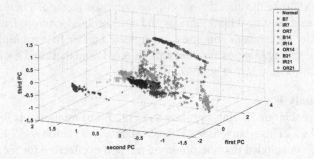

Fig. 6. Scatter plots of the first three principal components for the learned feature

5   Conclusion

Aiming to improve the diagnosis accuracy, this paper propose a novel data-driven diagnosis method. A deep learning method is designed to mine features automatically from the pre-processed data. Features are identified by a supervised learning technique – support vector machine. Two datasets are used to validate the proposed method. From the experimental results, we can draw the conclusions:

(1) The designed feature extraction method is effective to mine representative features automatically. It releases us from the work of designing features and overcomes the weakness that traditional data-driven methods depend on much prior knowledge.
(2) The mined features are linearly separable, so that simple classifiers are applicable.
(3) The proposed method is an effective fault diagnosis method.

Acknowledgments. This research is supported by the National Natural Science Foundation of China NO. 51435009, NO. 51375004 and NO. 51421062.

References

1. Gao, Z., Cecati, C., Ding, S.X.: A survey of fault diagnosis and fault-tolerant techniques-Part I: Fault diagnosis with model-based and signal-based approaches. IEEE Trans. Industr. Electron. **62**(6), 3757–3767 (2015)
2. Hong, S., et al.: Condition assessment for the performance degradation of bearing based on a combinatorial feature extraction method. Digit. Signal Proc. **27**, 159–166 (2014)
3. Grasso, M., et al.: A data-driven method to enhance vibration signal decomposition for rolling bearing fault analysis. Mech. Syst. Signal Process. **81**, 126–147 (2016)
4. Lei, Y., et al.: An intelligent fault diagnosis method using unsupervised feature learning towards mechanical big data. IEEE Trans. Industr. Electron. **63**(5), 3137–3147 (2016)
5. Chen, J., et al.: Wavelet transform based on inner product in fault diagnosis of rotating machinery: A review. Mech. Syst. Signal Process. **70–71**, 1–35 (2016)
6. Zhang, X., Liang, Y., Zhou, J.: A novel bearing fault diagnosis model integrated permutation entropy, ensemble empirical mode decomposition and optimized SVM. Measurement **69**, 164–179 (2015)
7. Soualhi, A., Medjaher, K., Zerhouni, N.: Bearing health monitoring based on hilbert-huang transform, support vector machine, and regression. IEEE Trans. Instrum. Meas. **64**(1), 52–62 (2015)
8. Ng, A.: Sparse autoencoder. CS294A Lecture notes **72**, 1–19(2011)
9. Zhang, L., Zhou, W.-D.: Fisher-regularized support vector machine. Inf. Sci. **343–344**, 79–93 (2016)
10. Zheng, W., et al.: Fast B-spline curve fitting by L-BFGS. Comput. Aided Geom. Des. **29**(7), 448–462 (2012)
11. Le Roux, N., Bengio, Y.: Representational power of restricted Boltzmann machines and deep belief networks. Neural Comput. **20**(6), 1631–1649 (2008)
12. Qiu, H., et al.: Wavelet filter-based weak signature detection method and its application on rolling element bearing prognostics. J. Sound Vib. **289**(4–5), 1066–1090 (2006)
13. Lou, X., Loparo, K.A.: Bearing fault diagnosis based on wavelet transform and fuzzy inference. Mech. Syst. Signal Process. **18**(5), 1077–1095 (2004)
14. Ben Ali, J., et al.: Application of empirical mode decomposition and artificial neural network for automatic bearing fault diagnosis based on vibration signals. Appl. Acoust. **89**, 16–27 (2015)
15. Shao, H., et al.: Rolling bearing fault diagnosis using an optimization deep belief network. Meas. Sci. Technol. **26**(11), 115002 (2015)

16. Gryllias, K.C., Antoniadis, I.A.: A Support Vector Machine approach based on physical model training for rolling element bearing fault detection in industrial environments. Eng. Appl. Artif. Intell. **25**(2), 326–344 (2012)
17. Yu, Y., Junsheng, C.: A roller bearing fault diagnosis method based on EMD energy entropy and ANN. J. Sound Vib. **294**(1), 269–277 (2006)
18. Liu, Z., et al.: Multi-fault classification based on wavelet SVM with PSO algorithm to analyze vibration signals from rolling element bearings. Neurocomputing **99**, 399–410 (2013)
19. Xu, H., Chen, G.: An intelligent fault identification method of rolling bearings based on LSSVM optimized by improved PSO. Mech. Syst. Signal Process. **35**(1), 167–175 (2013)

High Performance Computing

Simulated Precipitation and Reservoir Inflow in the Chao Phraya River Basin by Multi-model Ensemble CMIP3 and CMIP5

Thannob Aribarg[1,2(✉)] and Seree Supratid[3]

[1] College of Information and Communication Technology, Rangsit University,
Mueang Pathum Thani, Thailand
thannob.a@rsu.ac.th
[2] Climate Change and Disaster Center, Rangsit University,
Mueang Pathum Thani, Thailand
[3] Provincial Waterworks Authority, Bangkok, Thailand

Abstract. Climate Change caused by global warming is a growing public concern throughout the world. It is well accepted within the scientific community that an ensemble of different projections is required to achieve robust climate change information for a specific region. For this purpose we have compiled a Multi-Model Ensemble and performed statistical downscaling for 9 GCMs of CMIP3 and CMIP5. The observed precipitation data from 83 stations around the country were interpolated to grid data using the Inverse Distance Weighted method. The precipitation projection was downscaled by the Distribution Mapping for the near-future (2010–2039), the mid-future (2040–2069) and the far-future (2070–2099). The nonlinear autoregressive neural network with exogenous input (NARX) was used to forecast the mean monthly inflow to reservoirs. The projection inflow for the future periods are shown to increase in inflow in the wet season. A possibility of increase in hydrological extreme flood in the wet season may be indicated by these findings.

Keywords: CMIP3 · CMIP5 · Nonlinear autoregressive neural network with exogenous input · Statistical downscaling · Chao Phraya river basin

1 Introduction

Climate Change caused by global warming is a growing public concern throughout the world. Changes in temperature and precipitation due to climate change may affect to water resource of river basins around the world. Since the Coupled Model Intercomparison Project (CMIP) was launched in 1995, coupled ocean-atmosphere general circulation models developed in dozens of research centers around the world have been compared and analyzed extensively. The program has improved our scientific understanding of the processes of Earth's climate system and of our simulation capabilities in this field. CMIP also plays an important social role by contributing to the Intergovernmental Panel on Climate Change (IPCC). The CMIP phase three (CMIP3) provided the scientific base for the Fourth Assessment Report (AR4) of IPCC published in 2007.

© Springer International Publishing AG 2017
Y. Tan et al. (Eds.): DMBD 2017, LNCS 10387, pp. 455–463, 2017.
DOI: 10.1007/978-3-319-61845-6_45

CMIP phase 5 (CMIP5) was initiated in 2008, and the CMIP5 data are now available for analyses and are expected to provide new insights on our climate for the Fifth Assessment Report (AR5).

The Chao Phraya River basin, the largest basin in Thailand, is located in the centre of the northern part of the country. This basin has two large-scale reservoirs: the Bhumibol reservoir (capacity 13.5 billion m3) and the Sirikit reservoir (capacity 9.5 billion m3). Between June and October 2011, five tropical storms caused historical levels of flooding in the Chao Phraya river basin consequence in a total of 815 deaths with 13.6 million people affected. Due to limited capacity of the river, several overbank flows and broken dike were observed causing excessive flow to several communities. Total estimated economic damage was about 45.7 billion US$ [1]. In order to address this issue, the precipitation projection at the local scale is a key input for flood impact assessments. In this study, we focus on inflows into the two major reservoirs, Bhumibol and Sirikit reservoirs through downscaling precipitation of 9 GCMs of CMIP3 (B1 and A2) and CMIP5 (RCP4.5 and RCP8.5). The nonlinear autoregressive neural network with exogenous input (NARX) is used to forecast the mean monthly inflow of Bhumibol and Sirikit reservoirs. All these findings will carry implications related to the degree to which the region will need to adapt to projected changes in precipitation and reservoir inflow and flood and drought from the future warming.

2 Related Works

Santer et al. [2] and Knutti et al. [3] suggested about the value of using Multi-Model Ensemble (MME) versus a single model. The MME tends to be an improvement over any individual model, because the bias in one model is cancelled out by another. Supharatid [4] used nine GCMs from CMIP3 and CMIP5 under 4 emission scenarios: B1, A2, RCP4.5 and RCP8.5 to project precipitation change in the Chao Phraya River basin, Thailand. Many researchers have applied artificial neural network (ANN) in hydrology and Coulibaly et al. [5] reported that 90% of the experiments use of backpropagation neural network (BPNN). However, BPNN is calculated on a basis of static based learning. The static network is inferior to the dynamic ones because the inputs of the static network depend solely on observed data, whereas those of the dynamic networks incorporate observed data with time delay units through recurrent connections. Chang et al. [6] performed a dynamic nonlinear autoregressive neural network with exogenous input (NARX) indicated the best performance on multi-step-ahead water level forecasting for Taiwanese urban flood control, compared to the static BPNN and Elman's recurrent neural network (RNN).

3 Materials and Method

3.1 Study Area

The Chao Phraya river basin (see Fig. 1.), which is the largest basin in the country, covering an area of 159,000 km^2 or about 35% of the total land area of the country. There are 2 main reservoirs, namely Bhumibol and Sirikit located in the Ping and Nan rivers. The Chao Phraya river system consists of four principal tributaries: the Ping (36,018 km^2), the Wang (11,708 km^2), the Yom (24,720 km^2) and the Nan (34,557 km^2). These four tributaries flow southward to join each other in Nakhon Sawan to become the Chao Phraya River. The river flows southward through a large alluvial plain to the sea at the Gulf of Thailand.

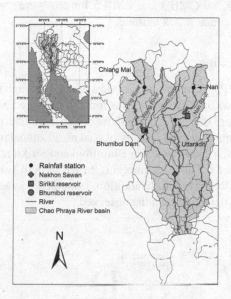

Fig. 1. The Chao phraya river basin in Thailand.

3.2 Data Processing

The daily precipitation data from 83 meteorological stations were collected from Thailand Meterological Department (TMD). In this study, we used 9 climate model pair

Table 1. List of IPCC CMIP3 and CMIP5 GCMs.

CMIP3	Resolution	CMIP5	Resolution	Country
CNRM-CM3	128 × 64	CNRM-CM5	128 × 64	France
CSIRO-Mk3.0	192 × 96	CSIRO-Mk3.6	192 × 96	Australia
GFDL-CM2.0	144 × 90	GFDL-CM3	144 × 90	USA
GFDL-CM2.1	144 × 90	GFDL-ESM2 M	144 × 90	USA
GISS-ER	72 × 46	GISS-E2-H	144 × 90	USA
INM-CM3.0	72 × 45	INM-CM4	180 × 120	Russia
IPSL-CM4	96 × 72	IPSL-CM5A-LR	96 × 96	France
MIROC3.2	128 × 64	MIROC5	256 × 128	Japan
MRI-CGCM2.3.2	192 × 96	MRI-CGCM3	320 × 160	Japan

based on IPCC CMIP3 and CMIP5 GCMs as shown in Table 1. These models have been selected based on availability and horizontal grid resolution. We have performed mean climatology for the historical period (1980–1999), for the near-future period (2010–2039), for the mid-future period (2040–2069) and the far-future period (2070–2099). Besides, the future monthly precipitation data has been taken from comparable greenhouse warming scenarios, SRES B1 and A2 from CMIP3 and RCP4.5 and RCP8.5 from CMIP5 models. All such 9 CMIP3 and CMIP5 models were regridded to a common grid of 0.5° or 720 longitude × 278 latitude.

3.3 Climate Downscaling

The 9 CMIP3 and CMIP5 GCMs models are downscaled through Distribution Mapping (DM) [7]. Figure 2 shows the Taylor diagrams for historical precipitation over Bhumibol and Sirikit reservoirs for DM. The models show both generations performed reasonably well in capturing the amplitude and phasing of past mean annual precipitation over both locations. The correlation coefficient over Bhumibol and Sirikit reservoirs from CMIP3 and CMIP5 lies between 0.7 and 0.8, implying MME of both generations simulate the mean precipitation reasonably well. In addition, both model generations give approximately lower standard deviation than the observed.

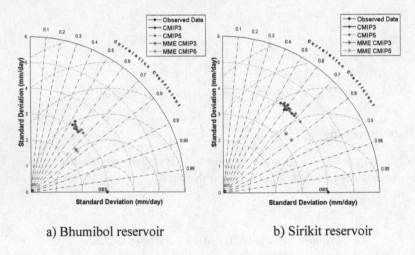

a) Bhumibol reservoir   b) Sirikit reservoir

Fig. 2. Taylor diagram of the annual mean precipitation for the historical period (1980-1999) at Bhumibol and Sirikit reservoirs.

3.4 Nonlinear Autoregressive Neural Network with Exogenous Factors (NARX)

The NARX model used in this study is a recurrent dynamic neural network [8]. The defining equation for the NARX model as (1).

$$y(n + 1) = f\left[y(n), y(n - 1), \dots, y\left(n - d_y\right); u_s(n), u_s(n - 1), \dots, u_s\left(n - d_{u_s}\right)\right] \tag{1}$$

Where $y(n)$ is regressed on previous values of the output signal and exogenous input discrete at time n; $y(n + 1)$ the next value of the output at discrete time $n + 1$; u_s represented by MME precipitation at a particular station s. The nonlinear mapping $f(\cdot)$ is generally an unknown smooth function; and can be approximated by a standard multilayer perceptron network. $d_{u_s} \in Z$ and $d_y \in Z$ consecutively are the lags of the exogenous input s and output regressors of the system, where $d_y \geq d_{u_s} \geq 1$. The output regressor $y(n)$ can be written in two different modes, depending on the training modes of the NARX network:

$$\hat{y}_{sp}(n + 1) = \hat{f}\left[y(n), y(n - 1), \ldots, y(n - d_y); \hat{u}_s(n), \hat{u}_s(n - 1), \ldots, \hat{u}_s(n - d_u)\right] \qquad (2)$$

$$\hat{y}_p(n + 1) = \hat{f}\left[\hat{y}(n), \hat{y}(n - 1), \ldots, y(n - d_y); \hat{u}_s(n), \hat{u}_s(n - 1), \ldots, \hat{u}_s(n - d_u)\right] \qquad (3)$$

where the series-parallel (SP) mode, $\hat{y}_{sp}(n + 1)$, $\hat{u}_s(n)$ and $\hat{f}(\cdot)$ are an estimators of $y(n + 1)$, $u_s(n)$ and $f(\cdot)$ respectively. The SP mode is applied for the historical period forecast due to data availability. According to (3), the parallel (P) mode refers to the actual values of the reservoir inflow which is a system's output are available during only the training of the NARX network. Besides, the testing performed outputs are feedback. Therefore, NARX using P mode utilized for future projection of reservoir inflow.

In this study, the NARX model was used to forecast mean monthly inflow of Bhumibol and Siriki reservoirs. From Fig. 1, the downscaled MME precipitation data were applied to the locations of 2 (Bhunmibol dam and Chiangmai) and 2 (Nan and Uttaradit) near-by rainfall stations regarding Bhumibol and Sirikit reservoirs respectively. However, the selection of the station is one of the most important steps. The stations are chosen by high correlation coefficient between precipitation and inflow. MME precipitation of CMIP3 and CMIP5 are found to be the highest correlation coefficient and it will be used for training by NARX. The characteristics relevant to the models are as follows: the tangent sigmoid transfer function is applied in the hidden layer and linear transfer functions in the output layers; the number of nodes in the hidden is 10; the learning method is Levenberg Marquard as explained in [9].

4   Results

4.1   Future Precipitation Projection

The precipitation projection in the Chao Phraya river basin was analysed for period 2010–2099. Figure 3 shows an annual cycle of precipitation projection of CMIP3 and CMIP5 models for all scenarios. The black line denote the observed data for the twentieth century. Both MME3 and MME5 give higher precipitation in the wet season than the observed data. However, we do not see any significant differences between MME3 and MME5. The projection precipitation for the near-future, mid-future, and far-future periods show seasonal variation with double peaks similar to the observed data for the twentieth century. Continuously increase of precipitation is found from the near-future to the far-future period except CMIP3 (B1 and A2) show decreasing precipitation in May. The maximum precipitation increase of 32% in October and 28% in September are seen for Bhumibol reservoir and Sirikit reservoir respectively.

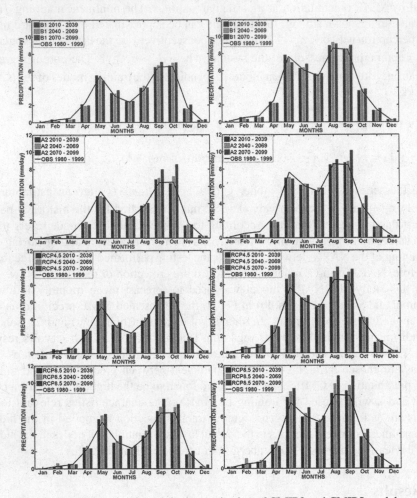

Fig. 3. Annual cycle of precipitation projection of CMIP3 and CMIP5 models.

4.2 Future Inflow Simulation

The comparison tests employ the NARX as well as those of other related forecasters such as BPNN and RNN. In this study, the experiment considers historical baseline for period 1980–1999, which divided into 90% for training and 10% for testing. NARX performs the best performance in terms of root mean square error (RMSE) and Pearson's correlation coefficient (r) as shown in Table 2.

Table 2. The performance measures in terms of RMSE and r.

Forecasting models	Bhumibol reservoir				Sirikit reservoir			
	RMSE		r		RMSE		r	
	CMIP3	CMIP5	CMIP3	CMIP5	CMIP3	CMIP5	CMIP3	CMIP5
NARX	**4.723**	**5.676**	**0.955**	**0.934**	**7.135**	**7.674**	**0.917**	**0.904**
BPNN	9.757	8.405	0.819	0.873	18.024	22.195	0.708	0.567
RNN	12.451	10.163	0.789	0.809	23.528	29.083	0.569	0.417

Figure 4, the future inflow simulation for the near-future (2010–2039), mid-future (2040–2069), and far-future (2070–2099) periods show seasonal variation similar to the observed inflow of the twentieth century (1980–1999). Continuously increase in inflow in the wet season is found from the near-future to the far-future period. However, there are decreasing trend in the dry period. The maximum inflow increase of 13% in September and 37% in August are seen for Bhumibol reservoir and Sirikit reservoir. The

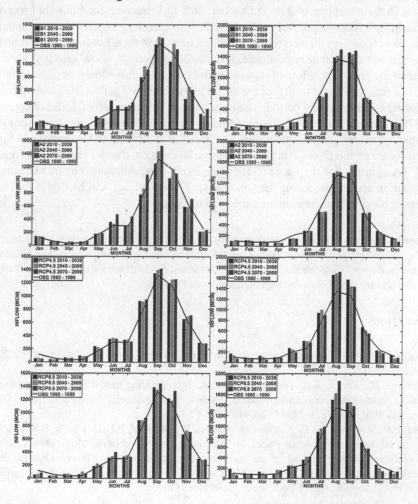

Fig. 4. Annual cycle of inflow simulation of CMIP3 and CMIP5 models.

future inflow of both reservoirs show a potential increase in future precipitation, implying more flood vulnerability. Similar trends are found for CMIP3 (A2) and CMIP5 (RCP8.5) models, but with different scales.

5   Conclusion

In this study, we have compiled a state-of-the-art multi-model multi-scenario ensemble of global (CMIP3 & CMIP5) and performed statistical downscaling for 9 GCMs. The precipitation projection was downscaled by the distribution mapping for the near-future (2010–2039), the mid-future (2040–2069) and the far-future (2070–2099). Both model generations perform reasonably well in capturing the amplitude and phasing of past mean annual precipitation. The correlation coefficient from all models lies between 0.6 – 0.8, implying reasonable simulation. The summer monsoon precipitation has an increase trend, with the maximum of 32% in October, 28% in September for Bhumibol reservoir and Sirikit reservoir respectively. The NARX was used to forecast the mean monthly inflow. We found very high correlation coefficients (>0.9) for all stations. The projection inflow for the near-future, mid-future, and far-future periods show seasonal variation similar to the observed inflow of the twentieth century. Continuously increase in inflow in the wet season is found from the near-future to the far-future period. However, there are decreasing trend in the dry period. The maximum inflow increase of 13% in September and 37% in August are seen for Bhumibol reservoir and Sirikit reservoir. The results indicate that the overall trend for the future precipitation and inflow at Bhumibol and Sirikit reservoirs are on the increase in the wet season. In contrast, the dry season is decreasing trend. These imply that the future climates are likely to be dominated by increases in the precipitation and inflow during the wet season. These findings will be useful for policy makers in pondering adaptation measures due to flooding.

Acknowledgement.   The authors wish to thank the Royal Irrigation Department of Thailand and the Thai Meteorological Department for providing their observed precipitation. The Electricity Generating Authority of Thailand for the Bhumibol and Sirikit reservoirs inflow characteristic. Also, the climate model datasets were obtained from PCMDI.

References

1. World Bank: The World Bank support Thailand's post-floods recovery effort, World Bank, Washington DC (2011)
2. Santer, B.D., Taylor, K.E., Glecker, P.J., et al.: Incorporating model quality information in climate change detection and attribution studies. Proc. National Acad. Sci. United States America **106**(35), 14778–14783 (2009)
3. Knutti, R., Abramowitz, G., Collins, M., Eyring, V., Gleckler, P.J., Hewitson, B.M., Mearns, L.: Good Practice Guidance Paper on Assessing and Combining Multi Model Climate Projections. Meeting Report of the Intergovernmental Panel on Climate Change Expert Meeting on Assessing and Combining Multi Model Climate Projections, IPCC Working Group I Technical Support Unit, University of Bern, Bern (2010)

4. Supharatid, S.: Skill of precipitation projection in the Chao Phraya river Basin by multi-model ensemble CMIP3-CMIP5. Weather and Clim. Extremes **12**, 1–14 (2016)
5. Coulibaly, P., Anctil, F., Bobee, B.: Daily reservoir inflow forecasting using artificial neural networks with stopped training approach. J. Hydrol. **230**(3), 244–257 (2000)
6. Chang, F.J., Chen, P.A., Lu, Y.R., Huang, E., Chang, K.Y.: Real-time multi-step-ahead water level forecasting by recurrent neural networks for urban flood control. J. Hydrol. **517**, 836–846 (2014)
7. Teutschbein, C., Seibert, J.: Bias correction of regional climate model simulations for hydrological climate-change impact studies: Review and evaluation of different methods. J. Hydrol. **456**, 12–29 (2009)
8. Nørgård, P.M., Ravn, O., Poulsen, N.K., Hansen, L.K.: Neural Networks for Modelling and Control of Dynamic Systems-A Practitioner's Handbook (2000)
9. Hagan, M.T., Demuth, H.B., Beale, M.H., De Jesús, O.: Neural network design, vol. 20, Boston (1996)

An IDL-Based Parallel Model for Scientific Computations on Multi-core Computers

Weili Kou[1]([⊠]), Lili Wei[1], Changxian Liang[1], Ning Lu[1],
and Qiuhua Wang[2]

[1] School of Computer and Information, Southwest Forestry University,
Kunming 650224, Yunnan, China
kwl_eric@163.com, lilywlwei@163.com, xnllcx@163.com,
swfuluning@163.com
[2] Faculty of Civil Engineering, and Yunnan Key Laboratory
of Forest Disaster Warning and Control, Southwest Forestry University,
Kunming 650224, Yunnan, China
qhwang2010@163.com

Abstract. Parallel computing is an efficient way to improve the efficiency of scientific computations. However, most of current parallel methods are implemented based on massage passing interface and very complicated for researchers. This study presents a new parallel model based on interactive data language. This paper specifically designed and described the parallel principles, strategies, architectures, and algorithms of the model. An experiment of a time series extraction was conducted to evaluate its performance, and the result illustrated that this model can significantly improve efficiency of scientific computations. Additionally, the model can be easily extended by third-part modules or toolkits. This study provides a general and upper-layer parallel model for scientists and engineers in scientific community.

Keywords: Parallel computing · Scientific computation · Multi-core computer

1 Introduction

The increasing volume of scientific data poses huge challenges for scientific research, so lots of computational scientists and engineers are concentrating on performance improvement of scientific applications [1–3]. For instance, much time of geospatial researchers is taken to process and analyze the large amount of remotely sensed images [4–6]. So, how to bring scientists and engineers out of tedious and boring data processing is a very important research topic that attracts more and more attentions of computational scientists. A simple resolution to solve these problems is to take advantages of multi-core computers and clusters. Parallel computing is an effective paradigm for analysis of huge volumes of data based on multi-core computers [7–13] or across clusters [14, 15], which presents new opportunities to solve these problems.

Parallel computing is generally used for scientific applications that are data-intensive, computing-intensive or high-real-time, and is able to coordinate and utilize multi-core computer resources [2, 16]. Multi-thread and multi-process are two

© Springer International Publishing AG 2017
Y. Tan et al. (Eds.): DMBD 2017, LNCS 10387, pp. 464–471, 2017.
DOI: 10.1007/978-3-319-61845-6_46

parallel computing models running on multi-core computers. Comparing to multi-process, multi-thread has two major advantages: (1) smaller computer resources are consumed to create a thread, and (2) communications among threads are simpler and faster than processes. But multi-process is easier than multi-thread to change serial codes into parallel programs in some scenarios.

Parallel computing is extensively implemented based on Message Passing Interface (MPI) [17, 18], which is always developed and maintained by professional personnel or teams, and is too complicated for researchers working in scientific communities. Interactive Data Language (IDL) is widely used to process and analyze scientific data, and can read and write almost any type of them although need to be supported by open database connectivity (ODBC) sometimes, has abilities to process array-based big data, and automatically supports parallel computing for built-in functions.

There are two typical kinds of IDL-based parallel computing solutions: data-based (such as FastIDL) and task-based (such as TaskDL and mpiDL). However, current parallel toolkits are too complex to parallelize serial codes for common scientists. The IDL_IDLBridge (IIB) is a parallel component encapsulated for high-level parallel applications, is able to simplify development processes and reduces maintenance costs [19]. Taking the IIB to implement parallel computing does not need to support by third-part toolkits or software modules with lower human and financial resources, and takes full advantages of multi-thread and multi-process programming models on multi-core computers.

The objective of this study is to develop a new parallel model for scientific computations based on the IDL IIB to improve computing efficiency on multi-core computers, and finds an easy-to-implementation manner for common researchers to realize parallel scientific computations in various scientific communities.

2   IDL Parallel Computing

The IDL provides parallel computing solutions for huge volumes of scientific data by using distributed computing resources in computer cluster systems, local network, or multi-core workstations. Third-part parallel toolkits, including FastDL, TaskDL, and mpiDL, are always used to implement parallel computations [20]. They have strong parallel ability but very expensive and complex for common scientists.

Many IDL built-in functions have parallel computing abilities, but non-built-in functions must be parallelized by user-defined methods. The IIB is a parallel computing component of the IDL, and is an object programming class to create and control IDL sessions in each of which program is independently executed as a computer process. An IIB instance corresponds to such a child process that is controlled by the IDL parent process. The IIB provides abundant attributes and methods to support parallel computing, and enables one or more the IIB child processes simultaneously execute programs written by IDL. Copying variables mechanism was employed to exchange data between parent and child processes of an IDL IIB instance. Child processes can execute arbitrary commands or programs that comply with the IDL syntax and grammars.

Synchronous and asynchronous are two manners for a child process to execute tasks within IDL IIB. Synchronous is that IDL waits for executing the next task until a child process completing the specified tasks. For asynchronous, the call to execute immediately, and the caller can track its progress status. This manner of overlapped execution can be used to perform extensive computations and visualizations in parallel with the main IDL process. Generally, all child processes cannot get the status of parent processes including data, compiled functions, system variables, and the current workspace.

3   Parallel Computing Model for Multi-core Computers

3.1   Presenting the Parallel Principles and Strategies

The IIB-based parallel principle in Fig. 1 was presented for researchers to improve scientific computing efficiency. First, a task is split into several subtasks $t_i \in \{t_1, t_2, t_3, \ldots, t_n\}$; then, the IIB object assigns a subtask t_i to a free CPU core $c_i \in \{c_1, c_2, c_3, \ldots, c_n\}$ for processing, and storing each intermediate result $r_i \in \{r_1, r_2, r_3, \ldots, r_n\}$ into memory temporarily; and finally, the parent process collects every result r_i from different cores, and combines them into one final result after all sub-processes complete their tasks and are in free of use.

This model takes different parallel strategies according to various requirements of processing and computing scenarios. There are two major parallel strategies are designed for third-part toolkits and others in Fig. 2. Some processes need to employ the third-part toolkits, such as Landsat Ecosystem Disturbance Adaptive Processing System (LEDAPS) and Fmask, that cannot be parallelized inside. To improve the processing efficiency, this study parallelizes those using parallel packages of Linux such as the parallel package based on IIB.

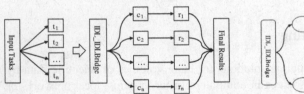

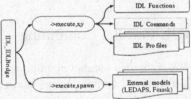

Fig. 1. Illustration   of   IIB-Based   parallel principles.

Fig. 2. The IIB-based parallel strategies.

3.2   Construction of the Parallel Architecture

The parallel architecture of the model is shown in Fig. 3. Users interact with the model through Secure Shell (SSH) or command lines in operating systems including Linux, Mac OS, or Windows. The IDL and the IIB component corporately control the computing process over the storage and computing servers. Third-part toolkits used for scientific computations are installed in multi-core computers or cluster systems.

3.3 Design of the Parallel Algorithm

This study designed a parallel algorithm based on IIB in Fig. 4. In the algorithm, the number of cores that involves in parallel computing is directly specified by users or automatically assigned by the "!cpu.hw ncpu" command. The algorithm partitions a task into several subtasks and assigns them into different cores to simultaneously execute in asynchronous mode.

4 Experiment and Results

4.1 The Workflow of Time Series Extraction

Time series profiles are very useful for remote sensing monitoring applications, such as land use and land cover change, disasters, forests, crops [21–23]. Extracting time series from temporal Landsat images across decades is time-consuming and computing-intensive. This study used the IIB-based parallel algorithm to improve the extraction efficiency of Landsat time series. Land surface water index (LSWI) were extracted from Landsat 4/5/7 TM/ETM+ images between 2000 and 2010 in Xishuangbanna Dai Nationality Autonomous Prefecture (XSBN), China. The workflow is show in Fig. 5, and is implemented by IIB-based parallel algorithm in Fig. 4. To parallelize this workflow, a function was created to implement this task and assigned to multiple computer cores to execute.

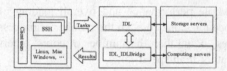

Fig. 3. The system architecture of the parallel model.

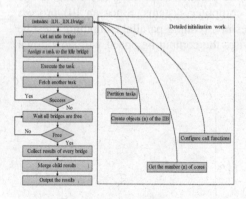

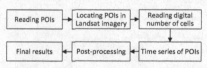

Fig. 4. The parallel computing algorithm.

Fig. 5. The time series extraction workflow. The POI is point of interest.

4.2 Results and Performance Testing

XSBN was selected as the study area. 248 Landsat 4/5/7 TM/ETM+ images (path/row 130/045) with Beijing 54 coordination system between 2000 and 2010 were downloaded from U.S. Geological Survey for time series extraction. Its total size is up to 1.0 TB. Computers with 16 cores of two processors (Intel® Xeon® E5-2650 2.00 GHz) were used to test the model. Based on the algorithm in Fig. 4, time series in Fig. 6 of natural forests LSWI at a point of interest (POI) (22.16,101.13) and rubber plantations at POI (21.98,100.24) were automatically extracted from Landsat TM/ETM+ Images between 2000 and 2010. Observing from these time series profiles, natural forests can be distinguished from rubber plantations by LSWI during about the Day of Year (DOY) between 0 and 125 in the study area.

Speedup calculated by formula (1) is an effective metric to evaluate performance of parallel algorithms and was taken to measure the performance of the model presented in this study. It is used to show how much speed of a parallel algorithm can be improved than a serial one. Under the situation of linear speedup ($S_p = p$), computing efficiency would double when the number of processors is doubled.

$$S_p = T_1/T_p \tag{1}$$

where S_p is the speedup, p is the number of CPU cores, T_1 and T_p is the computing time of the serial and parallel algorithms running on computers with p cores separately.

The performance testing result in Fig. 7 shows in a whole the speedup is continually increasing with the increase of cores involved in parallel computing. The speedup starts to slow down while the number of cores attending parallel computing is greater than 6. That is, the IIB-based parallel algorithm can obviously improve the computing rate, but limitations of computer system resources lead that the speedup slows down after core number reaches to a certain threshold in Fig. 7. Thereby, to optimize the scientific computing efficiency of the IIB-based parallel algorithms, the number of computer cores should be assigned to execute parallel computing according to the whole performance and overall loading of computer systems. Since an IIB process occupies one core, so the number of processes involving in parallel computing must be less equal than the number of cores; otherwise, the computing speed will slow down rather than speed up.

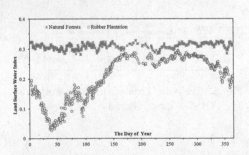

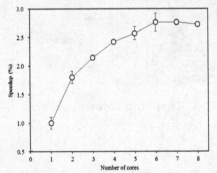

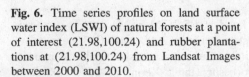

Fig. 6. Time series profiles on land surface water index (LSWI) of natural forests at a point of interest (21.98,100.24) and rubber plantations at (21.98,100.24) from Landsat Images between 2000 and 2010.

Fig. 7. The speedup is continually increasing with the number increasing of cores.

5   Discussion

This study presented a novel high-level parallel model for scientific computing on multi-core computers for researchers in scientific community based on the IDL IIB parallel component. It splits a dataset into multiple sub-datasets and executes them in different cores of a multi-core computer simultaneously. After all tasks completed, the results gathering from different cores are merged and outputted. It was designed under the popular map/reduce principle and easy to be understood and implemented for common researchers.

Integration of serval toolkits to do a research is very common in scientific community. Some toolkits without parallel ability are time-consume, and need to be parallelized to improve the computing efficiency. This model enables effective integration of third-part toolkits, such as Fmask and LEDAPS that are used for Landsat imagery processing, which increases the flexibility and extensibility of scientific applications. Current parallel models (such as mpiDL, fastDL and taskDL) do not provide parallel solutions for third-part toolkits or modules.

Taking the IIB into parallel computing algorithms can obviously improve efficiency of scientific computations, and take full use of multi-core computer system resources in an easy-to-use way. The performance testing result agrees that computing efficiency was gradually speeded up with the increasing number of CPU cores if the thread number is less equal than maximum of total CPU cores (Fig. 7), otherwise is unable to improve even slow down [18]. Since this parallel model took asynchronous mechanisms inside, it is of interest that the loose coupled between processes increases the scope for fault tolerance [24]. The IIB provides an easy way for scientists and engineers to implement parallel algorithms by automatically coordinating multi-core computers, even clusters. Thereby, researchers can easily improve research efficiency by parallel computations without helping with experts and professional teams.

Although the IIB-based parallel model has the advantages of taking full use of multi-core computer resources, it is limited by hardware configurations and memory capacity and still cannot support asynchronous interactions with users. The IDL can be embedded into other software toolkits or integrated development environment such as Visual Studio .NET, which should be properly choose according to the practical computing requirements. Additionally the IDL is able to run across complex clusters environments. Furthermore, data exchanging efficiency of GetVar and SetVar that use in the model is relative slow and needs to be improved by other technology such as shared memory (SHMMAP, etc.) that is 4–10 times faster than GetVar [25].

6   Conclusions

This study developed an IIB-based parallel algorithm for scientific computations, which can enhance the efficiency of data processing and analyzing and helps researchers and engineers speed up their research progresses. First, the algorithm presented in this study partitions a task into several subtasks that are separately simultaneously processed by multi-core computers. Second, it collects results from every core and merges them into one. Additionally, it is able to parallelize the third-part toolkits (such as LEDAPS and Fmask) through integrating the parallel ability of operating systems such as Linux parallel package. An experiment of extracting time series of temporal Landsat images was done and the result shows the parallel paradigm is capable of taking full use of multi-core computer resources based on multi-thread technology, and significantly improving processing and analyzing efficiency of huge volumes of scientific data. This algorithm can be also used in other similar research fields. The IIB-based algorithm is very fit to apply in middle-size laboratories or institutes, and should be further improved in distributed computing, deployment, visualization, and data sharing. Visualization can be realized through not only the IDL itself but also other integrated development environment including .NET and J2EE. This algorithm can be extended to large-scale laboratories or enterprises by using parallel solutions of the third-part toolkits (such as FastDL) or MPI clusters, even cloud-based solutions.

Acknowledgement. This work was supported by the National Natural Science Foundation of China (No. 31400493), the Science Research Fund of Yunnan Provincial Education Department, China (No. 2014Y324), and the Research Center of Kunming Forestry Information Engineering Technology (No. 2015FIB04).

References

1. Cai, X., Langtangen, H.P., Moe, H.: On the performance of the python programming language for serial and parallel scientific computations. Sci. Program. **13**, 31–56 (2005)
2. Fillmore, D., Galloy, M., Messmer, P.: Parallel IDL and python for earth and space science data analysis. In: AGU Fall Meeting (2007)
3. Commer, M., Kowalsky, M.B., Doetsch, J., Newman, G.A., Finsterle, S.: MPiTOUGH2: A parallel parameter estimation framework for hydrological and hydrogeophysical applications. Comput. Geosci. **65**, 127–135 (2014)
4. Paul, K., Mickelson, S., Dennis, J.M., Xu, H.: Light-weight parallel python tools for earth system modeling workflows. In: IEEE International Conference on Big Data (2015)
5. Canty, M.J.: Image Analysis, Classification, and Change Detection in Remote Sensing: With Algorithms for ENVI/IDL and Python, 2 edn. CRC Press/Taylor & Francis (2010)
6. Xiong, X., Wu, J.: IDL-based remote sensing image database design and implementation. In: The Second International Conference on Electric Information and Control Engineering, pp. 772–775. IEEE Computer Society (2012)
7. Alekseeva, E., Mezmaz, M., Tuyttens, D., Melab, N.: Parallel multi-core hyper-heuristic grasp to solve permutation flow-shop problem. Concurr. Comput. Pract. Exp. **29**(9), e3835 (2016)
8. Jannach, D., Schmitz, T., Shchekotykhin, K.: Parallel model-based diagnosis on multi-core computers (2016)
9. Laura, J., Rey, S.J.: Spatial data analytics on homogeneous multi-core parallel architectures (2016)
10. Liu, B., Liao, S., Cheng, C., Wu, X.: A multi-core parallel genetic algorithm for the long-term optimal operation of large-scale hydropower systems. In: World Environmental and Water Resources Congress (2016)

11. Liu, T., Xu, W., Yin, X., Zhao, X.: Multi-core parallel implementation of data filtering algorithm for multi-beam bathymetry data. In: Mechanical Engineering and Control Systems (2016)
12. Wang, J., Wang, B., Xiaohua, L.I., Yang, X.: Multi-core parallel substring matching algorithm using BWT (2016)
13. Wei, R., Murray, A.T.: A parallel algorithm for coverage optimization on multi-core architectures. Int. J. Geogr. Inf. Sci. **30**, 432–450 (2016)
14. Wasi-Ur-Rahman, M., Lu, X., Islam, N.S., Rajachandrasekar, R.: High-performance design of YARN MapReduce on modern HPC clusters with lustre and RDMA, pp. 291–300 (2015)
15. Chen, L., Huo, X., Agrawal, G.: A pattern specification and optimizations framework for accelerating scientific computations on heterogeneous clusters. In: IEEE International Parallel & Distributed Processing Symposium, pp. 591–600 (2015)
16. Liu, Z., Yang, J.: Remote sensing image parallel processing system. In: Proceedings of SPIE - The International Society for Optical Engineering, vol. 7497, p. 74970G (2009)
17. Shasharina, S.G., Volberg, O., Stoltz, P., Veitzer, S.: GRIDL: high-performance and distributed interactive data language. In: 2005 IEEE Proceedings of the International Symposium on High Performance Distributed Computing, HPDC 2014, pp. 291–292 (2005)
18. Xing-Qiang, L.U., An-Xi, Y.U., Liang, D.N.: Distributed parallel method for IDL. J. Syst. Simul. **18**(SUPPL.2), 256–258 (2006)
19. Bian, X., Zhang, D., Zhang, C., Wang, J.: Wind field retrieval from SAR images based on parallel computing technology in IDL. Comput. Eng. Appl. **50**(18), 261–264 (2014)
20. Nicter, C.: Improving between-shot fusion data analysis with parallel structures. Office of Scientific & Technical Information Technical reports (2005)
21. Zeng, L., Wardlow, B.D., Wang, R., Shan, J., Tadesse, T., Hayes, M.J., Li, D.: A hybrid approach for detecting corn and soybean phenology with time-series MODIS data. Remote Sens. Environ. **181**, 237–250 (2016)
22. Kastner, G.: Dealing with stochastic volatility in time series using the R package stochvol. J. Stat. Softw. **69**, 1–30 (2016)
23. Sulla-Menashe, D., Friedl, M.A., Woodcock, C.E.: Sources of bias and variability in long-term landsat time series over canadian boreal forests. Remote Sens. Environ. **177**, 206–219 (2016)
24. Bethune, I., Bull, J.M., Dingle, N.J., Higham, N.J.: Performance analysis of asynchronous Jacobi's method implemented in MPI, SHMEM and OpenMP. Int. J. High Perform. Comput. Appl. **28**, 97–111 (2014)
25. Coyote: Shared memory with IDL_IDLBridge, vol. 2017 (2016)

Accelerating Redis with RDMA Over InfiniBand

Wenhui Tang[(✉)], Yutong Lu, Nong Xiao, Fang Liu, and Zhiguang Chen

State Key Laboratory of High Performance Computing,
National University of Defense Technology, Changsha, China
wenhui_tang@yeah.net

Abstract. Redis is an open source high-performance in-memory key-value database supporting data persistence. Redis maintains all of the data sets and intermediate results in the main memory, using periodical persistence operations to write data onto the hard disk and guarantee the persistence of data. InfiniBand is usually used in high-performance computing domains because of its very high throughput and very low latency. Using RDMA technology over InfiniBand can efficiently improve network-communication's performance, increasing throughput and reducing network latency while reducing CPU utilization. In this paper, we propose a novel RDMA based design of Redis, using RDMA technology to accelerate Redis, helping Redis show a superior performance. The optimized Redis not only supports the socket based conventional network communication but also supports RDMA based high-performance network communication. In the high-performance network communication module of optimized Redis, Redis clients write their requests to the Redis server by using RDMA writes over an Unreliable Connection and the Redis server uses RDMA SEND over an Unreliable Datagram to send responses to Redis clients. The performance evaluation of our novel design reveals that when the size of key is fixed at 16 bytes and the size of value is 3 KB, the average latency of SET operations of RDMA based Redis is between 53 μs and 56 μs. This is about two times faster than IPoIB based Redis. And we also present a dynamic Registered Memory Region allocation method to avoid memory waste.

Keywords: Hybrid communication · InfiniBand · RDMA · Redis · Big data

1 Introduction

Because of the rapid development of Internet technology and the growing number of Internet users, huge amounts of data have been produced and we begin to enter the era of big data. These huge amounts of data are no longer a single structured relational data, but on this basis contain a large number of semi-structured and unstructured data. Due to the uncertainty of data structure of the semi-structured and unstructured data, it is difficult for the traditional relational database to carry on the effective management of data. In order to make up for the inadequacy of the traditional relational database, NoSQL arose at the historic moment. NoSQL systems [1] have shown that they have enormous scalability to satisfy the immense data storage and look-up requirements of data intensive web applications [2]. So NoSQL systems have been widely used as the

© Springer International Publishing AG 2017
Y. Tan et al. (Eds.): DMBD 2017, LNCS 10387, pp. 472–483, 2017.
DOI: 10.1007/978-3-319-61845-6_47

data center in various web applications such as the use of Redis [3] in WEIBO and the use of HBase [4] in Twitter and Facebook.

Database management systems based on memory primarily rely on main memory for data storage, saving the persistent files in the underlying persistent storage system, to avoid hard disks accesses and storage I/O operations when databases are processing clients' requests, eliminating the bottleneck storage I/O operations bring [20, 21]. These kinds of characteristics make in-memory stores provide lower data accesses latency and high-performance data processing performance and become the hot research spot in the big data area. What's more, the disappearance of hard disks accesses latency makes network latency, network bandwidth, CPU utilization and memory utilization become the main influences which affect the performance of in-memory stores.

The existing open-source Redis implementation uses traditional C (TCP) Sockets. This conventional implementation method provides a great degree of portability, but needs to endure the loss of performance.

Using high-performance networks such as InfiniBand [5] in high-performance computing clusters can provide advanced functions, such as Remote Direct Memory Access (RDMA), to achieve high throughput and low latency along with low CPU utilization. Therefore, combining high-performance networks with NoSQL systems, using advantages of high-performance networks to reduce access latency of NoSQL systems and increase throughput of NoSQL systems, has become the new research direction. A lot of RDMA based in-memory key-value stores have been generated such as Memcached [6], MICA [7], Pilaf [8], and HERD [9].

Therefore, in this paper, we hope on the basis of previous works, using RDMA technology to improve Redis, helping Redis show a superior performance. The major contributions of this paper are:

- A detailed analysis to reveal the performance of InfiniBand verbs, such as SEND/RECV, RDMA write and RDMA read. Provide data support for choosing high-efficiency RDMA verbs.
- A novel design for Redis communication module which supports both socket and RDMA. And make efficiently management of communication buffers.
- An extensive evaluation to study the performance implications of the new RDMA-capable module, in contrast to existing socket based module.

2 Background

2.1 Redis

Redis is an open source high-performance in-memory key-value database project supporting data persistence. It supports different data structures such as strings, hashes, lists, sets, sorted sets and so on to meet user's storage requirements under different scenarios. Redis maintains all of the data sets in the main memory. As a result of the memory read/write speed is obviously faster than the hard disk, compared with other databases based on the hard disk storage, Redis has significant performance advantages. Although the data sets stored in the memory are volatile, Redis has still realized the persistence of data sets. Redis generates snapshot files or log files, saving snapshot files or log files on the hard disk. When system crashes, using the snapshot files or log files on the hard disk to recover the system. So the persistence of data sets of Redis can be guaranteed.

2.2 RDMA

In the process of the conventional network transmission, data need to be transferred and copied among user space, kernel space and network equipment. This entire transmission process involves the consumption of memory bandwidth and CPU cycles. However, the Remote Direct Memory Access (RDMA) technology allows the local host to directly access the remote host's application buffer without impacting the remote host's operating system. Zero copy technique is adopted in the process, to avoid the copy of data between the user space and kernel space, reduces the time delay and CPU overhead.

Three RDMA technologies are in use today [10]: InfiniBand (IB), Internet Wide-Area RDMA Protocol (iWARP), and RDMA over Converged Ethernet (RoCE). Infini-Band [11, 12] defines a completely self-contained protocol stack, utilizing its own interface adapters, switches, and cables. It is worth noting that the InfiniBand in addition to support RDMA, it also provides an IPoIB (IP over InfiniBand) protocol. IPoIB allows existing socket based applications to run on InfiniBand with no modification. iWARP defines three thin protocol layers [13, 14] on top of the existing TCP/IP protocol [10]. Both iWARP and RoCE are permitted to be used over Ethernet physical devices which must support RDMA.

InfiniBand is a computer-networking communications standard used in high-performance computing, featuring very high throughput and very low latency [5]. This high-speed interconnection network is often used in high-performance computing clusters. Compared with traditional Ethernet, high-speed interconnection networks have lower latency and higher bandwidth.

Among the RDMA technology, User space programs access RDMA NICs directly using functions called verbs [9]. There are several types of verbs. Those most relevant to this work are RDMA read (READ), RDMA write (WRITE), SEND and RECEIVE (RECV) [9]. Applications post verbs to the queues of queue pairs which are maintained inside the RDMA NICs. The queue pair (QP) consists of send queue and receive queue. And every queue pair will be associated with a completion queue (CQ) to record the completion of the verbs operations.

Verbs operations can be divided into two types: one-sided operations and two-sided operations. SEND/RECV verbs are commonly regarded as two-sided operations because every local SEND operation requires a corresponding RECV operation at the remote side. Different with two-sided operations, one-sided operations such as RDMA reads and RDMA writes directly take read or write operations on application buffer of the remote host without notifying the remote host. And the remote host does not take any operation and will not be aware of local host's operations on its application buffer.

Similar to the conventional transmission protocol, RDMA transports can be connected or unconnected. A connected transport requires that a local queue pair only communicates exclusively with a remote queue pair, in a similar way, a remote queue pair can only communicates exclusively with a local queue pair. There are two main types of connected transports: Reliable Connection (RC) and Unreliable Connection (UC) [9]. Totally different with connected transports, unconnected transports do not require the local queue pair communicates exclusively with the remote queue pair, a local queue pair can communicate with any number of remote queue pairs. And the Unreliable Datagram (UD) is the only unconnected transport. Under different transport types, different verbs are available. The Table 1 shows the verbs which every transport type supports.

Table 1. The verbs which every transport type supports [9]

Verb	RC	UC	UD
SEND/RECV	Yes	Yes	Yes
WRITE	Yes	Yes	No
READ	Yes	No	No

3   Motivation

This section experiments use one x86-64 machine as a server and one x86-64 machine as the client. Every machine is equipped with 2.10 GHz Intel Xeon E5-2620L 6-core processors, featuring 32 KB for L1 instruction and data caches, 256 KB L2 and 15 MB L3 cache. 16 GB memory is installed on each node. Each machine is equipped with a Mellanox MT27500 ConnectX-3 40 Gbps InfiniBand HCA as well as an Intel gigabit Ethernet adapter. The machines run Red Hat Enterprise Linux Server 6 with kernel 2.6.32. And perftest which is an InfiniBand verbs performance test tool is used in this section experiments. The perftest package contains a set of bandwidth and latency benchmark such as ib_send_lat, ib_read_lat, ib_write_lat and so on.

The aim of this section experiments is to evaluate the latency and bandwidth of InfiniBand verbs over different transport types respectively: SEND/RECV over RC, UC and UD, RDMA write over RC and UC, RDMA read over RC.

What's more, we also discuss the SEND/RECV and RDMA write's performance under inlined mode. The inlined mode provides a feature that the small message up to 256 bytes can be inlined in the work request to avoid that the small message is fetched by the RDMA NIC via a DMA read [9]. This method can effectively avoid the DMA operations to reduce latency. But as we have mentioned, this mode only suits for small messages, the size of which is smaller than 256 bytes.

Figure 1 reveals that using inlined SEND/RECV operations can obviously reduce latency when the message size is less than 256 bytes. Without inlined mode, SEND/RECV operations over UD transport can slightly reduce latency, but the effect is not obvious.

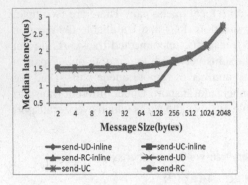

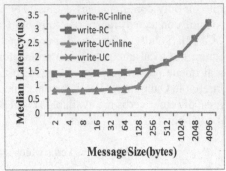

Fig. 1. The median latency of SEND/RECV operations

Fig. 2. The median latency of RDMA writes

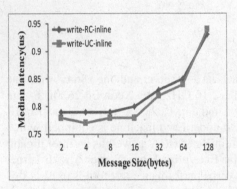

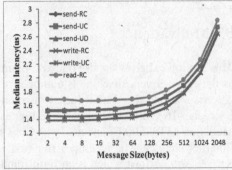

Fig. 3. The median latency of inlined RDMA writes

Fig. 4. The median latency comparison among SEND/RECV, RDMA write and RDMA read

Figure 2 indicates that when transferring small messages, using inlined mode can help RDMA writes have lower latency on RC transport and UC transport, similar to SEND/RECV operations. Although inlined RDMA writes can help to reduce latency no matter on RC transport or on UC transport, Fig. 3 gives us the comparison of median latency between inlined RDMA writes over RC transport and inlined RDMA writes over UC transport that inlined RDMA writes over UC transport have lower latency. Different with Fig. 3, RDMA writes without inlined mode over UC transport do not have better performance than RDMA writes without inlined mode over RC transport.

After the comprehensive comparison of the latency among SEND/RECV, RDMA write and RDMA read, it can be found that the latency of RDMA read operations is longer than SEND/RECV operations and RDMA write operations' latency, as Fig. 4 displays. This is because the RDMA read is a round-trip operation and the tested latency is round-trip time. But on the contrary, RDMA write and SEND/RECV are one-way operation and the tested latency is the half of the round-trip time.

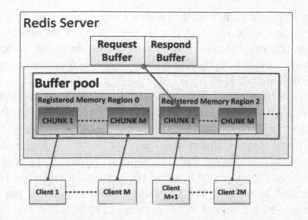

Fig. 5. Layout of server's buffer pool

As for the bandwidth of SEND/RECV, RDMA write and RDMA read, all of them attain the maximum of bandwidth when message size up to 4096 bytes.

4   Design and Implementation of Hybrid Communication Module

4.1   Hybrid Communication Framework

We present a novel hybrid communication module design for Redis to incorporate both the socket and advanced networks such as InfiniBand in this section. The hybrid communication module consists of two components: RDMA technology based high-performance network communication module and socket based conventional network communication module.

Redis server sets the parameters of the server through configuration file called redis.conf. Therefore, we add a new parameter Rdma in the configuration file to help Redis server to select network communication mode. When the value of Rdma is equivalent to yes, Redis server chooses high-performance network communication. When the value of Rdma is no, Redis server chooses conventional network communication. According to the service Redis server provides, Redis client chooses the corresponding communication module.

4.2   Communication Module Design for High Performance Networks

According to the design of HERD [9] and the performance evaluation of RDMA verbs in Sect. 3, we choose inlined and unsignaled RDMA writes to send clients' requests over UC transport. When the RDMA NIC completes the work request posted by the application, a completion event will be produced. And the completion event can be pushed to the queue pair's associated completion queue (CQ) by RDMA NIC via DMA write [9]. If we use unsignaled RDMA writes, DMA write can be avoided and extra overhead can be reduced. It is worth noting that even we use unsignaled operations, the transmission operations still consume completion queue resources. It does not generate work completion data structure or notification event [10]. So in order to avoid using up completion queue resources, we need use signaled mode transmission operations to handle the completion events which stay in the completion queue, using a selectively signaled send queue of size K, up to K−1 consecutive verbs can be unsignaled and the last verb operation is signaled [9].

Redis server maintains a buffer pool, when Redis server starts to initialize, a contiguous memory region is allocated and we call it Registered Memory Region 0. This contiguous memory region is divided into M Chunks, as showed in Fig. 5. Each Chunk consists of two parts: Request Buffer and Respond Buffer, as Fig. 5 displays. When Client i connects to the Redis server, Redis allocates a Chunk called Chunk i to the Client i. The clients' requests which are sent to the server's clients' associated Chunk's Request Buffer are formatted as follows. We show the details of clients' requests. Clients' requests are composed by request head, command and ending flag. The length of command and the content of ending flag are included in the request head. By adding ending flag at the tail of the request, we can directly rewrite Request Buffer without reclaiming and zeroing out Request Buffer after the Redis server extracts the request from the Request Buffer. And Redis server can use ending flag to poll the arriving of requests. What's more, ending flag makes the extraction of requests become easier. When the Redis server has already dealt with client's requests, the response messages are placed in the client's associated Chunk's Respond Buffer, and then the responses are sent back to the corresponding client by SEND operation.

When there are M clients keep connection with Redis server at the same time, this means that all of the Chunks in Registered Memory Region 0 are used up. So if the Redis server wants to deal with the connection of the M+1th client, it is necessary for Redis server to allocate another Registered Memory Region which includes M Chunk too. Such a dynamic Registered Memory Region allocation method avoid memory waste caused by allocating too much buffer at one-time, efficiently save the memory space.

5   Evaluation

In this section, we detailed evaluate the optimized Redis's performance and compare the original Redis based on socket over 1GigE network and IPoIB with optimized Redis based on RDMA over InfiniBand in terms of latency of SET and GET operations.

5.1   Experimental Setup

Use one x86-64 machine as a server and one x86-64 machine as the client. Every machine is equipped with 2.10 GHz Intel Xeon E5-2620L 6-core processors, featuring 32 KB for L1 instruction and data caches, 256 KB L2 and 15 MB L3 cache. 16 GB memory is installed on each node. Each machine is equipped with a Mellanox MT27500 ConnectX-3 40 Gbps InfiniBand HCA as well as an Intel gigabit Ethernet adapter. The machines run Red Hat Enterprise Linux Server 6 with kernel 2.6.32.We have allocated 16 MB memory to each Registered Memory Region of Redis server, and each Chunk possesses 2 MB memory, 1 MB memory is allocated to Request Buffer and another

1 MB memory is allocated to Respond Buffer. All the Registered Memory Region is registered to RDMA NIC when they are allocated. In all of our experiments, we use Redis version 3.0.5. And we do not use Redis default configuration. In order to get better performance, we have modified Redis's configuration file redis.conf. We have turned off the persistence function no matter based on snapshot or log.

5.2 Workloads

We use Yahoo! Cloud Serving Benchmark (YCSB) to examine the performance of Redis. We have implemented a C Redis Client to bridge the Redis with YCSB and the Redis Client is used to use workloads generated by YCSB. It is also worth noting that our Redis Client cannot batch local requests. Given that YCSB workloads generator can be highly CPU-intensive [15], we have generated workloads before we begin the tests. And the workloads are cached in the memory for Redis Client use. According to the previous work [16, 17] of workload analysis, we can clearly know that the number of GET operations and the number of SET operations is unequal at most of the time. Therefore, we have generated workloads following Zipf distribution and uniform distribution respectively. As for each kind of distribution, there are three kinds of GET/SET ratios which are 50%GET + 50%SET, 100%GET and 100%SET. Each workload haves 1 thousand operations. As for all the operations, the size of the key is fixed at 16 bytes. And the size of the value ranges from 4B to 3 KB. There are six workloads for each kind of value size.

And in our experiments, in order to explore the impact of the percentage of SET operations on the performance of Redis, we varied the set percentage for the experiments between 5%, 25%, 50%, 75% and 100%. And the value size is fixed at 3 KB.

5.3 Performance Analysis

Figure 6 exhibits the average latency of SET and GET operations under different workloads. From Fig. 6, we can know that no matter following what kind of distribution or what kind of GET/SET ratio, when value size ranges from 4 bytes to 512 bytes, the average latency of SET operations and GET operations of RDMA technology based Redis keeps in the area between 9 to 14 μs. Compared with RDMA based Redis, the average latency of SET operations and GET operations of socket based Redis keeps in the area between 95 to 104 μs which is much higher than RDMA based Redis when value size ranges from 4 bytes to 512 bytes. When value size is equal or greater than 1 KB, Fig. 6(a), (b), (d) and (e) indicate that because of huge amounts of SET operations, average latency begins to increase obviously. So through the Fig. 7, we can discover that no matter what kind of distribution we have used, with the increase of the set percentage, the average latency has increased.

What's more, Fig. 6 indicates that the average latency of the SET operations and GET operations over IPoIB of socket based Redis is just slightly lower than over 1GigE network when value size ranges from 4 bytes to 512 bytes. But when value size is greater than 1 KB, the gap between the average latency over IPoIB and the average latency over 1GigE network grows large. When value size is equal to 3 KB, the SET operations of RDMA based Redis is faster than IPoIB based Redis about 2 times.

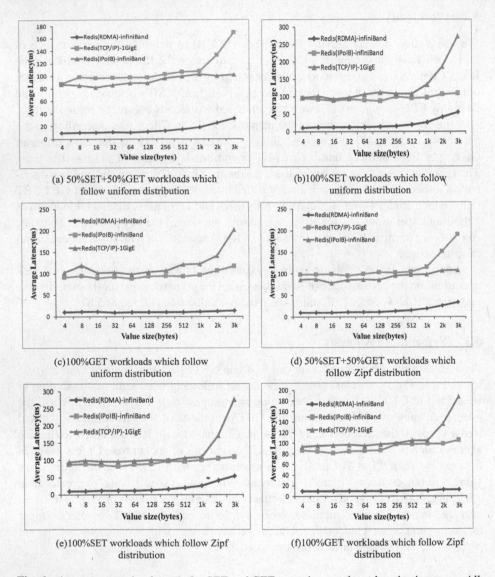

(a) 50%SET+50%GET workloads which follow uniform distribution

(b)100%SET workloads which follow uniform distribution

(c)100%GET workloads which follow uniform distribution

(d) 50%SET+50%GET workloads which follow Zipf distribution

(e)100%SET workloads which follow Zipf distribution

(f)100%GET workloads which follow Zipf distribution

Fig. 6. Average operation latency for SET and GET operations as the value size increases. All tests use workloads which follow uniform distribution and Zipf distribution and each distribution has different GET/SET ratios

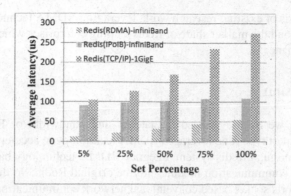

(a).   Average latency of varied set percentage with uniform distribution

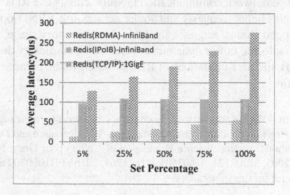

(b).   Average Latency of varied set percentage with Zipf distribution

Fig. 7. Average Latency of varied set percentage (a)Average latency of varied set percentage with uniform distribution (b)Average Latency of varied set percentage with Zipf distribution

6   Related Work

In the past few years, leveraging high-performance network to accelerate key-value store systems become a novel research trend. Some research work [2, 18] leveraging RDMA technology to accelerate HBase and Memcached have already existed. Some RDMA based key-value stores have been generated one by one such as Pilaf [8], HERD [9] and FaRM [19]. Pilaf is a RDMA based key-value store. Its clients use RDMA reads to realize GET operations to avoid CPU overhead in its server. HERD is a key-value store too which use RDMA writes over UC transport to send clients' requests and use SEND operations over UD transport to send server's response messages. FaRM is a RDMA based distributed share memory system. In addition to RDMA based key-value stores, C-Hint [15] has been generated, an efficient and reliable cache management system for in-memory key-value stores that leverage RDMA Read. C-Hint is utilized to solve the problem RDMA reads bring.

On the basis of existing research work, leveraging RDMA technology to accelerate Redis can expand the market share of Redis helping enterprises which use Redis gain a better performance.

7   Conclusion

In this paper, we have evaluated the latency and bandwidth of RDMA verbs over different transport types and leveraged RDMA technology to accelerate Redis. Without changing the original Redis system, integrate RDMA technology based high-performance network communication module into the original Redis. So the optimized Redis not only supports socket based conventional network communication but also supports RDMA based high-performance network communication. Within the optimized Redis's high-performance network communication module, clients use RDMA writes over UC transport to send clients' requests and Redis servers use SEND operations over UD transport to respond clients' requests. In order to save memory space, we design a buffer pool which can allocate buffer dynamically. Our design indicates that when value size is equal to 3 KB, the SET operations of RDMA based Redis is faster than IPoIB based Redis about 2 times.

Acknowledgment.   We are grateful to our anonymous reviewers for their suggestions to improve this paper. This work is supported by the National High-Tech Research and Development Projects (863) and the National Natural Science Foundation of China under Grant No. 2015AA015305, 61232003, 61332003, 61202121, 61402503, U1611261, 2016YFB1000302 and NSFC61402503. And we are grateful for this opportunity.

References

1. Cattell, R.: Scalable SQL and NoSQL data stores. J. ACM Sigmod Record. **39**(4), 12–27 (2010)
2. Huang, J., et al.: High-performance design of HBase with RDMA over InfiniBand. IEEE Int. Parallel Distrib. Process. Symp. **19**, 774–785 (2012)
3. Redis. http://redis.io
4. HBase. http://hbase.apache.org
5. InfiniBand Trade Association. http://www.infinibandta.org
6. Memcached: High-Performance, Distributed Memory Object Caching System. http://memcached.org
7. Lim, H., Han, D., Andersen, D.G., et al.: MICA: a holistic approach to fast in-memory key-value storage. J. Manage. **15**(32), 36 (2014)
8. Mitchell, C., Geng, Y.F., Li, J.Y.: Using one-sided RDMA reads to build a fast, CPU-efficient key-value store. In: Usenix Conference on Technical Conference, pp. 103–114. USENIX Association (2013)
9. Kalia, A., Kaminsky, M., Andersen, D.G.: Using RDMA efficiently for key-value services. J. ACM Sigcomm Comput. Commun. Rev. **44**(4), 295–306 (2014)

10. Macarthur, P., Russell, R.D.: A performance study to guide RDMA programming decisions. In: IEEE, International Conference on High PERFORMANCE Computing and Communication & 2012 IEEE, International Conference on Embedded Software and Systems, pp. 778–785. IEEE (2012)
11. Keeton, B.K., Patterson, D.A., Hellerstein, J.M.: InfiniBand Architecture Specification, Volume 1.0, Release 1.0. InfiniBand Trade Association (2010)
12. Grun, P.: Introduction to InfiniBand for End Users. InfiniBand Trade Association (2010)
13. Shah, H., et al.: Direct data placement over reliable transports. J. Heise Zeitschriften Verlag (2007)
14. Recio, R., Metzler, B., Culley, P., et al.: A remote direct memory access protocol specification. J. Neurophys. **93**(1), 467–480 (2007)
15. Wang, Y.D., et al.: C-Hint: an effective and reliable cache management for RDMA-accelerated key-value stores. In: ACM Symposium on Cloud Computing, pp. 1–13. ACM (2014)
16. Atikoglu, B., Xu, Y., Frachtenberg, E., et al.: Workload analysis of a large-scale key-value store. J. ACM Sigmetrics Perform. Eval. Rev. **40**(1), 53–64 (2012)
17. Cooper, B.F., et al.: Benchmarking cloud serving systems with YCSB. In: ACM Symposium on Cloud Computing, pp. 143–154. ACM (2010)
18. Jose, J., et al.: Memcached design on high performance RDMA capable interconnects. In: IEEE International Conference on Parallel Processing, pp. 743–752. IEEE (2011)
19. Dragojevi, A., Narayanan, D., et al.: FaRM: fast remote memory. In: Usenix Conference on Networked Systems Design and Implementation. USENIX Association (2014)
20. Yang, X.J., et al.: The TianHe-1A supercomputer: its hardware and software. J. Comput. Sci. Technol. **26**(3), 344–351 (2011)
21. Chen, Z.G., Xiao, N., Liu, F., Du, Y.M.: Reorder write sequence by hetero-buffer to extend SSD's lifespan. J. Comput. Sci. Technol. **28**, 14–27 (2013)

Knowledge Base and its Framework

Application of Decision Trees in the Development of a Knowledge Base for a System of Support for Revitalization Processes

Agnieszka Turek[✉]

Faculty of Geodesy and Cartography, Warsaw University of Technology, Warsaw, Poland
agnieszka.turek@pw.edu.pl

Abstract. The objective of the paper was to develop an efficient system supporting the management of degraded areas and their revitalization. The author developed a knowledge base for a system of assessment and support of revitalization processes through the application of selected data mining methods. The database included more than 100 objects for which approximately 100 attributes on different measurement scales were collected. The analysis of the collected data involved the application of decision trees. The intermediate goal was the determination of the applicative potential of properly transformed spatial data for the requirements of revitalization procedures. The studies carried out represent early work in this scientific field, and the author provides a methodological basis for further research.

Keywords: Spatial data mining · Revitalization · Decision trees · CART · Knowledge discovery · Data enrichment

1 Introduction

Revitalization is a long-term process involving planned measures aimed at transforming the functional spatial structure of degraded urban areas, and consequently their economic and social recovery. Its objective is to recover an area from a state of crisis, restoring its former functions, or to introduce new functions generating conditions for further development. Therefore, the process requires the coordination of comprehensive and inter-disciplinary activities [1].

The experience of the countries of Western Europe shows that solving problems related to degraded areas requires the development of an efficient system supporting the management of such areas and their revitalization. Relevant procedures regulating the course of action in the case of such measures have still not been developed in Poland. The development of these types of procedures or recommendations, however, requires not only the analysis of current solutions and examples, but more importantly the trans-formation of collected data into useful information and its translation into the form of a knowledge base supporting revitalization processes.

The dynamic development of information technologies permits access to an increasing number of data sets and tools which can be applied not only in designing

© Springer International Publishing AG 2017
Y. Tan et al. (Eds.): DMBD 2017, LNCS 10387, pp. 487–495, 2017.
DOI: 10.1007/978-3-319-61845-6_48

revitalization measures, but also in the assessment and forecasting of the potential effects of proposed investments, as well as the development of prediction models. Classic methods of data analysis, however, are largely insufficient when the objective of the analysis is not a simple comparison, but the extraction of knowledge or the formalization of decision-making rules. The data mining methods applied by the author offer new possibilities in the field of data analysis (and also spatially-distributed data) through "discovering" knowledge contained in the collected set of spatial and descriptive data [2].

Data mining is a relatively young, interdisciplinary scientific endeavour resulting from the integration of knowledge from the fields of database techniques, statistics, artificial intelligence, and social and economic sciences [3]. The term is defined as the process of research into, and analysis of, large amounts of data with the application of specified algorithms for the purpose of discovery of significant models and rules [4, 5]. Techniques of data exploration and data enrichment applied in the process of discovering knowledge permit the detection of patterns, correlations, schemes and models "hidden" within the observation set. Therefore, records collected in databases can be used for supporting the decision-making process. It should be emphasized that skilful "enrichment" of descriptive information by cartometric parameters resulting from the performance of spatial analyses, allows for the expansion of data mining techniques by incorporating spatial components [3, 6, 7]. Thus, defined spatial data mining permits simultaneous consideration of the analytical process of descriptive parameters characterizing particular objects, as well as their location, contiguity, and correlations of a geometric and/or topological character.

Methods of spatial data mining do not only offer the possibility of the analysis of source information, i.e., finding spatial models and the extraction of decision-making rules, but they also allow for forecasting conclusions. The objective of spatial data mining analyses is the development of methods permitting "indirect" analysis of objects, phenomena and correlations occurring in space, performed through the analysis of the model, which is a properly-designed spatial database [6, 8].

Data mining is used for many scientific and practical applications, such as economics, medicine, or business (understood broadly). However, it is still not used in revitalization procedures. The present paper is a first attempt to apply data mining techniques to a revitalization process. In order to do this, it is necessary to gather at least basic data, and due to the specificity of the revitalization process, the requirement for expert evaluation and the large number of parameters (attributes), this is an extremely laborious task. However, the author has taken up the challenge.

Many methods of spatial data mining could potentially be useful for the purposes of development of a knowledge base for the system of support for revitalization processes. In this article, however, the author consciously decided to limit the conducted research to the application of so-called decision trees. This decision is determined by pragmatic factors, i.e., the need to obtain forecast models with high credibility. The specificity of the applied approaches permits on the one hand the development of very clear, multi-level classifications of the analysed data, and on the other hand, the extraction of useful decision-making rules.

One important advantage of trees, among others, is a comprehensive sequence of decision-making rules permitting the classification of new objects based on changed values. Trees are also resistant to outliers. As explanatory variables, any combination of continuous (expressed on a quantitative scale) and category variables (expressed on a classification scale) can be used. In a tree, a single variable can be used multiple times in different parts. Decision trees permit discovery of the context of correlations and interactions between variables [6, 9, 10].

2 Analyses Performed

The objective of this stage of the research was the development of a knowledge base for the system of assessment and support for revitalization processes through the application of selected data mining methods. The analysis of the collected data involved the application of so-called decision trees. The intermediate goal was the determination of the applicative potential of properly-transformed spatial data for the needs of revitalization programmes. The development of the knowledge base, hierarchical data classification, and especially the development of credible decision-making rules, however, requires the collection of a large set of source data. Generally, the larger (and more credible) the set of so-called learning data is, the more valuable the prediction model will be. It is also important for analysis by means of spatial data mining techniques, to enrich descriptive data with parameters resulting from cartometric, geometric and topological analyses.

Taking into consideration the different descriptive parameters of the analysed objects and the geometric and topological relations characterizing their spatial surroundings, permits the development of a credible knowledge base for a system supporting the holistically-understood revitalization process, extracted from single examples.

Based on numerous examples of revitalization measures implemented in different parts of the country, a tabular knowledge base of the systems supporting the decision-making process concerning the type and scope of revitalization was developed.

The research involved collecting information concerning transformations of 102 industrial objects of different types distributed unevenly throughout the territory of Poland. At the preliminary stage of the work, as many as 99 attributes (parameters) were collected, expressed on the level of quantitative rank, as well as qualitative measurement scales characterizing given objects. Further in the analyses, the data were subject to reduction and generalization. Data mining analyses were performed for all the selected research objects. The distribution of the analysed objects is presented in Fig. 1.

The approach proposed in this paper combines the possibilities of GIS packages and advanced statistical software, permitting the performance of complex analyses. The research involved the application of the tool packages ArcGIS, MapInfo, MapBasic and Statistica Data Miner by Statsoft. The GIS tools permitted the collection and analysis of spatial data, and the data mining analyses were implemented in a specialized Data Miner tool environment, constituting a component of the Statistica package. In the GIS environment, a model of the spatial analyses was developed allowing for the selection of an optimal scenario of measures for revitalization, using spatial multi-criterion analyses. The minimum scope of information for the system of management and

Fig. 1. Distribution of the analysed research objects in Poland.

organization of this type of undertaking was also determined. Such a system should generate a preliminary assessment and valorization of the terrain for the user, calculated by means of a relevant algorithm with the application of spatial data mining methods.

The development of models describing significant correlations, and the proper analysis of such a complex research problem, required the collection of large amounts of data from different sources (public administration, scientific institutions, commune offices, literature review, internet search and field research). The integration of data was performed with reference to spatial data parameters (ArcGIS tool environment).

For the analysed objects, descriptive data were collected concerning, among other things, the original function, date of establishment and discontinuation of industrial activity, investment costs and problems occurring during transformations (23 parameters in total). Based on spatial data obtained from the Database of Topographic Objects (BDOT10 k), among other sources, the assessment of parameters such as attractiveness of the location of the area, transport accessibility, access to media, ownership structure or negative contiguity (22 attributes in total) was performed. The planning conditions and technical state of the existing building development was also analysed. The implemented projects for revitalization of post-industrial areas were subject to expert assessment after division into two categories – general and detailed. Each of the analysed issues was assessed on the rank scale defined by the author (−1, 0 and 1), and values belonging to particular categories were summed. In this way, in the first approach, values from −2 to +4 were obtained for the general category, and values from −5 to +8 for the detailed category. Data mining analyses were also performed for the sum of all the analysed variables (from −7 to +12).

The collected data required "cleaning". This involved, among other things, solving problems resulting from lack of data, incoherence of the data set, and false observations [11]. Some of the parameters required standardization, generalization and division into classes for the purpose of facilitating the generation of generalized knowledge.

Some of the objects and attributes characterized by a considerable lack of data (predominantly resulting from lack of access to data) or high variability of parameters were excluded from the analysis. The consequence of these measures was incomplete data. On the other hand, the collected data were enriched by means of spatial analyses which permitted the researcher to obtain quantitative measures such as distance from important objects or complexes, or their number within a radius of 1000 m.

The resulting database is of a complex character, and covers data on objects and their surroundings. The collected examples (of both good and bad practices in revitalization processes) contain "hidden" knowledge. Such knowledge is implicit in character, and is difficult to interpret. The application of spatial data mining methods permits the "extraction" or "disclosure" of such knowledge, and enables it to be shown it in an explicit way in the form of clearly-defined rules. From many spatial data mining methods, decision trees were selected, because the method allows for the rules to be shown in the model. Therefore, the results of such analyses can be used for forecasting.

2.1 Ranking of Predictors

Data mining methods permit consideration in the analyses of N dimensions in the non-metric space of parameters. The ranking of predictors allows for the reduction of the number of independent attributes (and therefore for transition to a lower number of dimensions) through the selection, from a large set of different predictors, of only those which are significant and have some (statistical) effect on the obtained result [6].

The analysis performed shows that the development of a credible model of assessment requires different descriptive and spatial attributes, whereas the ranking of predictors shows that the most significant ones include attributes belonging to spatial analyses, and therefore descriptive data alone are insufficient. The most significant effect of revitalization measures concerns the attribute describing the distance from building development. Other important parameters having a considerable impact on the final effect of revitalization measures are the distance from sporting and recreational objects, and the distance from industrial and economic objects. As it turns out, the attribute describing distance from sporting and recreational objects is equally important in the case of generalized values and their division into four classes (1 – near, 2 – medium distance, 3 – far and 10 – out of reach). With such a small sample of data, the choice of predictors cannot be based only on statistical tests – it requires expert evaluation. The data enrichment – extraction directly from spatial database parameters resulting from geometric and topological analysis is of key importance.

2.2 CART Classification Trees

Decision trees use data of both quantitative and qualitative character for building the model. Due to this fact, the applied spatial predictors do not require aggregation and classification [10]. This usually results in loss of part of the information. Any type of correlation can be directly recorded in an attribute. An attribute of qualitative character can contain, among other things, information on the type of topological relation between objects, and quantitative attributes can contain information concerning distance, direction or topological relations. Moreover, algorithms for the development of the model use tabular data [6].

Values for the dependent variable were determined based on the sum of particular elements of the expert assessment. In the first approach, the effect of the implementation of revitalization measures was regarded as "positive" when the sum of assessments from detailed categories for all the analysed elements was within a range from +3 to +8. The effect of revitalization was regarded as "negative" when the sum of assessments was within a range from −5 to +2. An example classification tree developed during the processing of the analysed data is presented in Fig. 2. In general, rules that are read directly off a decision tree are far more complex than necessary, and rules derived from trees are usually pruned to remove redundant tests [12]. The resulting CART tree shows that revitalization measures are successful if the analysed object is located at a distance of not more than 18 m from other buildings, and if objects with an industrial and economic function are located at a distance of more than 154.5 m from the revitalized object. Should other buildings be located at a distance of less than 18 m, the effect of the revitalization measures is also positive.

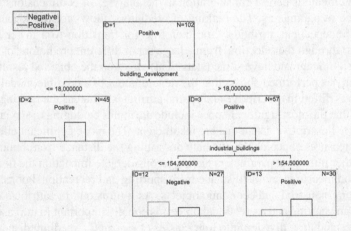

Fig. 2. CART classification tree. Leaf ID = 13: the effect of revitalization is positive, if building development is located at a distance of more than 18 m, and objects with industrial and economic function are located at a distance of more than 154.5 m from the revitalized object. Leaf ID = 2: the effect of revitalization is positive, if building development is located at a distance of less than 18 m. Leaf ID = 12: the effect of revitalization is negative, if building development is located at a distance of more than 18 m, and objects with industrial and economic function are located at a distance of less than 154.5 m from the revitalized object.

In the second approach, the classification was changed. "Positive" effect was ascribed to all objects for which the sum of assessments was equal to or higher than +4. The analysis included all quantitative predictors. As a result of the analysis, a tree was obtained according to which the effect of revitalization is positive if the object is located at a distance of less than or equal to 388 m, and when industrial and economic objects are located at a distance of more than 47.5 m.

In the next approach, all spatial attributes were divided into classes (Fig. 3). Generalization and division into four classes was performed as follows: 1- object located nearby, 2 – object located at a medium distance, 3 – object located far away and 10 – object located out of range (at a distance of more than 1000 m from the analysed parameter). Such defined data permitted a tree to be obtained where the effect of revitalization is negative when transport complexes are located far from (3) or at a medium distance from the objects (2). The effect of revitalization is positive when transport complexes are located near the object (1) (or are out of reach), and simultaneously commercial and service objects are located near the object (1).

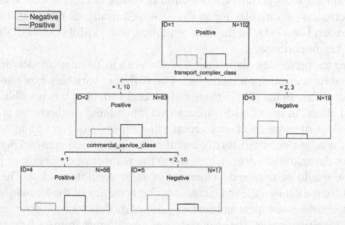

Fig. 3. Classification tree where spatial attributes were divided into four classes.

By including only selected predictors in the analysis, different correlations were obtained. For example, the effect of revitalization was negative when "object category I" was equal to 7 (only clearing the area) or 2 (restoration of the industrial function and preservation of the existing objects with simultaneous introduction to the area of new functions and capacities).

3 Conclusion

As mentioned before, data mining methods are not being used in revitalization procedures at present. This research constitutes a preliminary stage of exploration of data mining techniques, which could be useful in the revitalization process. Revitalisation is an interdisciplinary issue, and therefore the preparation of data for analysis was a long and time-consuming process. Both spatial and descriptive data were collected, based on many different sources and databases. The collected data required cleaning, generalization, relevant arrangement, and classification. Relying solely on expert knowledge, finding key correlations in a database which includes more than 100 objects for which approximately 100 attributes were collected on different measurement scales, was a difficult goal to achieve.

The author has performed the first approach to such research. Despite the large amount of work, the collected data set is relatively small, and so the results are quite simple. Further research, however, is expected to deepen the analysis, and explore new rules and patterns. However, further studies will require the enlargement of the existing data set, by using examples of revitalization activities from other countries.

The analysis of all the collected data with the application of data mining tools without any expert control and supervision may provide results in which the correlation between particular predictors can be of spurious character, providing false results and false conclusions. The example shows that only the synergy of expert knowledge and skilful use of data mining tools permits the detection of valuable correlations in the data set. Unskilled selection of variables for analysis caused irrational results to be obtained. Therefore, expert knowledge in the field combined with skilful control of the obtained results is of key importance.

The analyses performed show that it is possible to develop models imaging the correctness of revitalization measures based on different variables expressed on many measurement scales. In the case of the use of several variables, it is possible to forecast the potential effects of revitalization measures. Data mining methods can contribute to the development of more functional geoinformation systems, providing credible and complex knowledge that is particularly useful in the area of decision-making processes. In the case of the analysed issue, an increase in the number of observations by an order of magnitude would permit the development of more credible models. The expansion of the set of source data implies an increase in the accuracy of the knowledge base.

In the case of the assessment and development of models for revitalization measures, additional difficulties appear. The analysed cases are different from each other, and often not comparable. Some areas require the application of individual indicators adjusted to the specificity of the analysed area. The greatest problem occurred in the case of environmental indices, due to difficulties with obtaining this type of data. Another problematic aspect is the fact that the success or failure of an undertaking can often be determined only after several years from completion of the investment. Moreover, data analysis is a complex process involving problems related to the dimensions of the data, the dynamic structure of the data, non-systematic errors (noise) and missing values.

The proposed guidelines, developed with respect to a system of support for the decision-making process as "revitalization rules", and considering the spatial context for subjects implementing revitalization, can support the launch of an efficient and effective process of renewal and management of areas with potential functional and spatial value, the transformation of which is hindered by unfavourable development conditions. The methodical approach to the problem of revitalization and management of post-industrial areas has not been described in the literature to date.

References

1. Act on Revitalization of October 9, 2015, Dz. U. 2015, item 1777
2. Olszewski, R., Turek, A.: Application of the spatial data mining methodology and gamification for the optimisation of solving the transport issues of the "Varsovian Mordor". In: Tan, Y., Shi, Y. (eds.) Data Mining and Big Data DMBD 2016. LNCS, vol. 9714, pp. 103–114. Springer, Cham (2016). doi:10.1007/978-3-319-40973-3_10
3. Migut, G.: Czy stosowanie metod data mining może przynosić korzyści w badaniach naukowych? Statsoft Polska, 49–65 (2009)
4. Berry, M.J.A, Linoff, G.S.: Mastering data mining. In: The Art and Science of Customer Relationship Management. Wiley, New York (2000)
5. Fayyad, U.M., Piatetsky-Shapiro, G., Smyth, P.: From data mining to knowledge discovery in databases. AI Mag. 17(3), 37–54 (1996). American Association for Artificial Intelligence
6. Fiedukowicz, A., Gąsiorowski J., Olszewski R.: Selected methods of exploration of spatial data analysis (Spatial Data Mining). Faculty of Geodesy and Cartography, Warsaw University of Technology, Warsaw (2015)
7. Zakrzewicz, M.: Data Mining i odkrywanie wiedzy w bazach danych. In: Materiały konferencyjne III Konferencji Polskiej Grupy Użytkowników Systemu Oracle, Zakopane (1997). http://www.cs.put.poznan.pl/mzakrzewicz/pubs/ploug97.pdf
8. Gotlib D., Iwaniak A., Olszewski R.: GIS. Obszary zastosowań. Wydawnictwo Naukowe PWN, Warszawa (2007)
9. Migut, G.: Data mining methods, Training materials, Statsoft Poland (2016). http://www.statsoft.pl/Portals/0/Downloads/Czy_stosowanie_metod.pdf
10. Breiman, L., Friedman, J. Olshen, R., Stone, C.: Classification and Regression Trees. Chapman & Hall/CRC Press, Boca Raton (1984)
11. Pyle, D.: Data preparation for Data Mining. Morgan Kaufmann Publishers Inc., Burlington (1999)
12. Written, I.H., Frank, E.: Data Mining: Practical Machine Learning Tools and Techniques. Morgan Kaufmann Publishers, Burlington (2005)

ABC Metaheuristic Based Optimized Adaptation Planning Logic for Decision Making Intelligent Agents in Self Adaptive Software System

Binu Rajan[1(✉)] and Vinod Chandra[2]

[1] Department of Computer Science and Engineering, SCT College of Engineering,
Thiruvananthapuram, India
bin_rajan@yahoo.com
[2] Computer Centre, University of Kerala, Thiruvananthapuram, India
vinod@keralauniversity.ac.in

Abstract. The potential of machine intelligence is enormously increasing with a vision of computing systems that can act as good decision making and self managing entities. This led to the introduction of systems that are more intelligent with self* properties and are known as Self Adaptive Software Systems (SAS). Intelligent Agents which has a high adaptation capability forms the main component of such systems. These self adaptive systems are provided with the ability of self–configuring based on the run time environmental changes which guarantee the overall system functional and QoS goals. This paper proposes an optimized decentralized adaptation logic for modeling SAS which exploits the multi-agent concept. Each subsystem has an objective and uses an Artificial Bee Colony metaheuristic to achieve local optimization which in turn leads to the optimization of the whole distributed system.

1 Introduction

The complexity of modern software is increasing both in its size and varying nature of its behavior during runtime. This evolution in software led to the development more manageable systems that are capable of managing and modifying itself after the system is released for use. These systems are adaptive to its runtime environment satisfying certain goals and conditions. They are able to accommodate new policy changes during runtime [1]. The effect of adaptation can viewed as appearance or disappearance of some components or they may change their behavior in an unpredictable manner. Such systems called Self-Adaptive Systems (SAS) need more significant innovations and progresses in order to cope with the challenges associated with the design and management. It has also made it imperative to investigate new approaches for designing such self adaptive systems with highly dynamic optimized adaptation with respect to environmental changes to avoid the high cost of system downfalls.

© Springer International Publishing AG 2017
Y. Tan et al. (Eds.): DMBD 2017, LNCS 10387, pp. 496–504, 2017.
DOI: 10.1007/978-3-319-61845-6_49

1.1 Self-adaptive Software System

A formal definition for Self Adaptive system is *"Systems that are capable of adjusting their behavior at runtime to achieve certain functional or Quality of Service Goals"*. The main features associated with SAS are dynamicity, uncertainty, adaptivity and complexity. Various methodologies have been developed to address the requirements such as modeling and self organization of SAS, but optimized approaches are little known. The self organization strategy used for adaptation holds the main challenge in designing such high performance autonomic systems.

Feedback Loop Control and MAPE-K Architecture

The key element of control theory is feedback loop control, which provides well defined mathematical models, tools and techniques for verification and validation in order to measure their performance and various other factors. The basic design of self-adaptive systems is derived from feedback loop control in control theory system. The system is continuously monitored and feedbacks are analyzed. The feedback may be positive or negative. Based on the positive and negative feedback a new action will be initiated without affecting the overall system behavior.

MAPE-K feedback loop model, the architectural blue print by IBM [3] is one of the common reference model for modeling a dynamic and distributed self adaptive systems. It is a well organized model with a loop back control which helps to realize the real time behavior of a self adaptive system. It comprise a sequence of four coordinating operations–*Monitor, Analyze, Plan and Execute along with a Knowledge Base*.

1.2 Agent Based Modeling

The complex autonomic systems can be considered to be a combination of heterogeneous components that interact with each other to achieve the optimal behavior. In order to accomplish this, system may undergo changes like adding and deleting of components without affecting the overall goals. During change, the system may evolve its behavior to an optimized one by means of online learning about the current behavior and then moving towards appropriate path. This action–reaction property is the main feature of a dynamic system. The most conventional methods for implementing this uses the conventional **if..else** statements based on some local parameter monitoring. But this is not seemed to be an optimal solution in online systems. The system behavior will have to change rigourously. In modeling such systems dynamic selection of action on various inputs and parameters are required. Thus an Agent-Based modeling is proposed for modeling such systems which makes it convenient for local parameter optimization which leads to global behavior. Agents are the main entities in the runtime system that has some observable state, actions–reactions and a set of rules to perform these actions and reactions. The basic concept of intelligent agents and the practical issues in its design are explained in [6]. In agent-based modeling the dynamic behavior of the real time system is decided by the collective behavior of the agents. The agents interact with each other in a coordinated way and reach an optimized solution. Many nature inspired behaviors like swarm intelligence, ant colony and bee colony along with the bottom up design methodology has also given a large contributions in the evolving agent based modeling.

1.3   Multi-agent Multi-zone System

More complex self adaptive systems include various subsystems or zones each with a large number of autonomous intelligent agents which may communicate, cooperate and negotiate with each other which is illustrated in Fig. 1. These agents are grouped on the basis of the subsystem to which they belongs. The agents of different groups may interact with each other. Also these individual agents continuously monitor the changing environment and act according to the perceived information from the environment. The optimized behavior of the agents of each zone and the sharing of information between them lead to the overall system optimization.

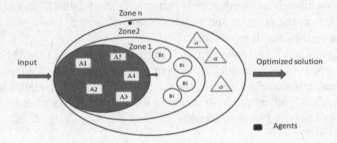

Fig. 1.   Multi-agent multi-zone system

1.4   Background

Various studies have been undergone in the modeling and development of complex adaptive systems with high ambience of intelligence. Agent Based modeling concept [6] has also influence the autonomic system development. The need of more intelligent systems with self* properties requires the design of various intelligent agent models with different functionalities. These agents may have high adaptation capability. These SAS are provided with the ability of self–configuring based on the run time environmental changes which guarantee the overall system functional and QoS goals. It has also made it imperative to investigate new approaches for designing such self adaptive systems with highly dynamic optimized adaptation with respect to environmental changes to avoid the high cost of system downfalls. This paper proposes an ABC metaheuristic based optimized decentralized adaptation logic for modeling SAS which also exploits the multi-agent concept. It also allows more agile agent handling while improving the overall system goals. As a proof of study, we demonstrated this strategy in restaurant simulation software. It is a distributed system with different system components each with different behaviors. These components are individual agents with some computational capability. The complete system is divided into various subsystems called zones. Each zone contains agents of same or different behavior and are synchronized.

These agents interact with each other and are able to respond to environmental triggers by taking appropriate decisions in order to optimize the whole system following certain rules. Each zone may use optimized adaptation logic to cause local optimization which in turn may lead to global optimization of the whole system. But solving such optimization problem is a current area of research. Several nature inspired meta-heuristics has been defined for solving such problems. Swarm intelligence is a group of metaheuristic methods designed based on the intelligent behavior of self-adaptive agents to obtain global optimization. The collective behavior of swarms adjusts the solutions based on each individual solution and then reaches the global best solution. Artificial Bee Colony (ABC) Algorithm is one of the most popular bioinspired metaheuristic based on a particular intelligent behavior of honey bee swarms. Our objective in this work is to minimize the average waiting time of each clients in the restaurant without affecting the normal functioning of the system and not overloading the employees. This work proposes some modified ABC algorithms that can optimize various subsystems of the defined distributed system.

2   Restaurant of Dining Philosophers Problem

An enhanced version of Classical Dining Philosophers Problem, named Restaurant of Dining Philosophers problem is used as the case study by adding some dynamic behaviour and constraints to the Classical Dining Philosophers synchronization problem. The restaurant system is divided into three subsystems: Dining subsystem, Service subsystem and Kitchen subsystem. The Problem is defined as follows:

(i) A Restaurant is open for service which offers **k** number of meals $(m_1, m_2 ... m_k)$

(ii) There is one table with **n** chairs surrounding it with one for each seat (n forks) in the dining hall.

(iii) The restaurant has p number of philosophers as clients for that restaurant who visit randomly carrying a fixed amount of money.

(iv) The philosopher Pm entering the restaurant can be in any of the three states: *Thinking, Eating, and Leaving*

(v) There are w waiters $(W_1, W_2 .... W_w)$ and Ch Chefs $(C_1, C_2 ... C_{ch})$ appointed in the restaurant.

(vi) The philosopher entered will be allotted a seat if available; else he goes to the thinking state. If the philosopher has to wait for a long time, he will leave the restaurant and never visit again.

(vii) Once the seat is allotted the philosopher can order his meals to the available chef.

(viii) The waiter places the order to the less loaded chef.

(ix) Each chef maintains two queues of size **s**, input queue for accepting the order and output queue for placing the cooked meal. (each location of the queue holds one meal)

(x) The chef places the cooked food to the output tray and the waiter will serve it to the client.

(xi)   The philosopher can choose another seat(if any) if he is not able to eat after the delivery of meals.
(xii) All the constraints in classical dining philosophers for synchronization are also included in this system

The problem is to find an optimal solution (more number of philosophers are serviced without much delay) with Less Average waiting time for the clients.

<u>Assumptions</u>

1. All the waiters and chefs have the same speed for their assigned work.
2. The time to cook all meals takes equal time.

3   Artificial Bee Colony Algorithm for Optimization

In the basic ABC algorithm, high-quality nectar sources can be found by communication among three groups of foraging artificial bees, namely, employed bees, onlookers, and scouts. The algorithm begins with a number of food sources (candidate solutions for optimization problems) that are randomly generated. Then, the following three steps are repeated until a termination criterion is met [13–15, 16]. First, the employed bees are sent to each food source and the amounts of nectar are then measured (evaluated by fitness of solutions), with the highest-quality food source retained using the greedy selection mechanism. Second, the food sources are selected by the onlookers after the employed bees share information according to the bees in the hive, and the retained food sources are determined. Third, if a food source is not improved within a limited number of repetitions, a scout is sent out to generate a new possible food source randomly.

Three control parameters should be set in the basic ABC algorithm: (1) SN: the number of food sources; (2) Limit: the number of repetition cycles to activate a scout bee. If a food source cannot be improved further in "limit" cycles, then it will be abandoned and replaced by a new food source generated by scout bee; this is a particular phase of bee-based algorithm to skip out of local optimum; and (3) MCN: the number of maximum cycle iterations, which is a termination criterion.

The probability of each food source chosen by the onlookers is as follows:

$$P_j = \frac{fit_j}{\sum_{n=1}^{SN} fit_n} \tag{1}$$

Where fit_j denotes the fitness of solution j.

3.1   Proposed ABC Algorithms

In the dining subsystem of the restaurant system, the main objective is to allocate an optimized seat for the incoming philosopher without delay. This allocation is subjected to various constraints like, no philosophers should wait more than a particular time. The control parameters that affect the waiting time of the customer includes the seat available time (ts1) and the forks available time (fr1). The philosophers are allocated seats in their arriving order. The main objective of this subsystem is to lessen the waiting time of each philosophers considering the constraints also. The modified part of the algorithm is specified below

Algorithm 1. Pseudo-code of the ABC algorithm for Seat Allocation

---

Input: List of Available seats and a candidate Philosopher
Output: The best solution for seat allocation for the candidate Philosopher

1. Set list of incoming philosophers
2. Allocate seat to the philosopher and update Available Seat List
3. Put any incoming philosopher in the Philosopher List in the order of their arriving time
4. while (Philosopher List!=NULL)
        philosophe r=get philosopher with earliest arriving time from Philosopher
            List
        seatavail[] =get existing Available Seat List
        seatindex (best solution)=ABC algorithm(philosopher,seatavail[])
        Allocate philosopher to corresponding seat
        Update the table of seat availability
        Remove Philosopher from Philosopher List
5. end while

In the kitchen subsystem the main objective is to assign the order to an available chef so that it is delivered as early as possible. The control parameters that affect the delivery time are the availability and the work load of each chef. The constraint in this is that no chef should be overloaded. The orders are kept in a list according to their ordering time. The main objective is to place these orders ed to a chef in an optimized way so as to reduce the delivery time subjected to the constraints. The main part of the algorithm used is specified below.

The time complexity of the proposed method mainly consists of five parts: the initialisation, the search operation of employed bees, the calculation of the probability of food sources, and the search operation of scouts and onlookers. The computational cost of these five parts are: $O(Hnm)$, $O(H(nm + nC))$, $O(H)$, $O(nm)$, and $O(H(H + (nm + nC) + nm))$, respectively. Here n is the number of populations in the solution set P; m is the number of control parameters; H is the number of employed bees or food sources; C is the total number of categories for all attributes where:

$$C = \sum_{j=1}^{m} t_i \tag{2}$$

Therefore the overall time complexity of the proposed approach is $O(Hmn + s(nm + H(H + nm + nC)))$. Here, s is the number of iterations.

Algorithm 2. Pseudo-code of the ABC algorithm for Load Balanced Scheduling

---

Input: List of available Chefs and Order list
Output: The best solution for placing order to the optimized chef
    1. Set list of orders given by seated philosophers
    2. Place order to the chef and update available Chef List
    3. Put any incoming orders in the Order List in order of their ordering time
    4. while (Order List!=NULL)
            order=get order with earliest ordering time from Order List
            chefavail[]=get existing Available Chef List
            chefindex (best solution)=ABC algorithm(order,chefavail[])
            Place order to corresponding chef
            Update the table of chef availability
            Remove order from Order List
    5.end while

---

4   Results and Discussions

The Restaurant of Dining Philosophers system shows that the optimized decision making behavior of individual intelligent agent in the system results in the overall optimized system behaviour. By using the relevant optimization strategy in each zone of the system, the system is able to attain maximum advantage in the subsystem and also led to the optimization of overall system behavior without affecting the QoS goals. The components of the system are modeled as agents which has the ability to adapt dynamically that can make decisions according to the change in the environment. The interaction between the various agents and its coordinated behavior in different subsystems lead a way to optimize the systems functional and QoS goals. The working of the system exhibits the self-adaptivity of its component agents and thereby reaches an optimized solution. In this problem it is expected that the philosophers should not need to wait for a long time after entering the restaurant. Also no philosopher should be allowed to leave the restaurant without being serviced. So the main objective in the dining subsystem is to allocate the incoming philosopher with an optimized seat so that he got easily serviced. In the kitchen subsystem the objective is to schedule the orders to an optimized chef so as to minimize the delivery time as well as to ensure that no chefs are overwhelmed. Moreover, the overall goal is to reduce the average waiting time of each philosopher. The results outperformed the system that works without using any adaptation and it shows that the average waiting time of the philosophers is highly reduced by using the proposed framework. Comparison result of the system with and without adaptation capability is plotted in Fig. 2. The rate of philosophers in each state (Thinking, Eating, Leaving) at various time slots is also presented in Fig. 3. It can thus be concluded that the proposed methodology provides a better adaptability for any software with multi components and that can be modelled as a multi-agent system with single or multiple objectives. The system is also evaluated on the basis of the rate of lost clients.

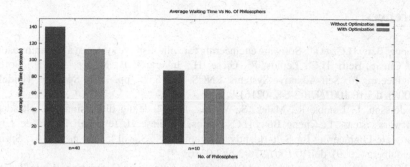

Fig. 2. Number of philosophers vs average waiting time

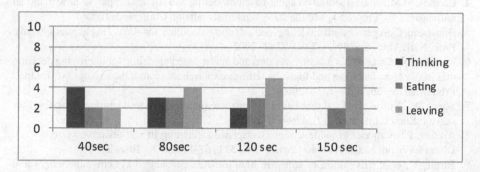

Fig. 3. Resource utilization and state graph at various time slots

5 Conclusion

A framework for intelligent agents based self-adaptive software system is proposed. The intelligent agents are designed to adapt its behavior according to the changing environment by perceiving information from it. As a proof of study, the proposed method is demonstrated by simulating the Restaurant of Dining Philosophers problem. The concurrently executing subsystems are modeled with a set of agents which are autonomous and communicating with each other. These agents have the capability of choosing an optimized option locally, thereby optimizing the overall system behavior. The ABC algorithm is used for optimized seat allocation in dining subsystem and for optimized load balancing and scheduling in kitchen subsystem. The results shows that using these algorithms dynamically improves the self-adaptivity of the agents in the system which in turn highly optimizes the whole system by lessening the average waiting time of the philosophers and reducing the number of lost customers. It allows the system entities to select the best adaptation logic and propel the system to act and react in an optimized way based on the current system configuration and the runtime environmental triggers. Thus it is inferred that the zone wise optimization in a multi-agent system in which the agents communicate with each other and which has self–decision making capability will optimize the overall system behavior without affecting the QoS goals of the system.

References

1. Cheng, Betty H.C., et al.: Software engineering for self-adaptive systems: a research roadmap. In: Cheng, Betty H.C., Lemos, R., Giese, H., Inverardi, P., Magee, J. (eds.) Software Engineering for Self-Adaptive Systems. LNCS, vol. 5525, pp. 1–26. Springer, Heidelberg (2009). doi:10.1007/978-3-642-02161-9_1
2. Andersson, J., Lemos, R., Malek, S., Weyns, D.: Modeling dimensions of self-adaptive software systems. In: Cheng, Betty H.C., Lemos, R., Giese, H., Inverardi, P., Magee, J. (eds.) Software Engineering for Self-Adaptive Systems. LNCS, vol. 5525, pp. 27–47. Springer, Heidelberg (2009). doi:10.1007/978-3-642-02161-9_2
3. IBM Corporation: An Architectural Blueprint for Autonomic Computing. White Paper. 4th edn., IBM Corporation (2005)
4. Charles, M.M., et al.: Tutorial on agent-based modeling and simulation part 2: how to model with agents. In: Proceedings of the 2006 Winter Simulation Conference (2006)
5. Bonabeau, E.: Agent–based modeling methods and techniques for simulating human systems. Proc. Natl. Acad. Sci. **99**(3), 7280–7287 (2002)
6. Fredrick, N.: On complex adaptive systems and agent-based modeling for improving decision making in manufacturing and logistics setting: experiences from a packaging. Int. J. Oper. Prod. Manage. **26**, 1351–1373 (2006)
7. Shen, W., Norrie, D.H.: Agent-based system for intelligent manufacturing: a state-of-art survey. Knowl. Inf. Syst. **1**(2), 129–156 (2013)
8. Andres, P.: A model to guide dynamic adaptation planning in self-adaptive systems. Sci. Direct Electron. Notes Theoret. Comput. Sci. **321**, 67–88 (2016). Elsevier
9. Birgit, V., et al.: Evolution of software in automated production systems: challenges and research directions. J. Syst. Softw. **110**, 54–84 (2015). Elsevier
10. Krik, S., et al.: A unified algorithm for load balancing adaptive scientific simulation. In: Proceedings of the 2000 ACM/IEEE Conference on Supercomputing Article, No. 59, IEEE Computer Society, Washington DC, USA (2000)
11. Wolf, T., Holvoet, T.: Design patterns for decentralised coordination in self-organising emergent systems. In: Brueckner, S.A., Hassas, S., Jelasity, M., Yamins, D. (eds.) ESOA 2006. LNCS, vol. 4335, pp. 28–49. Springer, Heidelberg (2007). doi:10.1007/978-3-540-69868-5_3
12. Weys, D., Malek, S.M., Anderson, J.: FORMS: unifying reference model for formal specification of distributed self-adaptive systems. ACM Trans. Auton. Adapt. Syst. **7**, 8 (2012)
13. Saritha, R., Vinod, C.: A novel algorithm based on honey bee foraging principle for transportation problems. In: ACCIS, Proceedings of Elsevier, 26–28 June 2014, Kollam, India (2014)
14. Karaboga, D., Basturk, B.: Artificial Bee Colony (ABC) optimization algorithm for solving constrained optimization problems. In: Melin, P., Castillo, O., Aguilar, L.T., Kacprzyk, J., Pedrycz, W. (eds) IFSA 2007. LNCS, vol. 4529, pp. 789–798. Springer, Heidelberg (2007). doi:10.1007/978-3-540-72950-1_77
15. Karaboga, D., Akay, B.: A comparative study of artificial bee colony algorithm. Appl. Math. Comput. **214**(1), 8–32 (2009)

A Knowledge-Based Framework for Mitigating Hydro-Meteorological Disasters

Pg. Hj. Asmali Pg. Badarudin$^{(\boxtimes)}$, Thien Wan Au, and Somnuk Phon-Amnuaisuk

School of Computing and Informatics, Universiti Teknologi Brunei,
Jalan Tungku Link, Gadong, Brunei Darussalam
salamdamit@gmail.com, {twan.au,somnuk.phonamnuaisuk}@utb.edu.bn

Abstract. A knowledge-based framework for mitigation of hydro-meteorological disasters in Brunei is proposed where a data mining process is used to predict anomalous intense rainfalls that were causing destructive floods and landslides. A previous study pointed the causes to anomalous oceanographic and atmospheric conditions. This expert knowledge and satellite data is used to create the model. Interoperable collaborative platforms are also crucial. This approach can potentially alter the prevailing disaster management where reactive response dominated the proactive bottom-up approach of disaster mitigation based on expert knowledge-based prediction from data mining.

Keywords: Data mining · Hydro-meteorological disasters · Mitigation framework · Collaboration platforms · Knowledge management

1 Introduction

With a recent scientific study that determined the causes of anomalous episodic intense rainfalls in Brunei Darussalam, and with the availability of data mining tools, there is a great opportunity to create a new approach for hydro-meteorological disaster management in the country. This can be achieved by putting together various pertinent best practices and putting data mining at the heart of this new arrangement.

The existing institutional and legal frameworks by default entail a mostly top-down approach such as giving directions from the top using command-and-control style. This approach mostly suits disaster response efforts; however, a more proactive way driven by domain experts' knowledge can greatly assist in prevention and mitigation.

2 Background Studies

Disaster management (DM) aims to reduce potential losses from hazards, assure prompt and appropriate assistance to victims of disaster, and achieve rapid and effective recovery. It is an 'ongoing process by which governments, businesses, and civil society plan for and reduce the impact of disasters, react during and immediately following a disaster, and take steps to recover after a disaster has occurred.' [1].

© Springer International Publishing AG 2017
Y. Tan et al. (Eds.): DMBD 2017, LNCS 10387, pp. 505–513, 2017.
DOI: 10.1007/978-3-319-61845-6_50

In Brunei, the Incident Command System (ICS) is activated in disaster response and recovery while the current framework for managing disasters locally, NaSOP [4] serves as a legal framework and guideline. As very high level documents, they do not fully address interoperability and collaboration for group decision support. There are no knowledge management (KM) concepts used like community of practices (CoP), a platform that can enable collective action and responsibility in the management and sharing of knowledge among experts across various agencies and disciplines.

A group decision-support system (GDSS) is a platform that facilitates collaborative decision-making. Though more suitable during mitigation and preparedness, GDSS can potentially complement ICS during catastrophic events. Data mining, in turn, can enhance the usefulness of GDSS by providing an element of predictive analytic capabilities along with pattern and new knowledge discovery. Besides lacking collaboration and interoperability, current framework also lacks inter-agency data collection of previous disasters, an issue often cited in local DM circles. Thus a new framework should provide input and output mechanisms that include real-time and online data capture, processing and visualization which can be built-in into the CoP and GDSS, as shown in a proposed mitigation framework in Fig. 1 below.

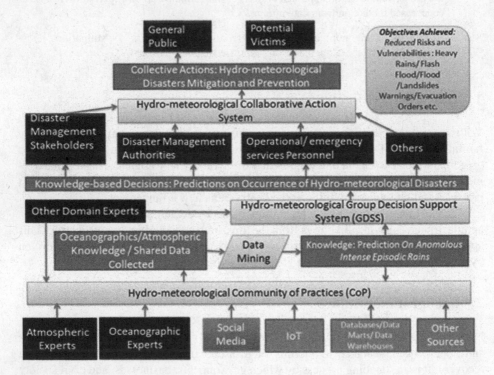

Fig. 1. A proposed data mining knowledge-based framework for mitigation of hydro-meteorological disasters in Brunei Darussalam

The local DM lags behind in adopting knowledge management practices. Knowledge management (KM) is the process of systematically and actively managing and leveraging stores of knowledge in an organization [5]. An ideal framework should thus incorporate the design, development and governance of KM framework, policies and procedures and guidelines. The knowledge products such as prediction from data mining, and services such as new videos on a new algorithm, should be identified, and shared with all concerned including the communities through the use of the ubiquitous social media and KM tools that can capture, process and store knowledge.

With the increased use of KM and the rise of big data, knowledge extraction is paradoxically becoming more difficult. By encouraging CoP however, regular sharing and learning, based on common interests and expertise, can lessen and eliminate the costs of knowledge acquisition and extraction. Experts interested in hydro-meteorological risks, for example, can discuss many issues emanating from a data mining process such as weather prediction, with a view of iteratively improving the accuracy of the model that predicts the results.

The data preparation or preprocessing stage is the most time-consuming in the data mining process. For hydro-meteorological DM, many domain knowledge experts are required to share knowledge towards understanding issues before arriving at accurate solutions and decision-making. Many types of data are thus required to undergo various sub-processes to make them well-formed before delving into data mining. This calls for an integrated and effectively coordinated hydro-meteorological risks management team [3] and platform. Domain-specific CoP fits the bill as depicted in Fig. 1 above. CoP can thus alleviate the difficulties in data preparation.

In order to fully appreciate how CoP creates organizational value, it is worth thinking of a community as an engine for generating social capital. Social capital found in CoP leads to changes in behavioral which lead to greater sharing of knowledge. This will in turn positively influence performance [6].

While a flourishing hydro-meteorological CoP leads to collaboration among various domain knowledge experts related to field, interoperability serves as an important enabler of inter-agency collaboration [7]. Decision makers need information from various resources [7, 8] and require multiple agencies to work together, and information needs to change rapidly as the disaster event evolves [7, 9].

A novel perspective to interoperability issues takes into account the advent of Internet technology feeding the emergence of the 'Internet of Things' (IoT), enabling the different artefacts to sense, process, share and act in an ubiquitous, Internet-like environment [10]. Potentially, an application of a novel IoT-aware 'interoperability as a property' (IaaP) paradigm can assist in efficient preparation by DM organisations and agile, adaptive response delivered by synergic task force and rescue teams [10].

For hydro-meteorological DM, the use of pervasive real-time IoT sensors and data-capture devices provide DM professionals with timely information for which to make a more informed decision-making. As shown in Fig. 1 above, IoT feeds data into the CoP. Near real-time meteorological data obtained from remote-sensing satellites are also available for the public to download from various internet sites at no costs.

3   Data Mining Methodology and Model

3.1   Introduction

In this paper, data mining is used for predicting anomalous weather conditions that caused intense episodic rainfall during the months of January in 2009, 2011 and 2014. The weather anomalies during the said periods posed serious hydro-meteorological hazards that caused massive flash-floods, floods and landslides. This section highlights the knowledge from domain knowledge experts can be used for building a model that will be used for predicting anomalous intense rainfall.

3.2   Sources of Experts' Knowledge

Using the conceptual framework in Fig. 2, a bottom-up knowledge-based approach of building a disaster mitigation framework takes effect. Using meteorological data obtained from remote-sensing satellites which are available on various websites, [3] focused on the causes of intense anomalous rainfall that triggered flash-floods, floods and landslides in 2009, 2011 and 2014. The real underlying reasons were revealed in the shape of anomalous atmospheric conditions and anomalous oceanographic conditions, dismissing the long-held belief that such events were always the result of anomalous monsoons or climate change [3].

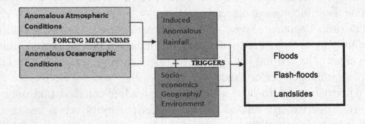

Fig. 2.   A simplified framework on forcing mechanisms of anomalous rainfalls [3]

Using meteorological data obtained from remote-sensing satellites, Ndah et al. focused on the causes of intense anomalous rainfall as follows (see Fig. 2 above):

1. Strong variability and unpredictability of anomalous rainfall events
2. Variability of anomalous rain-fall forcing or inducing mechanisms
3. Socio-geographical and environmental factors that aid in triggering floods and landslide disasters [9].

From their analysis of past data, the intensity of rainfall in Brunei was found to be greater than in wider Borneo, with the country's average daily January rainfall being highly variable [9]. The variance of daily January rainfall on the other-hand reveals periods of absolutely strong peaks in Brunei and Borneo. The major peaks represent periods of anomalous daily rainfall, which generally trigger flashfloods and landslides [9].

To emphasise the importance of their study, it should be noted that the 2009, 2011 and 2014 events were among the worst the country had seen, one caused the death of a woman due to flood and another in a landslide, while also causing massive and widespread economic losses.

Based on results of the study by Ndah et al., they stated that a window of opportunity is opened to attempt the prediction of anomalous episodic rainfall events and consequently enhance early detection capabilities for floods and landslides [9].

3.3 Data Preprocessing

Two choices of data sets from Brunei Meteorological Department (BDMD) and satellite data are considered. The average rainfall data from BDMD are from one location in one district only and do not represent the whole country. Satellites data were then analyzed. APDRC [13] provides many rainfall data sets and one was selected. Oceanographic and atmospheric data were also analyzed but later omitted due to complexity. For this, earlier study by [3] was used to represent TRUE and FALSE.

The average daily rainfall data sets were filtered to include only January data for 2009, 2011 and 2014. Some missing values are assigned to the dates other than the following periods when analysis were either not performed or mentioned by [3].

The data sets in Table 1 below are then partitioned into training (70%) and test data (30%). The purpose of test data is to check the accuracy of the prediction.

Table 1. Available data adapted from [3] and remote sensing satellite data [13]

Year	Atmospheric anomalies	Oceanographic anomalies	Daily average rainfall
2009	1st-12th January	January monthly average	January time-series
2011	1st -12th January	January monthly average	January time-series
2014	1st -12th January	January monthly average	January time-series

3.4 Model

A simple logical model in Table 2 below is derived from Fig. 2 and [3].

Table 2. A simple logical model based on scientific knowledge

Anomalous atmospheric conditions	Anomalous oceanographic conditions	Anomalous intense rainfall
TRUE	TRUE	Yes
TRUE	FALSE	Yes
FALSE	TRUE	Yes
FALSE	FALSE	No

CRISP-DM [11], the most popular methodology [12] is adopted. CHAID and CART, two popular algorithms that classify class label (anomalous intense rainfall).

4   Experiment and Results

January data from 2009, 2011 and 2014 from APDRC [13] are used. IBM SPSS Modeler is used for running the data mining processes.

Table 3 below shows the results using CHAID algorithm, showing the prediction and accuracy of anomalous intense rainfall ($R-Anomalous Rainfall and $RC-Anomalous Rainfall). The confidence level is either 85% or 100% correct.

Table 3. Results of anomalous rainfall prediction modeling using CHAID algorithm.

Jan	Anomalous atmospheric conditions	Anomalous oceanographic conditions	Average daily rainfall	Anomalous rainfall	Partition	$R-anomalous rainfall	$RC-Anomalous rainfall
1	FALSE	TRUE	5.46	Yes	Training	No	0.857
2	FALSE	TRUE	27.51	Yes	Training	Yes	1.000
3	FALSE	TRUE	23.39	Yes	Training	Yes	1.000
4	FALSE	TRUE	25.43	Yes	Testing	Yes	1.000
5	FALSE	TRUE	6.11	No	Training	No	0.857
6	FALSE	TRUE	19.98	No	Training	No	0.857
7	FALSE	TRUE	4.23	No	Training	No	0.857
8	FALSE	TRUE	16.56	No	Training	No	0.857
9	FALSE	TRUE	27.68	Yes	Training	Yes	1.000
10	FALSE	TRUE	5.94	No	Training	No	0.857
11	FALSE	TRUE	18.12	No	Training	No	0.857

Table 4 below shows the CART results showing the prediction and its accuracy of the anomalous intense rainfall. The confidence level is 72%, 94% or 100% correct.

Table 4. Results of anomalous rainfall prediction modeling using CART algorithm

Jan	Anomalous atmospheric conditions	Anomalous oceanographic conditions	Average daily rainfall	Anomalous rainfall	Partition	$R-anomalous rainfall	$RC-anomalous rainfall
1	FALSE	TRUE	5.46	Yes	Training	No	0.722
2	FALSE	TRUE	27.51	Yes	Training	Yes	0.944
3	FALSE	TRUE	23.39	Yes	Training	Yes	0.944
4	FALSE	TRUE	25.43	Yes	Testing	Yes	0.944
5	FALSE	TRUE	6.11	No	Training	No	0.722
6	FALSE	TRUE	19.98	No	Training	Yes	0.944
7	FALSE	TRUE	4.23	No	Training	No	0.722
8	FALSE	TRUE	16.56	No	Training	No	0.722
9	FALSE	TRUE	27.68	Yes	Training	Yes	0.944
10	FALSE	TRUE	5.94	No	Training	No	0.722
11	FALSE	TRUE	18.12	No	Training	No	0.722

Table 5 below analyzed the accuracy of the prediction model for both training and testing data that shows 90% correct for CHAID for training data, in actual numbers are 54 out of 60. For testing data, 88.46% or 23 out of 26 are correct and the corresponding wrong results in actual number of records in this case 6 out of 60 and the percentage of wrong prediction in training data is 10%.

Table 5. Results of anamolous rainfall prediction modeling using CHAID algorithm: comparing the accuracy of the predicted anomalous rainfall

Partition	Training		Testing	
Correct	54	90%	23	88.46%
Wrong	6	10%	3	11.54%
Total	60		26	

In Table 6 breakdown results are also shown for CART for both the training and testing which shows a slightly better result for predicting correct training data at 91.67% or 55 out of 60 while giving the same results as CHAID for testing data. Hence, it can be concluded that the model give a good accuracy in predicting anomalous intense rainfall.

Table 6. Results of anamolous rainfall prediction modeling using CART algorithm: comparing the accuracy of the predicted anomalous rainfall

Partition	Training		Testing	
Correct	55	91.67%	23	88.46%
Wrong	5	8.33%	3	11.54%
Total	60		26	

5 Comparison with Another Platform

For comparison purposes, the data set is run into a free machine learning platform called WEKA. During data preprocessing, a CSV file was created converted to ARFF, the format accepted by WEKA. WEKA's date attribute format was also followed (Table 7).

Table 7. Results of anamolous rainfall prediction using WEKA's J48 and REPTree

Classifier	J48				REPTree			
Test methods	Cross-validation 10 folds		30% split for Test Data		Cross-validation 10 folds		30% split for Test Data	
Correct	93	100%	28	100%	92	98.9247%	28	100%
Wrong	0	0%	0	0%	1	1.0753%	0	0%
Total	93		28		93		28	

Comparing IBM SPSS Modeler results against WEKA reveals that WEKA is better in correctly classifying instances for predicting January anomalous intense rainfall.

6 Discussions and Lessons Learnt

The experiment shows the feasibility of predicting using experts' knowledge [3] and satellite data. By using data mining, previously unpredictable anomalous intense rainfall conditions in Brunei can be predicted. The following are the talking points:

1. Business understanding and knowledge is very important in data mining. Understanding of the aims and objectives of mitigation and preparedness can lead to better risk management while expert knowledge can enhance insight.
2. Knowing the underlying causes presents a bottom-up approach in mitigation where benefits will reach higher authorities and eventually potential disaster victims. A valuable GDSS can assist in realizing the real benefits.
3. The understanding of data is a perquisite for quality predictions. Experts' guidance can give insights. One data sample can be more representative to use.
4. Having taxonomy in data mining, GDSS, CoP and knowledge systems in a framework can help to avoid potential misinterpretation and miscommunications.
5. For big data, consolidation, extraction and integration will be harder and require cooperation and collaboration. Using CoP and other KM tools are essential.
6. From the results, the best algorithms can be ascertained. Comparing different platforms is also useful. CoP and GDSS can be used for sharing findings.
7. In model testing and evaluation, experts can meet up and discuss the accuracy and whether objectives are met. CoP and GDSS can facilitate these activities.
8. The sharing of findings and knowledge gains from above can lead to major future projects. CoP and GDSS enforce the best rules and create self-learning system.
9. Visibility of assets in disaster response and recovery areas by using CoP and GDSS such as current resources available to deal with flood-causing rainfall, can improve preparedness and mitigation and overall resilience towards disasters.
10. A comprehensive framework for disaster mitigation will help in the development and use of knowledge-based analytics tool like data mining.

7   Conclusion

The paper proposes a framework in tackling and managing hydro-meteorological disaster risks in Brunei that had posed the greatest danger in the past. It shows the importance of the data mining in enabling a knowledge-based approach. Together with GDSS and CoP as the main platforms the prevailing top-down mindset of focusing primarily on response and recovery in disaster management and on devoting more resources to them can be altered. By formalizing data mining predictive capabilities and aligning actions in the proposed framework can potentially address many issues in the country's vision for better disaster resilience, particularly as part of disaster mitigation for two of the most prevalent and destructive disasters in recent years.

Acknowledgements. The authors gratefully acknowledge the kind assistance given by Mr. Anthony Banyouko Ndah, one of the authors of [3] for explanations and guidance, in giving permission for use of parts of their findings for the works above and the use of conceptual framework in Fig. 2. The authors also acknowledge and grateful for the valuable helpful comments and suggestions of the reviewers, which have improved the content and presentation of this paper.

References

1. Wisner, B., Adams J. (eds.): Environmental Health in Emergencies and Disasters: A Practical Guide. World Health Organization (2002)
2. Ismail, I.: Brunei to host regional disaster response exercise (2016). http://www.bt.com.bn
3. Ndah, A.B., Kumar, D.L., Becek, K.: Dynamics of hydro-meteorological disasters: revisiting the mechanisms and drivers of recurrent floods and landslides in Brunei Darussalam. In: Int. J. Earth Atmos. Sci. (2016)
4. NDMC Brunei NaSOP.ppt (2016). http://122.155.1.145/upload/filecenter/263/files/Brunei %20NaSOP.ppt
5. Laudon, K., Laudon, J.: Management Information Systems: Managing the Digital Firm, 13th edn. Pearson, Upper Saddle River (2014)
6. Lesser, L.E., Storck, J.: Communities of Practice and Organizational Performance. IBM Syst. J. **40**(4), 831–841 (2011)
7. Andargoli, A., Bernus, P., Kandjani, H.: Analysis of interoperability in the queensland disaster management system. In: 15th International Conference on Enterprise Information Systems (ICEIS) Proceedings (2013)
8. Gollery, S.J., Pohl, J.G.: The Role of Discovery in Context-Building Decision-Support Systems Collaborative Agent Design (CAD) Research Center, p. 55 (2002)
9. Janssen, M., Lee, J.K., Bharosa, N., Cresswell, A.: Advances in multi-agency disaster management. Inf. Syst. Frontiers **12**, 1–7 (2010)
10. Noran, O., Zdravković, M.: Interoperability as a property: enabling an agile disaster management approach. In: ICIST 2014 (2014)
11. Chapman, P., Clinton, J., Kerber, R., Khabaza, T., Reinartz, T., Shearer, C., Wirth, R.: CRISP-DM 1.0 Step-by-step data mining guides (2000)
12. KDnuggets: What main methodology are you using for your analytics, data mining, or data science projects? Poll. http://www.kdnuggets.com/polls/2014/analytics-data-mining-data-science-methodology.html
13. Asia-Pacific Data: Research Center of the IPRC in the School of Ocean and Earth Science and Technology at the University of Hawaii at Manoa. http://apdrc.soest.hawaii.edu/data/data.php

Fuzzy Control

Improved Stability and Stabilization Criteria for T-S Fuzzy Systems with Distributed Time-Delay

Qianqian Ma[1], Hongwei Xia[1(✉)], Guangcheng Ma[1], Yong Xia[2], and Chong Wang[2]

[1] Harbin Institute of Technology, Harbin 150001, China
xiahongwei@hit.edu.cn
[2] The First Natural Gas Plant, Petro China Changqing Oilfield Company, Jingbian 718500, China

Abstract. This paper investigates the stability and stabilization analysis problem of nonlinear systems with distributed time-delay. The T-S fuzzy model is employed to describe the nonlinear plant. When designing fuzzy controller, the novel imperfect premise matching method is adopted, which allows the fuzzy model and the fuzzy controller to use different premise membership functions and different number of rules. As a result, greater design flexibility can be obtained. By applying a new tighter integral inequality which involves information about the double integral of the system states, and introducing the information of membership functions, less conservative stability and stabilization conditions are derived. Finally, a numerical example is provided to clarify the effectiveness of the proposed approach.

Keywords: T-S fuzzy model · Distributed time-delay · Lyapunov stability theory · Imperfect premise matching

1 Introduction

Time-delay is a common phenomenon in practical control systems, which can deteriorate the system performance and cause instability. Therefore, the research for control systems with time-delay is crucial and has received considerable attention [1, 2]. When analyzing the stability conditions of control systems with time-delay, a Lyapunov-Krasovskii functional (LKF) approach [3] is usually applied, which can denote the conditions in terms of linear matrix inequalities (LMIs). Though the LKF approach can handle the stability analysis problem for time-delay systems, the criteria derived from LKF method are conservative. To reduce conservatism, much research has been done: the free-weighting matrix [4] technique was introduced to obtain more tighter bound of the derivative of Lyapunov function; the wirtinger-based integral inequality [5] was developed to deal with integral term $\int_{t-h}^{t} \dot{\mathbf{x}}^T(s)\mathbf{R}\dot{\mathbf{x}}(s)$; especially, a new tighter integral inequality was proposed in [6], which yields less conservative stability conditions.

© Springer International Publishing AG 2017
Y. Tan et al. (Eds.): DMBD 2017, LNCS 10387, pp. 517–526, 2017.
DOI: 10.1007/978-3-319-61845-6_51

Besides, the Takagi-Sugeno (T-S) fuzzy model [8] is normally used to describe fuzzy control systems. It can represent the dynamics of a complex nonlinear system as a weighted sum of some local linear models, which will facilitate the stability analysis. When the fuzzy controller is also denoted as a weighted sum of some linear sub-controllers, and is connected with the T-S fuzzy model in a closed loop, a Takagi-Sugeno fuzzy-model-based (TSFMB) control system is formed. When dealing with the stabilization analysis problem for TSFMB system, an efficient technique named parallel distribution compensation (PDC) [9] is usually adopted, which requires the fuzzy controller and the T-S fuzzy model have the same premise membership functions and the same number of rules. Such design can make the stabilization analysis easier. However, PDC method also bears some inevitable drawbacks such as limiting the design flexibility of the fuzzy controller and complicating the fuzzy controller structure unnecessarily in some cases. For the sake of solving these problems, some non-PDC design techniques were developed [7,8]. Among them, an effective methodology called imperfect premise matching method [8], which allows the fuzzy controller and the fuzzy model use different premise membership functions and different number of rules. Hence, this approach can avoid the limitations of PDC method.

Currently, the research of stability and stabilization analysis for TSFMB systems with time-delay has yet been thorough enough. First, most of literature applied PDC method to conduct stabilization analysis which would limit the design flexibility of fuzzy controller. Second, the existing stability and stabilization conditions can be further relaxed. This is because the bound of integral term $\int_{t-h}^{t} \dot{\mathbf{x}}^T(s)\mathbf{R}\dot{\mathbf{x}}(s)$ can be estimated more accurate, and the information of the membership functions has not been applied fully. Therefore, in this paper, we aim to investigate the improved stability and stabilization conditions for TSFMB systems with time-delay.

To achieve our goal, we will apply the imperfect premise matching method to design stable fuzzy controller. In order to relax the results, we will employ the novel integral inequality technique in the stability analysis, and take the information of the membership functions into account. The analyses results will be presented as theorems in terms of LMIs. Finally, a numerical example will be given to illustrate the advantages and superiority of the proposed approach.

2   Preliminaries

A TSFMB nonlinear control system with distributed time delay is considered.

Fuzzy Model: Construct a p-rule polynomial fuzzy model to represent the nonlinear system with time-delay:

$$\dot{\mathbf{x}}(t) = \sum_{i=1}^{p} \omega_i(\mathbf{x}(t))\big(\mathbf{A}_i\mathbf{x}(t) + \mathbf{A}_{1i}\mathbf{x}(t-h) + \mathbf{A}_{2i}\int_{t-h}^{t}\mathbf{x}(s)ds + \mathbf{B}_i\mathbf{u}(t)\big), \qquad (1)$$

where $\mathbf{x}(t) \in \mathbb{R}^{n \times 1}$ denotes the vector of the system state; $\mathbf{A}_i \in \mathbb{R}^{n \times n}, \mathbf{A}_{1i} \in \mathbb{R}^{n \times n}, \mathbf{A}_{2i} \in \mathbb{R}^{n \times n}$ and $\mathbf{B}_i \in \mathbb{R}^{n \times m}$ represent the system matrices and the system input matrices; $\mathbf{u}(t) \in \mathbb{R}^{m \times 1}$ is system input; the time-delay h is a constant satisfying $h \in [h_{min}, h_{max}]$; $\omega_i(\mathbf{x}(t))$ stands for the normalized grade of membership, and satisfying: $\omega_i(\mathbf{x}(t)) \geq 0$ for all i, and $\sum_{i=1}^{p} \omega_i(\mathbf{x}(t)) = 1$.

Fuzzy Controller: Motivated by the imperfect premise matching method [8], we consider a c-rule polynomial fuzzy controller in this paper:

$$\mathbf{u}(t) = \sum_{j=1}^{c} m_j(\mathbf{x}(t))\mathbf{K}_j\mathbf{x}(t), \tag{2}$$

where $\mathbf{K}_j \in \mathbb{R}^{m \times n}$ represents the feedback gain of jth rule; $m_j(\mathbf{x}(t))$ stands for the normalized grade of membership, and satisfying: $m_j(\mathbf{x}(t)) \geq 0$ for all i, and $\sum_{j=1}^{c} m_j(\mathbf{x}(t)) = 1$.

According to (1) and (2), the closed-loop fuzzy control system can be easily acquired:

$$\dot{\mathbf{x}}(t) = \sum_{i=1}^{p} \sum_{j=1}^{c} \omega_i(\mathbf{x}(t))m_j(\mathbf{x}(t))\big(\mathbf{G}_{ij}\mathbf{x}(t) + \mathbf{A}_{1i}\mathbf{x}(t-h) + \mathbf{A}_{2i}\int_{t-h}^{t} \mathbf{x}(s)ds \tag{3}$$
$$+ \mathbf{B}_i\mathbf{u}(t)\big),$$

where $\mathbf{G}_{ij} = \mathbf{A}_i + \mathbf{B}_i\mathbf{K}_j, \quad i = 1,2,...,p, \quad j = 1,2,...,c.$

Lemma 1 is given to facilitate the proof of the main results in the following sections.

Lemma 1 [6]. *It is assumed that* $\mathbf{x}(t)$ *is a differentiable function:* $[\alpha, \beta] \to \mathbb{R}^n$. *For* $\mathbf{N}_1, \mathbf{N}_2, \mathbf{N}_3 \subset \mathbb{R}^{4n \times n}$, *and* $\mathbf{R} \subset \mathbb{R}^{n \times n} > 0$, *the following inequality holds:*

$$-\int_{\alpha}^{\beta} \dot{\mathbf{x}}^T(s)\mathbf{R}\dot{\mathbf{x}}(s)ds \leq \xi^T\Omega\xi, \tag{4}$$

where

$$\Omega = \tau\Phi_2 + \Phi_3, \Phi_2 = \mathbf{N}_1\mathbf{R}^{-1}\mathbf{N}_1^T + \frac{1}{3}\mathbf{N}_2\mathbf{R}^{-1}\mathbf{N}_2^T + \frac{1}{5}\mathbf{N}_3\mathbf{R}^{-1}\mathbf{N}_3^T,$$

$$\Phi_3 = Sym\{\mathbf{N}_1\Delta_1 + \mathbf{N}_2\Delta_2 + \mathbf{N}_3\Delta_3\}, \mathbf{e}_k = [\mathbf{0}_{n \times (k-1)n} \quad \mathbf{I}_n \quad \mathbf{0}_{n \times (4-k)n}],$$

$$k = 1,2,3,4, \Delta_1 = \mathbf{e}_1 - \mathbf{e}_2, \Delta_2 = \mathbf{e}_1 + \mathbf{e}_2 - 2\mathbf{e}_3, \Delta_3 = \mathbf{e}_1 - \mathbf{e}_2 - 6\mathbf{e}_3 + 6\mathbf{e}_4,$$

$$\xi = [\mathbf{x}^T(\beta) \quad \mathbf{x}^T(\alpha) \quad \frac{1}{\tau}\int_{\alpha}^{\beta} \mathbf{x}^T(s)ds \quad \frac{2}{\tau^2}\int_{\alpha}^{\beta}\int_{\alpha}^{s} \mathbf{x}^T(u)duds]^T, \quad \tau = \beta - \alpha.$$

In addition, to simplify the computational complexity, $\omega_i(\mathbf{x}(t))$ and $m_j(\mathbf{x}(t))$ will be denoted has ω_i and m_j, respectively in the following sections.

3 Stability Analysis

Firstly, the stability condition of TSFMB autonomous system, i.e., the system (1) with $\mathbf{u}(t) = \mathbf{0}$, will be investigated, which can be described by

$$\dot{\mathbf{x}}(t) = \sum_{i=1}^{p} \omega_i(\mathbf{A}_i\mathbf{x}(t) + \mathbf{A}_{1i}\mathbf{x}(t-h) + \mathbf{A}_{2i}\int_{t-h}^{t} \mathbf{x}(s)ds). \tag{5}$$

The inferred stability analysis results are summarized in the following theorems.

Theorem 1. *For TSFMB autonomous system (1) and a prescribed constant $h \in [h_{min}, h_{max}]$, if there exist positive definite matrices $\mathbf{P} = \mathbf{P}^T \in \mathbb{R}^{3n \times 3n}, \mathbf{R} = \mathbf{R}^T \in \mathbb{R}^{n \times n}, \mathbf{Q} = \mathbf{Q}^T \in \mathbb{R}^{n \times n}$, such that LMIs (6) are satisfied, then the system (1) is asymptotically stable.*

$$\begin{bmatrix} \mathbf{\Phi}_{1i} + \mathbf{\Phi}_3 + \mathbf{\Phi}_{4i} & \sqrt{h}\mathbf{N}_1 & \sqrt{h}\mathbf{N}_2 & \sqrt{h}\mathbf{N}_3 \\ * & -\mathbf{R} & 0 & 0 \\ * & * & -3\mathbf{R} & 0 \\ * & * & * & -5\mathbf{R} \end{bmatrix} < 0, \tag{6}$$

where

$$\mathbf{\Phi}_{1i} = Sym\{\Delta_4^T \mathbf{P} \Delta_{5i}\} + \mathbf{e}_1^T \mathbf{Q} \mathbf{e}_1 - \mathbf{e}_2^T \mathbf{Q} \mathbf{e}_2 + h\mathbf{\Gamma}_i^T \mathbf{R} \mathbf{\Gamma}_i, \Delta_4 = \begin{bmatrix} \mathbf{e}_1^T & h\mathbf{e}_3^T & \frac{h^2}{2}\mathbf{e}_4^T \end{bmatrix}^T,$$

$$\mathbf{P} = \begin{bmatrix} \mathbf{P}_{11} & \mathbf{P}_{12} & \mathbf{P}_{13} \\ * & \mathbf{P}_{22} & \mathbf{P}_{23} \\ * & * & \mathbf{P}_{33} \end{bmatrix}, \Delta_{5i} = \begin{bmatrix} \mathbf{\Gamma}_i^T & \mathbf{e}_1^T - \mathbf{e}_2^T & h\mathbf{e}_3^T - h\mathbf{e}_2^T \end{bmatrix}^T, \mathbf{\Gamma}_i = \begin{bmatrix} \mathbf{A}_i & \mathbf{A}_{1i} & h\mathbf{A}_{2i} & 0 \end{bmatrix}$$

$$\mathbf{\Phi}_{4i} = \mathbf{T}_i - \mathbf{F}_i + \sum_{r=1}^{p} \bar{h}_r \mathbf{F}_r - \sum_{k=1}^{p} \underline{h}_k \mathbf{T}_k, i = 1, 2, ..., p,$$

and $\Delta_1, \Delta_2, \Delta_3$ are defined in Lemma 1. Besides, in order to decrease the computational complexity, we will eliminate the free matrices $\mathbf{N}_1, \mathbf{N}_2, \mathbf{N}_3$ by assuming $\mathbf{N}_1 = \frac{1}{h}\begin{bmatrix} -\mathbf{R} & \mathbf{R} & 0 & 0 \end{bmatrix}^T, \mathbf{N}_2 = \frac{3}{h}\begin{bmatrix} -\mathbf{R} & -\mathbf{R} & 2\mathbf{R} & 0 \end{bmatrix}^T, \mathbf{N}_3 = \frac{5}{h}\begin{bmatrix} -\mathbf{R} & \mathbf{R} & 6\mathbf{R} & -6\mathbf{R} \end{bmatrix}^T.$

Proof. To investigate the stability of the TSFMB autonomous system (1), the following Lyapunov-Krasovskii functional candidate is considered.

$$V(t) = \begin{bmatrix} \mathbf{x}^T(t) & \eta_1^T(t) & \eta_2^T(t) \end{bmatrix} \mathbf{P} \begin{bmatrix} \mathbf{x}^T(t) & \eta_1^T(t) & \eta_2^T(t) \end{bmatrix}^T$$
$$+ \int_{t-h}^{t} \mathbf{x}^T(s)\mathbf{Q}\mathbf{x}(s)ds + \int_{-h}^{0} \int_{t+\theta}^{t} \dot{\mathbf{x}}^T(s)\mathbf{R}\dot{\mathbf{x}}(s)dsd\theta, \tag{7}$$

where $\eta_1(t) = \int_{t-h}^{t} \mathbf{x}(s)ds, \quad \eta_2(t) = \int_{t-h}^{t} \int_{t-h}^{s} \mathbf{x}(u)duds.$
Therefore, the derivative of $V(t)$ can be presented as

$$\dot{V}(t) = \sum_{i=1}^{p} \omega_i \left(\zeta^T(t)\mathbf{\Phi}_{1i}\zeta(t) - \int_{t-h}^{t} \dot{\mathbf{x}}^T(s)\mathbf{R}\dot{\mathbf{x}}(s)ds \right), \tag{8}$$

where $\zeta(t) = \begin{bmatrix} \mathbf{x}^T(t) & \mathbf{x}^T(t-h) & \frac{1}{h}\eta_1^T(t) & \frac{2}{h^2}\eta_2^T(t) \end{bmatrix}^T$, and $\mathbf{\Phi}_{1i}$ is defined in (6).
Moreover, according to Lemma 1, we can obtain

$$-\int_{t-h}^{t} \dot{\mathbf{x}}^T(s)\mathbf{R}\dot{\mathbf{x}}(s)ds \le \zeta^T(t)(\mathbf{\Phi}_2 + \mathbf{\Phi}_3)\zeta(t), \tag{9}$$

where Φ_2, Φ_3 are defined in (4). So we have

$$\dot{V}(t) \leq \sum_{i=1}^{p} \omega_i \zeta^T(t)(\mathbf{\Phi}_{1i} + \mathbf{\Phi}_2 + \mathbf{\Phi}_3)\zeta(t) = \sum_{i=1}^{p} \omega_i \zeta^T(t)\mathbf{\Phi}_i\zeta(t) \qquad (10)$$

After some algebraic manipulations, we get

$$\dot{V}(t) = \sum_{i=1}^{p} \omega_i \zeta^T(t)\mathbf{\Phi}_i\zeta(t) \leq \sum_{i=1}^{p} \omega_i \zeta^T(t)\mathbf{\Phi}_i\zeta(t) + \sum_{i=1}^{p} (\bar{\omega}_i - \omega_i)\zeta^T(t)\mathbf{F}_i\zeta(t) + \sum_{i=1}^{p}$$

$$(\omega_i - \underline{\omega}_i)\zeta^T(t)\mathbf{T}_i\zeta(t) = \sum_{i=1}^{p} \omega_i \zeta^T(t)(\mathbf{\Phi}_i - \mathbf{F}_i + \mathbf{T}_i + \sum_{r=1}^{p} \bar{w}_r\mathbf{F}_r - \sum_{k=1}^{p} \underline{\omega}_k\mathbf{T}_k)\zeta(t),$$

$$(11)$$

where $\underline{\omega}_i$ is the lower bound of ω_i, and $\bar{\omega}_i$ is the upper bound of ω_i, $\mathbf{F}_i$ and $\mathbf{T}_i$ are positive semi-definite matrics. Hence, if

$$\mathbf{\Phi}_i - \mathbf{F}_i + \mathbf{T}_i + \sum_{r=1}^{p} \bar{w}_r\mathbf{F}_r - \sum_{k=1}^{p} \underline{\omega}_k\mathbf{T}_k < \mathbf{0}, \qquad (12)$$

then $\dot{V}(t) < \mathbf{0}$. By using Schur Complement theorem, the condition (12) is equivalent to condition (6). In other words, if condition (6) holds, the system (1) is asymptotically stable. Thus, the proof of Theorem 1 is accomplished.

Remark 1. It can be seen from the proof that a novel integral inequality (9) is adopted to deal with integral term $-\int_{t-h}^{t} \dot{\mathbf{x}}^T(s)\mathbf{R}\dot{\mathbf{x}}(s)ds$. The advantage of this integral inequality is that it is tighter than other similar integral inequalities, which can relax the results. Additionally, since the stability condition derived from it has a simpler structure, the implementation costs could be lowered.

4   Stabilization Analysis

Based on Theorem 1, we will mainly investigate how to design a fuzzy controller (2) to stabilize TSFMB control system (3) in this section.

Theorem 2. *For TSFMB control system (3) and the given constants σ, $t_i, i = 2, 3, .., 6$, $h \in [h_{min}, h_{max}]$, if there exist positive definite matrices $\bar{\mathbf{P}} = \bar{\mathbf{P}}^T \in \mathbb{R}^{3n \times 3n}, \bar{\mathbf{R}} = \bar{\mathbf{R}}^T \in \mathbb{R}^{n \times n}, \bar{\mathbf{Q}} = \bar{\mathbf{Q}}^T \in \mathbb{R}^{n \times n}$, and positive semi-definite matrices $\bar{\mathbf{F}}_i = \bar{\mathbf{F}}_i^T \in \mathbb{R}^{4n \times 4n}, \bar{\mathbf{T}}_i = \bar{\mathbf{T}}_i^T \in \mathbb{R}^{4n \times 4n}$, such that LMIs (13) are satisfied, then the system (3) is asymptotically stable.*

$$\begin{bmatrix} \Pi_1 & \sqrt{h}\bar{\mathbf{N}}_1 & \sqrt{h}\bar{\mathbf{N}}_2 & \sqrt{h}\bar{\mathbf{N}}_3 & \sqrt{h}\bar{\mathbf{\Gamma}}_{ij}^T \\ * & -\bar{\mathbf{R}} & 0 & 0 & 0 \\ * & * & -3\bar{\mathbf{R}} & 0 & 0 \\ * & * & * & -5\bar{\mathbf{R}} & 0 \\ * & * & * & * & \frac{1}{\sigma^2}\bar{\mathbf{R}} - \frac{2}{\sigma}\mathbf{X} \end{bmatrix} < \mathbf{0}, \qquad (13)$$

where

$$\Pi_1 = \bar{\Phi}_{1ij} + \bar{\Phi}_3 + \bar{\Phi}_{4ij}, \bar{\Phi}_{1ij} = Sym\{\Delta_4^T \bar{P} \Delta_{5ij}\} + e_1^T \bar{Q}_i e_1 - e_2^T \bar{Q}_i e_2, \bar{\Phi}_2 =$$

$$h\bar{N}_1 \bar{R}^{-1} \bar{N}_1^T + \frac{h}{3}\bar{N}_2 \bar{R}^{-1}\bar{N}_2^T + \frac{h}{5}\bar{N}_3\bar{R}^{-1}\bar{N}_3^T, \bar{\Phi}_3 = Sym\{\bar{N}_1 \Delta_1 + \bar{N}_2 \Delta_2$$

$$+ \bar{N}_3 \Delta_3\}, \bar{\Phi}_{4ij} = -\bar{F}_{ij} + \bar{T}_{ij} + \sum_{r=1}^{p}\sum_{s=1}^{c}\bar{h}_{rs}\bar{F}_{rs} - \sum_{k=1}^{p}\sum_{l=1}^{c}\underline{h}_{kl}\mathbf{T}_{kl}.$$

And the state feedback gain can be denoted as $\mathbf{K}_j = \bar{\mathbf{K}}_j \mathbf{X}^{-1}$. $i = 1, 2, ..., p, \quad j = 1, 2, ..., c.$

Proof. Substitute $\mathbf{A}_i$ for $\mathbf{G}_{ij} = \mathbf{A}_i + \mathbf{B}_i \mathbf{K}_j$, and by following the same line of analysis of Theorem 1, we can obtain

$$\dot{V}(t) = \sum_{i=1}^{p}\sum_{j=1}^{c}\omega_i m_j\Big(\zeta^T(t)\mathbf{\Phi}_{1ij}\zeta(t) - \int_{t-h}^{t}\dot{x}^T(s)\mathbf{R}\dot{x}(s)ds\Big). \tag{14}$$

Applying Lemma 1, we have

$$\dot{V}(t) \leq \sum_{i=1}^{p}\sum_{j=1}^{c}\omega_i m_j \zeta^T(t)(\mathbf{\Phi}_{1ij} + \mathbf{\Phi}_2 + \mathbf{\Phi}_3)\zeta(t). \tag{15}$$

Since

$$\zeta^T(t)\mathbf{\Phi}_{1ij}\zeta(t) = \zeta^T(t)Sym\{\mathbf{M}_{1ij}\}\zeta(t) + \zeta^T(t)Sym\{\mathbf{M}_2\}\zeta(t) + \zeta^T(t)Sym\{$$

$$\mathbf{M}_{3ij}\}\zeta(t), \zeta^T(t)(\mathbf{\Phi}_2 + \mathbf{\Phi}_3)\zeta(t) = \zeta^T(t)\mathbf{M}_4\zeta(t) + \zeta^T(t)\mathbf{M}_5\zeta(t) + \zeta^T(t)\mathbf{M}_6\zeta(t), \tag{16}$$

where

$$\mathbf{M}_{1ij} = \begin{bmatrix} \mathbf{P}_{11}\mathbf{G}_{ij} + \mathbf{P}_{12} & \Pi_2 & h\mathbf{P}_{11}\mathbf{A}_{2i} + \mathbf{P}_{13}h & 0 \\ 0 & 0 & 0 & 0 \\ h\mathbf{P}_{12}^T\mathbf{G}_{ij} + h\mathbf{P}_{22} & \Pi_3 & h^2\mathbf{P}_{12}^T\mathbf{A}_{2i} + h^2\mathbf{P}_{23} & 0 \\ \frac{h^2}{2}\mathbf{P}_{13}^T\mathbf{G}_{ij} + \frac{h^2}{2}\mathbf{P}_{23}^T & \Pi_4 & \frac{h^2}{2}\mathbf{P}_{13}^T\mathbf{A}_{2i} + \frac{h^2}{2}\mathbf{P}_{33} & 0 \end{bmatrix}, \mathbf{M}_2 = \begin{bmatrix} \mathbf{Q} & 0 & 0 & 0 \\ 0 & \mathbf{Q} & 0 & 0 \\ 0 & 0 & 0 & 0 \\ 0 & 0 & 0 & 0 \end{bmatrix},$$

$$\mathbf{M}_{3ij} = \begin{bmatrix} \mathbf{G}_{ij}^T\mathbf{R}\mathbf{G}_{ij} & \mathbf{G}_{ij}^T\mathbf{R}\mathbf{A}_{1i} & \mathbf{G}_{ij}^T\mathbf{R}(h\mathbf{A}_{2i}) & 0 \\ \mathbf{A}_{1i}^T\mathbf{R}\mathbf{G}_{ij} & \mathbf{A}_{1i}^T\mathbf{R}\mathbf{A}_{1i} & \mathbf{A}_{1i}^T\mathbf{R}(h\mathbf{A}_{2i}) & 0 \\ (h\mathbf{A}_{2i}^T)\mathbf{R}\mathbf{G}_{ij} & (h\mathbf{A}_{2i}^T)\mathbf{R}\mathbf{A}_{1i} & (h\mathbf{A}_{2i}^T)\mathbf{R}(h\mathbf{A}_{2i}) & 0 \\ 0 & 0 & 0 & 0 \end{bmatrix}, \mathbf{M}_4 = \frac{1}{h}\begin{bmatrix} -\mathbf{R} & \mathbf{R} & 0 & 0 \\ \mathbf{R} & -\mathbf{R} & 0 & 0 \\ 0 & 0 & 0 & 0 \\ 0 & 0 & 0 & 0 \end{bmatrix}$$

$$\mathbf{M}_5 = \frac{3}{h}\begin{bmatrix} -\mathbf{R} & -\mathbf{R} & 2\mathbf{R} & 0 \\ -\mathbf{R} & -\mathbf{R} & 2\mathbf{R} & 0 \\ 2\mathbf{R} & 2\mathbf{R} & -4\mathbf{R} & 0 \\ 0 & 0 & 0 & 0 \end{bmatrix}, \mathbf{M}_6 = \frac{5}{h}\begin{bmatrix} -\mathbf{R} & \mathbf{R} & 6\mathbf{R} & -6\mathbf{R} \\ \mathbf{R} & -\mathbf{R} & -6\mathbf{R} & 6\mathbf{R} \\ 6\mathbf{R} & -6\mathbf{R} & -36\mathbf{R} & 36\mathbf{R} \\ -6\mathbf{R} & 6\mathbf{R} & 36\mathbf{R} & -36\mathbf{R} \end{bmatrix}.$$

$$\Pi_2 = \mathbf{P}_{11}\mathbf{A}_{1i} - \mathbf{P}_{12} - h\mathbf{P}_{13}, \Pi_3 = h\mathbf{P}_{12}^T\mathbf{A}_{1i} - h\mathbf{P}_{22} - h^2\mathbf{P}_{23},$$

$$\Pi_4 = \frac{h^2}{2}\mathbf{P}_{13}^T\mathbf{A}_{1i} - \frac{h^2}{2}\mathbf{P}_{23}^T - \frac{h^3}{2}\mathbf{P}_{33},$$

So we have $\zeta^T(t)(\mathbf{\Phi}_{1ij} + \mathbf{\Phi}_2 + \mathbf{\Phi}_3)\zeta(t) = \zeta^T(t)(\mathbf{M}_{1ij} + \mathbf{M}_2 + \mathbf{M}_{3ij} + \mathbf{M}_4 + \mathbf{M}_5 + \mathbf{M}_6)\zeta(t)$. Therefore, if

$$\mathbf{M}_{1ij} + \mathbf{M}_2 + \mathbf{M}_{3ij} + \mathbf{M}_4 + \mathbf{M}_5 + \mathbf{M}_6 < 0, \tag{17}$$

we can obtain $\dot{V}(t) < 0$.

Define some new variables as

$$\mathbf{P}_{12} = t_2\mathbf{P}_{11}, \quad \mathbf{P}_{13} = t_3\mathbf{P}_{11}, \quad \mathbf{P}_{22} = t_4\mathbf{P}_{11}, \quad \mathbf{P}_{23} = t_5\mathbf{P}_{11}, \quad \mathbf{P}_{33} = t_6\mathbf{P}_{11},$$

$$\mathbf{X} = \mathbf{P}_{11}^{-1}, \quad \bar{\mathbf{R}} = \mathbf{X}\mathbf{R}\mathbf{X}, \quad \bar{\mathbf{Q}} = \mathbf{X}\mathbf{Q}\mathbf{X}, \quad \bar{\mathbf{K}}_j = \mathbf{K}_j\mathbf{X}, \quad \bar{\mathbf{G}}_{ij} = \mathbf{A}_i\mathbf{X} + \mathbf{B}_i\bar{\mathbf{K}}_j,$$

$$\bar{\mathbf{A}}_{1i} = \mathbf{A}_{1i}\mathbf{X}, \quad \bar{\mathbf{A}}_{2i} = \mathbf{A}_{2i}\mathbf{X}, \quad \bar{\mathbf{B}}_i = \mathbf{B}_i\mathbf{X}, \quad \bar{\mathbf{N}}_1 = \frac{1}{h}\left[-\bar{\mathbf{R}}\ \bar{\mathbf{R}}\ 0\ 0\right]^T,$$

$$\bar{\mathbf{N}}_2 = \frac{3}{h}\left[-\bar{\mathbf{R}}\ -\bar{\mathbf{R}}\ 2\bar{\mathbf{R}}\ 0\right]^T, \quad \bar{\mathbf{N}}_3 = \frac{5}{h}\left[-\bar{\mathbf{R}}\ \bar{\mathbf{R}}\ 6\bar{\mathbf{R}}\ -6\bar{\mathbf{R}}\right]^T,$$

$$\bar{\mathbf{\Gamma}}_{ij} = [\bar{\mathbf{G}}_{ij}\ \bar{\mathbf{A}}_{1i}\ h\bar{\mathbf{A}}_{2i}\ 0], \quad \bar{\mathbf{\Delta}}_{5ij} = \left[\bar{\mathbf{\Gamma}}_{ij}^T\quad \mathbf{X}(\mathbf{e}_1^T - \mathbf{e}_2^T)\quad \mathbf{X}(h\mathbf{e}_3^T - h\mathbf{e}_2^T)\right]^T. \tag{18}$$

Let Eq. (17) be pre-multiplied and post-multiplied by $diag\left[\mathbf{X}\ \mathbf{X}\ \mathbf{X}\ \mathbf{X}\right]$ and its transpose, then we can obtain

$$Sym\{\mathbf{\Delta}_4^T\bar{\mathbf{P}}\bar{\mathbf{\Delta}}_{5ij}\} + \mathbf{e}_1^T\bar{\mathbf{Q}}_i\mathbf{e}_1 - \mathbf{e}_2^T\bar{\mathbf{Q}}_i\mathbf{e}_2 + h\bar{\mathbf{N}}_1\bar{\mathbf{R}}^{-1}\bar{\mathbf{N}}_1^T + \frac{h}{3}\bar{\mathbf{N}}_2\bar{\mathbf{R}}^{-1}\bar{\mathbf{N}}_2^T$$
$$+ \frac{h}{5}\bar{\mathbf{N}}_3\bar{\mathbf{R}}^{-1}\bar{\mathbf{N}}_3^T + Sym\{\bar{\mathbf{N}}_1\mathbf{\Delta}_1 + \bar{\mathbf{N}}_2\mathbf{\Delta}_2 + \bar{\mathbf{N}}_3\mathbf{\Delta}_3\} + h\mathbf{\Gamma}_{ij}^T\mathbf{R}\mathbf{\Gamma}_{ij} < 0. \tag{19}$$

So we have

$$\dot{V}(t) \leq \sum_{i=1}^{p}\sum_{j=1}^{c}\zeta^T(t)\omega_i m_j \mathbf{\Phi}_{ij}\zeta(t)$$
$$= \sum_{i=1}^{p}\sum_{j=1}^{c}\omega_i m_j \zeta^T(t)\left(\bar{\mathbf{\Phi}}_{1ij} + \bar{\mathbf{\Phi}}_2 + \bar{\mathbf{\Phi}}_3 + h\mathbf{\Gamma}_{ij}^T\mathbf{R}\mathbf{\Gamma}_{ij}\right)\zeta(t) < 0, \tag{20}$$

where $\bar{\mathbf{\Phi}}_{1ij}, \bar{\mathbf{\Phi}}_2, \bar{\mathbf{\Phi}}_3, i = 1, 2, ..., p, j = 1, 2, ..., c$ are defined in (13).

By denoting $\omega_i m_j$ as $h_{ij}, i = 1, 2, ..., p, j = 1, 2, ...c$, and using similar algebraic manipulations as in the proof of Theorem 1, we can get

$$\dot{V}(t) \leq \sum_{i=1}^{p}\sum_{j=1}^{c}h_{ij}\zeta^T(t)\bar{\mathbf{\Phi}}_{ij}\zeta(t)$$
$$\leq \sum_{i=1}^{p}\sum_{j=1}^{c}h_{ij}\zeta^T(t)(\bar{\mathbf{\Phi}}_{ij} - \bar{\mathbf{F}}_{ij} + \bar{\mathbf{T}}_{ij} + \sum_{r=1}^{p}\sum_{s=1}^{c}\bar{h}_{rs}\bar{\mathbf{F}}_{rs} - \sum_{k=1}^{p}\sum_{l=1}^{c}\underline{h}_{kl}\mathbf{T}_{kl})\zeta(t), \tag{21}$$

where $\bar{h}_{ij} \geq h_{ij}$ is the upper bound of h_{ij}, $\underline{h}_{ij} \leq h_{ij}$ is the lower bound of h_{ij}, $\bar{\mathbf{F}}_{ij} = \bar{\mathbf{F}}_{ij}^T \geq 0$, and $\bar{\mathbf{T}}_{ij} = \bar{\mathbf{T}}_{ij}^T \geq 0$.

So if

$$\bar{\Phi}_{ij} - \bar{\mathbf{F}}_{ij} + \bar{\mathbf{T}}_{ij} + \sum_{r=1}^{p}\sum_{s=1}^{c} \bar{h}_{rs}\bar{\mathbf{F}}_{rs} - \sum_{k=1}^{p}\sum_{l=1}^{c} \underline{h}_{kl}\mathbf{T}_{kl} < \mathbf{0}, \tag{22}$$

we can obtain $\dot{V}(t) < 0$, which means the closed-loop TSFMB control system (3) is asymptotically stable.

Applying Schur Complement theorem, inequality (22) can be denoted as

$$\begin{bmatrix} \Pi_1 & \sqrt{h}\bar{\mathbf{N}}_1 & \sqrt{h}\bar{\mathbf{N}}_2 & \sqrt{h}\bar{\mathbf{N}}_3 & \sqrt{h}\bar{\mathbf{\Gamma}}_{ij}^T \\ * & -\bar{\mathbf{R}} & \mathbf{0} & \mathbf{0} & \mathbf{0} \\ * & * & -3\bar{\mathbf{R}} & \mathbf{0} & \mathbf{0} \\ * & * & * & -5\bar{\mathbf{R}} & \mathbf{0} \\ * & * & * & * & -\mathbf{R}^{-1} \end{bmatrix} < \mathbf{0}, \tag{23}$$

where $\Pi_1 = \bar{\mathbf{\Phi}}_{1ij} + \bar{\mathbf{\Phi}}_3 + \bar{\mathbf{\Phi}}_{4ij}, i = 1, 2, ..., p, j = 1, 2, ..., c.$

Moreover, as $\mathbf{R}$ is symmetric and positive definite matrix, for any scalar σ, the following inequality holds:

$$\left(\sigma\mathbf{R}^{-1} - \mathbf{X}\right)\mathbf{R}\left(\sigma\mathbf{R}^{-1} - \mathbf{X}\right) > \mathbf{0}. \tag{24}$$

Then we have

$$-\mathbf{R}^{-1} < -\frac{2}{\sigma}\mathbf{X} + \frac{1}{\sigma^2}\bar{\mathbf{R}}, \tag{25}$$

which means inequalities (23) is true, if LMIs (13) holds. Thus, the proof of Theorem 2 is completed.

5   Numerical Examples

In this section, a numerical example is given to demonstrate the effectiveness of the proposed methods.

Example 1. Consider a TSFMB autonomous system in the form of (1) with

$$\mathbf{A}_1 = \begin{bmatrix} -2.1 & 0.1 \\ -0.2 & -0.9 \end{bmatrix}, \mathbf{A}_2 = \begin{bmatrix} -1.9 & 0 \\ -0.2 & -1.1 \end{bmatrix}, \mathbf{A}_{11} = \begin{bmatrix} -1.1 & 0.1 \\ -0.8 & 0.9 \end{bmatrix},$$

$$\mathbf{A}_{12} = \begin{bmatrix} -0.9 & 0 \\ -1.1 & -1.2 \end{bmatrix}, \mathbf{A}_{21} = \begin{bmatrix} 0 & 0 \\ 0 & 0 \end{bmatrix}, \mathbf{A}_{22} = \begin{bmatrix} 0 & 0 \\ 0 & 0 \end{bmatrix},$$

$$\omega_1(x_1(t)) = 1 - \frac{0.5}{1 + e^{-3-x_1(t)}}, \quad \omega_2(x_1(t)) = 1 - \omega_1(x_1(t)).$$

Using the stability conditions introduced in literature [1–3,9,10] and Theorem 1 of this paper, respectively to calculate the maximum allowable time-delays. The results are presented in the following table.

From Table 1, we can see that Theorem 1 of this paper can yield larger maximum allowable time-delays h than literature [1–3,9,10], which means the proposed method in this paper is less conservative than the ones in [1–3,9,10].

Table 1. The maximum allowable time-delays h for Example 1

Method	[3]	[1]	[2]	[9]	[10]	Theorem 2
h	3.37	4.28	4.61	4.64	5.58	7.62

Remark 2. There are two reasons for the less conservative results in Example 1. First, the introduction of the new integral inequality (4), which is tighter than other existing integral inequalities. Second, Theorem 1 of this paper takes the information of membership functions into consideration while the methods in other literature are membership functions independent.

6   Conclusion

This paper concerns the stability and stabilization analysis issue of TSFMB control systems with distributed time-delay. The imperfect premise matching methodology has been employed to design fuzzy controller, which allows the fuzzy controller to use different premise membership functions and different number of rules from the fuzzy model. Hence, greater design flexibility can be achieved. Besides, a tighter integral inequality has been applied, and the information of the membership functions has been introduced in the criteria as well. As a consequence, less conservative stability conditions and stabilization criteria have been developed. Finally, a numerical example has been provided to illustrate the effectiveness of the proposed approach.

References

1. Zhao, Y., Gao, H., Lam, J., Du, B.: Stability and stabilization of delayed T-S fuzzy systems: a delay partitioning approach. IEEE Trans. Fuzzy Syst. **17**(4), 750–762 (2009)
2. Mozelli, L.A., Souza, F.O., Palhares, R.M.: A new discretized Lyapunov-Krasovskii functional for stability analysis and control design of time-delayed T-S fuzzy systems. Int. J. Robust Nonlinear Control **21**(1), 93–105 (2011)
3. Wu, H.N., Li, H.X.: New approach to delay-dependent stability analysis and stabilization for continuous-time fuzzy systems with time-varying delay. IEEE Trans. Fuzzy Syst. **15**(3), 482–493 (2007)
4. Wu, M., He, Y., She, J.H., Liu, G.P.: Delay-dependent criteria for robust stability of time-varying delay systems. Automatica **40**(8), 1435–1439 (2004)
5. Seuret, A., Gouaisbaut, F.: Wirtinger-based integral inequality: application to time-delay systems. Automatica **60**, 189–192 (2015)
6. Hong, B.Z., Yong, H., Min, W., Jinhua, S.: New results on stability analysis for systems with discrete distributed delay. Automatica **60**, 189–192 (2015)
7. Pan, J.T., Guerra, T.M., Fei, S.M., Jaadari, A.: Nonquadratic stabilization of continuous T-S fuzzy models: LMI solution for a local approach. IEEE Trans. Fuzzy Syst. **20**(3), 594–602 (2013)

8. Lam, H.K., Narimani, M.: Stability analysis and performance design for fuzzy-model-based control system under imperfect premise matching. IEEE Trans. Fuzzy Syst. **17**(4), 949–961 (2009)
9. Zhao, L., Gao, H., Karimi, H.R.: Robust stability and stabilization of uncertain T-S fuzzy systems with time-varying delay: an input-output approach. IEEE Trans. Fuzzy Syst. **21**(5), 883–897 (2013)
10. Zhang, Z., Lin, C., Chen, B.: New stability and stabilization conditions for T-S fuzzy systems with time delay. Fuzzy Sets Syst. **263**, 82–91 (2015)

Adaptive Neuro-Fuzzy Inference System: Overview, Strengths, Limitations, and Solutions

Mohd Najib Mohd Salleh$^{(\boxtimes)}$, Noureen Talpur, and Kashif Hussain

Faculty of Computer Science and Information Technology,
Universiti Tun Hussein Onn Malaysia, Batu Pahat, Johor, Malaysia
najib@uthm.edu.my

Abstract. Adaptive neuro-fuzzy inference system (ANFIS) is efficient estimation model not only among neuro-fuzzy systems but also various other machine learning techniques. Despite acceptance among researchers, ANFIS suffers from limitations that halt applications in problems with large inputs; such as, curse of dimensionality and computational expense. Various approaches have been proposed in literature to overcome such shortcomings, however, there exists a considerable room of improvement. This paper reports approaches from literature that reduce computational complexity by architectural modifications as well as efficient training procedures. Moreover, as potential future directions, this paper also proposes conceptual solutions to the limitations highlighted.

Keywords: ANFIS · Fuzzy logic · Neural network · Neuro-fuzzy · Big data

1 Introduction

According to Lofti Zadeh (informally, the fuzzifier of crisp domain) "in human cognition almost all classes have unsharp (fuzzy) boundaries" [1]. Hence, coupling, embedding or meshing fuzzy ingredients into neural networks with bivalent logic will enable us to comply with Zadeh's statement. This marriage of learning capability of neural network and knowledge representation ability of fuzzy logic has given birth to fuzzy neural networks. As a result, the drawback of neural network "black box" – inability to explain decision (lack of transparency), and weakness of learning in fuzzy logic have been conquered.

According to literature, fuzzy neural networks are able to approximate any plant with high degree of accuracy; be it engineering, medicine, transportation, or business and economics, etc. [2]. This success has led to significantly conspicuous literature comprising of improvements and modifications [3], applications [4], and surveys or reviews [2] of fuzzy neural networks. As one of the prominent neuro-fuzzy systems, adaptive neuro-fuzzy inference system (ANFIS), introduced by Jang in 1993 [5], has gained remarkable attention from researchers. Nevertheless, ANFIS faces major limitations such as curse of dimensionality and training complexity which restrict applications on problems with large datasets. As [6]

© Springer International Publishing AG 2017
Y. Tan et al. (Eds.): DMBD 2017, LNCS 10387, pp. 527–535, 2017.
DOI: 10.1007/978-3-319-61845-6_52

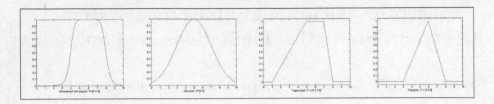

Fig. 1. Basic shapes of membership functions

rightly pointed out a major problem that it is again a problem on making decisions about number, type and initial values of membership functions, initial rule-base, and input space clustering or partitioning method.

In this study, emphasis is placed on basic concepts and architectural aspects of ANFIS. The rest of the paper is organized as follows: the subsequent section explains ANFIS architecture, its strengths and limitations. Solutions to limitations from existing literature are reported in Sect. 3, whereas the conceptual solutions to the shortcomings are proposed in Sect. 4. Finally, Sect. 5 concludes the study and highlights future prospects in this area of research.

2   Adaptive Neuro-Fuzzy Inference System (ANFIS)

In machine learning area, FNN – also referred to as fuzzy inference system – is an effective hybrid of fuzzy logic and neural network which has achieved significant success in approximation and control models. Unlike ANN, FNN maps inputs through input membership functions to the desired output(s) via output membership functions, and this mapping generates rule-base in the course of learning. These rules in FNN are directly mapped into the neural structure of the network. The accuracy of rules depends on appropriateness of type (Fig. 1) and parameters of membership functions.

ANFIS is based on Takagi-Sugeno-Kangmodel (TSK), or simply Sugeno fuzzy model, proposed by [7] where a rule R_k can be represented as:

$$R_k : \text{IF} \mu_{A_i}(x) \text{AND} \mu_{B_i}(y) \text{THEN} f = p_k x + q_k y + r_k \tag{1}$$

where k is the number of rules, A_i and B_i are n fuzzy membership functions of any shape i.e., gaussian, triangular, triopzoidal, etc., denoted by μ in the antecedent part of the rule R_k, and p_k, q_k, r_k are the linear parameters of consequent part of the kth rule. The parameters of membership functions (antecedent or premise parameters) and consequent part of the rule (consequent parameters) are tuned during the training process.

ANFIS five-layers architecture comprises of two types of nodes: fixed and adaptable (Fig. 2). The nodes in membership function layer and consequent layer are tunable, the rest of the nodes are fixed.

In Layer 1, the node i is a membership function i.e., triangle, trapezoidal, or gaussian, etc. For example, if μ_{A_1}, μ_{A_2} and μ_{B_1}, μ_{B_2} are the membership

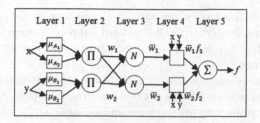

Fig. 2. ANFIS architecture

functions of gaussian shape with two parameters center (c) and width (σ). Layer 2 calculates the firing strength of a rule via product $\prod$ operation. Layer 3 is normalized firing strength of a rule from previous layer. In Layer 4, each node represents consequent part of fuzzy rule. The linear coefficients of rule consequent are trainable. Nodes in Layer 5 perform defuzzification of consequent part of rules by summing outputs of all the rules. Further detail on computation performed in ANFIS can be found in the related paper [7].

ANFIS learns by tuning all its tunable parameters (c, σ and p_k, q_k, r_k) in order to map input to the desired output with minimum error. The default ANFIS learning algorithm employs gradient descent (GD) for tuning membership functions and least square estimation (LSE) for training the consequent parameters. It is a two pass learning process as presented in Table 1.

Table 1. Two-pass ANFIS learning algorithm

	Forward pass	Backward pass
Antecedent parameters	Fixed	GD
Consequent parameters	LSE	Fixed
Signals	Node outputs	Error signals

The two pass learning algorithm tunes consequent parameters by LSE in forward pass and while back-propagating error back to first layer, it updates membership functions using GD. Since, GD is influenced by back propagation (BP) algorithm of ANN, which has the drawback to be likely trapped in local minima [8]. On the other hand, the convergence of gradient method is also very slow; depending on initial parameter values.

2.1 Strengths and Limitations of ANFIS

The success of ANFIS can be attributed to the robustness of results it provides [2]. ANFIS has as highly generalization capability as neural networks and other machine learning techniques [4]. ANFIS is able to take crisp input, represent in the form of membership functions and fuzzy rules, as well as, again generates crisp output out of fuzzy rules for reasoning purpose. This provides room for applications that involve crisp inputs and outputs. It is exceptionally potential tool yet to be explored in various other non-linear and complex approximation and control problems.

The computational cost of ANFIS is high due to complex structure and gradient learning. This is a significant bottleneck to applications with large inputs. Broadly, the limitations are: (a) the type and number of membership functions; (b) the location of a membership function; and (c) the curse of dimensionality [9]. Additionally, the trade-off between interpretability and accuracy is considered as crucial problem.

In ANFIS, tunable parameters consist of membership function parameters and consequent parameters. This demands efficient training mechanism that can tune the parameters more effectively. The parameter complexity is directly related to computational cost. Therefore, the more the parameters in ANFIS architecture, the more is the training and computational cost. This is explained in Table 2 as case study of Iris Classification Dataset [10]. Here, ANFIS-1 is created by genfis1 command in MATLAB which generates ANFIS with grid partitioning method. ANFIS-2 is the one generated with subtractive clustering using genfis2 command in MATLAB. And, ANFIS-3 is the network structure generated with fuzzy c-mean clustering approach using genfis3 command in MATLAB [8].

Table 2. Computational complexity of different ANFIS networks: Iris as case study

ANFIS type	Inputs	MF type (params)	MFs	MFs params.	Total rules	Conseq. params.	Total params.	Training RMSE
ANFIS-1	n=4	Trapezoidal (p=4)	m=3	48	r=81	405	453	0.00062849
ANFIS-1	n=4	Bell (p=3)	m=3	36	r=81	405	441	5.9994e-05
ANFIS-1	n=4	Triangle (p=3)	m=3	36	r=81	405	441	8.5363e-05
ANFIS-1	n=4	Guassian (p=2)	m=3	24	r=81	405	429	5.8176e-05
ANFIS-2	n=4	Guassian (p=2)	m=11	88	r=11	55	143	0.0043667
ANFIS-3	n=4	Guassian (p=2)	m=15	120	r=15	75	195	0.00011801

It is obvious from total number of parameters 453 in Table 2 that maximum computational complexity is in case of ANFIS-1 generated using grid partitioning methods, as it involves maximum number of tunable parameters. This also influences the computational time as well to reach its peak. In other cases of ANFIS-1, the one which employs gaussian membership function, involves least number of parameters, however, it still maintain high computational complexity as compared to ANFIS-2 and ANFIS-3. On the other hand, in terms of accuracy, ANFIS-1 generates best results among all the other types of ANFIS models as grid partitioning produces all possible rules to interpret the problem in hand. ANFIS2 and ANFIS3 cut computational complexity but suffer from the loss of accuracy.

Moreover, in terms of interpretability, ANFIS with grid partitioning produces a large number of rules which indeed cannot be easily understood by model users.

Hence, interpretability is highly compromised, even though, the large number of rules contribute to improvement in model accuracy. This is expressed via Fig. 3 that more rules tend to produce better accuracy. Though, on the other hand, it is difficult to interpret the designed model. Contrarily, reducing the rule-base may result in low accuracy. Therefore, this tradeoff is difficult to solve [11]. In this paper, other than accuracy and interpretability mentioned in [11], computational complexity is additionally expressed in Fig. 3. Other than grid methods, in clustering approaches, the partitioning of input space is performed with the clustering input-output data. This raises concerns of how to label the partitioned membership functions as partitioning is performed on data with non-linear and complex characteristics.

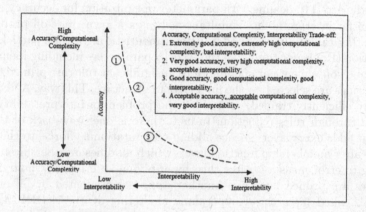

Fig. 3. Interpretability vs. accuracy/computational complexity in fuzzy system

ANFIS is good when the number of inputs is not more than five. The more the inputs, the higher will be computational expense of ANFIS-based model. Majority of the applications have been found with inputs not more than 5 i.e., [12,13], etc. However, a few studies are found with slightly greater than 5 inputs; e.g., [14] used 6 inputs. It seems that ANFIS is difficult to implement in Big Data paradigm.

3 Solutions to Limitations from Literature

The limitations of ANFIS motivated researchers to find better alternatives. In this section, such significant contributions from the existing literature are reported.

Mainly, regarding parameter training and computational effort, there are three approaches proposed in literature: reducing rule-base (discussed below), reducing number of parameters, and efficient training methods. The ANFIS rules are minimized to shrink the number of parameters as well as to reduce computational effort along with achieving acceptable accuracy. These approaches are discussed below. A novel approach to addressing the excessive computational effort in ANFIS was proposed in [15]. In this research, third layer which normalizes the rule strength is removed. However, this approach still uses hybrid learning algorithm as typical ANFIS. In order to propose efficient training methods, many researchers trained ANFIS parameters using metaheuristic algorithms either in the hybrid with least square or gradient decent, or training all the parameters with metaheuristic algorithm alone. [16] employed cat swarm optimization (CSO) algorithm with gradient descent to training membership function parameters and consequent parameters, respectively. [17] proposed a hybrid of particle swarm optimization (PSO) and least square method to tune premise and consequent parameters, respectively. [18] proposed a variant of artificial bee colony (ABC) algorithm to tune all the ANFIS parameters.

According to [11], dealing with parameter optimization for accuracy and rule generation for interpretability simultaneously is a known trade-off problem. To overcome this, this study proposed using metaheuristic algorithm, particle swarm optimization (PSO), for tuning all the ANFIS parameters including membership function and consequent parameters. Here, insignificant rules are pruned to select only best rules are selected for the final ANFIS structure. This way, ANFIS parameters are efficiently trained, as well as, the problem of interpretability is also addressed through rules reduction. In-fact, there is one draw-back of this approach as it adds extra layer – hence adding computational effort – to the ANFIS structure after membership function layer, which modifies membership functions according to error measure. After this, by applying an error threshold, insignificant rules are pruned later on. Another approach presented by [3] employed a method, called hierarchical hyperplane clustering synthesis (HHCS), which increases rules to the ANFIS rule-base constructively until the desired accuracy is achieved. This research achieves interpretability by producing the optimum rule-base. However, this approach still relies on typical parameter tuning algorithms like gradient descent and least square estimation. Moreover, similar to previously discussed technique, this research also makes ANFIS structure more complex. Other than clustering techniques, in study [19], ANFIS rule-base is reduced using Karnaugh Map while designing traffic signal controller. In this method, rules are mapped into K-Map so that minimum map that represents reduced rules with high accuracy.

As mentioned earlier, due to input constraints, ANFIS applications found in literature involve 5 to 7 input variables. However, input selection techniques are integrated with ANFIS models to first reduce the number of inputs to select best suitable inputs, and then apply ANFIS on the selected inputs. Such as, [20] first selected 4 out of 8 inputs and then employed ANFIS to approximate the desired model. More of such tactics can be found in [21].

4   Proposed Conceptual Solutions

In response to limitations identified previously, this paper proposes conceptual solutions that can act as potential future directions in research related to ANFIS performance improvement.

Table 2 reveals that ample number of rules produces better accuracy of the ANFIS output. In this connection, grid partitioning is useful method that generates maximum number of rules. That said, it also increases computational cost as consequent part of rules contains most of the parameters. Therefore, fourth layer which holds linear coefficients shares most of the computational cost of training algorithm. The removal of fourth layer may contribute to reduction of computation hence ANFIS architecture can be reduced to four layers. Furthermore, instead of gradient based learning mechanism in typical ANFIS, metaheuristic algorithms can be employed to train all the parameters. This will not only reduce computational complexity of ANFIS but also apply efficient training using metaheuristic paradigm.

Additional to layer reduction approach discussed before, rules can also be reduced by selecting potential rules for producing the most suitable rules set to address the trade-off of lessening computational cost, increasing accuracy and also enhancing interpretability. This approach, may apply threshold for error tolerance on fourth layer to filter the rules that best meet the error criteria. Rule minimization by selecting suitable rule-set for producing powerful rule-base, that contains only the rules that contribute the most in accuracy of the model, thus can be achieved.

5   Conclusion

In this research work, an overview of ANFIS architecture has been presented in order to highlight the computational complexity of the network. The number of parameters and rules play crucial part in increasing the computational cost of ANFIS-based models. Moreover, different limitations such as curse of dimensionality, interpretability of rules, and parameter training are the major hurdles that need to be overcome for the implementation in problems with larger number of inputs. This is the reason, ANFIS is often integrated with additional techniques for input selection, rule reduction and parameter tuning, which again increases the complexity of the designed model. Various structural and parameters optimization techniques have been proposed in literature, however, there is enough room of improvement in ANFIS architecture so that applications in larger problems can be achieved easily. To overcome these issues, conceptual solutions have been proposed in this research, which reflect interesting future research directions.

Acknowledgments. The authors would like to thank Universiti Tun Hussein Onn Malaysia (UTHM), Malaysia for supporting this research under Postgraduate Incentive Research Grant, Vote No.U728.

References

1. Zadeh, L.A.: Fuzzy logica personal perspective. Fuzzy Sets Syst. **281**, 4–20 (2015)
2. Kar, S., Das, S., Ghosh, P.K.: Applications of neuro fuzzy systems: a brief review and future outline. Appl. Soft Comput. **15**, 243–259 (2014)
3. Panella, M.: A hierarchical procedure for the synthesis of anfis networks. Adv. Fuzzy Syst. **2012**, 20 (2012)
4. Zamani, H.A., Rafiee-Taghanaki, S., Karimi, M., Arabloo, M., Dadashi, A.: Implementing anfis for prediction of reservoir oil solution gas-oil ratio. J. Nat. Gas Sci. Eng. **25**, 325–334 (2015)
5. Jang, J.-S.R.: Anfis: adaptive-network-based fuzzy inference system. IEEE Trans. Syst. Man Cybern. **23**(3), 665–685 (1993)
6. Alizadeh, M., Lewis, M., Zarandi, M.H.F., Jolai, F.: Determining significant parameters in the design of anfis. In: Fuzzy Information Processing Society (NAFIPS), 2011 Annual Meeting of the North American, pp. 1–6. IEEE (2011)
7. Sugeno, M., Tanaka, K.: Successive identification of a fuzzy model and its applications to prediction of a complex system. Fuzzy Sets Syst. **42**(3), 315–334 (1991)
8. Inc., The MathWorks. anfis (2017)
9. Ciftcioglu, O., Bittermann, M.S., Sariyildiz, I.S.: A neural fuzzy system for soft computing. In: Fuzzy Information Processing Society, NAFIPS 2007, Annual Meeting of the North American, pp. 489–495. IEEE (2007)
10. Lichman, M.: Uci machine learning repository (2013)
11. Rini, D.P., Shamsuddin, S.M., Yuhaniz, S.S.: Balanced the trade-offs problem of anfis using particle swarm optimisation. TELKOMNIKA (Telecommun. Comput. Electron. Control) **11**(3), 611–616 (2013)
12. Barati-Harooni, A., Najafi-Marghmaleki, A., Mohammadi, A.H.: Anfis modeling of ionic liquids densities. J. Mol. Liq. **224**, 965–975 (2016)
13. Tatar, A., Barati-Harooni, A., Najafi-Marghmaleki, A., Norouzi-Farimani, B., Mohammadi, A.H.: Predictive model based on anfis for estimation of thermal conductivity of carbon dioxide. J. Mol. Liq. **224**, 1266–1274 (2016)
14. Taylan, O., Karagözoğlu, B.: An adaptive neuro-fuzzy model for prediction of student's academic performance. Comput. Ind. Eng. **57**(3), 732–741 (2009)
15. Peymanfar, A., Khoei, A., Hadidi, K.: A new anfis based learning algorithm for cmos neuro-fuzzy controllers. In: 14th IEEE International Conference on Electronics, Circuits and Systems, ICECS 2007, pp. 890–893. IEEE (2007)
16. Orouskhani, M., Mansouri, M., Orouskhani, Y., Teshnehlab, M.: A hybrid method of modified cat swarm optimization and gradient descent algorithm for training anfis. Int. J. Comput. Intell. Appl. **12**(02), 1350007 (2013)
17. Zuo, L., Hou, L., Zhang, W., Geng, S., Wu, W.: Application of PSO-adaptive neural-fuzzy inference system (ANFIS) in analog circuit fault diagnosis. In: Tan, Y., Shi, Y., Tan, K.C. (eds.) ICSI 2010. LNCS, vol. 6146, pp. 51–57. Springer, Heidelberg (2010). doi:10.1007/978-3-642-13498-2_7
18. Karaboga, D., Kaya, E.: An adaptive and hybrid artificial bee colony algorithm (aabc) for anfis training. Appl. Soft Comput. **49**, 423–436 (2016)
19. Soh, A.C., Kean, K.Y.: Reduction of anfis-rules based system through k-map minimization for traffic signal controller. In: 12th International Conference on Control, Automation and Systems 2012 (ICCAS), pp. 1290–1295. IEEE (2012)

20. Polat, K., Güneş, S.: An expert system approach based on principal component analysis and adaptive neuro-fuzzy inference system to diagnosis of diabetes disease. Digit. Signal Proc. **17**(4), 702–710 (2007)
21. Güneri, A.F., Ertay, T., YüCel, A.: An approach based on anfis input selection and modeling for supplier selection problem. Expert Syst. Appl. **38**(12), 14907–14917 (2011)

Acquisition of Knowledge in the Form of Fuzzy Rules for Cases Classification

Tatiana Avdeenko[✉] and Ekaterina Makarova

Novosibirsk State Technical University, Novosibirsk, Russia
avdeenko@corp.nstu.ru, katmc@yandex.ru

Abstract. We consider an approach to automatic knowledge acquisition through machine learning based on integration of two basic paradigms of reasoning – case-based and rule-based reasoning. Case-based reasoning allows to use high-performance database technology for storing and accumulating cases, while rule-based reasoning is the most developed technology for creating declarative knowledge on the basis of strong logical inference. We also propose an improvement of classification algorithm through extraction of fuzzy rules from cases. We have obtained higher classification accuracy for various membership functions and for sequentially reducing amount of training sample by application of special strategies for expanding the scope of fuzzy rules in the control sample classification.

Keywords: Knowledge-based systems · Case-based reasoning · Rule-based reasoning · Fuzzy rules · Classification

1 Introduction

An important stage in building knowledge-based systems (KBS) is the choice of knowledge representation model for creating knowledge base. The first KBS appeared in 1970-s were based on rules. When implementing rule-based KBS, developers faced many challenges. The major one is the problem of knowledge elicitation and formulating it in the form of a set of rules. Most often, the experts intuitively make decisions based on their vast experience, without hesitation, what rules they apply in this or that case. Partitioning a specific behavior of an expert into separate building blocks such as rules is a very complicated problem requiring high skilled specialists. Thus, acquisition of knowledge is the key problem of rule-based reasoning (RBR) systems.

However, since 1980s an alternative reasoning paradigm has increasingly attracted more and more attention. Case-based reasoning (CBR) solves new problems by adapting previously successful solutions to similar problems, just as a human does it. The paper [1] of Shank is widely cited to be the origins of CBR. In this work it has been proposed to generalize knowledge about previous situations and store them in the form of scripts that can be used to make conclusions under similar conditions. Later Shank continued to explore the role that dynamic memory about previous situations (cases), represented as a knowledge container, plays both in the process of decision making and in the process of learning [2]. The model of dynamic memory became a basis for the creation of a

© Springer International Publishing AG 2017
Y. Tan et al. (Eds.): DMBD 2017, LNCS 10387, pp. 536–544, 2017.
DOI: 10.1007/978-3-319-61845-6_53

number of other systems: MEDIATOR [3], CHEF [4], PERSUADER [5], CASEY [6] and JULIA [7].

CBR allowed to overcome a number of restrictions inherent to the rule-based systems [8]. The acquisition of knowledge in CBR is reduced to identification of essential features and accumulation of decision making stories (cases) described by the features, that is much easier task than building an explicit knowledge model of the application domain. There are different ways of presenting and storing cases – from simple (linear) to complex hierarchical ones.

At the same time there are two essential shortcomings of traditional CBR. The first one is that description of the cases does not usually take into account the deeper knowledge of the application domain. The second one reveals itself when the number of cases accumulated in the knowledge base becomes great. The large case base results in reduced system performance. Very often the search dictionaries and algorithms for determining similarity are needed to debug manually. These shortcomings could neutralize the benefits of case-based approach.

In order to overcome these disadvantages CBR has been widely integrated with other methods in various application domains [9, 10]. Some systems (ADIOP, CADRE, CADSYN, CHARADE, COMPOSER, IDIOM, JULIA) integrated CBR with constraint satisfaction problem (CSP). Some systems (ANAPRON, AUGUSTE, CAMPER, CABARET, GREBE, GYMEL and SAXEX) combined CBR with RBR. It is worth to note that the first prototype of the system, integrating CBR with RBR was CABARET [11]. In [12] it is proposed possible connection of CBR with RBR and its application to the financial domain implemented in the MARS prototype system.

In present paper, an original approach is proposed in which it is possible the dynamic interaction of both CBR and RBR models of knowledge representation and its application to the task of cases classification. Section 2 provides a hybrid model of dynamic interaction of CBR and RBR. In Sect. 3 we give the method of transformation of a set of cases into the system of fuzzy classification rules and discuss its improvement. Section 4 contains some experimental results and their explanation.

2   Dynamic Integration of CBR and RBR

The interaction between the employees of the organization on solving the important tasks generates creation of new knowledge. In the process of knowledge transformation both its forms are used: non-formalized and formalized knowledge. While creating new knowledge the formalized (explicit) knowledge and non-formalized (tacit) knowledge interact in four ways: socialization, externalization, combination, internalization according to SECI model (Socialization, Externalization, Combination, Internalization [13]. Sequential alternation of four processes - socialization, externalization, combination, internalization - creates a spiral of knowledge. The process is developing by spiral consistently through these four stages.

In the previous section we described two basic approaches to knowledge representation - case-based and rule-based approaches. Both models have their advantages and disadvantages. Therefore, in practice it would be reasonable to join advantages of both approaches. The combination of these approaches could be represented as cyclic knowledge transformation from case-based form to the rule-based and vice versa, as shown in Fig. 1, by the analogy with Nonaka-and-Takeuchi spiral [13].

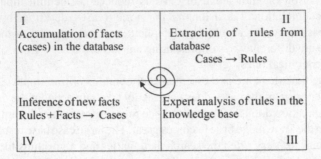

I Accumulation of facts (cases) in the database	II Extraction of rules from database Cases → Rules
Inference of new facts Rules + Facts → Cases IV	Expert analysis of rules in the knowledge base III

Fig. 1. Transformation of knowledge between case base and rule base

At the initial stages of the KBS life cycle, when there is no insight into the application domain, it is advisable to use a case-based model in which the knowledge is represented by relevant precedents (cases) of decision making (stage I). To create a precedent it is necessary to determine a set of features that uniquely determine the situation and the specific solution made in this situation. A simple linear form of the case can be presented as $(n + 1)$ - dimensional tuple

$$CASE = (x_1, x_2, x_3, \ldots, x_n, s), \tag{1}$$

where $x_1, x_2, x_3, \ldots, x_n$ are the features identifying the situation, s is a solution to the problem defined in the case. Subsequently, with the deepening into the problem domain, possible complication of the case structure is possible, through, for example, the introduction of hierarchy and other relationships between the features.

Already at the earliest stages of the KBS development it is possible to extract, adapt and apply cases in solving the current problems. After applying the case from the knowledge base to the current situation, a new precedent is recorded to the case base for future use. Note that in terms of Nonaka-and-Takeuchi cycle, we do not carry out the formalization of knowledge here. Moreover, we do not try to analyze why the solution is made (i.e. we do not try to transform tacit knowledge into the formal representation), but we simply fix the fact of tacit knowledge in a particular case.

As soon as we accumulate sufficient volume of decision-making cases and deepen our knowledge on the basis of analyzing the case base, it is possible to carry out the formalization of knowledge, contributing to the transformation of tacit knowledge into the explicit one (stage II). Machine learning offers a variety of approaches to extract knowledge from data (simple cases). In this article we consider the problem of obtaining a set of formal rules for classification of cases. Classification allows, on the one hand, to arrange a space of cases in order to improve the retrieving performance. On the other hand, classifying rules are formal knowledge of higher level than simple cases. They can be meaningfully analyzed by an expert who can get possibly new information for decision-making in the future.

If we have simple cases of the form (1), we could apply, for example, decision tree method [14] to this sample of cases, in which the intermediate nodes of the tree correspond to the features $x_1, x_2, x_3, \ldots, x_n$, and terminal nodes are the classes to which solutions belong. Another method of forming a set of fuzzy rules for cases classification is given in the next section and based on the original papers [15–17].

The transition to a rule-based model means that we obtain the explicit (formalized) form of knowledge, able to explain the cause-and-effect relationships in the application domain. Such reasoning could be presented to experts for analysis and interpretation (stage III of knowledge transformation cycle). The important point at this stage is the resolution of conflicts arising in a system of rules (conflicting rules lead to different conclusions for the same premises).

At the stage IV the accumulation of new cases takes place using the received version of a decision support system. New cases can be obtained from application of logical inference methods to the knowledge base consisting of rules and facts (cases), and also as a result of CBR-cycle, adapting old cases from the knowledge base to new problems. Getting new cases means the use of formal (explicit) knowledge and turning them into the implicit (tacit) form.

The considered cycle of knowledge transformation is repeated further on the next level. As soon as our knowledge about application domain deepen it is possible to make the structure of cases more complex by transition from the linear parametric form to the hierarchical or more complex logical form. At each stage of the proposed cycle, the conversion is made of the implicit knowledge into the explicit form, or vice versa, and each new turn of the spiral brings additional new aspects and dimensions to the knowledge of the application domain.

3   Method of Cases Classification Through Fuzzy Rules

Here we consider the method of obtaining fuzzy rules from cases in application to the problem of cases classification. The idea of the method is as follows. Consider arbitrary case of the form (1), where the features $x_1 \in X_1, x_2 \in X_2, x_3 \in X_3, \ldots, x_n \in X_n$, and the solution variable s belongs to a certain set of classes $D = \{d_1, .d_2, \ldots, d_m\}$. Without loss of generality we assume that the features are numerical and sets $X_1, X_2, X_3, \ldots, X_n$ are continuous numerical intervals.

Define linguistic variables V_i corresponding to the features $x_i, i = \overline{1, n}$. Let each variable V_i has J term-values $A_i^{(j)}$ that are fuzzy sets determined by membership functions $\mu_{A_i^{(j)}}$, on the universal sets $X_i, j = \overline{1, J}$. Then the rule *Rule* corresponding to the case (1), is formulated as follows:

$$Rule: \quad IF \ V_1 = F(x_1) \& \ldots \& V_n = F(x_n) \ THEN \ R = d \qquad (2)$$

where R is the classifying variable, $d \in D, F(x_i)$ is a fuzzy term-value which the linguistic variable V_i accepts as a result of fuzzification of the feature x_i:

$$F(x_i) = A_i^{(j^*)}, \text{ where } j^* = \arg \max_{|j=\overline{1,J}} \mu_{A_i^{(j)}}(x_i).$$ (3)

Then, knowledge base consisting of a set of rules from the initial case base is formed in accordance with the following algorithm.

An algorithm of obtaining fuzzy classification rules from case base

Step 1.   Define the number and the type of membership functions and universal sets X_i for each linguistic variable V_i corresponding to the feature x_i.

Step 2.   Determine fuzzy rule $Rule_i$ of the form (2) for each case of the form (1).

Step 3.   Assign truth degree $TD(Rule)$ to each rule. The simplest way to do this is to compute minimum of all $\mu_{A_i^{(j^*)}}(x_i)$ for every x_i and fuzzy sets $A_i^{(j^*)}$ defined by the formula (3) as a result of fuzzification for each rule $Rule$ of the form (2). The result is obtained in the following way

$$TD(Rule) = \min_{i=\overline{1,n}} \mu_{A_i^{(j^*)}}(x_i) = \min_{i=\overline{1,n}} \max_{j=\overline{1,J}} \mu_{A_i^{(j)}}(x_i).$$ (4)

Step 4.   Resolve conflicts between the rules. After the step 2 we obtain a set of rules uniquely corresponding to the set of cases. But this set can contain subsets of the rules with the same premises. These rules can have the same or different conclusions (solutions $d \in D$). If for all rules in such subset we have the same conclusion we simply remove all duplicate rules from the set except one. For the rules with conflicting conclusions, we propose two different strategies for classification problem. In each case we leave only one rule in the subset with conflicted conclusions in the rule base. Let we have a subset of rules $Rule_1, \ldots,$ $Rule_m$ with the same premise. Divide the set of indexes $I = \{1, 2, \ldots, m\}$ into K subsets $I_1, I_2, \ldots, I_K$ corresponding to the classes of solutions $d_1, d_2, \ldots, d_K$ in the conclusions of the rules (3), $\dim I_k = m_k$, $\sum_{k=1}^{K} m_k = m$.

The first strategy of resolving conflicts is to obtain the optimal rule

$$l^* = \arg \max_{l \in I} TD(Rule_l),$$ (5)

and then to choose a solution d corresponding to the conclusion of this optimal rule. The second strategy of resolving conflicts is a bit more complex. The optimal solution d_{k^*} that is assigned to the conflicting set of rules could be obtained as follows

$$k^* = \arg \max_{k} \left(\sum_{i \in I_k} TD(Rule_i) \right).$$ (6)

4   Experimental Results

In this section we investigate the proposed approach with the data set Iris (available at http://archive.ics.uci.edu/ml/) with 150 cases, characterized by 4 features ($n = 4$), which are classified into 3 classes ($K = 3$) through classifying attribute.

In the first experimentation, we randomly selected training sample of given size (120 cases) from Iris set, while the rest of the data (30 cases) was used as a control sample. Then, for each feature, minimum and maximum values were determined from the training sample. The obtained intervals were used as the universal sets for corresponding linguistic variables. Then we divided each universal set X_i into equal intervals for three variants of choosing the number of term-values $A_i^{(j)}$, $j = \overline{1, J}$: $J = 3, 5$ and 7 (J does not depend on i) for membership functions (3MF, 5MF and 7MF). We also considered two classes of MF – linear (class T) and quadratic (class S).

Then a set of fuzzy classifying rules obtained as a result of machine learning algorithm was applied to control sample to compute an accuracy of classification as a ratio of correctly classified cases to the general volume of the control sample. In Table 1 we presented the results of classification accuracy for two strategies of control sample classification with linear membership functions (3MF, 5MF and 7MF).

Table 1. Classification accuracy for two strategies of control sample classification with linear membership functions (3MF, 5MF and 7MF)

Number of trial	Control accuracy for strategy 1			Control accuracy for strategy 2		
	3 MF	5 MF	7 MF	3 MF	5 MF	7 MF
1	0,90	0,83	0,83	0,97	0,88	0,88
2	0,87	0,80	0,63	0,93	0,92	0,78
3	0,97	0,77	0,73	0,95	0,83	0,83
4	0,80	0,63	0,63	0,97	0,83	0,83
5	0,83	0,73	0,57	0,95	0,88	0,77
6	0,87	0,83	0,60	0,95	0,90	0,82
7	0,80	0,70	0,63	0,97	0,88	0,83
8	0,90	0,77	0,70	0,95	0,85	0,78
9	0,93	0,83	0,73	0,93	0,85	0,83
10	0,87	0,67	0,80	0,97	0,87	0,83
Average	**0,88**	**0,76**	**0,69**	**0,95**	**0,87**	**0,82**

From Table 1 we can see that the best results were obtained for 3MF. Reducing in the accuracy for 5MF and 7MF could be explained by the fact that the number of possible combinations of preconditions in the rules increases significantly, so classification on the control sample fails because of the lack of appropriate rule for the classified case. Therefore we compare two types of control cases classification. For the first strategy we apply each rule only to exactly corresponding control cases. If there are no appropriate rules, incorrect classification is registered. For the second strategy, if there is no appropriate rule to classify the control case, we exclude one feature and try to find classifying rule for reduced number of features (Here we assume that the excluded feature is not important for classification). The first appropriate rule is applied for classification of the corresponding case. The second strategy permits to improve the results of cases classification. The experiments were performed ten times, each time having arbitrary set of cases for training.

In Table 2 we give classification accuracy for linear and quadratic membership functions (class T and S for 3MF). Here we can see that for smoother membership functions (class S) the classification accuracy is better.

Table 2. Classification accuracy for two strategies of control sample classification with linear and quadratic membership functions (class T and S for 3MF)

Number of trial	Control accuracy for strategy 1		Control accuracy for strategy 2	
	Class T	Class S	Class T	Class S
1	0,90	0,97	0,97	0,98
2	0,87	0,97	0,95	0,97
3	0,93	0,93	0,98	0,95
4	0,87	0,97	1,00	0,93
5	0,87	0,83	0,98	0,95
6	0,85	0,92	0,95	1,00
7	0,87	0,88	0,95	0,93
8	0,82	0,90	0,95	0,92
9	0,87	0,87	0,98	0,98
10	0,85	0,90	0,97	0,95
Average	**0,87**	**0,91**	**0,97**	**0,96**

Figure 2 gives visual illustration of comparison between the three classes of linear membership functions (3MF, 5MF and 7MF) for sequentially reducing volume of training sample. Here we can observe that the classification accuracy for sequentially

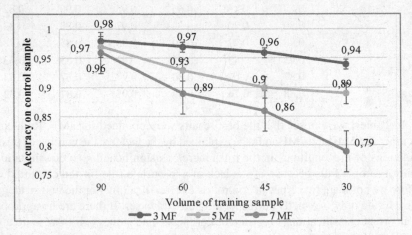

Fig. 2. Classification accuracy for sequentially reducing volume of training sample for linear membership function (T) for 3MF, 5MF and 7 MF

reducing volume of training sample (from 90 cases to 30) remains quite high. The results could be explained by the fact that the number of rules obtained from the initial training sample of 150 cases was not more than 17 in all the experiments. If we randomly choose less amount of cases for training, the information content (or representativeness) of the sample remains still high.

Thus, we have proposed rather simple method of generating classification rules from case base that can be used in the cycle of knowledge transformation in the KBS. Despite the simplicity we have achieved quite high accuracy of classification even for sequentially reducing volume of training samples. We believe that accuracy of classification can be improved further through the extension of the rule structure (2) by introducing disjunctions into a conjunctive premises as well as considering the possibility of reducing the number of the rule premises by leaving the most informative ones, that is the subject of further research.

Acknowledgments. The reported study was funded by Russian Ministry of Education and Science, according to the research project No. 2.2327.2017/4.6

References

1. Schank, R.C., Abelson, R.P.: Scripts, Plans, Goals and Understanding. Erlbau, New York (1977)
2. Schank, R.C.: Dynamic Memory: A Theory of Reminding and Learning in Computers and People. Cambridge University Press, New York (1982)
3. Simpson, R.L.: A computer model of case-based reasoning in problem solving: an investigation in the domain of dispute mediation. Technical report GIT-ICS-85/18, Georgia Institute of Technology, School of Information and Computer Science (1985)
4. Hammond, K.J.: CHEF: a model of case-based planning. In: American Association for Artificial Intelligence, AAAI 1986, Philadelphia, PA (1986)
5. Sycara, E.P.: Resolving adversarial conflicts: an approach to integrating case-based and analytic method. Technical report GIT-ICS-87/26, Georgia Institute of Technology, School of Information and Computer Science (1987)
6. Koton, P.: Using experience in learning and problem solving. Ph.D. thesis, Massachusetts Institute Technology, Laboratory of Computer Science (1989)
7. Hinrichs, T.R.: Problem Solving in Open Worlds. Lawrence Erlbaum, Hillsdale (1992)
8. Watson, I., Marir, F.: Case-based reasoning: a review. Knowl. Eng. Rev. **9**(4), 327–354 (1994)
9. Marling, C., Sqalli, M., Rissland, E., Hector, M.A., Aha, D.: Case-based reasoning integrations. AI Mag. **23**(1), 69–86 (2002)
10. Yang, H.L., Wang, C.S.: Two stages of case-based reasoning - integrating genetic algorithm with data mining mechanism. Expert Syst. Appl. **35**, 262–272 (2008)
11. Rissland, E.L., Skala, D.B.: Combining case-based and rule-based reasoning: a heuristic approach. In: Eleventh International Joint Conference on Artificial Intelligence, IJCAI 1989, Detroit, pp. 524–530 (1989)
12. Dutta, S., Bonissone, P.P.: Integrating case- and rule-based reasoning. Int. J. Approx. Reason. **8**(3), 163–203 (1993)
13. Nonaka, I., Takeuchi, H.: The Knowledge Creating Company: How Japanese Companies Create the Dynamics of Innovation. Oxford University Press, New York (1995)

14. Quinlan, J.R.: Induction of decision trees. Mach. Learn. **1**, 81–106 (1986). Kluwer Academic Publishers
15. Xiong, N.: Learning fuzzy rules for similarity assessment in case-based reasoning. Expert Syst. Appl. **38**, 10780–10786 (2011)
16. Xiong, N.: Fuzzy rule-based similarity model enables learning from small case bases. Appl. Soft Comput. **13**, 2057–2064 (2013)
17. Avdeenko, T.V., Makarova, E.S.: Integration of case-based and rule-based reasoning through fuzzy inference in decision support systems. Procedia Comput. Sci. **103**, 447–453 (2017)

Author Index

Printed in the United States
By Bookmasters